Measuring Climate Change to Inform Energy Transitions

Measuring Climate Change to Inform Energy Transitions

Carbon Footprint Calculations

Sunny E. Iyuke
University of the Witwatersrand
Johannesburg, South Africa

WILEY

Published by John Wiley & Sons, Inc., Hoboken, New Jersey.
Published simultaneously in Canada.

Library of Congress Cataloging-in-Publication Data Applied for:

Hardback ISBN: 9781394207312

Cover Design: Wiley
Cover Images: © Olena Ruban/Getty Images

Set in 9.5/12.5pt STIXTwoText by Straive, Chennai, India

Contents

Preface

In this century of COP27, almost all delegates, including the United Nations, have requested rich countries to stop using coal by 2030 and poor countries by 2040. Otherwise, before now, climate professionals have shown that heat waves, coastal floods and tropical cyclones will continue to produce devastation that is more violent if the world does not rapidly effect policies, build infrastructure and develop skills and workforce to make clean electricity, transportation and other energy applications. It is convenient to remind humans that in 2015, more than 200 governments agreed to stop global warming by 1.5 °C (2.7 °F) by the end of the century. However, seven years later, as they meet again for a climate summit in COP27 in Egypt, policies that will enable heating the planet by less than this value are being pursued.

It therefore becomes obvious that for world leaders to honour their promises, they must sharply cut the amount of heating on the planet by immediately transiting the use of coal, petroleum oil and gas or fossil energy sources to cleaner energies such as hydrogen, fuel cells, solar, wind, biomass and renewables into the air each year. Greenhouse gas (GHG) emissions such as carbon dioxide (CO_2) methane (CH_4) would have to fall by 45% by the end of the decade and reach net zero by 2050. The main driver for optimistic shift would be to generate clean electricity, clean fuels for transportation and other stationary applications instead of using fossil fuels, and electrifying activities that involve burning coal, petroleum oil, gas, etc.

Most of the world now generates power for grid, households and transportation from coal, the dirtiest source of energy, and fossil oil and gas, which are cleaner but still polluting. Combusting these fuels releases GHGs into the Earth, heating the planet and making extreme weather worse as is currently observed around the world, whether developed or underdeveloped. About a year after the previous COP26, less than 40% of the world's electricity still comes from low-carbon sources like solar, wind, nuclear and hydropower, to mention but a few.

This book therefore presents transition mechanisms and model (Figure 1) that the world can adopt to move from fossil fuels as energy sources to renewables and other clean energy technologies, as mentioned above. The manuscript can thus be used by energy policymakers, trainers and educationists in both elementary and tertiary education, including research and development experts.

The main objective of compiling the author's previous works and reports with the students and colleagues into this book is therefore to provide sustainable development plans for both policymakers and end users, while the book is aimed at developing an innovative planning tool to assess the effect of energy consumption and production on the overall carbon emissions. It is also to measure, calculate and benchmark against set standards and government policies. Technologies to compact or control GHG emissions are also proposed and applied in case studies distributed across the chapters. This tool is used for both modelling of imminent global warming and assessment of current situations, supported with examples and case studies, which may involve the use of triangulation tool to convert gradients of CO_2 emissions to spatial coordinates and bearings.

Policymakers and educators with technical backgrounds will find the book useful. Individuals and environmentalists, engineers, scientists, students, etc., will buy the book. In addition, because of the inclusion of a chapter on policy with reference to the preceding chapters on calculations, measurement and decision-making, it will be a reference textbook accessed through libraries serving academic researchers and teachers for classroom teaching and learning resource materials.

Figure 1 Effect of carbon footprint on global warming: (a) modified image (Gambrill 2021), (b) drying lake bed at the University of the Witwatersrand, Johannesburg, South Africa as of 2023. (c) The Ocean Foundation highlighted that seagrass habitants are up to 35× more effective than Amazonian rainforests in their carbon uptake and storage abilities (Owens et al. 2019).

1 Topics

The book comprises the following seven chapters: (1) 'Introduction, Carbon Footprint and Climate Change' that establishes the background and implementation of the tool to determine carbon footprint and its impacts on climate change. (2) 'Vegetative Sinks for Carbon Capture' that introduces and applies carbon sequestration as the process of capturing molecules of carbon dioxide (CO_2), which is one of the leading GHGs in the atmosphere. Trees, forests and even microalgae are used as vegetative sinks that sequester carbon through photosynthesis, including food gardens. (3) 'Carbon Transition for the Petrochemical Sector' involves transformative actions to decarbonise the chemical, petrochemical, oil and gas sectors in finding alternative non-fossil carbon feedstocks and provides overall understanding of the environmental implications, scientific, engineering resources and equipment. (4) 'Energy Transition and Power Reforms from Coal'. Since power supply is a major component to the social and economic development of any nation and is a component of the Sustainable Development Goals of the United Nations, energy demands increase with increasing population and in turn more coal, oil and gas are burned for power generation. This exacerbates GHG emissions and environmental concerns, whereas abundance of renewable energy in Africa in general calls for new energy policies that will promote renewable energy-sourced power generation and ensure a cleaner environment for South Africa and the continent as a whole. (5) 'Carbon Footprint of Internal Combustion Engines and Mitigations' involves the transportation sector as a major contributor to CO_2 release into the atmosphere, which in turn acts as the key culprit in increasing climate change. In order to meet the long-term climate change mitigation targets, as set by the Paris Agreement and for individual countries, policies towards decarbonising road, air and ocean transportation should be built around more efficient transport, society and larger shares of renewable fuels and faster introduction of electric vehicles and engines. (6) 'Application of Carbon Footprint to Climate Change Solutions' presents that investments in development and use of renewable energy will facilitate the transiting from fossil fuels to clean and renewable energy sources such as solar, wind, wave, tidal and geothermal power and hydrogen and help with climate change solutions since global warming and climate change are inseparably the world's most topical issues in the current environment-economy dialogue. Thus, since CO_2 emission is the main GHG responsible for global warming and climate change, many scholars and policymakers are beginning to tune the relationship between climate change, energy production, and sustainable economic growth. (7) 'Climate Change Policy and Skills Education' involves skills development policies and green job creation plans. It then presents an elegant, thought-provoking inquiry on how the engineering discipline can impact and battle climate change to build a more sustainable world and ensure that engineering should be the key to delivering a sustainable future.

2 Acknowledgements

This excellent compilation entitled *Measuring Climate Change to Inform Energy Transitions: Carbon Footprint Calculations* would not have been possible without the support from the University of the Witwatersrand (Wits), Johannesburg, in South Africa, which has given permission to use the University's policy on climate change and its associated works. The current Head of School of Chemical and Environmental Engineering, Professor Josias van der Merwe, permitted the use of the School's Quinquennial Review Report (QQR) for the period from 2017 to 2022 on the curriculum for skills training and education of experts on climate change, carbon transition and global warming.

The Wits students' design reports and postgraduate research, which were supervised by the author and other colleagues, became handy as the author received permission to include extracts of these reports in this textbook. The examples and case studies from these students provide the credibility and resourcefulness the reader will find in the work, which are gratefully appreciated. It is therefore with heartfelt gratitude that the author gives recognition to the following former students who have magnanimously allowed their hard work to be included in this manuscript.

Dr. Osayi Ilawe Julius, whose research work is entitled Production and characterisation of biocrude from used tyres and natural rubber. Miss Thandeka Lukhele and Mr. Thembalethu Sibuye who researched into evaluation of membrane electrode assembly performance based on hydrogenated and sulfonated polystyrene-butadiene rubber in a single-cell PEM fuel cell and electrolyser and process design and integration of proton exchange membrane electrolyser for 500 kW proton exchange membrane fuel cell to power the Richard Ward building at Wits University, respectively. The excellent design reports presented as case studies were co-authored by Mr. Darren Bouttell, Ishaam Ismail and Ketalya Reddy on their design work entitled Balance of Plant Design for Simultaneous DME and Methanol Production and Power Generation. Mr. Michael Landgrebe's design report on a 4000 Barrel Per Day Modular Refinery and Ms. Nyiko Ngxola's design report on Student Food Garden at the Wits University, which was supervised by the author's colleague, Dr. R.T. Paton as a final year design project at the School of Mechanical, Industrial and Aeronautical (MIA) Engineering at Wits University.

My gratitude also goes to the South African government for the interns provided for this work through the Presidential Youth Employment Intervention (PYEI) Graduate Programme. Finally, to my family and the entire Iyuke family: without their love and enormous help and support, the project would never have been possible.

3 Synopsis

The book proposes temperature/CO_2 and temperature/time gradients as measures of global warming and climate change, respectively. These gradients recommend that CO_2 should be reduced to the pre-industrial revolution of the 1800s gradient of $0.0551 \leq 0.5000\Delta°C/\Delta CO_2$ and $-35.64 \leq 0.500\Delta°C/\Delta Yr$ from 2010 to 2035, respectively. A Triangulation tool that converts gradient bearings to spatial coordinates is hereby employed to interpret and avoid the worst impacts of climate change. Its application provides the world with the urgent need to reduceGHGs such as carbon dioxide (CO_2), methane (CH_4), nitrous oxide (N_2O), dichlorodifluoromethane (CCl_2F_2) especially CO_2 emissions. This is a responsibility between regions, countries and individuals, which possess endless points of contention at the international arena, such as the Conference of the Parties (COP). Thus, the role of governments in these efforts to control or stop climate change is to encourage governments and businesses to adopt the following strategies for predictable energy transition goals presented in the table below. As shown, CO_2 concentration increases at a gradient of 76.76 from 1990–2010, followed by a negative gradient of −35.64 from 2010–2035, while the 3rd spline has a gentle slope of 5.4, which is still higher than the minimum value of $0.500\Delta°C/\Delta Yr$. Upon the analysis of the measured atmospheric temperature against CO_2 concentration, again a negative gradient of $-0.0004C_{CO2}$ is obtained, which supports the earlier negative value $-0.0004 \leq 0.500\Delta°C/\Delta CO_2$ during the second spline 2010–2035. In both cases, therefore, the trigonometric triangulation projection supports the fact that the 'net annual change in CO_2 equivalent model adopted by the UK woodlands' is presenting positive results to global warming and, in turn, should assist in taking decisions for the efforts to transit carbon economy as an example. It is therefore expedient to quote the UK Forests and Climate Change future predictions that carbon uptake may assume that commercial conifer plantations should be replanted when felled. That, also planting of new woodland to support the current Rural Development Plan should be encouraged because the CO_2 release is still on the increase, as shown by the gradient of $5.4 \geq 0.500\Delta°C/\Delta Yr$ from 2035 to 2050.

Emission (ppmv CO_2)	Period (year)	Gradient of global warming	Energy transition $\geq$277.04 ppmv $\geq 0.500\Delta°C/\Delta CO_2$ $\geq 0.500\Delta°C/\Delta Yr$
Predictive modelling			
277.04–277.10	1647–1756	$0.5000\Delta°C/\Delta CO_2$	Maintain CO_2
284–287	1856–1870	$0.0380\Delta°C/\Delta CO_2$	Reduce CO_2
298–314	1915–1940	$0.0201\Delta°C/\Delta CO_2$	Reduce CO_2
335–374	1970–2005	$0.0142\Delta°C/\Delta CO_2$	Reduce CO_2
Application of modelling on UK Forests and Climate Change data			
4050–5670	1990–2010	$76.76 \geq 0.500\Delta°C/\Delta Yr$	Reduce CO_2
5751–5022	2010–2035	$-35.64 \leq 0.500\Delta°C/\Delta Yr$	Maintain CO_2
5049–5103	2035–2050	$5.4 \geq 0.500\Delta°C/\Delta Yr$	Reduce CO_2

However, in the international community, the importance of education and training to address climate change is currently receiving attention because education is crucial to promoting climate action. It certainly helps people to understand and address the impacts of the climate crisis, thus empowering them with the right knowledge, skills, values and attitudes needed to act as agents of change as buttressed by a case study from the University of the Witwatersrand in Johannesburg. It is therefore a common belief that the provision of the right skills, however, paves the way for a productive structural transformation towards a greener economy and decent job creation. This also means that effective anticipation and development of skills are fundamentals for a just and inclusive transition. Such skill development to create green jobs also serves as a "buffer" against the effects of transitory disruptions and emerging challenges that will surface as economies become green. A well-coordinated policy approach will ease the transition to a greener future, which again should be inclusive.

Keywords

climate change; energy transitions; students' food garden; triangulation tool; modular refinery; power reforms from coal; electric vehicle; CO_2 as feedstocks; skills training and education

School of Chemical & Metallurgical Engineering
University of the Witwatersrand
Johannesburg, South Africa

Jan 27, 2024
Sunny E. Iyuke

References

Gambrill, D. (2021). Aviva Canada wants to talk to you about reducing your carbon footprint. Canadian Underwritter, 8 March 2021. https://www.canadianunderwriter.ca/insurance/aviva-canada-wants-to-talk-to-you-about-reducing-your-carbon-footprint-1004204790 (accessed 8 June 2023).

Owens, A., Voros, B., Fore, P., and Ruballo, G. (2019). The Ocean Foundation. Global, 16 April 2019. Source: https://oceanfdn.org/how-to-reduce-your-carbon-footprint-when-you-travel/ (accessed 8 June 2023).

1

Introduction, Carbon Footprint and Climate Change

1.1 Introduction

A carbon footprint can be defined as the measure of the impact of human activities have on the environment due to direct or indirect release of greenhouse gases. It estimates the greenhouse gases in pounds or tons of carbon dioxide (CO_2). Within the context of total Ecological Footprint, the tonnes of carbon dioxide emissions are expressed as the amount of productive land area required to sequester those carbon dioxide emissions. This refers to biocapacity that is necessary to neutralise the emissions from burning fossil fuels. Thus, measuring the carbon footprint in land area shows how much biocapacity is needed to take care of untreated carbon waste and avoid a carbon build-up in the atmosphere. This method of measurement allows addressing climate change challenge in a holistic manner that does not simply shift the burden from one natural system to another because the planet does not provide enough biocapacity to neutralise all the carbon dioxide from fossil fuel and other-natural CO_2 sequestration medium.

Levels of CO_2 in the atmosphere are moderately normal, because it helps to keep the planet warm and plays an integral role in many key biological processes, which includes photosynthesis and other biogenic transformations by nature. These earthly transformations produce the so-called Global Carbon Cycle. However, human activities have altered this natural cycle by adding more concentration of CO_2 to the atmosphere, which has overburdened natural ability to remove it. The major culprit in the increased CO_2 concentrations in the atmosphere is the burning of fossil fuels (oil, coal and natural gas), as well as changes in land use from deforestation.

The ocean is not left out in sustaining the stability of the carbon cycle by absorbing excess CO_2 from the atmosphere. As seawater absorbs the excess CO_2, chemical reactions known as ocean acidification occur, which increases the acidity of the water. The resultant acids make life very difficult for corals to build and maintain skeletons, shellfish, lobsters, scallops and clams to build shells and function normally or even die. The oceans also experience warmer temperatures and sea levels increase due to the warming of the atmosphere.

Ordinarily, carbon footprint may be determined from the average number of gallons an engine consumes per hour, then multiply this by the total number of hours the engine has run during the period in question.

For example:

Over a year, if a car consumes 10 gallons of fuel per hour and the engine runs for 204 hours. The gallons consumed are,

$$10 \text{ gallons} \times 24 \text{ hours} = 2040 \text{ gallons}$$

In case of marine diesel, the CO_2 released is:

$$20\,140 \text{ gallons} \times 21{,}24 \text{ lb of } CO_2/\text{gallon}$$
$$= 43329 = 43\ 329 \text{ lb of } CO_2, \text{ equivalent to 21.7 US tons.}$$

where marine fuel consumption can be given as:

Measuring Climate Change to Inform Energy Transitions: Carbon Footprint Calculations, First Edition. Sunny E. Iyuke.

One gallon of fuel	Pounds (lb) of CO_2 per gallon
Marine diesel	21.24
Marine unleaded 93	19.88
Marine unleaded 91	19.51
Marine unleaded 89	19.52
Jet A	21.10
Biodiesel	5.02

Climate change, on the other hand, is the long-term shifts in temperatures and weather patterns. The shifts may be due to natural variations in the solar cycle. However, since the Industrial Revolution in 1800s, the main causes of climate change have been due to human activities, mainly due to burning fossil fuels like coal, petroleum oil and gas. It has been proven that the greenhouse emissions from burning of fossil fuels wrapped around the earth like blankets, trapping the sun's heat and raise temperatures of the earth. Although, clearing land and forests, landfills for garbage are sources of CO_2 and methane (CH_4) emissions, respectively, greenhouse gas emissions that are causing climate change come from using gasoline and diesel for transportation, heating buildings, industry, transport, agriculture and land.

It is also expedient to mention that in the last Conference of the Parties (COP) 26 in Glasgow the last decade (2011–2020) has experienced the warmest weather on record. Also, the greenhouse gas concentrations are at their highest levels in about 2 million years, and the emissions continue to rise. Such that the earth is now 1.1 °C warmer than the temperature in the late 1800s.

1.2 Global Warming Due to Carbon-Cycle Feedbacks in a Coupled Climate Model

In the 1992 Supplementary Report, the Intergovernmental Panel on Climate Change (IPCC) Special Report on Emissions published Six alternative IPCC scenarios (IS92a to f). The scenarios contained assumptions affecting how future greenhouse gas emissions would affect the climate policies that are already in place. As a result, Cox et al. (2000), simulated atmospheric CO_2, which diverged very rapidly from the standard IS92a concentration scenario in the future as presented in Figure 1.1. They observed that vegetation carbon in South America was declining due to the drying and increase in temperature of Amazonia that started loss of forest, which was a result of climate change. It was also proposed that a critical point will be reached by 2050, when the land biosphere may switch from being a weak sink to a strong source for CO_2 (Figure 1.1). Such that reduction in terrestrial carbon by 2050 onward may be associated with a widespread climate-driven loss of soil carbon. On the other hand, rise in concentration of atmospheric CO_2 will tend to promote increasing rate of photosynthesis and terrestrial carbon storage. Since the rate of plant maintenance and soil respiration will increase with temperature, terrestrial

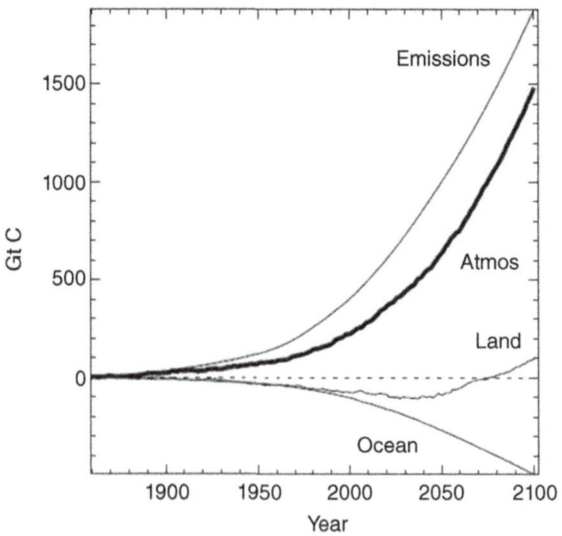

Figure 1.1 CO_2 simulation, where the thick line shows simulated change in atmospheric carbon. The thinner lines are the integrated impact of the emissions of land and ocean fluxes. The terrestrial biosphere takes up CO_2 at decreasing rate from 2010 and become a net source by 2050. These reversing rates balance themselves by 2010 on land and oceanic sinks, which would result into increasing atmospheric carbon content at a fraction of 1. Source: Cox et al./Springer Nature.

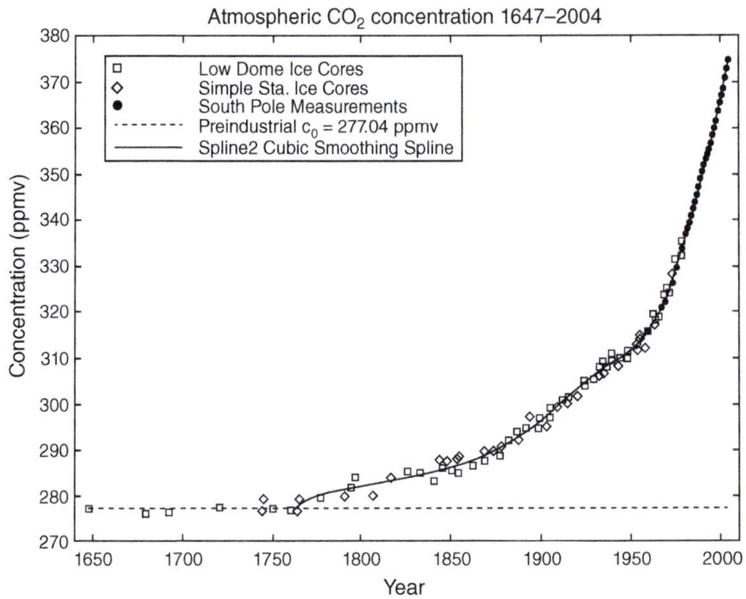

Figure 1.2 Atmospheric CO_2 concentrations measured in units of parts per million by volume Rust and Thijsse/National Institute of Standards and Technology/https://math.nist.gov/~BRust/pubs/CSC07/ReprintCSC-07.pdf/Public Domain.

carbon storage will reduce. This scenario will be most observed in the warmer regions where an increase in temperature does not necessarily support photosynthesis because, at this point, CO_2 concentration saturates photosynthesis whilst soil respiration increases with temperature. This means that a transition will come in year 2050 whereby carbon stored on land would decrease by about 170 Gt C from 2000 to 2100, which in turn accelerates the rate of atmospheric CO_2. This is a scary scenario if urgent governments of nations do not take urgent actions to give adequate control on the evolving threat of climate change as enormously and better presented elsewhere (Cox et al. 2000) (Figure 1.2).

1.3 A Mathematical Model of Man-Made CO_2 Emissions

This section discusses the analysis by Rust (2011) who proposed that atmospheric CO_2 concentrations $c(ti)$ can be related to man-made CO_2 emissions $F(ti)$ by a linear regression model whose solution vector gives the unknown retention fractions $\gamma(ti)$ of the $F(ti)$ in the atmosphere. In this estimation of $\gamma(t)$, enough accuracy was obtained to establish the connection between emissions from fossil fuel and land uses and increase in global average annual temperature anomalies.

It was explained that uncertainties in global climate models are often doubted on the basis that global warming is too complicated to estimate climate models. Rust and Thijsse (2007) and Thijsse and Rust (2008) were able to use the measurements of global average temperatures and atmospheric CO_2 concentrations to confirm that global warming is principally caused by increasing concentrations of the greenhouse of CO_2. By combining atmospheric measurements from 1958–2004 at the South Pole with measurements from 1647–1978 at Antarctic ice cores gave the plot in Figure 1.3 as the record of atmospheric CO_2 concentrations obtained for these regions and periods of time.

Thus, Rust and Thijsse (2007) assumed that changes in the temperature anomaly are linearly proportional to changes in the atmospheric concentration of CO_2 as:

$$\frac{dT}{dt} = \eta \frac{dC}{dt}, \tag{1.1}$$

where T is temperature, C is concentration of CO_2, t is period of time and η is a constant to be determined by curve fitting. The time scale should be chosen so that $t = 0$ at epoch 1856.0 and integrating the above equation gives:

$$T(t) = T_0 + \eta\,[C(t) - c_0] \tag{1.2}$$

where Rust and Thijsse (2007) assumed that $c_0 = 277.04$ is the preindustrial concentration illustrated in Figure 1.3 and T_0 is the corresponding temperature anomaly (not the temperature anomaly at $t = 0$) which is also estimated by curve fitting.

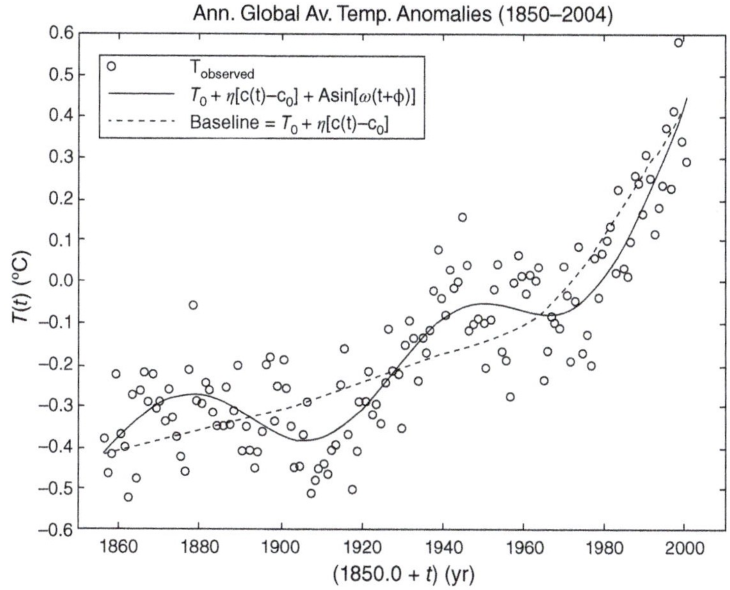

Figure 1.3 Annual global average temperature anomalies. The anomaly for any given year was obtained by subtracting a reference temperature from the average temperature for that year. The reference temperature used was the average temperature for the years 1961–1990. The solid curve is the fit of Eq. (1.2) to the data, and the dashed curve is a plot of $T_0 + \eta[C(t) - 277.04]$ where T_0 and η are the parameter estimates obtained from that fit. Source: Rust and Thijsse/National Institute of Standards and Technology/https://math.nist.gov/~BRust/pubs/CSC07/ReprintCSC-07.pdf/Public Domain.

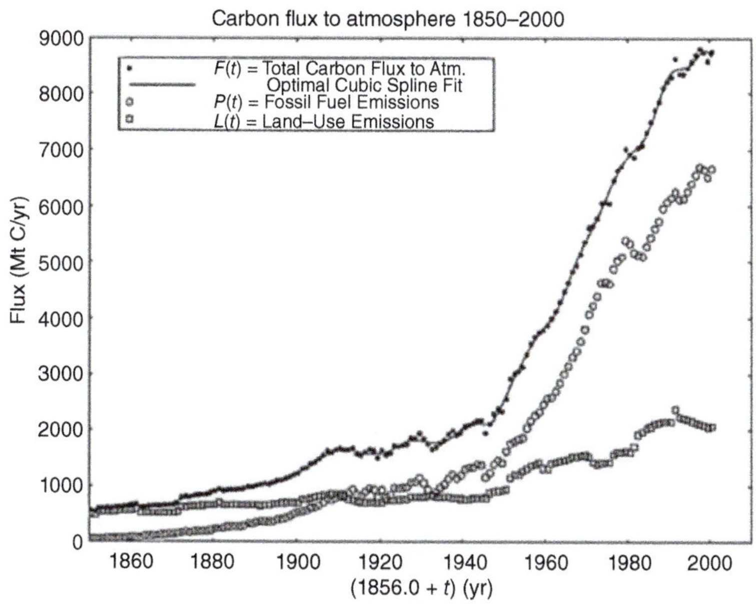

Figure 1.4 Annual man-made carbon emissions to the atmosphere. The flux units are megatons of carbon per year (Rust 2011).

Figure 1.4 was derived from the fitting of Eq. (1.2) with measured data that gave a curve very similar to the dashed line at 278 ppmv of CO_2 leading to Eq. (1.3).

$$T(t) = T_0 + \eta[C(t) - 277.04] + A\ \sin\left[\frac{2\pi}{\tau}(t + \emptyset)\right] \tag{1.3}$$

The solid curve in Figure 1.4 is the nonlinear least squares fit of the model (3) to the data. The authors (Rust and Thijsse 2007) confirmed that: (1) there is a linear relationship between global warming and increasing CO_2 in the atmosphere, (2) warming since 1856 has been ≈0.9 °C and (3) that the warming is accelerating at a faster rate. Again, from their modelling, the annual total man-made carbon emissions to the atmosphere for 1850–2000 were also presented and shown in the upper curve of Figure 1.4.

The application of the model (Eq. (1.3)) will be used as the carbon footprint barometer of the measurement of climate change for energy transition in all the chapters of the book.

1.4 Estimation of Global Warming for Energy Transition

With reference to Eq. (1.3), the circular functions sin x can be defined by the power series expansions (James 1999) as:

$$\sin x = x - \frac{x^3}{3!} + \frac{x^5}{5!} + \frac{x^7}{7!} + \ldots + (-1)^n \frac{x^{2n+1}}{(2n+1)!} + \ldots \tag{1.4}$$

Thus, the linear approximation of sin x at $a = 0$ of Figure 1.5 and Table 1.1 gives:

$$\sin x \approx x \tag{1.5}$$

Therefore, Eq. (1.3) becomes:

$$T(t) = T_0 + \eta[C(t) - 277.04] + A\left[\frac{2\pi}{\tau}(t + \emptyset)\right] \tag{1.6}$$

Thus, a plot of $T(t)$ against t of Eq. (1.6) will give the constants A, τ and $\emptyset$. Once these constants are obtained, they can be fed into Eq. (1.3) to predict the carbon footprint as function of global warming and energy transition in this and subsequent chapters.

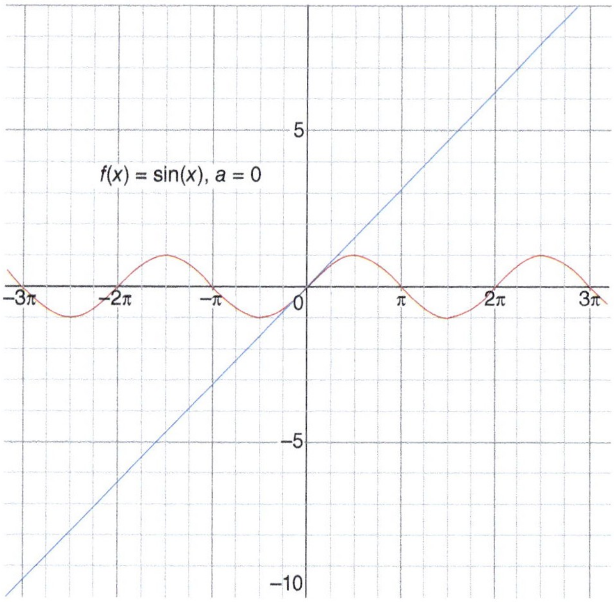

Figure 1.5 Linearised sine function at $a = 0$.

Table 1.1 Trigonometric table.

θ	0°	30°	45°	60°	90°	180°	270°	360°
sin θ	0	$\frac{1}{2}$	$\frac{1}{\sqrt{2}}$	$\frac{\sqrt{3}}{2}$	1	0	−1	0
cos θ	1	$\frac{\sqrt{3}}{2}$	$\frac{1}{\sqrt{2}}$	$\frac{1}{2}$	0	−1	0	1
tan θ	0	$\frac{1}{\sqrt{3}}$	1	$\sqrt{3}$	Not defined	0	Not defined	0
cosec θ	Not defined	2	$\sqrt{2}$	$\frac{2}{\sqrt{3}}$	1	Not defined	−1	Not defined
sec θ	1	$\frac{2}{\sqrt{3}}$	$\sqrt{2}$	2	Not defined	−1	Not defined	1
cot θ	Not defined	$\sqrt{3}$	1	$\frac{1}{\sqrt{3}}$	0	Not defined	0	Not defined

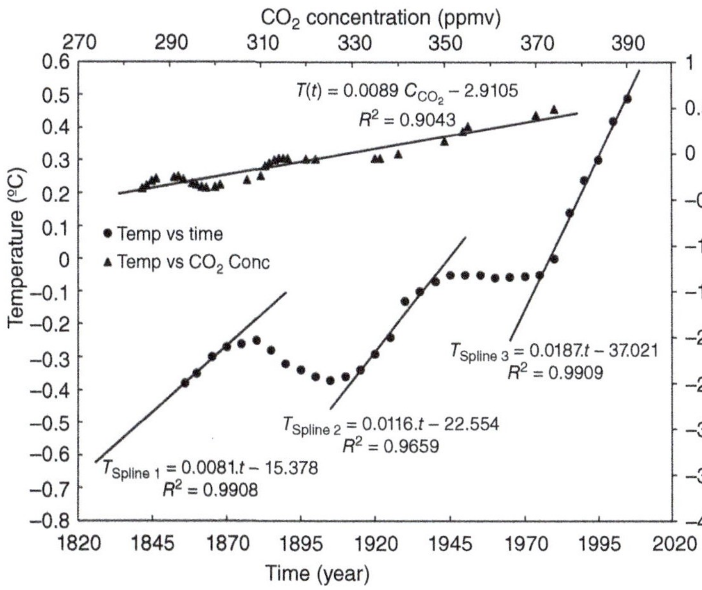

Figure 1.6 Plots of temperature versus time and CO_2 concentration against time to determine global warming from increasing CO_2 and environmental temperature at the Antarctic.

However, since Thijsse and Rust (2008) referred to T_0, η, A, ϕ and τ, as free parameters and nonlinear constants, which they determined from least-square fit to the temperature data, Eq. (1.6) is rewritten as Eq. (1.7):

$$T(t) = T_0 + \eta \cdot C(t) - \eta \cdot 277.04 + \frac{A}{\tau} 2\pi t + A(\emptyset) \tag{1.7}$$

which implies that:

$$T(t) = T_0 + \eta \cdot C(t) - \eta \cdot 277.04 + S(T) \cdot t + L(t) \tag{1.8}$$

where, $S(T) = \frac{A}{\tau} 2\pi$ and $L(t) = A\emptyset$ are the temperature variation and sums of greenhouse emissions due to changes in land and sea uses, respectively. Thus, the plots of Eqs. (1.6) and (1.3) in Figure 1.6, whilst substituting Eq. (1.9) gave approximately same values as the free parameters of Thijsse and Rust (2008), which were tested statistically:

$$\begin{aligned} \hat{T}_0 &= -0.507 \pm 0.016\,^\circ\text{C} \\ \hat{\eta} &= 0.01039 \pm 0.00042\,^\circ\text{C/ppmv} \\ \hat{A} &= 0.099 \pm 0.012\,^\circ\text{C} \\ \hat{\tau} &= 71.5 \pm 2.2\,\text{years} \\ \hat{\emptyset} &= -1.0 \pm 1.4\,\text{years} \end{aligned} \tag{1.9}$$

The sinusoidal gradients of the increasing temperatures and CO_2 concentration present measure of global warming and recommendation for control of CO_2 emission into the atmosphere as presented in Figure 1.7 in order to effect energy transition policies either to maintain or reduce Carbon Footprint. The figure supports the increasing risk of global warming with recommendations as tabulated in Table 1.2, which were interpreted from the data in Thijsse and Rust (2008) at different splines 1–4. Such analytical procedure can be integrated into CO_2 sensor and magnetic compass to detect CO_2 in the environment (Arena 2013).

1.4.1 Application of Temperature/CO_2 Gradient as a Measure of Global Warming

Although the International Panel of Climate Change reported that for five decades (1951–2012) the earth has experienced warming at a rather high rate from 0.08 to 0.14 °C per decade, it has become worse than that in the last decade (2012–2022). As a matter of fact, UNICEF echoed during the COP27 climate conference in Sharm et Sheik, Egypt that this year has brought devastating flooding to 27.7 million children in 27 countries worldwide. This trend has been observed over both land and over the ocean. In the records by IPCC (IPCC 2013), this increasing temperature is attributed to the total increase

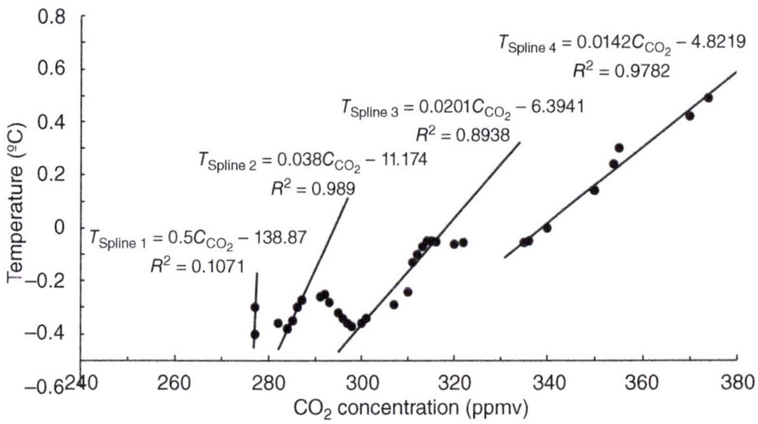

Figure 1.7 Temperature/CO_2 gradient as measure of global warming.

Table 1.2 Increasing global warming with temperature and CO_2 concentration.

Emission (ppmv CO_2)	Period (year)	Correlation coefficient (R^2)	Gradient of global warming	Energy transition ≥277.04 ppmv ≥0.500Δ°C/ΔCO_2 ≥0.500Δ°C/ΔYr
277.04–277.10	1647–1756	0.1071	0.5000Δ°C/ΔCO_2	Maintain CO_2
284–287	1856–1870	0.9890	0.0380Δ°C/ΔCO_2	Reduce CO_2
298–314	1915–1940	0.8938	0.0201Δ°C/ΔCO_2	Reduce CO_2
335–374	1970–2005	0.0142	0.0142Δ°C/ΔCO_2	Reduce CO_2

primarily caused by the increase in the atmospheric concentration of CO_2 during the last 200 years. As discussed in the preceding section, the warming rate changes with time, raising questions regarding the causes underlying the observed trends. Of course, greenhouse gases and other anthropogenic forcing are prevalent that more precise quantification of the temperature changes. Detection and quantification of this attribution are still regarded as key priorities in climate change research. However, IPCC defines detection as the process of demonstrating that climate has changed in some statistical sense, which means that the likelihood of occurrence by chance due to internal variability alone is small (Stips et al. 2016).

Thus, the application of gradient of change in temperature versus change in CO_2 concentration as a measure of global warming to the 2004 Global Temperature Anomalies (NASA 2004), Figure 1.8 was obtained, which recommends that CO_2 should be reduced since the gradient, 0.0551 ≤ 0.5000Δ°C/ΔCO_2 (Table 1.2).

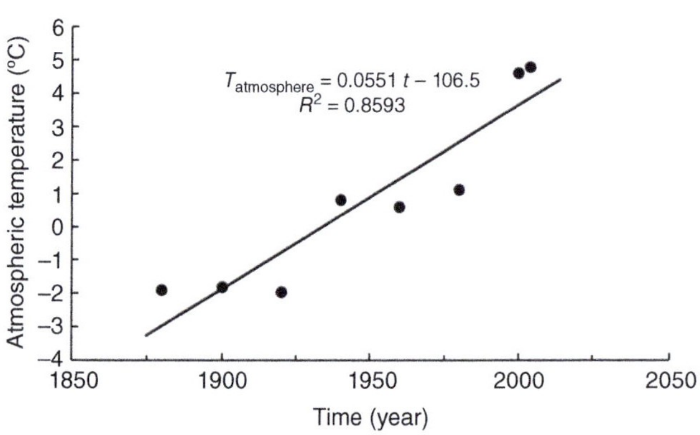

Figure 1.8 2004 global temperature anomalies.

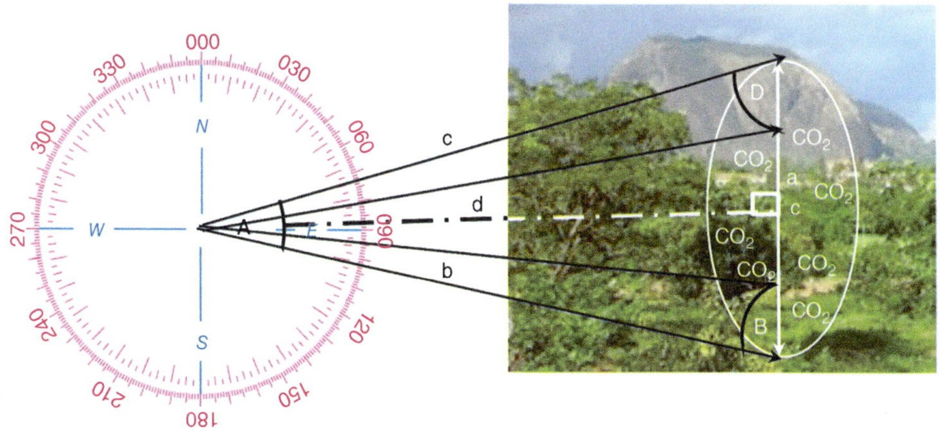

Figure 1.9 Triangulation of concentration of CO_2 gradient compass, angles *A* and *B* are CO_2 sensors or observer with geographic information system (GIS) data capturing, storing, checking and displaying capacity and related to positions on Earth's surface (Arena, 2013).

1.4.2 Triangulation Tool to Convert Gradient Bearings to Spatial Coordinates

In remote and inaccessible locations such as the wilderness, cliffs or geological platforms where measurements of features may be required to calculate the distance to a faraway objects and greenhouse CO_2 concentrations. In the example and model presented in Figure 1.9, a compass and basic trigonometry triangulation calculation has been used, where CO_2 concentration can easily be estimated from distance faraway distance. Thus, if any three of the sides or angles of a triangle are known, the remaining angles and sides are calculated. Wilderness Arena Newsletter 2013 has used the following illustrations, adapted as:

(1) Adopt two distances *b* and *c* to the remote arena,
(2) Sight the faraway pollutants in a compass from both points,
(3) Record the angle for both paths to be used in the distance calculations.

Take a known distance as if it were walked in steps and the angles measured at each of the two points as to the intended venue. Using a basic trigonometry of tangent, the trigonometric method will produce the most accurate measurement to calculate the associated angles using a standard magnetic compass (Figure 1.9) by triangulation technique.

Nightingale (2020) has provided various definitions for triangulation as a technique used for analysing the results from multiple-method research designs. It has often been to compare degree of overlap of research results. The three obvious types of triangulation are:

- Convergence: this type refers to strong degree of overlap between the datasets and serves to help ensure methods are appropriate and the data are accurate,
- Complementarity: to build a richer picture of the research results by allowing the results from different methods to complement one another,
- Divergence: indicate that the results maybe be flawed but can be treated as new data and analysed to look for new insights.

Triangulation is also a technique to analyse results of the same study using different methods of data collection for three main purposes to enhance validity, create more in-depth picture of a research problem and to interrogate different ways of understanding a research problem.

It can also be used to interrogate inconsistencies and data that are not expected to align. The methodological framework used would determine how the degree of overlap between methods is conceptualised.

Complementarity builds a richer picture of the research results by allowing the results from different methods to inform each other. Divergence presents a different set of challenges within the methods, and how it is interpreted depends on the conceptual framework for the research. Divergence can either indicate the methods or the results are flawed or be treated as new data and analysed to look for new insights (Nightingale 2009).

Thus, adopting the divergent triangulation, the basic geometry calculations define from Figure 1.9, the trigonometric functions for the angle *A* of the right-angle triangle *abd* (Figure 1.10).

The three sides of the triangle are: *a*, opposite, *c*, hypotenuse and *d*, adjacent, which is the distance between the compass and the platform BCD, whilst angle *C* is a right angle and angle *A* is of interest in this case. Since in ordinary Euclidean

Figure 1.10 Trigonometric triangulation projections from Figure 1.9 to remote platform.

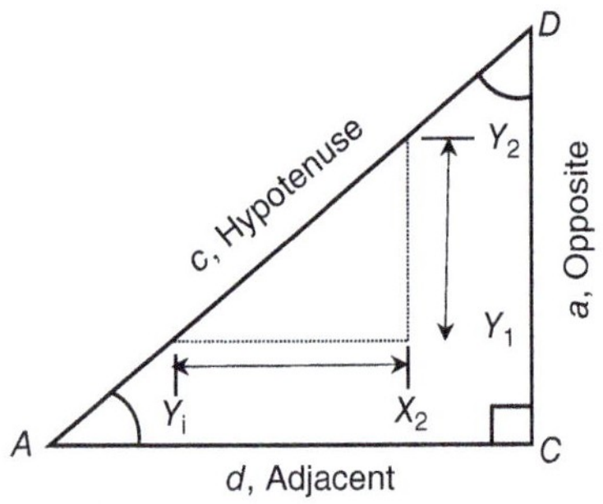

geometry, the sum of the internal angles is equal to 180°, the sum of angles A and D 90°, with the following definitions:

$$\sin A = \frac{\text{Opposite}}{\text{Hypotenuse}} = \frac{a}{c} \tag{1.10}$$

$$\cos A = \frac{\text{Adjacent}}{\text{Hypotenuse}} = \frac{d}{c} \tag{1.11}$$

$$\tan A = \frac{\text{Opposite}}{\text{adjacent}} = \frac{a}{d} \tag{1.12}$$

The definitions for sin, cos and tan are important in calculating the angles, the lengths of the sides in Figures 1.9 and 1.10. Thus, the distance, d in the trigonometric triangulation from Eq. (1.13) gives:

$$d = \frac{a}{\tan A} = \frac{a}{\text{gradient (Figs. 1.6 to 1.8)}} = \frac{a}{Y_{2-Y_1}/X_2-X_1} \tag{1.13}$$

where Eq. (1.14) confirms the use of triangulation to achieve the principles of trigonometry and gradient of change of temperature versus time and/or CO_2 concentrations in interpreting global warming as a result of CO_2 emission (Figures 1.6–1.9).

1.4.3 Algorithm for the Application of Trigonometric Triangular Tool

Posavec et al. (2017) discussed the general concept of similar algorithm, which consists of an Excel Visual Basic for Applications (VBAs) algorithm utilising trigonometry laws in time series and create master recession curves (MRCs). The Excel VBA algorithm for modelling MRC using the trigonometry approach was implemented into a spreadsheet tool (MRC Tools v3.0). In this section, the trigonometry approach-based algorithm is developed as shown in Figure 1.11 in the following steps.

1.4.3.1 Step 1 – Data Processing

Data on the effect of CO_2 emission over time on global warming as measured by temperature variations was collected over five centuries (1647–2005) as reported by Thijsse and Rust (2008), which were fitted into Eq. (1.3).

1.4.3.2 Step 2 – Gradient Calculation on Temperature versus Time and CO_2

By graphical calculations, the data in step 1 was used to obtain the constants in the sinusoidal Eq. (1.3) by adopting gradient or least square correlations on the repeating splines in Figures 1.6 and 1.7. The highest gradient which occurred in the last century of spline 3 (1975–2005) was used to obtain the parameters requited to calculate Eq. (1.3).

1.4.3.3 Step 3 – Trigonometric Triangulation Projections

In step 3, the solved parameters from the highest to the lowest vertexes are used to project into remote geographic platforms in trigonometric triangulation. Absolute times are converted into relative times in a date/time conversion where concentrations of CO_2 are measured by sensors or observers with GIS data capturing, storing, checking and displaying capacity and related to any appropriate positions on earth's surface.

1.4.3.4 Step 4 – Energy Transition Limits

With relative time, being equal to zero at the beginning of each spline CO_2 concentration is set to the minimum of 277.04 ppmv, which was observed in about 1647 before the industrial revolution era (Thijsse and Rust 2008). Minimum gradients of temperature change with change in time and CO_2 concentrations are set to $0.500\Delta°C/\Delta CO_2$ and $0.500\Delta°C/\Delta Yr$ (Table 1.2), respectively.

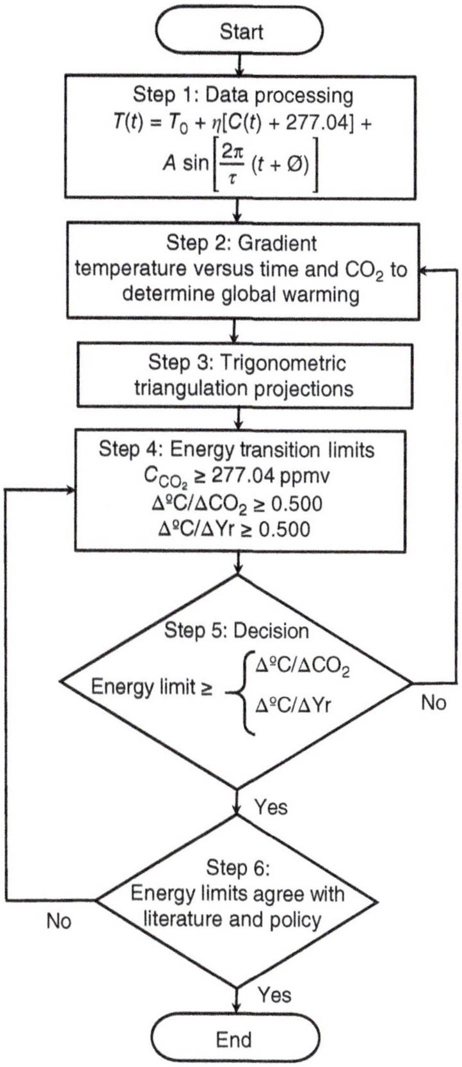

Figure 1.11 Flowchart of a trigonometry approach-based algorithm.

1.4.3.5 Step 5 – Decision to Assess Level of Global Warming

Decision to assess the level of global warming by comparing gradients to the limits set in Table 1.2 is taken. If the gradient or tan A is higher than the limits, the process is sent back to Step 2 where CO_2 emission should be controlled until allowable limit is obtained, which will in turn affect the global warming at that time. If the assessment is equal or less than the limit, then the process continues to step 6. This loop is repeated until the allowable environmental temperature and the corresponding CO_2 concentration are achieved.

1.4.3.6 Step 6 – Energy Limits Agree with Literature and Policy

In step 6, it is checked if environmental temperature, the CO_2 concentration and the gradients agree with literature and standards set by government policies. If the answer is no, the process is set back to step 4, where the limits are interrogated with the Tan *A* and gradient limitations. If on the contrary, the outcome is affirmative, compliant to government policy and the environment is deemed safe for human inhabitation.

1.5 Conclusion and Future Direction

It is confirmed repeatedly the implication of carbon dioxide emissions as the primary driver of global climate change. It is also widely recognised that to avoid the worst impacts of climate change, the world needs to urgently reduce greenhouse gases (carbon dioxide, CO_2), methane (CH_4), nitrous oxide (N_2O), dichlorodifluoromethane (CCl_2F_2), etc., especially CO_2

emissions. However, sharing this responsibility between regions, countries and individuals has been an endless point of contention at international arena, such as the COP.

1.5.1 Sustainable Development Plan

Ligardo-Herrera et al. (2022) contributed to the discussions on carbon emission that its reduction in buildings and whole districts is one of the main objectives of sustainable development plans for both policymakers and end-users. As a result, they developed an innovative planning tool to assess the effect of energy consumption on the overall carbon emissions of districts, buildings and house units. Whilst remote access is also considered in the case presented in this manual, it comes to the end-users and help them know the actual environmental impact of their homes, the environment and sound financial decisions can be made before investing in new equipment.

This compilation concludes that providing calculation tools for assessing both global warming and the CO_2 concentration and the duration of emission is pivotal to planning carbon transition to near zero emission and towards pure renewables.

Thus, imagining an imminent 'double carbon' era projects into industrial carbon reduction with high energy consumption and high carbon emission is imperative. With foregoing carbon emission is influenced by the following five variables, viz: energy intensity, energy structure, industrial structure, economic efficiency and employee scale (Ligardo-Herrera et al. 2022). As discussed above therefore, the manual provides reference and theoretical basis for the government to adjust the industrial structure, control industrial overcapacity, and realise the 'double carbon' goal as soon as possible, which attract optimised energy layout, ensure energy security and implement the new energy strategy.

Double carbon is beginning to attract government organisations and the private companies in planning new energies from wind power, photovoltaic power generation, energy storage and strategies in transition from carbon have introduced relevant policies to promote industrial development to high quality, high efficiency and low cost.

1.5.2 The Warning Signs and Prospective Solutions

CO_2 emission due to human activities is a small fraction of the total global climate load of greenhouse gas (GHG) infected earth. The oceans contain about 37 400 billion tons of carbon dioxide; land biomass contains some 2000 to 3000 billion tons; the atmosphere holds some 720 billion tons (Ouellette et al. 2016). At about 6 billion tons, the major sources of carbon include the burning of fossil fuels such as coal, oil and natural gas, industrial processes activities and production, deforestation, decomposition, ocean release and respiration. The natural sources include animals and plants respiration, soils respiration and decomposition, volcanic eruptions and the earth's oceans, whilst the natural sources on the other hand of the carbon cycle are the air, ground, oceans, plants, animals and fossil fuels. With the disruption of this carbon cycle therefore, human is now faced with but not limited to the following calamities: hotter temperatures, more severe storms, increased drought, warming and rising oceans, loss of species, unavailability of food, more health risks, poverty and displacements.

However, mitigation pathways that stabilise global warming to 1.5 °C would include deployment of land-based solutions such as increasing carbon sequestration in soils and biomass more now than before. Although this practice will call for additional demands for land, which will compete with food security and other ecosystem services, a carefully implemented sustainable intensification of agriculture and aquaculture should provide the potential to facilitate meeting demands for food and fibre, whilst simultaneously meeting environmental and ecological goals.

The next Chapter therefore presents vegetative sinks for carbon capture and sequestration as the processes of capturing molecules of CO_2, which is one of the leading greenhouse gases in the atmosphere. A Case Study and series of examples have also been cited on smallholder agriculture, food farmers, food garden, roof top farm, etc. in Africa.

References

Arena, W. 2013. *Using a Magnetic Compass and Triangulation to Calculate the Distance of a Faraway Object*. https://www.wildernessarena.com/environment/navigation/use-magn-etic-compass-triangulation-to-calculate-distance-of-object.

Cox, P.M., Betts, R.A., Jones, C.D. et al. (2000). Acceleration of global warming due to carbon-cycle feedbacks in a coupled climate model. *Nature* 408: 184–187.

IPCC (2013). Summary for policymakers. In: *Climate Change 2013: The Physical Science Basis. Contribution of Working Group I to the Fifth Assessment Report of the Intergovernmental Panel on Climate Change* (ed. T.F. Stocker et al.). Cambridge, United Kingdom and New York, NY, USA: Cambridge University Press.

James, G. (1999). *Advanced Modern Engineering Mathematics*, 2ee, 49. Harlow, England: Addison-Wesley.

Ligardo-Herrera, I., Quintana-Gallardo, A., Wolfgang Stascheit, C.W., and Gómez-Navarro, T. (2022). Make your home carbon-free. An open access planning tool to calculate energy-related carbon emissions in districts and dwellings. *Energy Rep.* 8: 11404–11415.

NASA (2004). *Global Temperature Anomalies*. NASA Earth Observatory https://earthobservatory.nasa.gov/images/5236/2004-global-temperature-anomalies.

Nightingale, A.J. (2009). Triangulation. In: *International Encyclopedia of Human Geography*, 489–492. Elsevier Ltd.

Nightingale, A.J. (2020). Triangulation. In: *International Encyclopedia of Human Geography*, 2ee, 477–480. Elsevier Ltd.

Ouellette, R.P., John, F., Munro, M.A. (2016). *The Politics of Climate Change*. https://www.researchgate.net/publication/299560962_The_Politics_of_Climate_Change.

Posavec, K., Giacopetti, M., Materazzi, M., and Birk, S. (2017). Method and excel VBA algorithm for modeling master recession curve using trigonometry approach. *Groundwater* 55 (6): 1–8.

Rust, B.W. (2011). A mathematical model of atmospheric retention of man-made CO_2 emissions. *Math. Comput. Simul.* 81: 2326–2336.

Rust, B.W. and Thijsse, B.J. (2007). Data-based models for global temperature variations. In: *Proceedings of the 2007 International Conference on Scientific Computing* (ed. H.R. Arabnia, J.Y. Yang, and M.Q. Yang), 10–16. Las Vegas, NV: CSREA Press.

Stips, A., Macias, D., Coughlan, C. et al. (2016). On the causal structure between CO_2 and global temperature. *Sci. Rep.* 6: 21691. https://doi.org/10.1038/srep21691.

Thijsse, B.J. and Rust, B.W. (2008). Freestyle data fitting and global temperatures. *Comput. Sci. Eng.* 10: 49–59.

2

Vegetative Sinks for Carbon Capture

Carbon sequestration is the process of capturing molecules of carbon dioxide (CO_2), which is one of the leading greenhouse gases (GHGs) from the atmosphere, thereby reducing the total concentration in the atmosphere. Trees, forests and even microalgae are, therefore, vegetative sinks that sequester carbon through photosynthesis. These sinks use light, the captures CO_2 molecules and combining them with water molecules to create sugar and other biomass that are stored in the forest within living biomass, soil and litter and contribute to the forest carbon stock. Oxygen (O_2) is then released into the atmosphere as a by-product of this process. This process happening in forests is part of the carbon cycle that helps regulate the climate. Increasing planting of trees, therefore, will increase carbon sequestration, which will in turn remove additional carbon released through combustion of fossil fuels. The young trees because of their high rate of growth either in number or in size are preferred in a carbon market context due to their corresponding high rate of carbon capture.

The forest carbon sink describes the natural reservoir that accumulates and stores sequestered carbon, which is dynamic, whereby carbon is gained by sequestration through photosynthesis and it lose by emission through forest fires (UK Forests and Climate Change 2018).

Ciais et al. (2014) reported that from the 1980s to date about ¼ of annual fossil fuel and land-use CO_2 emissions could have been sequestered by natural terrestrial ecosystems, which in turn assist in curtailing the rate and magnitude of global climate change. However, current environmental devastations such as water flooding, heat waves, wildfires and meteorological droughts seen in many countries the sustainability of this natural CO_2 sink has progressively become inadequate.

As alluded in Chapter 1, if increasing atmospheric carbon content at a fraction of 1 remains unabated through 2100, the semi-arid regions of the planet, which account for about 41% of Earth's land surface (Dass et al., 2018) are expected to become even drier (Collins et al., 2013). Many recent studies support the trend that CO_2 uptake is dominated by the response of semi-arid vegetation to precipitation and temperature (Ahlstrom et al., 2015). Tree mortality has also been related to drought and heat stress coming from dry climates, which continues to pose threats to sustained carbon storage by forests in such regions (Allen et al., 2010, 2015; Anderegg et al., 2015; McDowell and Allen, 2015).

Grasslands, on the other hand, sequester most of their carbon in the roots and soil underground making them more adaptive to climate change, while forests store their carbon mostly in woody biomass and leaves. Such that, when wildfires cause trees to go up in flames, the burned carbon they formerly stored is released back to the atmosphere, when fire burns grasslands, the carbon is fixed and remains underground. Thus, promoting reliable rates of carbon sequestration in plant agricultural practices could help readily meet carbon emission–reduction goals.

2.1 Plant Agricultural Practices as Carbon Dioxide Sink

The understanding of photosynthesis informs that most life on Earth depends on this process called photosynthesis. The process involves plants, algae and some types of bacteria, which capture energy from sunlight to produce oxygen (O_2) and chemical energy, stored in glucose as sugar. Herbivores obtain their energy by eating plants, and carnivores obtain it by eating herbivores, while these living things burn the O_2 within them and release CO_2 to complete the carbon cycle as partly shown in Figure 2.1.

Measuring Climate Change to Inform Energy Transitions: Carbon Footprint Calculations, First Edition. Sunny E. Iyuke.

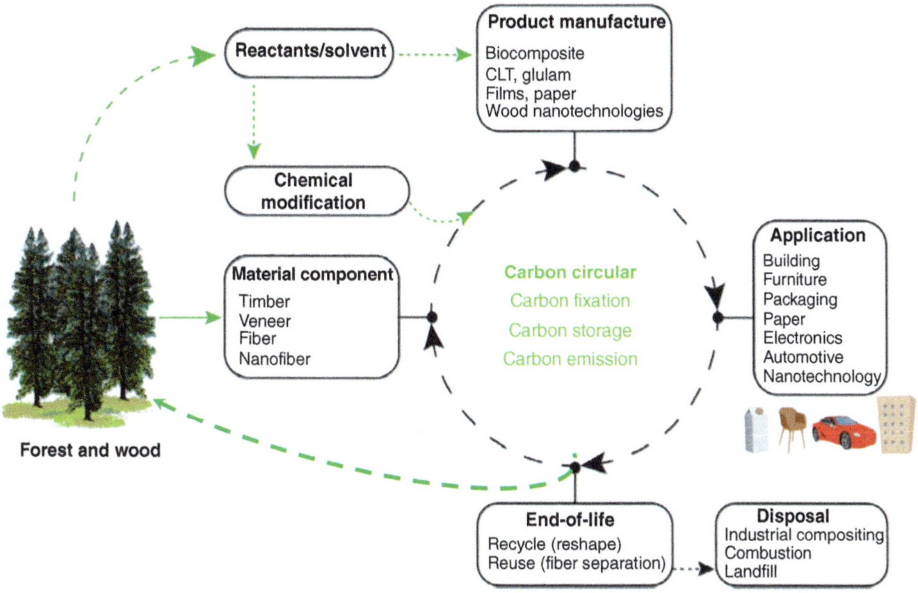

Figure 2.1 Modified carbon cycle in forest and wood for sustainable development of wood products towards extended carbon storage and CO_2 capture. Source: Montanari et al./Frontiers Media S.A./CC BY 4.0.

The forests are known to be the most important carbon sinks that contribute to CO_2 removal from the atmosphere. Carbon naturally and can artificially get sequestered and locked in trees through these CO_2 photosynthesis reactions, which give forests a critical role in tailoring the atmospheric carbon concentration during the global industrial age (Thomas and Martin, 2012). Human actions have been breaking the natural balance between carbon absorption and emission because the non-sustainable fossil fuels and non-biodegradable petroleum-based plastics are consumed very rapidly. The high rate of use leads to increasing release of CO_2 into the atmosphere resulting into global warming, which needs to be controlled efficiently as discussed in the preceding chapter. It, therefore, becomes worthwhile to implement management strategy for the forest carbon sink and utilisation of wood products in order to positively mitigate the emission of CO_2 that is resulting into climate change.

Harvesting wood and wood products on the other hand also influence the carbon cycle of the forests depending on the level of deforestation. At present, human beings have high demand for timber products at a higher rate than how fast trees are being planted. Thus, the replacement cycle of 15–25 years rotation period for trees or wood resources is longer than their use for supplying energy, burning, pulping and paper production (Geng et al., 2017). To abate the situation, new technologies for carbon capture and fixation of wood products over longer periods will have to be developed. Buchanan and Levine (1999) have shown that in New Zealand that 17% increase in wood used in the built environment has a direct impact on carbon emissions, reducing them by 20%. This estimation was based on using wood instead of brick, aluminium and other building products that require more process energy than wood. In abatement efforts, researchers are developing functional wood materials for the extension and increase of carbon storage and CO_2 capture (Montanari et al., 2021). Wood, being a highly porous material with a hierarchical cell wall structure that provides water and nutrient transportation and mechanical support functions (Cai et al. 2022), is often subject to cracking, degradation, aging and dimensional instability, which leads to the reduction in its service lifetime resulting into easy return of the carbon to the atmosphere (Santori et al., 2018). Thus, in order to extend service life span for wood. Its drying, preservation and modification are applied to reduce the moisture content and treat or alter the cell wall to also improve the quality of wood products and extend carbon storage. However, emerging technologies are now being developed in the use of wood for capture CO_2 from the air, e.g. membrane adsorption, electrocatalytic and photocatalytic reactions.

2.1.1 Kyoto Protocol Requirement for Net Annual Rate of Carbon Dioxide Accumulation

In the case of the United Kingdom for instance, the net annual rate of CO_2 accumulation by the forests is projected to fall from around 21 million tonnes CO_2 in total in 2020 to around 19 million tonnes of CO_2 by 2030, as presented in Table 2.1 (UK Forests and Climate Change 2018). They also reported that under the Kyoto protocol, additional woodland planted since 1990 contributes to the UK's CO_2 emissions target.

Table 2.1 Net annual $CO_2 \times 10^6$ t/per year sequestered by UK forests from 1990 to 2050 due to land use, land use change and forestry, produced by the Centre for Ecology and Hydrology.

Year	Living biomass	Total carbon	Afforested since 1990	Emissions due to land deforested	Total CO_2 (ppmv × 10^3)
1990	12.3	15.0	−0.1	−0.2	4.05
1995	13.4	17.9	−0.5	−0.2	4.83
2000	14.0	19.8	−0.6	−0.8	5.35
2005	12.7	20.1	−0.4	−1.1	5.43
2010	11.6	21.0	0.4	−1.4	5.67
2015	10.9	21.3	1.9	−1.3	5.75
2020	10.9	20.5	3.7	−1.1	5.54
2025	10.0	20.2	5.5	−1.0	5.45
2030	8.9	19.3	7.0	−0.9	5.21
2035	8.9	18.6	8.2	−0.7	5.02
2040	9.5	18.7	8.1	−0.6	5.05
2045	9.8	18.9	7.3	−0.6	5.10
2050	9.4	18.9	6.6	−0.5	5.10

Source: Adapted from UK Forests and Climate Change.

Thus, the application of Eq. (1.3) to determine temperature changes with respect to sequestration of CO_2 Eqs. (1.6) and (1.9) are rearranged into;

$$T(t) = T_0 + \eta \cdot C(t) - \eta \cdot 277.04 + \frac{A}{\tau} 2\pi \cdot t + A\emptyset \tag{2.1}$$

$$\therefore C(t) = \frac{T(t) - T_0}{\eta} + 277.04 - \frac{A}{\eta}\left[\frac{2\pi}{\tau}(t + \emptyset)\right] \tag{2.2}$$

Thus, a plot of the concentration of CO_2, $C(t)$ against t in Eq. (2.1), gives a negative gradient as a measure of rate of CO_2 sequestration. While introducing the parameters from Eq. (1.10) into Eq. (2.1), as instructed by the scheme in Figure 1.11. produces the temperature versus CO_2 concentration of Figure 2.2 as a measure of global warming.

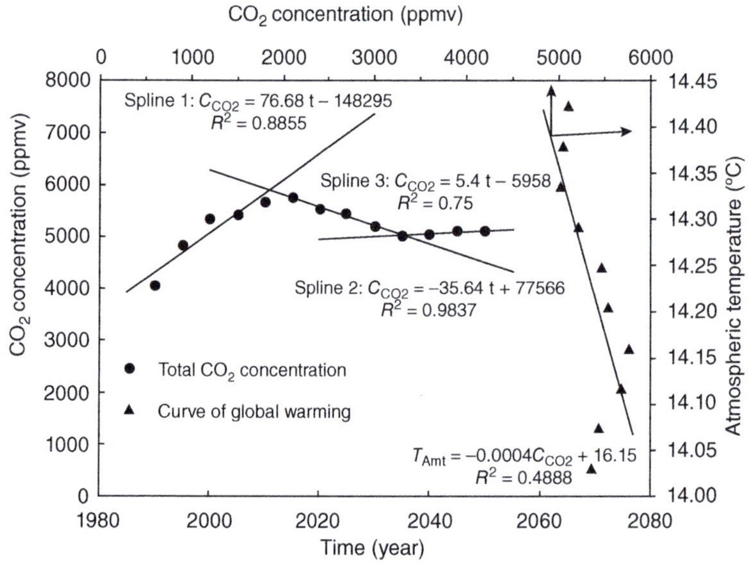

Figure 2.2 Determination of global warming by triangulation of the net annual data of CO_2 concentration sequestered per year by the UK forests projected from 1990 to 2050 due to land use, land use change and forestry, produced by the Centre for Ecology and Hydrology.

As seen from Figure 2.2, increasing CO_2 concentration from 1990 to 2010 at a gradient of 76.76 and then followed by a negative gradient of −35.64 (2010–2035), while the 3rd spline has a gentle slope of 5.4, which is still higher than the minimum value of $0.500\Delta^0C/\Delta Yr$ proposed in Chapter 1. Upon the analysis of the measured atmospheric temperature against CO_2 concentration, again a negative gradient of $-0.0004C_{CO2}$ is obtained, which supports the earlier negative value $-0.0004 \leq 0.500\Delta^0C/\Delta CO_2$ during the second spline 2010–2035. In both cases, therefore, the trigonometric triangulation projection supports the fact that the 'Net annual change in CO_2 equivalent model adopted by the UK woodlands' is presenting a positive results to global warming and in turn should assist in taking decisions for the efforts to transit carbon economy.

It is, therefore, expedient to quote the UK Forests and Climate Change future predictions that carbon uptake may assume that commercial conifer plantations should be replanted when felled. That, also planting of new woodland to support the current Rural Development Plan should be encouraged because the CO_2 release is still on the increase as shown by the gradient of $5.4 \geq 0.500\Delta^0C/\Delta Yr$ from 2035 to 2050 of Figure 2.2.

2.2 Microalgae as a Biological Capture for CO_2

It is obvious from the preceding discussion that the amount of GHG emissions to the atmosphere is resulting in the 1.1 °C rise in average global temperatures from the levels before industrial revolution in 1800s. The Paris Climate Agreement was aimed to limit global temperature increase to 1.5 °C, which has also been confirmed in the 2022 COP27 in Egypt. The main GHG contributor is CO_2, mainly emitted from energy generation, transport and industrial sectors (including food, drink, cement and other manufacturing industries).

To reduce carbon dioxide emissions from power generation of fossil fuels- and biofuels-based power plants, CO_2 capture is an indispensable technology. The three main methods to capture carbon dioxide from power plants are pre-, post- and oxy-fuel combustion capture. The pre-combustion carbon dioxide capture is the removal of CO_2 from the fossil fuel before completion of the combustion. In gasification and reforming processes, for instance, the coal/biomass is partially oxidised in steam and oxygen/air under high temperature and pressure to form synthesis gas before the actual combustion. The process is commonly referred to as integrated gasification combined cycles (IGCC), whereby gasification and partial oxidisation of the fuel produce CO_2 and hydrogen, which are further separated by absorption processes (Park et al., 2021).

A water-gas-shift (WGS) reaction -$CO + H_2O \rightarrow H_2 + CO_2$- and the pre-combustion capture involve some efficiency penalties compared to post-combustion technology, which separates dilute CO_2 around 5–15% in concentration in the flue gas after the fossil fuel has undergone combustion, generating CO_2 in the flue gas as;

$C + \frac{1}{2}O_2 \rightarrow CO$	gasification with oxygen
$C + O_2 \rightarrow CO_2$	combustion with oxygen
$C + CO_2 \rightarrow 2CO$	gasification with carbon dioxide
$C + H_2O \rightarrow CO + H_2$	gasification with steam
$C + 2H_2 \rightarrow CH_4$	gasification with hydrogen
$CO + H_2O \rightarrow H_2 + CO_2$	water-gas shift reaction

Although the pre-combustion capture proves to be more efficient and the hydrogen produced can be utilised for transportation fuel, the capital costs of the base gasification are higher than conventional pulverised coal power plants. The advantages of post-combustion technology, however, can easily be adapted to already existing power plant. One of the disadvantages for both technologies include large sizes of equipment as a result of high volume of flue gases and the low partial pressure of CO_2, energy-intensive cooling systems, pre-treatment to remove impurities and considerable cost of capture (Jiang et al., 2019; Nessi et al., 2021; Omoregbe et al., 2020).

Sharma and Maréchal (2019) proposed an on-board CO_2 capture and storage unit for an internal combustion engine for the transportation sector. The adsorption–desorption isotherms on vehicle with CO_2 liquefaction is an alternative CO_2 capture strategy in the transportation sector, which was applied to truck transport. Microalgae can also be used in carbon capture for a sustainable environment, especially carbon footprint emanating from the transportation sector. In this case, CO_2 emission in the transportation sector can be reduced by capture and store CO_2 onboard. This is buttressed by the fact that terrestrial plants assimilate CO_2 through photosynthesis to support respiration, growth and reproduction. Such that, atmospheric CO_2 is transformed into plant biomass through photosynthetic processes (Figure 2.3).

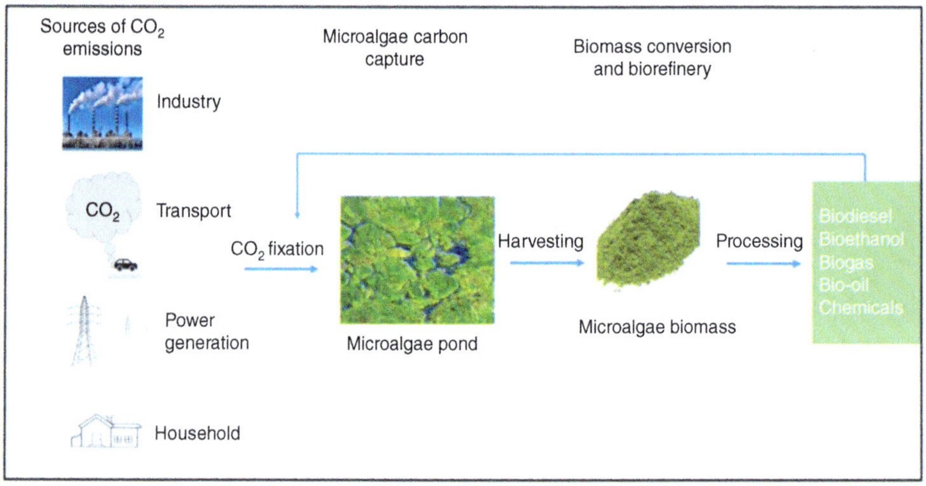

Figure 2.3 Schematic representation of microalgae cultivation, photosynthetic fixation of CO_2 into microalgae biomass and biomass processing. Source: Onyeaka et al./Elsevier/CC BY 4.0.

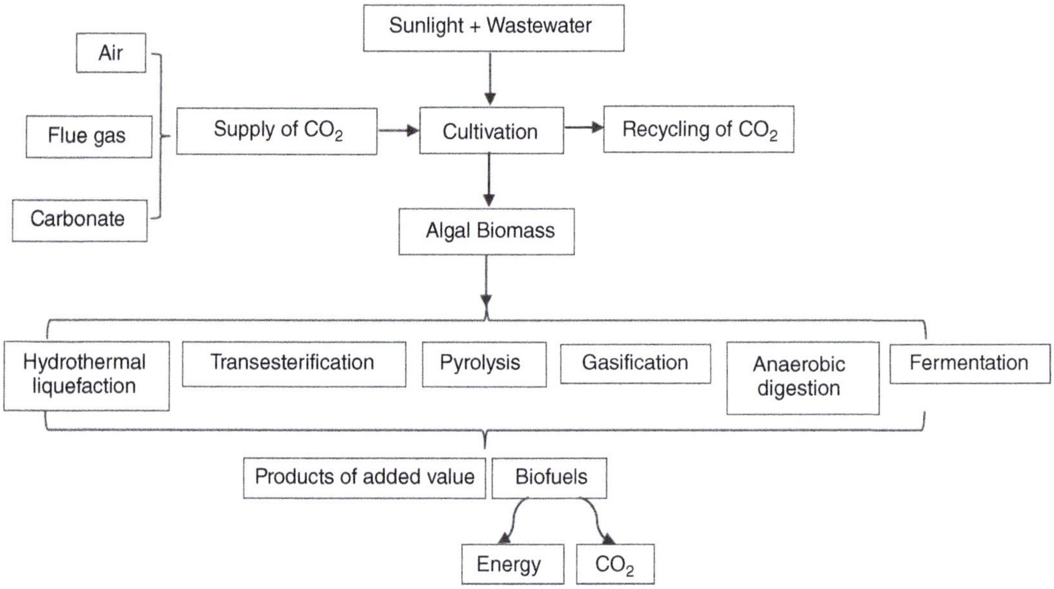

2.3 Cultivation and Adaptation of Microalgal Communities to 100% Coal-Fired Flue Gas

Aslam et al. (2017) constructed and studied an experimental setup of photobioreactors (PBRs) that was connected to a coal-fired flue gas in an outdoor culture system, which consisted of twelve 30 L photobioreactors (PBRs) (Figure 2.4). The set up was equipped with conical transparent polyethylene (PE) bags (50 cm height and 40 cm diameter) containing 15 L culturing volume and 15 L gas space, each as shown in Figure 2.4. Although a detail description of the photobioreactors is fully presented and discussed in the literature (Aslam et al., 2017), the bag sleeves were sealed at the bottom in a conical shape where also an inlet for aeration was located in order to prevent cell settling. PBRs were each hung on a steal rod supported by an iron frame. This PBR system contained twelve individual bags, as each of the two microalgal communities had three replicates with and without flue gas supplementation.

The proximate and constituent compositions were obtained from the flue gas of a typical 4 MW coal-fired boiler including 11.24% CO_2, 8.26% O_2, 388.8 mg/Nm^3 CO, 423.9 mg/Nm^3 NO_2, 781.8 mg/Nm^3 SO_2, 2.8 mg/Nm^3 SO_3 and 0.4 mg/Nm^3 while the concentrations of NO_2 and SO_2 were obtained by proportion (Aslam et al., 2017). Since the photosynthetic reactions

Figure 2.4 Experimental setup of photobioreactors that was designed and connected to coal-fired flue gas by Aslam et al./with permission of Elsevier.

Table 2.2 Flue gas composition, assuming 90% CO_2 uptake efficiency and 20 g DW/m^2 day productivity, 2.73 ha would be required to capture 1 ton CO_2/day, or 224 km^2, where a typical 500 MW coal power plant emits 3 million ton CO_2 and 1 kg of dry algal biomass utilises about 1.83 kg of CO_2 per year.

Percentage composition of CO_2 (%)		Component concentration (ppmv)					
		Replete			Deplete		
Proximate	Constituent	CO_2	NO_2	SO_2	CO_2	NO_2	SO_2
10	1.1	984.1	423.9	781.8	837.5	396.8	781.8
30	3.3	8857.2	1271.7	2345.4	7537.5	3571.3	2345.4
50	5.5	24 603.3	2119.5	3909.0	20 937.4	9920.3	3909.0
100	11.0	98 413.3	4239.0	7818.0	83 749.7	39 681.3	7818.0

Source: Adapted from Aslam et al.

where CO_2, NO_2 and SO_2 were consumed are too complicated to resolve within the scope of this book, the amounts of these constituents consumed are estimated from the initial proportion by proportion as presented in Table 2.2. It was also reported that only 7.6–14.9% of the flue-derived CO_2 was absorbed after 10 min, thus, the display of rapid growth and high biomass productivity, which is directly related to CO_2 uptake and low pH/high SO_2 and NO_2 tolerance.

Based on their reports, overall biomass productivities were from 50% flue gas supplementation during the nitrogen-replete or deplete conditions. The final biomass optical density of 450 nm and pH conditions after supplementation with 100% flue gas for 2 weeks proved that the microalgae consumed 7.6–14.9% of CO_2, but the concentrations of NO_2 and SO_2 remained unconsumed at the period of 10 minutes in the presence or absence of phosphate buffer compared to air-only control cultures.

Thus, the application of triangular plot for the three GHGs of CO_2, NO_2 and SO_2 provides the comparison of the constituent compositions before and after treatment with mixed culture of the microalgae (Figure 2.5). As seen from Figure 2.5, the concentrations of CO_2, NO_2 and SO_2 in the raw flue gas give gradient of 6.7376w_{ppmv}, while after treatment with microalgae, the gradient becomes 1.0269, which is higher than the minimum value of 0.500Δ^0C/ΔYr proposed in Chapter 1. Although the gradient value after the treatment with the microalgae is much less than that of the raw flue gas composition, it means the treatment with the mixed culture microalgae could not achieve the set standards to combat global warming or climate change. It is, therefore, necessary to engage hydroponic agricultural technologies to resolve climate change and its effects on food security and CO_2 sink as discussed in the next.

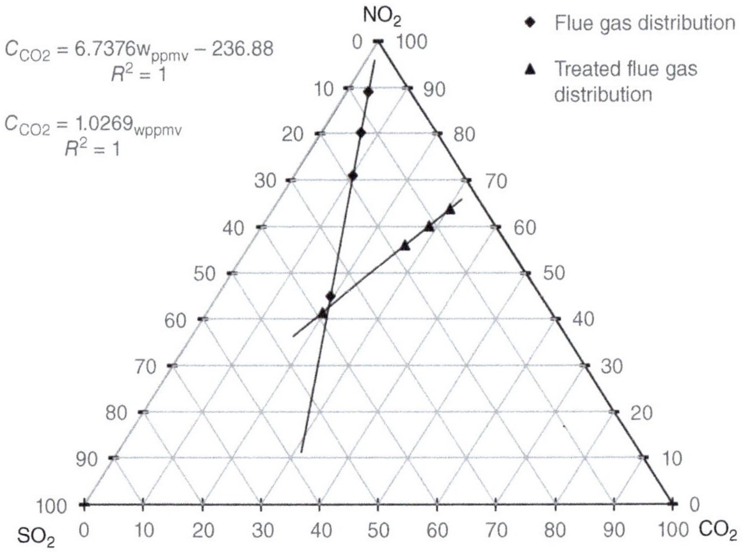

Figure 2.5 Mitigation of industrial CO_2 emission from the unfiltered smoke stack of a 4 MW coal-fired power plant through biotransformation into microalgal biomass.

2.4 Status of Food System, Food Insecurity and Climate Change

Mbow et al. (2019) defined food system as a system that encompasses all the activities and actors in the production, transport, manufacturing, retailing, consumption and waste of food, and their impacts on nutrition, health and well-being and the environment. food security, therefore, is an outcome of the food system that in turn generates impacts on the environment. This implies that environmental impacts and food systems contribute to GHG emissions and climate change. It is understandable that climate change invokes complex interactions within food systems that lead to food insecurity which consists food availability, access, utilisation and stability (Table 2.3). It is interesting to recognise that demand for food and its supply are interlinked such that they should be jointly assessed in order to identify the challenges of mitigation and adaptation to climate change. Since agricultural production of crops may contribute to both GHG emission and sink simultaneously, the demand for food shapes what is grown, where it is grown and how much is grown. Thus, mitigation and adaptation of agricultural activities should involve modifying production, supply chain and demand practices such as dietary choices, market incentives and trade relationships. FAO (2001) highlighted that food security is a situation that exists when all people, at all times, have physical, social and economic access to sufficient, safe and nutritious food that meets their dietary needs and food preferences for an active and healthy life. The highlight includes that – all people at all times' implies the need for equitable and stable food distribution, however, it has increasingly been recognised that it also covers the need for inter-generational equity and, therefore, 'sustainability' in food production –. Safe and nutritious food for a healthy life implies that food insecurity can occur if the diet is not nutritious, when there is consumption of an excess of calories, unsafe food and when it is not free from harmful substances. The overwhelming impact of food insecurity, therefore, is malnourishment, which can also be referred to as malnourishment or deficiencies, excesses and imbalances in a person's intake of energy and/or nutrients.

In another instance, FAO, IFAD, UNICEF, WFP and WHO (2018) defined undernourishment as when an individual's habitual food consumption is insufficient to provide the amount of dietary energy required to maintain a normal, active and healthy life. In terms of insufficient calories or hunger, undernourishment occurs when individual experiences nutritional deficiencies in vitamins and minerals (e.g. vitamin A, B and C, and iron, zinc and iodine respectively), which are times referred to as 'hidden hunger'

In countries with high levels of undernourishment, hidden hunger is prevalent (Muthayya et al., 2013), while micronutrient deficiency does occur in countries with low prevalence of undernourishment. In other words, it has been observed that teenage girls suffer from iron deficiency and calcium deficiency is common in Western-style diets (Aslam and Varani 2016) in many parts of the world. It, therefore, means that food security is related to nutrition, and conversely, food insecurity is related to malnutrition. Therefore, effects of climate change on food security may be projected to negatively impact the four

Table 2.3 Relationships between food security, the food system and climate change.

Food security pillar	Examples of observed and projected climate change impacts	Examples of adaptation and mitigation
Availability: Production of food and its readiness for use through storage, processing, distribution, sale and/or exchange	Reduced yields in crop and livestock systems	Development of adaptation practices
	Reduced yields from lack of pollinators; pests and diseases	Adoption of new technologies, new and neglected varieties
	Reduced food quality affecting availability (e.g. food spoilage and loss from mycotoxins)	Enhanced resilience by integrated practices, better food storage
	Disruptions to food storage and transport networks from change in climate, including extremes	Reduction of food demand by reducing waste, modifying diets
		Closing of crop yield and livestock productivity gaps
		Risk management, including marketing mechanisms, financial insurance
Access: Ability to obtain food, including effects of price	Yield reductions, changes in farmer livelihoods, limitations on ability to purchase food	Integrated agricultural practices to build resilient livelihoods
	Price rise and spike's effects on low-income consumers, in particular women and children, due to lack of resources to purchase food	Increased supply chain efficiency (e.g. reducing loss and waste)
	Effects of increased extreme events on food supplies, disruption of agricultural trade and transportation infrastructure	More climate-resilient food systems, shortened supply chains, dietary change, market change
Utilisation: Achievement of food potential through nutrition, cooking, health	Impacts on food safety due to increased prevalence of microorganisms and toxins	Improved storage and cold chains
	Decline in nutritional quality resulting from increasing atmospheric CO_2	Adaptive crop and livestock varieties, healthy diets, better sanitation
	Increased exposure to diarrheal and other infectious diseases due to increased risk of flooding	
Stability: Continuous availability and access to food without disruption	Greater instability of supply due to increased frequency and severity of extreme events; food price rises and spikes; instability of agricultural incomes	Resilience via integrated systems and practices, diversified local agriculture, infrastructure investments, modifying markets and trade, reducing food loss and waste
	Widespread crop failure contributing to migration and conflict	Crop insurance for farmers to cope with extreme events
		Capacity building to develop resilient systems
Combined Systemic impacts from interactions of all four pillars	Increasing undernourishment as food system is impacted by climate change	Increased food system productivity and efficiency (e.g. supply-side mitigation, reducing waste, dietary change)
	Increasing obesity and ill health through narrow focus on adapting limited number of commodity crops	Increased production of healthy food and reduced consumption of energy-intensive products
	Increasing environmental degradation and GHG emissions	Development of climate-smart food systems by reducing GHG emissions, building resilience, adapting to climate change
	Increasing food insecurity due to competition for land and natural resources (e.g. for land-based mitigation)	Governance and institutional responses (including food aid) that take into consideration gender and equity

Source: Adapted from Mbow et al.

Figure 2.6 Wits Food Sovereignty Centre in SA Food Sovereignty Campaign (Wits, 2023). Source: Jane Cherry/South African Food Sovereignty Campaign.

pillars of food security – availability, access, utilisation and stability – and their interactions (FAO, IFAD, UNICEF, WFP and WHO 2018).

Table 2.3 on the relationships between food security, the food system and climate change has been prepared to strengthen the understanding of how climate change affects each of these pillars across the full range of food system activities and relevant examples, under the following five criteria:

- Production of food and its readiness for use through storage, processing, distribution, sale and/or exchange
- Ability to obtain food, including effects of price
- Achievement of food potential through nutrition, cooking, health
- Continuous availability and access to food without disruption
- Combined Systemic impacts from interactions of all four pillars

Thus, in response to local food security issues amongst students on the campus of the University of the Witwatersrand in Johannesburg, five food gardens (Figure 2.6) were established within the University of which, two provide students with access to nutritious food, presented as case study in the next sections (Wits, 2023).

2.5 A Case Study: Student Food Garden at the Wits University

(Reprinted with permission from Ms. Nyiko Ngxola of the Wits University)

Groups of students signed up to be a part of a society to work on the ground in the season before harvest. A survey was therefore conducted towards obtaining insights of how students in the School of Mechanical, Industrial and Aeronautical Engineering (MIA, which is subsequently referred to as 'School' in the rest of the text) of Wits University access nutritious food groups in their day-to-day lives. It is, however, realised that there is an even larger underlying issue at hand, such that although only 100 responses were received from a school of just over 1000 students, 10% of the students that responded are currently receiving meals or have received meals before from the schools' feeding schemes. In some of the responses, students shared that they noticed that some of their fellow students may be suffering from hunger, but they do not know how to approach them to lend a helping hand. It is disheartening to learn that there are still fellow students suffering in silence although there is help on campus and within the school. However, there is not enough that is being done by students in an effort to solve the problems of the students because of lack of awareness. Of the 100 students who participated in the survey, 75 of them said that they could be interested in being involved in the school food garden initiative. This implies that there are at least 75 pairs of hands from this school who are willing to work the ground so that fellow students have access to nutritious food on campus.

This design project, therefore, involves developing a nutritious food garden that is aimed at providing 1000 cal of nutritious food groups to at least 25 students and to estimate the quantity of CO_2 required for a 6 m^2 of land. The design focuses on developing an external structure to protect the food crops from environmental loads and weather conditions experienced in Johannesburg. The design also focuses on creating the ideal climate for the food crops inside the structure while ensuring that the crops received sufficient irrigation and nutrients. The food garden is determined to be able to provide just under 30 000 cal (per 80-day grow period) of nutritious food crops at an energy cost of R0.053c/cal. The structure weighs 160.69 kg/m^2; therefore, it can be placed on a concrete roof of office buildings.

2.5.1 Background and Motivation

As human population increases, the demand for food is also increasing, while the arable land per capita is being reduced throughout the world (Gruda, 2019). By 2050, therefore, it is expected that the global population will reach approximately 10 billion people and it is estimated that by this time, 66% of the world's population will live in urban areas (United Nations, 2014). While population growth in cities creates a demand for fresh produce to ensure a healthy population, agricultural produce still has to travel a long way to reach the consumer, the produce loses quality and nutritional value along the way and a significant fossil fuel cost for transportation and storage is required. The average plate of food will have to travel over 2400 km before it reaches the final consumer. In such situation, vertical hydroponic farming (VHF) offers many advantages over conventional farming including more efficient water and land use, which will reduce transport costs if it is undertaken in the urban area. The model becomes economically sustainable if it is integrated into existing urban infrastructure. It also becomes mutually beneficial if VHFs are integrated as feedback into District heating (DH) areas during winter, to lower the return temperature in line with fourth-generation DH guidelines, and reduce CO_2 emissions in food and energy production (Gentry, 2019).

The World Health Organisation (WHO), on the other hand, has defined good nutrition as the intake of food that meets the body's dietary requirements. WHO further states that an adequate, well-balanced diet, coupled with physical activity is 'the cornerstone to good health'(WHO, 2018). The consumption of nutrient-dense foods, which is food that is high in nutrients but low in calories, improves immunity, decreases susceptibility to disease and promotes physical and mental development and productivity (Anon, 2018).

As defined earlier, food insecurity refers to the lack of access to nutritious foods for short or prolonged periods, resulting in malnutrition and the physiological state termed hunger. According to Statistics South Africa, in 2017, 20% of South African household's food access status ranged from inadequate to severely inadequate (Maboko, 2014). The 2017 Wits Food Bank statistics partially reflected South African statistics, as it was revealed that monthly food parcels were provided to over 3000 (of the 38 000) enrolled students (Ryan, 2020). A growing number of students at Wits are receiving support from the Wits Food Garden (Figure 2.7); however, the demand is not sufficiently met due to low food crop yields and high maintenance requirements. There are other intermittent supplies of fresh vegetables and cooked meals on campus, mainly focusing on the June examination period, provided by the Wits Siyakhana Food Project and the Masidleni Meal Project, which fall under the Wits Food Sovereignty Centre (Wits, 2023). However, the outreach falls short because of the limited catering capacity of only 800 students. This further emphasises the growing need for sustainable and accessible means of providing students with nutritious food crops on campus.

2.5.1.1 Design Motivation

The 3000 students at the University of the Witwatersrand have seemed to be deprived of the nutritional requirements that constitute a balanced diet. This translates to about 50 students in the MIA. The lack of a student's access to a nutritious diet impairs their cognitive abilities and serves as a hindrance in applying themselves sufficiently in their pursuit to completing the degree's requirements. This the calls for design of vertical farms to be used by students throughout the university for the growth of fresh produce. It is, therefore, necessary to design a sustainable urban farm that will provide students in the school with nutritious food crops. The fresh produce must be grown locally on campus and must be accessible to students throughout the year. The food crop yields must be enough to supplement the student's recommended daily requirements of fruits and vegetables, to form part of a balanced diet. The food farm is operated and maintained by students for the students.

Figure 2.7 Wits food garden. Source: Ngxola/with permission from University of the Witwatersrand, Johannesburg.

This arrangement results into the Wits Inala Food Garden Program, which is currently running three gardens on campus. The first garden is located on west campus at Wargames, and the other two are located at the International House and Sunnyside residence. The programme is also running two gardens off campus in Sunninghill and at the National School of Arts. The gardens are about 360 m^2 in area and the main crop produce includes leafy greens (spinach, lettuce, cabbage and herbs) as well as beetroot and carrots. The gardens mainly supply fresh food crops to the students and staff members involved in the planting and maintenance of the gardens. The rest of the produce is 3 donated to the Wits Masidleni Daily Meal Project and to the community outreach programmes run by the Wits Citizenship and Community Outreach (WCCO).

2.5.1.2 Johannesburg Food Farmers Rooftop Farm

On the rooftops of the building located at 77 Juta Street Braamfontein, a rooftop garden is developed. The garden floor space is only 66 m^2; however, the food farmers are able to produce leafy greens (spinach, lettuce, cabbage and mint), strawberries, lemons and limes, cacti plants, green beans, violas and lavender. The Johannesburg Food Farmers (JFF) Rooftop Farm (Figure 2.8) is developed to provide city dwellers with fresh produce grown organically and locally as well as a botanical environment for locals to enjoy coffee and tea.

The food crops are planted in wooden raised beds for better drainage, weed growth prevention and better exposure to the sunlight. The rooftop garden currently has no irrigation systems in place, the JFF staff water the plants using water cans. The JJF Food Garden managers provide workshop sessions to customers interested in growing their own miniature gardens, as well as provide additional materials required to plant the food crops such as grow media (peat, vermiculite and succulent mix) and worm castings. There are plans to expand the programme throughout the city; however, the roof space rent serves as a major hurdle to the expansion plans.

2.5.1.3 Nutrition and Recommended Dietary Allowance

Enteral nutrition is the intake of food sources, necessary for growth, metabolism and repair. A balanced diet ensures that all the necessary food groups that provide the certain aspects of nutrition are included in one's daily intake of food. Figure 2.9 depicts the classification of food groups that form part of a balanced diet.

Micronutrients are the chemical elements required to perform a host of bodily functions essential for growth and development. Thus, a diet richly supplemented with micronutrients improves immune system function and prevents against cardiovascular disease while reducing the production of free radicals in the body (Sloss, 2011). Although micronutrients are only required in trace amounts, deficiencies often lead to malnutrition resulting in impaired physical and cognitive function and development. The deficiency disorders are devastating and mostly irreversible, for instance, iodine deficiency disorder. The human body requires only 150 μg of Iodine daily, however, if it is consistently lacking in one's diet, it leads to reduced cognitive ability and significantly lower IQ levels (Sloss, 2011). Micronutrients are largely sourced from fruit and vegetable food crops and are sometimes provided as additional supplements, i.e. the iodisation of salt as recommended by the WHO and the United Nations Children's Fund (de Benoist, 2009).

Phytonutrients, antioxidants and vitamins are generally grouped together because of the common function of significantly improving the body's immunity by increasing resistance to disease and promoting the healthy function of systems. Antioxidants combat the presence of free radicals in the body. The free radicals are a result of exposure to radiation and polluted environments and abnormal functioning of metabolic processes in the body. The antioxidants neutralise the free radicals and exist in two forms: as antioxidant enzymes such as catalase (which protects cells from oxidative damage) or

Figure 2.8 Johannesburg food farmers rooftop farm. Source: Ngxola/with permission from University of the Witwatersrand, Johannesburg.

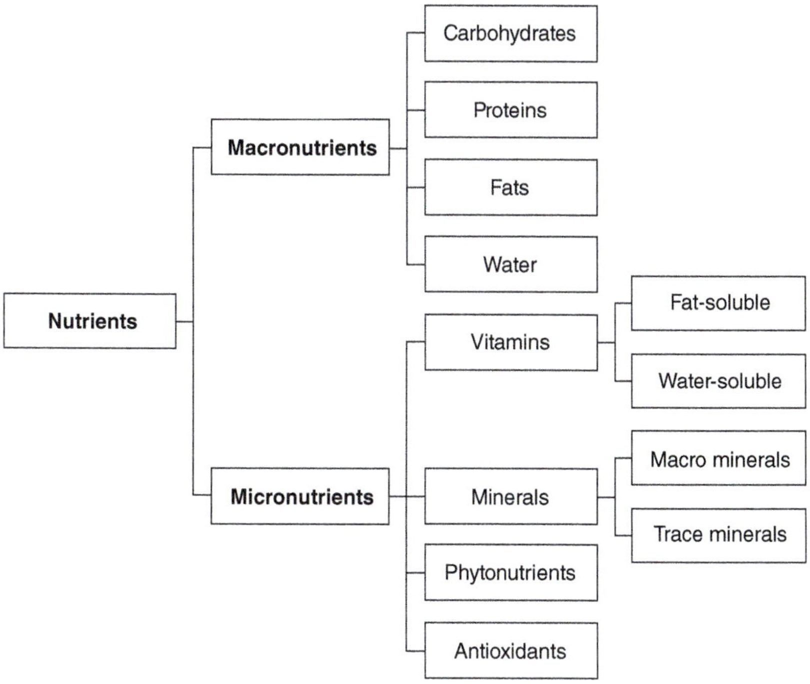

Figure 2.9 Nutrient classification tree.

as nutrient-derived antioxidants from phytonutrients and vitamins (such as vitamin C and flavonoids). Phytonutrients can convert to vitamins, for instance, alpha-carotene and beta-carotene are easily converted to vitamin A in the body (Sloss, 2011).

2.5.2 Conceptualisation of Design

The conditions of the Johannesburg climate in Table 2.4 can sustain seasonal plant growth with occasional stress periods on the food crops. However, in pursuit of providing students with food crops at considerable yields throughout the entire year, the low night and winter temperatures along with the low relative humidity are a hindrance. Therefore, a microclimate has to be developed around the food crops. It is for this reason coupled with the presence of environmental loads that the

Table 2.4 Climatic and soil condition requirements for selected food crops.

Food crop	Growth duration (days)	Climatic temperature (°C)	Soil pH levels	Plant spacing (m × m)	Companion crops
Cabbage	80–180	7–24	6.5–6.8	0.5 × 0.6	Corn, lettuce, spinach, cucumber
Cauliflower	75–85	10–24	6.5–6.8	0.5 × 0.6	Broccoli, spinach, cucumber, chard
Broccoli	100–150	10–24	6.0–6.8	0.3 × 0.6	Celery, onions, beetroot, garlic
Carrots	70–80	10–24	6.0–6.8	0.5 × 0.6	Chillies, onions, parsley, leeks
Beetroot	45–65	<27	6.5–7.0	0.1 × 0.25	Cauliflower, peas, onion, mint, garlic
Spinach	30–44	2–24	>7.0	0.3 × 0.3	Carrot, celery, cucumber, peppers
Lettuce	54–55	15–24	6.0–6.8	0.3 × 0.3	Aubergines, cucumbers, celery
Swiss chard	40	5–35	6.0–6.8	0.3 × 0.3	Tomatoes, beans, cabbage, radishes
Kale	55–70	−5–27	5.5–6.8	0.3 × 0.3	Artichokes, beetroot, celery
Butternut	110–120	15–29	6.0–6.5	0.4 × 0.5	Beans, peas, mint, sunflowers
Pumpkin	95–120	15–29	5.5–6.5	0.4 × 0.5	Thyme, sage, lavender, mint
Bell peppers	70–90	18–27	5.8–6.5	0.5 × 0.6	Carrots, cucumbers, radishes
Chilli peppers	10	12–27	6.0–6.8	0.1 × 0.1	Onions, garlic, leeks, chillies

Source: Adapted from Earheasy.

food garden concepts include external structures. Table 2.4 summarises the amount of time required to grow selected food crops, together with the required climatic temperature, the pH of the soil and the plant spacing. The table also includes the companion plants of the selected food crops.

The main food crop that determines the minimum depth of the growth beds is carrots and beetroot that require about 60 cm depth length (Earheasy, 2016). The plant spacing requirements need to be adhered to in order to prevent excessive competition between plants for nutrients and water.

The best range air temperature should be maintained between 18 and 26 °C, at a relative humidity between 75 and 85% during light hours, and between 90 and 95% during hours of darkness to ensure that the transpiration rate does not exceed water uptake rate at the roots and inhibit plant growth. The crop growing process also requires a minimum of 1500 ppm of carbon dioxide (Earheasy, 2016), which is easy to maintain naturally in open environments, however in closed systems the acceptable levels of carbon dioxide must be maintained using artificial means, which represents the quantity of CO_2 sequestrated from any CO_2 source to account for carbon banking when necessary.

2.5.3 Food Garden External Structure Concepts

The food garden 2 m by 3 m units is designed to yield enough produce to meet the fruit and vegetables Recommended Dietary Allowance of 1000 cal for 25 people. This implies that two of these units are required to meet the objective of supplementing the diets of 50 School students with 1000 cal of fresh produce. The units are able to operate separately or in tandem when attached to each other. The height of the greenhouse in the range between 2 and 2.7 to ensure that the internal environment of the greenhouse is conducive to crop growth. The exact requirements are summarised in Table 2.4. The main climate properties that require regulation are ambient temperature, relative humidity and carbon dioxide levels, which are effects of climate change and global warming and discussed in the proceeding sections.

2.5.3.1 Ambient Temperature

The air temperature should be maintained between 18 and 26 °C. Devices are required to cool the air within the greenhouse during peak summer and devices are required to heat the greenhouse air during the night and during the winter season. Active methods and passive methods can be used.

Passive methods involve using the suns direct heat and absorbing the incident rays through a black surface into a heat sink, a passive heat exchanger. The heat sink will release the energy later on in the night when temperature drops. This method of heating is viable because it is mostly self-sustaining unless insufficient sunlight is received that particular day. It requires minimal maintenance and the only heat sink takes up space and contributes to the weight of the greenhouse. The heat sink can be soil, pebbles, water or any other material with a considerable heating capacity. The passive methods of cooling involve natural convection through widows, the windows can be manually operated or use automatic window openers that operate using springs and the thermal expansion properties of expandable wax and the use of shade cloths.

Active methods of providing heating:

- Heat exchanger that circulates heated steam (uses gas or wood furnace)
- GHG heater
- Radiant tube heaters (electrical)

Active cooling and ventilation methods:

- Greenhouse fans with shutters
- Electrically controlled windows

The main issue with active methods is the requirement of a power source and the cost associated with running the devices. Heat exchanges generally take up considerable amounts of space and are heavy. Since the greenhouse only has a floor area of 6 m^2, a heat exchanger is not a viable option.

2.5.3.2 Relative Humidity

The relative humidity is expected to remain between 70 and 85%. This can be achieved using passive and active systems. The passive system would make use of the available heat from the sun to evaporate water within the greenhouse and provide the high amounts of humidity required. The active system would be a humidifier that would take up minimum space and require low wattages for operation. Since the air in Johannesburg is dry, and only reaches peaks of 60% during the rainy season, there is no considerable need for a dehumidifier

2.5.3.3 Carbon Dioxide Levels

At most, plants require a carbon dioxide concentration of 1500 ppm The global average of carbon dioxide concentration since 2018 is 407.4 ppm (Glandorehydro, 2016). In order to enrich the greenhouse carbon dioxide levels during light periods, a CO_2 generator, sprays CO_2 into the greenhouse. The carbon dioxide is intermittently released in controlled amounts from a canister as may be needed.

2.5.4 Nutrient Sources and Other Resources

The plants require a consistent supply of the important plant macro and micronutrients. The nutrients can be administered through the irrigation systems by dissolving nutrients in the water or the nutrients can be directly administered onto the growing media of the plants, where the roots are anchored. The nutrients can be sourced from:

- Nutrient solutions (in the case of hydroponic systems)
- Organic fish waste (aquaponics systems)
- Fertilisers and fertile soils in the case of crop growth using soil as a growth medium.

The selection of the nutrient source is dependent on the system compatibility, cost restrictions and preference.

2.5.4.1 Power Sources

The main sources of power considered are electricity from nearby buildings and solar power. These power sources are easily obtainable and will take up less space and require the least amount of maintenance once installed. The running costs of these power sources are comparable to the yield produced.

2.5.4.2 Morphological Diagram

Considering all the above-mentioned features in order to perform the various functions required for crop growth, there are tens of thousands of combinations of features. However, most of the combinations cancel because of incompatibility. The possible solutions for functions are summarised in Table 2.5. The possible solutions are used to develop the concepts in the next section.

2.5.5 Food Garden External Structure Concepts

2.5.5.1 Concept 1: The Gable House

The Gable House food garden structure is illustrated in Figure 2.10. The even span design has a floor area of 6 m^2 as required. The apex height of the gable roof is 2.7 m. The structure can be placed on any form of flat foundation. The foundation is

Table 2.5 Food garden morphological resources and solution.

Functions	Possible solutions		
Food garden external structure			
Glazing materials	Tempered glass	Polycarbonate	
Framing materials	Galvanised steel	Aluminium	PVC
Shape	Quonset	Even span	Saw-tooth
Water circulation and irrigation			
Method	Watering cans	Nutrient film technique	Flood and drain
Power source	Solar	Electricity	
Climate regulation			
Heating	Radiant tube heater	Passive heating	
Cooling	Windows	Fans with shutters	
Relative humidity	Passive humidification	Humidifier	
CO_2	CO_2 generator		

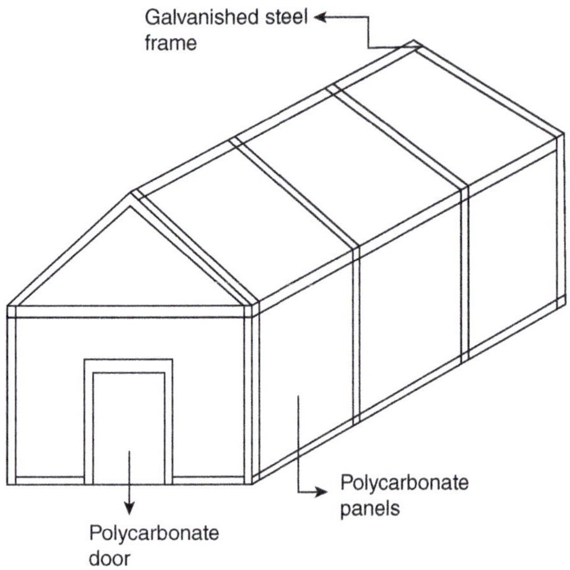

Figure 2.10 External structure concept 1.

Table 2.6 Advantages and disadvantages of external structure concept 1.

Advantages	Disadvantages
The gable roof provides additional space for overhanging plants and accommodates circulation fans	The galvanised steel structure might have to be assembled offsite
The frame structure does not require recoating throughout lifetime	Polycarbonate sheets require replacement after a period of 5 years, depending on the rate of discolouration
The galvanised steel flat bars provide high strength and resistance to environmental loads, with a low specific weight	High combined weight of the structure because of the galvanised steel frame
Polycarbonate sheets are lightweight with high light transmissibility at a comparable strength	

covered with protective baseboard and a layer of weed barrier fabric. The weed barrier fabric will ensure that unwanted weeds do not creep into the greenhouse. The greenhouse frame is made from galvanised steel flat bars, welded or riveted together at the joints. The galvanised steel provides high strength but with a low specific strength. The frames of the greenhouse are polycarbonate sheets. The door is also made from polycarbonate to ensure that sunlight passes through. Depending on the type of climate control mechanisms that are used, the greenhouse may have window panels inserted as well as fans from the roof space. Table 2.6 summarises the advantages and disadvantages of the even span food garden structure.

2.5.5.2 Concept 2: Curved Food Garden Structure

The curved food garden structure illustrated in Figure 2.11 has a floor area of 6 m^2 and a height of 2 m. The foundation of the greenhouse is protected with a base board, with a layer of weed liner. The frame of the structure is constructed from arched polyvinyl chloride (PVC) pipes that are anchored into the ground by slotting the PVC pipes into galvanised steel rods in the ground. The structure is glazed with polycarbonate sheets/film. The door is made from PVC. The PVC sheets are able to be rolled open in order to allow for circulation if that method of cooling is used. The advantages and disadvantages of this concept are summarised in Table 2.7.

2.5.5.3 Concept 3: The Glass House

The glass house illustrated in Figure 2.12 has a 6 m^2 floor space and features a roof structure with differing slopes and an apex height of 2.7 m. The shallower slope is North-facing to allow more sunlight into the structure. The framing material that is selected in aluminium as it is resistant to corrosion providing the necessary resistance to environmental loads. The joints of the structure may be welded using a gas-fed torch or riveted. The structure glazing is tempered glass, which provides

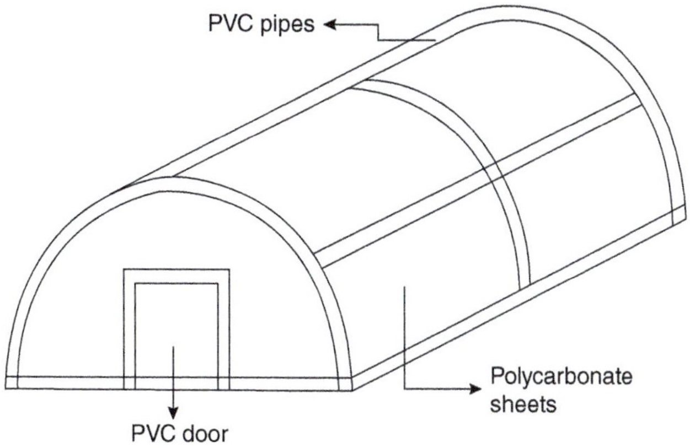

Figure 2.11 External structure concept 2.

Table 2.7 Advantages and disadvantages if the external structure concept 2.

Advantages	Disadvantages
The entire structure is lightweight because of the PVC pipes and the polycarbonate sheets	The structure requires annual maintenance of recoating PVC pipes to protect against UV rays
Low-cost structure	Cannot accommodate overhanging weight
Relatively easy assembly that can be completed on-site	The PVC pipes require anchorage into the ground using steel supports

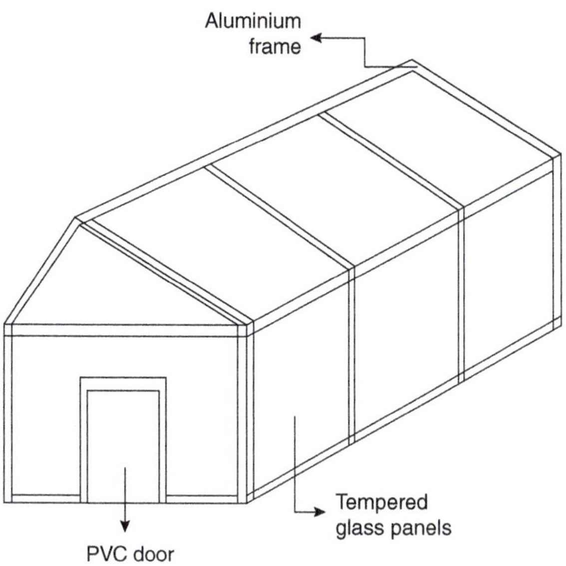

Figure 2.12 External structure concept 3.

excellent light transmissibility and impact strength. The door of the structure is constructed from PVC which is lightweight and opens outwards to maximise the usage of space within the structure. The roof structure is constructed using a truss arrangement for support. Table 2.8 includes the advantages and disadvantages of this external structure concept.

2.5.6 Concept Evaluation: Food Garden External Selection Concept

The concept selection process is based on the following user requirements specifications (URS) and criteria:

Food garden external structure URS requirements

- The food garden external structure must be resistant to environmental loads.
- The food garden should allow for sunlight to stream into the structure.

Table 2.8 Advantages and disadvantages of external structure concept 3.

Advantages	Disadvantages
Both aluminium and tempered glass are highly resistant to corrosion	Aluminium and tempered glass are expensive
Low maintenance frequency because materials have long life	The overall weight of the structure will be high
High resistance to environmental loads	Offsite assembly will be required
The higher roof structure provides additional space for overhanging plants	

Constraints

- The food garden external structure would have a floor area of 6 m^2.
- The materials used to construct the food garden structure must be compatible with the required internal climate.
- The combined load of the garden structure and supporting systems must not exceed 350 kg per square meter.
- The food garden external structure would have a minimum height of 2 m and a maximum height of 2.7 m.

Criteria

- The cost of the structure materials should be minimised.
- The structure must require minimal maintenance frequency.
- The material of the external food garden must be fire-resistant.
- The food gardens shell structure should be constructed from recyclable materials.
- The structure should be sufficiently insulated.
- The food garden external structure must be resistant to environmental degradation.
- The food garden external structure should have a low specific weight.

The criteria are ranked in order of importance using a weighting system. The weighting system of the criteria ranges from 1 to 5, with 5 – being of utmost importance and 1 – of minimal importance. The criterion that is of utmost importance (with a weighting of 5) is maintenance frequency since the structure will be maintained by the School volunteering students who, according to the survey conducted are willing to participate if the maintenance requirements are minimal. Environmental degradation resistance also carries a weighting of 5 because it relates to maintenance requirements and the structure is exposed to the elements throughout its lifetime; therefore, it is a criterion of utmost importance. The criteria with a weighting of 4 are 1, the cost of structural material and 2, Insulation. The greenhouse structure must be sufficiently insulated in order to ensure controlled interaction between the system and the surroundings.

The low specific weight criteria has a weighting of 3 because while it is important to use materials that provide the same strength at a reduced weight, it does not necessarily imply that the strength the material can provide is considerable. Two materials can have the same specific weight but one provides much better mechanical strength than the other. The recyclability also has a weighting of 3 because it is important to consider the full lifecycle of the design. For the plastic materials, the Design Guide for Recyclability issued by the Association of Plastic Recyclables (Design for Recycling, 2017) is used as an aid to determine the degree of recyclability of materials. Fire resistance has a weighting of 2 because it is important to protect the structure from fires however, the likelihood of the greenhouse being exposed to fire hazards is not considered.

The criteria that the concepts are compared against are cost of structural material, environmental degradation resistance, maintenance frequency, insulation, fire resistance, recyclability and the specific weight. The highest possible score a concept can obtain for the level of fulfilment of a criteria is 10. The weighting process is summarised in Table 2.9.

2.5.6.1 Selection Justification by Cost of Structural Material

Concept 2 has the lowest cost of materials because of the PVC frame. As illustrated in Table 2.10, the cost of PVC is 70% lower than the aluminium material. Comparison of framing materials to food garden external structure criteria. It can also be seen that the only material that presents flammable properties is wood which will not be included in the morphological diagram. All the selected materials are corrosion-resistant and recyclable. The main deciding factor between the framing materials is the cost of the materials (short term and long term) as well as the material strength properties. It can thus be seen that the best framing material is stainless steel because the mechanical properties are balanced. It provides a high ultimate tensile strength and specific weight, while the cost is half of the most expensive material. However, galvanised steel, aluminium, polycarbonate, and PVC are still viable options for the structural framing and are considered in the morphological diagram.

Table 2.9 Weighted decision matrix for the food garden external structure.

Criteria	Criteria weighting	Concept 1		Concept 2		Concept 3	
		Outcome	Total	Outcome	Total	Outcome	Total
Cost of structural material	4	7	28	8	32	3	12
Maintenance frequency	5	7	35	3	15	8	40
Insulation	4	7	28	6	24	4	16
Fire resistance	2	7	14	7	14	8	16
Recyclability	3	6	18	4	12	6	18
Specific weight	3	6	18	8	24	4	12
Environmental degradation resistance	5	7	35	6	30	8	40
Total			162		151		154

Table 2.10 Comparison of framing materials of the food garden external structure.

Frame material	Material cost	Maintenance frequency (throughout period of 5 years)	Ultimate tensile strength (MPa)	Specific strength (kNm/kg)	Flammable	Recyclable	Corrosion resistant
Aluminium	1	0	310	115	No	Yes	Yes
Stainless steel	0.58	0	505	63.1	No	Yes	Yes
Galvanised steel	0.33	0	350	46.4	No	Yes	Yes
Wood	0.24	5	3.2	130	Yes	Yes	Yes (if coated)
Polyvinyl chloride (PVC)	0.21	5	25	18.12	No	Yes	Yes

However, since the PVC structure will require recoating for UV resistance and frequent maintenance of the polycarbonate film, it means that in the long run, the costs will start being comparable. That is why the weighting for the cost of concept 3 is not exactly 70% higher than that of concept 2. Concept 1 also has a high cost of materials because of the galvanised steel.

2.5.6.2 Maintenance Frequency

Concept 3 has the lowest maintenance frequency of all three concepts because tempered glass and aluminium are long-lasting materials that are not affected by weathering and prolonged exposure to UV light. Concept 2 requires the most maintenance, as illustrated in Table 2.9, PVC requires recoating or replacement almost annually due to constant exposure to UV light to retain the mechanical properties of the material.

2.5.6.3 Insulation

Concepts 1 and 2 provide the best insulation properties because of the polycarbonate glazing, it has a thermal conductivity that is only 21% that of tempered glass. Although PVC has a lower thermal conductivity than galvanised steel, frames only take up minimal surface area of the entire structure and the sealing between the galvanised steel and polycarbonate panels is better than that of the PVC poles and the polycarbonate films.

2.5.6.4 Fire Resistance

All the concepts are fire resistant; however, concept 3 is most likely to retain structural integrity after being exposed to a fire because of the tempered glass panels.

2.5.6.5 Recyclability

Concepts 1 and 3 scored the same because the frame materials are completely recyclable. The glazing materials are also recyclable; however, polycarbonate has a rating of 7 in the Design Guide for Recyclability which implies that it is hard to recycle. The tempered glass also is a type of glass that is hard to recycle. Concept 2 scored an even lower score because PVC has a recyclability rating of 3, implying that it is seldom recycled because it releases toxic fumes during the process.

2.5.6.6 Specific Weight

Concept 2 scored the highest with regards to specific weight because PVC and polycarbonate both have low specific weight as compared to metals. However, plastic materials do not provide considerable mechanical strength. Concept 1 was somewhat comparable because galvanised steel has a specific weight that is lower than most metals as seen in Table 2.10.

2.5.6.7 Environmental Degradation Resistance

Concept 3 has the highest resistance to environmental degradation because aluminium and tempered glass have a good tolerance of environmental conditions and UV rays do not adversely affect the structure's integrity. PVC thins out and becomes brittle when exposed to sunlight for prolonged periods, but the polycarbonate frames have high tolerance to environmental conditions. After prolonged periods, the galvanised steel zinc protective layer may be broken and the steel be susceptible to corrosion; however, the zinc will oxidise and form another protective layer to prevent against further environmental degradation.

Concept 1 obtained the overall highest score, as it met all of the criteria. The concept provides a viable solution for the external structure of the food garden, as it provides great resistance against environmental weathering. The structure provides high mechanical strength to be able to resist environmental loads with a specific weight that is comparable to that of the plastic materials. Most importantly, it requires minimal maintenance frequency with a moderate cost of materials initially but with low running costs because of the maintenance frequency.

2.5.7 Climate Regulation Concepts for Carbon Credit

2.5.7.1 Concept 1

This concept will make the use of natural ventilation through windows, passive heating and humidification and an organic carbon dioxide generator. The organic CO_2 generator is a device that generates carbon dioxide by means of an alkaline reaction. Once opened, the carbon dioxide gas is gradually released from a 4 L bottle, which provides enough carbon dioxide to sustain the carbon dioxide levels inside a 2 m by 4 m by 2 m enclosure at 1500 ppm for 6 weeks. This additional supply of CO_2 is assumed the amount of CO_2 that is sequestrated from sources such as coal power plants, coal gasification, natural gas reformation to produce hydrogen, petroleum refining and fine chemical production.

The windows are operated using a thermal actuator that relies on the thermal expansion of mineral wax, to allow a piston to gradually expand when the temperature increases. When temperatures drop, a spring aids the gradual closing of the window.

Passive heating is used to heat the system. This is achieved by using the method detailed above in this section. Water is used as a heat sink, which is housed in black plastic tanks to allow for maximised heat absorption, while passive humidification is used. An open source of water is exposed to direct sunlight and the process can be made efficient by surrounding the water source with reflective materials. Table 2.11 summarises the advantages and disadvantages of the concept.

2.5.7.2 Concept 2

Concept 2 makes use of windows and a fan to aid the circulation of air for increased cooling rates. The windows are operated using the thermal actuator. For heating, both passive and active heating are used. The passive heating uses water as a

Table 2.11 Advantages and disadvantages of climate control for carbon credit: concept 1.

Advantages	Disadvantages
The climate regulation concept requires no energy input besides solar radiant heat	Overcast weather would result in reduced humidity and temperatures
The system is cost-efficient because the only additional equipment required is the organic CO_2 generator, which is a measure of carbon credit	The water tanks required as a heat sink will occupy space required for plant growth
The concept requires minimal maintenance, only the organic CO_2 generator will require replacement every 6 weeks	Not much control over the internal climate that has been created

Table 2.12 Advantages and disadvantages of climate control concept 2.

Advantages	Disadvantages
The concept combines both passive and active systems in order to compensate for sunless days	The climate system requires an energy source in order to run the fan and the radiant heater
The cooling system operates on both natural and active convection to provide sufficient cooling	The passive heating water tanks will take up space that must be utilised for food crops
The concept requires minimal maintenance, only the replacement of the organic CO_2 generator after 6 weeks	The fan and radiant heater require control either manually or by means of sensors

heat sink in plastic tank. The active heating uses radiant heaters to provide heating during the winter season and during cold nights. The radiant heaters are suspended in the water to provide both heating and humidity in the system. Passive humidification is still used on days when the system receives radiant heat from the sun; otherwise, it is supplemented by the radiant heater inducing further evaporation of the open source water. The organic CO_2 generator is used to increase the CO_2 levels in the system. Table 2.12 provides the advantages and disadvantages of concept 2.

2.5.7.3 Concept 3

Concept 3 makes use of fans and shutters in order to provide cooling into the system. The fans will require a sensor in order to start turning when the temperatures get too high. The heating of the system will be provided by means of a radiant heater that will also be connected to a sensor that switches the heater on when the temperatures drop too low. A humidifier will be installed in order to maintain the humidity in the room at a consistent level. The humidifier has a built-in sensor to assist in its operation. The organic CO_2 generator will still be used. The advantages and disadvantages of this concept are outlined in Table 2.13.

2.5.8 Concept Evaluation: Climate Regulation Concepts Selection

The same concept selection method by means of a weighted matrix outlined in Section 2.5.6 is used. The criteria that the concepts are evaluated against are from the Climate regulation URS in Section 2.5.6. The criteria with an utmost importance rating of 5 are the low maintenance frequency and system reliability. The criteria that has a weighting of 4, which are of high importance are the minimal energy usage, the minimised running costs and the minimal area occupied by system components. If the system has components that draw significant amounts of energy to perform their function, the operating cost of that system will outweigh the food crop production. For the minimal area occupied by the components, the internal structure of the greenhouse has limited space that should be used to maximise the amount of yield of food crops. The perfect balance should be obtained where the crop yields can be high but with reduced cost and energy requirements to sustain the crop growing process. The respective concepts are evaluated against each of the criteria on a scale of 1 to 10, 10 being the highest level of fulfilment of the specific criteria. Table 2.14 contains the weighted matrix of the climate regulation concepts.

2.5.8.1 Selection Justification by Criteria

2.5.8.1.1 Minimal Energy Consumption Concept 1 obtained the highest score for minimal energy consumption because it requires no additional energy input except the radiant heat from the sun. Concept 2 received a good score because it is

Table 2.13 Advantages and disadvantages of the climate regulation concept 3.

Advantages	Disadvantages
The climate system provides the greenhouse with more consistent and controlled levels of humidity and temperature	The cost of installation and the running cost of the concept is considerable
The concept is automated and only requires that operating points be provided into the systems	The energy consumption will be high because the humidifier, radiant heater and fan require power
The system will require minimal maintenance, except for the replacement of the CO_2 generator as carbon sink	The humidifier will take up space that is intended for crop growth

Table 2.14 Weighted matrix of the climate regulation concepts.

		Concept 1		Concept 2		Concept 3	
Criteria	Criteria weighting	Outcome	Total	Outcome	Total	Outcome	Total
Minimal energy consumption	4	9	36	8	32	4	16
Minimal running cost	4	9	36	8	32	5	20
Area occupied by components minimised	4	7	28	7	28	8	32
Low maintenance frequency	5	7	35	7	35	6	30
System reliability	5	4	20	8	40	9	45
Total			155		167		143

highly reliant on the sun's energy too but on overcast days, energy is required to operate the radiant heater. The fan is required to provide cooling on extremely hot days. Concept 3 scored the lowest because it constantly requires an energy input to sustain its functionality.

2.5.8.1.2 Minimal Running Cost Concept 1 had the highest score in this criterion because it functions on the heat from the sun, there are no additional running cost requirements except for the common cost of the organic CO_2 generator. Concept 2 also has a low running cost because it is a predominantly passive system unless it is cloudy. Concept 3 scored the lowest because the continuous energy consumption requirement is accompanied by running costs.

2.5.8.1.3 Area Occupied by Components Minimised Concept 3 scored the highest because it only requires a humidifier that takes up floor space. Concepts 1 and 2 scored slightly less because the sink water in a plastic container that is required to absorb heat takes up floor space.

2.5.8.1.4 Low Maintenance Frequency Concepts 1 and 2 require the least amount of maintenance because the main components required are the water–plastic containers. Concept 3 has various components that even require sensors to function; this ultimately increases the risk of components requiring replacement and repair.

2.5.8.1.5 System Reliability Concept 3 provides the best system reliability because of the consistent amount of control of the inside climate because the system is automated, even when the weather is overcast for prolonged periods. However, concept 2 provides comparable reliability because if the power goes out, the greenhouse can be reliant on the passive heating system. Concept 1 obtains the lowest reliability reading because if it is overcast for prolonged periods, the greenhouse climate is not sufficiently regulated.

In conclusion, concept 2 provides the best climate control solution because it combines the use of passive systems when there is sunlight availability and is aided by active systems in cloudy conditions. Although the system is not as automated as solution provided by concept 3, and it requires more power, space and running costs than concept 1, it provides a sufficiently functional solution.

2.5.9 Climate Regulation Design Development

Two cases of climate regulation are considered. The first one is the amount of heating the food garden will receive on summer days, and if the passive heating system stored enough energy to keep the garden temperature between 18 and 26 °C during the cooler evenings. The second consideration is the amount of heating required inside the food garden if the winter temperatures dropped to 0 °C or increased abruptly due to incidental climate change from CO_2 concentration. The passive heating/cooling system uses the water in the exposed grow beds as a solar collector. The grow beds are lined with black pond liner in order to provide sufficient sunlight absorption. The water acts as both heat and CO_2 sink, and as it circulates to the tank, the larger surface area of the tank will ensure that at night, the released heat is evenly distributed in the food garden. However, because this is a design project, the concentration profile of the CO_2 is not monitored. The wall panels installed directly behind the tank are lined with reflective Mylar sheets to ensure that excess sunlight is directed towards the tank in an effort to retain the stored heat in the tank water until it is required at night.

Table 2.15 Johannesburg climatic conditions.

Weather property	Lower limit	Upper limit
Summer temperature (°C)	13	30
Winter temperature (°C)	0	18
Rainfall (mm/month)	10	140
Wind (km/h)	0	15.6
Relative humidity (%)	47	70
Sun period (hours/day)	10.5	13.75

2.5.9.1 Summer Conditions

Johannesburg's climate is categorised as sub-tropical highland with dry winter, with a warm temperate dry forest biome. Based on the day-to-day weather conditions recorded between 1 January 1980 and 31 December 2016 the average ambient temperature is 19.9 °C (Vincenta et al., 2013; Past Weather in Johannesburg, 2016). Table 2.15 details the climatic conditions in Johannesburg, averaged over the period of 36 years. The Johannesburg climate allows for crop growing for 9 months of the year and depending on how long the extremely dry and cold conditions of winter last, this period can be reduced (Past Weather in Johannesburg, 2016).

From Table 2.15, the minimum and maximum temperatures observed in summer Johannesburg are 13 °C at night and 30 °C during the day. The plants are tolerant of temperatures between 18 and 26 °C. The critical cases are assessed as follows:

2.5.9.1.1 Required Heat Removal Rate in Summer The extreme case will occur when the outside temperature is 30 °C and the inside temperature of the food garden has reached the maximum temperature of 26 °C. The main assumption is that the water was not absorbing heat during this heating process. A standard heat, q, transfer Eq. (2.3) is used to determine the excess heat that entered the food garden as;

$$\dot{q} = h\,A\,\Delta T \tag{2.3}$$

where h is the thermal conductivity of the polycarbonate panels since they occupy the majority of the surface area of the food garden. The thermal conductivity used is 0.22 W/m k. A is the exposed surface area of the polycarbonate panels, which is 24.72 m^2, while ΔT is the change in temperature is 4 K. The resulting heat flux q is 21.75 J/s. For the entire garden to be at 26 °C, it would imply that the water in the passive system has also reached that temperature. Therefore, to determine the absorbed heat of the water that would raise its temperature from 26 to 30 °C, Eq. (2.4) is used as follows:

$$Q = mCp\Delta T \tag{2.4}$$

where m is the mass of water, which is 481.29 kg (from the maximum volume of the tank), Cp is the heating capacity of the water taken as 4181 J/kg k and the change in temperature remains 4 K. If a sun period of 10.5 hours per day is considered, the amount of excess heat, Q, of the water received during that period is 8.05 × 106 J, at a heat flux of 212.94 J/s. This implies that to ensure that temperatures do not rise beyond 26 °C, a cooling system must remove heat from the food garden at a combined rate of 234.72 J/s or more. This is also assuming that all of the other surfaces in the food garden are good insulators.

2.5.9.1.2 Usage of Passive Heat for Summer Nights Equation (2.3) is used to determine the amount of heat exiting the food garden during the remaining nighttime of 13.5 hours. The temperature drops to the summer minimum of 13 from 26 °C during that time. The calculated heat flux is 70.70 J/s. The passive heating is able to provide 165.64 J/s, which will suffice but will require cooling at a rate of 94.94 J/s so the temperatures remain below 26 °C.

2.5.9.2 Winter Conditions

Minimum amount of heating required in winter to sustain the food garden at its minimum of 18 °C and if the outside temperature drops to its winter minimum of 0 °C is determined using Eq. (13). The required heat flux will be 97.89 J/s.

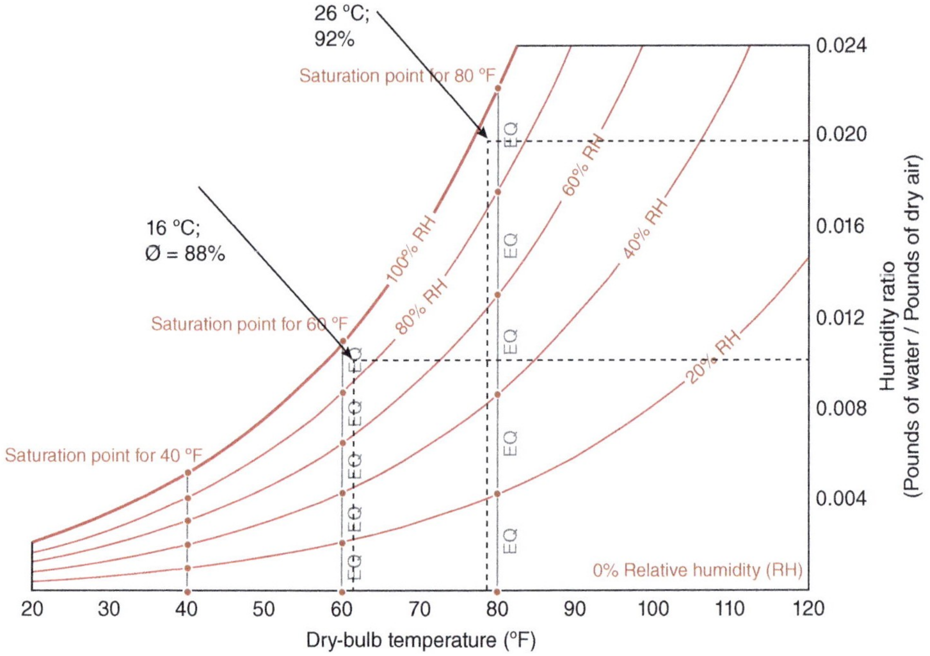

Figure 2.13 Psychrometric chart used to determine relative humidity. Source: Adapted from Perry and Green.

2.5.9.3 Relative Humidity

In order to check if the internal relative humidity requirements could be possible to achieved inside the food garden, a psychometric analysis is conducted using charts. At the minimum (16 °C) and maximum temperatures (26 °C) of the food garden, the corresponding humidity ratios are 0.011 and 0.020, respectively. When the humidity ratios and the dry bulb temperatures are projected onto the psychometric chart in Figure 2.13, the maximum relative humidity ratios available are 88–92%. The required relative humidity range of 70–95% has been sufficiently met without the requirement of a humidifier.

Both the clear and translucent panels in Figure 2.14 are made from polycarbonate polymer. The translucent panels are lined with Mylar reflective sheets to assist in retaining heat from the passive solar collector in the water and to control CO_2 or carbon load. While Figure 2.15 presents the detailed engineering drawings of the food garden for the school at Wits University.

Figure 2.14 The food garden to grow agricultural produce, control heat and CO_2 load.

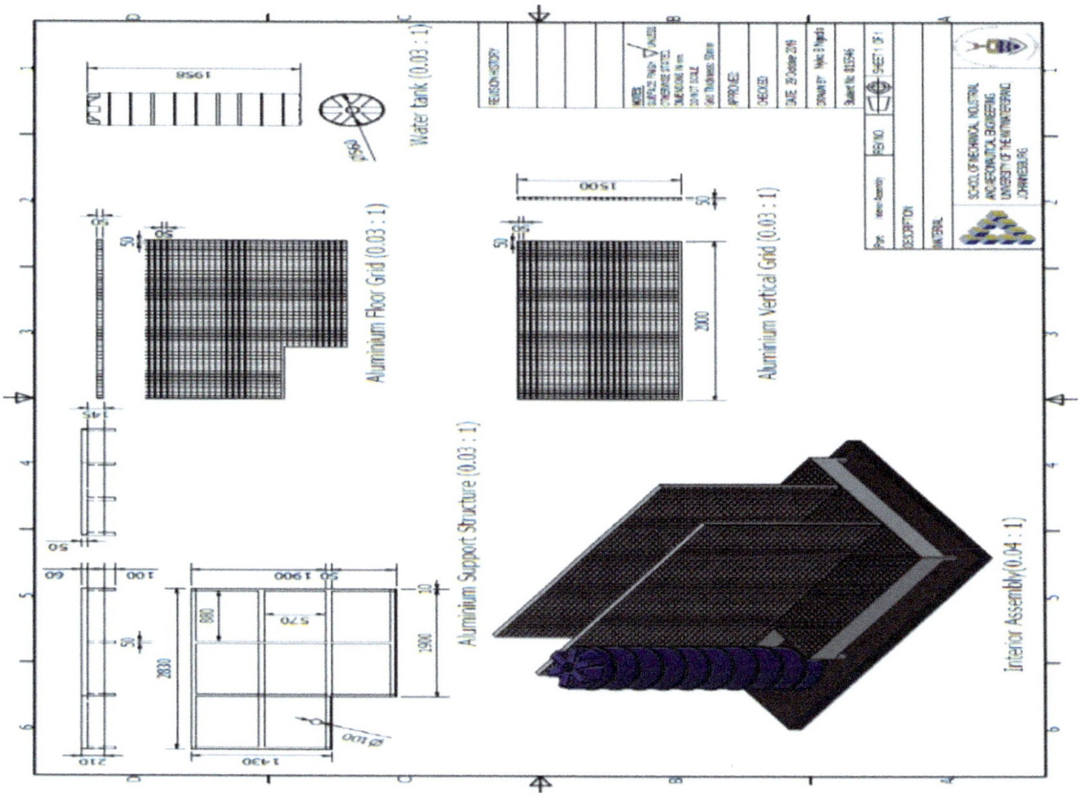

Figure 2.15 Detailed engineering drawing of food garden at Wits University.

2.6 Biomass Towards Extended Carbon Storage and CO_2 Capture

In conclusion, biomass from vegetation (microalgae, catch crops, food crops, trees, etc.) are the most abundant renewable resources and sustainable materials that can lock up carbon and CO_2 naturally. The use of the nanostructure of wood products with a long service life will substantially contribute to extended carbon storage and advanced functionalities. Several emerging technologies focussed on modifying and functionalising biomass cell walls and lumen targeting extended carbon storage and CO_2 capture would provide the so-called net-zero offerings of CO_2 sequestration. For industrial engineering purposes, a set of green and sustainable agricultural processes should contribute to climate change mitigation by increasing CO_2 photosynthetic consumption, reducing crop and livestock emissions, sequestering carbon in soils and biomass and decreasing emissions intensity within sustainable production systems. Total technical mitigation potential from crop and agroforestry is estimated as 2.3–9.6 $GtCO_2$-eq per year by 2050. Thus, the more diverse the food systems are, the more resilient they are in enhancing food security in the face of biotic and abiotic stresses. Diverse production systems are, therefore, important in providing regulatory ecosystem services such as nutrient cycling, carbon sequestration, soil erosion control, reduction of GHG emissions and control.

A conceptual design has also provided the best climate control solution because it combines the use of passive systems when there is sunlight availability that is aided by active systems in cloudy conditions for CO_2 sequestration into food crops to achieve a sustainability.

However, since land availability for food security might remain issue and the conventional processes as discussed in the preceding discussion for each of these solutions often have high-capital costs associated and kinetic limitations in different process steps, it becomes imperative to integrate process intensification technologies for CO_2 capture and conversion as presented in the next chapters. As a result, Chapter 3 presents transformative actions to decarbonise the chemical sector in finding alternative non-fossil carbon feedstock for more robust environmental implications, scientific, engineering resources and equipment.

References

Ahlstrom, A., Raupach, M.R., Schurgers, G. et al. (2015). The dominant role of semi-arid ecosystems in the trend and variability of the land CO_2 sink. *Science* 348: 895–899.

Allen, C.D., Macalady, A.K., Chenchouni, H. et al. (2010). A global overview of drought and heat-induced tree mortality reveals emerging climate change risks for forests. *For. Ecol. Manag.* 259: 660–684.

Allen, C.D., Breshears, D.D., and McDowell, N.G. (2015). On underestimation of global vulnerability to tree mortality and forest die-off from hotter drought in the Anthropocene. *Ecosphere* 6: 1–55.

Anderegg, W.R.L., Schwalm, C., Biondi, F. et al. (2015). Pervasive drought legacies in forest ecosystems and their implications for carbon cycle models. *Science* 349: 528–532.

Anon. (2018). *Dietary recommended for health: Fact sheet.* [Online] Available at: http://www.health.uct.ac.za/sites/default/files/image_tool/images/116/diet-debate/2014%20Fact%20Sheet%20-%20Dietary%20Guidelines%20for%20Health%20Final.pd [Accessed 15 October 2019].

Aslam, N.M. and Varani, J. (2016). The Western-style diet, calcium deficiency and chronic disease. *J. Nutr. Food Sci.* 6: 1–6. https://doi.org/10.4172/2155-9600.1000496.

Aslam, A., Thomas-Hall, S.R., Mughal, T.A., and Schenk, P.M. (2017). Selection and adaptation of microalgae to growth in 100% unfiltered coal-fired flue gas. *Bioresour. Technol.* 233 (2017): 271–283.

de Benoist, B. (2009). *Preventing and controlling micronutrient deficiencies in populations affected by an emergency. Joint statement by the World Health Organization, the World Food Programme and the United Nations Children's Fund.* Source: https://cdn.who.int/media/docs/default-source/nutritionlibrary/preventing-and-controlling-micronutrient-deficiencies-in-populations-affected-by-an-emergency.pdf?sfvrsn=e17f6dff_4&download=true (accessed 19 March 2023).

Buchanan, A.H. and Levine, S.B. (1999). Wood-based building materials and atmospheric carbon emissions. *Environ. Sci. Policy* 2 (6): 427–437.

Cai, T.M., Wang, Y., Jia, J. et al. (2022). Assessing the accumulation efficiency of various microbial carbon components in soils of different minerals. *Geoderma* 407: 115562.

Ciais, P., Sabine, C., Bala, G. et al. (2014). Carbon and other biogeochemical cycles. In: *Climate Change 2013: The Physical Science Basis: Working Group I Contribution to the Fifth Assessment Report of the Intergovernmental Panel on Climate Change*, 465–570. Cambridge: Cambridge University Press.

Collins, M., Knutti, R., Arblaster, J. et al. (2013). Long-term climate change: projections, commitments and irreversibility. In: *Climate Change 2013: The Physical Science Basis: Working Group I Contribution to the Fifth Assessment Report of the Intergovernmental Panel on Climate Change*, 1029–1136. Cambridge: Cambridge University Press.

Dass, P., Houlton, B.Z., Wang, Y., and Warlind, D. (2018). Grasslands may be more reliable carbon sinks than forests in California. *Environ. Res. Lett.* 13: 074027.

Design for Recycling (2017). Packaging SA. Source: https://www.packagingsa.co.za/wp-content/uploads/2019/11/PACSA-Recyclability-by-Design-WEB.pdf (accessed 20 March 2023).

Earheasy (2016). Raised Beds: Soil Depth Requirements. Source: https://learn.eartheasy.com/guides/raised-beds-soil-depth-requirements/ (accessed 19 March 2023).

FAO (2001). *Food Insecurity in the World 2001.* www.fao.org/3/a-y1500e.pdf, 8. Rome, Italy: Food and Agriculture Organization of the United Nations.

FAO, IFAD, UNICEF, WFP and WHO (2018). *The State of Food Security and Nutrition in the World 2018. Building Climate Resilience for Food Security and Nutrition.* Rome, Italy: Food and Agriculture Organization of the United Nations 202 pp.

Geng, A., Yang, H., Chen, J., and Hong, Y. (2017). Review of carbon storage function of harvested wood products and the potential of wood substitution in greenhouse gas mitigation. *For. Policy Econ.* 85: 192–200. https://doi.org/10.1016/j.forpol.2017.08.007.

Gentry, M. (2019). Local heat, local food: Integrating vertical hydroponic farming with district heating in Sweden. *Energy* 174: 191–197. https://doi.org/10.1016/j.energy.2019.02.119.

Glandorehydro (2016. CO_2 Explanation. A Guide to CO_2 in the Grow Room. Source: https://glandorehydro.com/a-guide-to-c02-in-the-grow-room (accessed 19 March 2023).

Gruda, N.S. (2019). Increasing sustainability of growing media constituents and stand-alone substrates in soilless culture systems. *Agronomy* 9: 298.

Jiang, L., Gonzalez-Diaz, A., Ling-Chin, J. et al. (2019). Post-combustion CO_2 capture from a natural gas combined cycle power plant using activated carbon adsorption. *Appl. Energy* 245: 1–15.

Maboko, M.M. (2014). *Guide to Hydroponic Vegetable* (ed. J.G. Niederwieser and C.P. du Plooy), 632–636. Agricultural Research Council, Roodeplaat Vegetable and Ornamental Plant Institute. ISBN: 13 978-1-86949 Source: https://www.arc.agric.za/arc-vopi/Pages/Crop%20Science/Hydroponic-Vegetable-Production.aspx (accessed 18 March 2023).

Mbow, C., Rosenzweig, C., Barioni, L.G. et al. (2019). Food security. In: *Climate Change and Land: An IPCC Special Report on Climate Change, Desertification, Land Degradation, Sustainable Land Management, Food Security, and Greenhouse Gas Fluxes in Terrestrial Ecosystems* (ed. P.R. Shukla, J. Skea, E.C. Buendia, et al.). https://doi.org/10.1017/9781009157988.007.

McDowell, N.G. and Allen, C.D. (2015). Darcy's law predicts widespread forest mortality under climate warming. *Nat. Clim. Change* 5: 669–672.

Montanari, C., Olsén, P., and Berglund, L.A. (2021). Sustainable wood nanotechnologies for wood composites processed by *in-situ* polymerization. *Front. Chem.* 9: 1–12. https://doi.org/10.3389/fchem.2021.682883.

Muthayya, S., Rah, J.H., Sugimoto, J.D. et al. (2013). The global hidden hunger indices and maps: an advocacy tool for action. *PLoS One* 8: e67860. https://doi.org/10.1371/journal. pone.0067860.

Nessi, E., Papadopoulos, I.A., and Seferlis, P. (2021). A review of research facilities, pilot and commercial plants for solvent-based post-combustion CO_2 capture: packed bed, phase-change and rotating processes. *Int. J. Greenhouse Gas Control* 111: 103474.

Ngxola, N.B. (2019). Student Food Garden. A Design Project at the University of the Witwatersrand, Johannesburg, unpublished.

Omoregbe, O., Mustapha, N.A., Steinberger-Wilckens, R. et al. (2020). Carbon capture technologies for climate change mitigation: a bibliometric analysis of the scientific discourse during 1998–2018. *Energy Rep.* 6: 1200–1212.

Onyeaka, H., Miri, T., Obileke, K. et al. (2021). Minimizing carbon footprint via microalgae as a biological capture. *Carbon Capture Sci. Technol.* 1: 100007.

Park, J., Cho, Y.S., Jung, M. et al. (2021). Efficient synthetic approach for nanoporous adsorbents capable of pre- and post-combustion CO_2 capture and selective gas separation. *J. CO_2 Util.* 45: 101404.

Past Weather in Johannesburg (2016). Source: https://www.timeanddate.com/weather/south-africa/johannesburg/historic?month=1&year=2016 (accessed 20 March 2023).

Perry, R.H. and Green, D.W. (2008). *Perry's Chemical Engineering Handbook*, 8e. New York: McGraw-Hill.

Ryan, J. (2020). How Aquaponics Benefits Children and the Community. Source: https://www.farmersweekly.co.za/agri-business/empowerment/how-aquaponics-benefits-children-and-the-community/ (accessed 18 March 2023).

Santori, G., Charalambous, C., Ferrari, M.-C., and Brandani, S. (2018). Adsorption artificial tree for atmospheric carbon dioxide capture, purification and compression. *Energy* 162: 1158–1168. https://doi.org/10.1016/j.energy.2018.08.090.

Sharma, S. and Maréchal, F. (2019). Carbon dioxide capture from internal combustion engine exhaust using temperature swing adsorption. *Front. Energy Res.* 7: 143. https://doi.org/10.3389/fenrg.2019.00143.

Sloss, W. (2011). The Chemistry and Microbiology of Aquaponics. Source: http://kohalacenter.org/HISGN/pdf/Lesson8ChemistryandMicrobiologyofAquaponics.pdf (accessed 19 March 2023).

Thomas, S.C. and Martin, A.R. (2012). Carbon content of tree tissues: a synthesis. *Forests* 3: 332–352.

UK Forests and Climate Change (2018). Carbon sequestration. Tools and Resources. https://www.forestresearch.gov.uk/tools-and-resources/statistics/forestry-statistics/forestry-statistics-2018/uk-forests-and-climate-change-5/carbon-sequestration/ (accessed 19 December 2022).

United Nations (2014). *World Urbanization Prospects: the 2014 Revision, Highlights.* Department of Economic and Social Affairs https://doi.org/10.4054/DemRes.2005.12.9.

Vincenta, C.J., Hicks, N., Arenstein, G. et al. (2013). The proposed CO_2 test injection project in South Africa. *Energy Procedia* 37: 6489–6501. https://doi.org/10.1016/j.egypro.2013.06.579.

Wits (2023). SA Food Sovereignty Campaign. Source: https://www.safsc.org.za/food-commons-projects/wits-food-sovereignty-centre/ (accessed 18 March 2023).

3

Carbon Transition for the Petrochemical Sector

Although the present environmental calamities and devastations and of course pressure groups demanding immediate downgrading of use of fossil fuels for power and chemical productions, care should be exercised to avoid immediate reverse of the industrial revolutions achieved since the 1800s. Transformative actions to decarbonise the chemical sector in finding alternative non-fossil carbon feedstock would need overall understanding of the environmental implications, scientific, engineering resources and equipment.

In their study, Gala-Martın et al. (2021) tried replacing fossil feedstocks with renewable carbon to reduce CO_2 emissions by 25–100% through technology pathways that gave the highest CO_2 mitigation in the petrochemical sector. Being a sector with a significant carbon footprint, it provides everyday materials and developing low-carbon technologies to support transition to decarbonised environment. They also estimated that the sector contributes 1.24 Gt CO_2-eq annually, accounting for about 10% of global total final energy consumption and 7% of greenhouse gas (GHG) emissions associated with industry. Majority of the emissions come from production of chemicals based on carbon feedstocks whereby, electrifying the energy source or involving carbon capture and storage (CCS) would not fully prevent the sector's strong reliance on fossil carbon as a feedstock. While the petrochemical industry is the raw material source for various fine and bulk chemical products, including ethylene, propylene, benzene, toluene, xylenes, and methanol, one-third of the total energy demand of the sector is produced from fossil carbon of oil and gas, coal through cracking of naphtha, ethane, and/or liquefied petroleum gas as schematically presented in Figure 3.1. Intermediate green methanol could be produced from renewable biomass or by the hydrogenation of captured CO_2. On the other hand, H_2 can be produced by fermentation with microbes and electrolysis of water. While the latter consumes vast amounts of energy supplied by renewable power sources of wind and solar, the bioenergy of H_2, the CO_2 released could be captured either at sources or directly from the air and then converted into economically valuable chemicals.

3.1 Carbon Intensity of Global Crude Oil Refining

The petroleum refining and petrochemical industry is seen as the third-largest global source of stationary greenhouse gas (GHG) emissions after power and the petroleum and natural gas systems. It should account for 40% of emissions from the oil and gas supply chain and 6% of all industrial GHG emissions (World Energy Outlook 2018; Tollefson 2018). The key role of the industry is maintaining a stable global energy market by providing fuels and feedstock from crude oil. They also reported that because the industry continues to grow, the global refining capacity doubled from about 45 million barrels per day (Mbbl d^{-1}) in 1970 to about 95 Mbbl d^{-1} in 2015. As a result, hundreds of marketable crude oil networks are created around the world, which consist of crude oil produced in one oil field in a country but processed in many refineries in different multiple destination countries (Crude Oil Pretreament in Petroleum Refining 2007). A very good example is the Arab light from Saudi Arabia is the world's largest network that supplies crude oil to 153 refineries located in 35 countries for refining and processing. Although the climate implications from the operation of these complex and dynamic networks are not well understood knowing and establishing carbon abatement and investment status will bring a relieving climate-guild and decisions. In order for most or all the jurisdictions to implement government GHG emissions regulations that affect the refining sector (Andress et al. 2010; Christensen and Hobbs 2016), private investors should also abide by emission stands, which will in turn inform new refineries while trending towards more complex configurations. However, while the refineries attempt to preserve tight profit margins (Oil 2018), the industry is faced with challenges in making climate-sensitive decisions with regards to a volatile heavy crude supply and increasing excess global refining capacity.

Measuring Climate Change to Inform Energy Transitions: Carbon Footprint Calculations, First Edition. Sunny E. Iyuke.

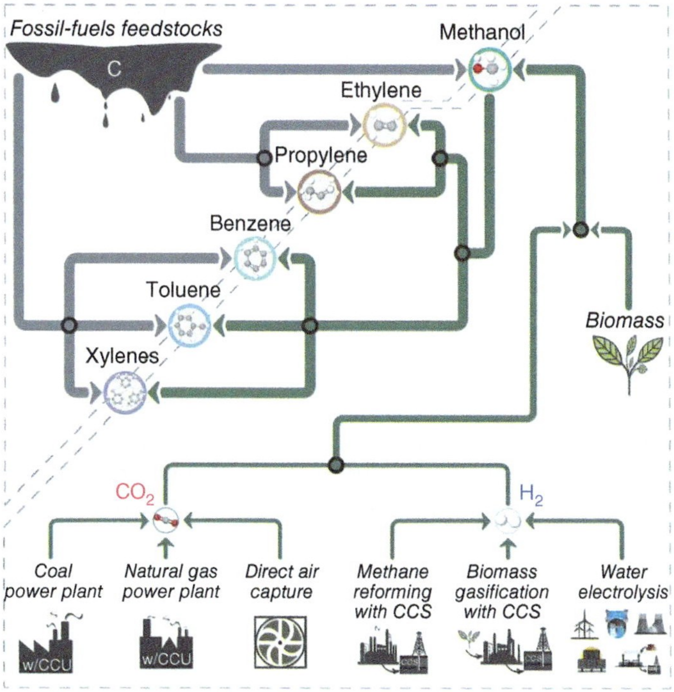

Figure 3.1 Schematics for sustainable carbon footprints of transition from fossil to renewable carbon. Source: Gala-Martín et al./Elsevier/CC BY 4.0.

Moretti et al. (2017) have expressed that the petroleum industry plays an important role in adopting and showcasing new technologies that could result in large GHG emissions reductions, which involves reducing methane intensity of aggregated upstream operations to below 0.25% by 2025 with the ambition of achieving 0.2% (OGCI 2018). It is relevant to express that such climate commitments, which might be lack detailed inventories of GHG emissions, should result into an industry-wide transition strategy and updating to evaluate and track mitigation actions.

The global crude oil refining throughput of 2015 (World Energy Outlook, IEA 2016), adopted in this study, gave the global crude oil refinery operations of about 74.4 Mbbl d^{-1} of crude oil. This is equivalent to 93% of the 10 countries with the highest throughput that accounts for 65% of the world's total volume of crude oil refined among the top 20 countries as presented in Table 3.1. Four types of refineries are categorised by refinery complexity and how heavy fractions of crude oil are processed through hydroskimming, medium conversion, deep conversion (coking) and deep conversion (hydrocracking) refineries, which are deployed in 37, 59, 34 and 8 countries, which share volume percentages of 7, 44, 45 and 4%, respectively (Jing et al. 2020).

Although a more detailed discussion has been presented by Jing et al. (2020), it is summarised here that deep conversion (coking) is dominant globally due to the heavy influence of the United States and China, which are together responsible for one-third of global crude oil refining throughput.

In order to meet the increasing need for more unconventional resources and increasing demand for transportation fuels while accommodating lower-quality feedstock, over 60% of US and 75% of Chinese refineries are deep conversion refineries. As for the growing number of newer refineries in Asia, Latin America and other areas, a substantial deep conversion refining capacity is adopted. On the other hand, in refineries, in the Middle East, Russia, Japan, South Korea and Europe, hydroskimming technology is mainly used, which results into medium conversion, possibly due to their historical processing of light crude slates and due to increasing demand for petrochemicals (The Future of Petrochemicals, IEA 2018).

It is observed that upstream oil production is dominated by a small number of countries that are comparatively more diversified to meet different market demands and product quality specifications. These refiners are seen to employ various strategies that in some instances lead to operational conditions that are suboptimal in terms of energy efficiency and emissions (Bredeson et al. 2010).

Table 3.1 Top 20 refining countries ranked by volume of crude oil refined in 2015. Shown are country, total volumetric crude oil refined, number of refineries, shares and refinery type breakdown.

Country	2015 Crude oil refined (Mbbl d^{-1}) (Ref. 22)	Number of refineries	2015 Share in crude oil refined (vol%)	Refinery type breakdown (vol%)			
				Hydroskimming	Medium conversion	Deep conversion (coking)	Deep conversion (hydrocracking)
United States	15.5	95	20.9	1	22	74	3
China	9.1	75	12.3	1	20	79	–
Russian Federation	5.3	31	7.2	21	46	33	–
India	4.5	18	6.1	–	29	71	–
Japan	3.1	22	4.2	–	68	17	15
South Korea	2.6	5	3.5	7	52	14	27
Saudi Arabia	2.5	8	3.4	8	63	29	–
Germany	1.9	11	2.6	–	65	35	–
Brazil	1.8	10	2.4	–	15	85	–
Iran	1.7	8	2.3	2	98	–	–
Canada	1.6	13	2.2	–	79	21	–
Italy	1.4	9	1.9	12	55	–	33
Spain	1.3	8	1.8	–	44	56	–
United Kingdom	1.2	6	1.6	–	84	16	–
Thailand	1.1	6	1.5	–	100	–	–
France	1.1	7	1.5	–	100	–	–
Mexico	1.1	6	1.5	–	59	41	–
Netherlands	1.1	5	1.5	6	44	17	33
Singapore	1	3	1.4	–	100	–	–
Venezuela	0.9	3	1.2	–	32	68	–
Global	74.4			7	44	45	4

Data reported here do not fully cover all refineries in each country. See main text for details.
Source: Adapted from Jing et al.

3.2 GHG Reduction Measures for the Petroleum Refining Industry

Among the top 20 petroleum refineries mentioned above, there are more than a thousand oil refineries worldwide that can be categorized broadly as hydroskimming or larger conversion refineries. Distillation and a series of process units make up the hydroskimming refineries to produce petrol, jet fuel and middle distillates. The larger conversion refineries, on the other hand, include hydrocrackers and catalytic cracking units to further reduce heavier crude fractions into lighter products (Jing et al. 2020). About 70% of all refineries in Europe and the United States belong to these groups, which together emit around four times as much CO_2 per barrel of oil as the simpler conversion refineries (Lei et al. 2021). The power stations in the refinery plant constitute about 29% of total emissions as the main emission sources in larger conversion refineries. This is followed by the fluid catalytic cracking (FCC) unit, atmospheric distillation units and steam methane reformer for hydrogen production at 19, 19 and 11%, respectively (IEAGHG 2017). A typically simplified flow diagram of a refinery is presented in Figure 3.2, which shows the key conversion processes and outputs. Nowadays, however, the power station often uses natural gas combined cycle (NGCC) plant with additional gas-fired boilers to produce the overall power requirements for the refinery, while the heat requirements by the atmospheric and vacuum distillation units are met by burning the fuel oil and gases. Even though the steam methane reformer uses natural gas both as a feedstock and a fuel, it produces two separate sources of CO_2 to release about 10 $kgCO_2$,eq for each kg of hydrogen produced and so on (Jing et al. 2020).

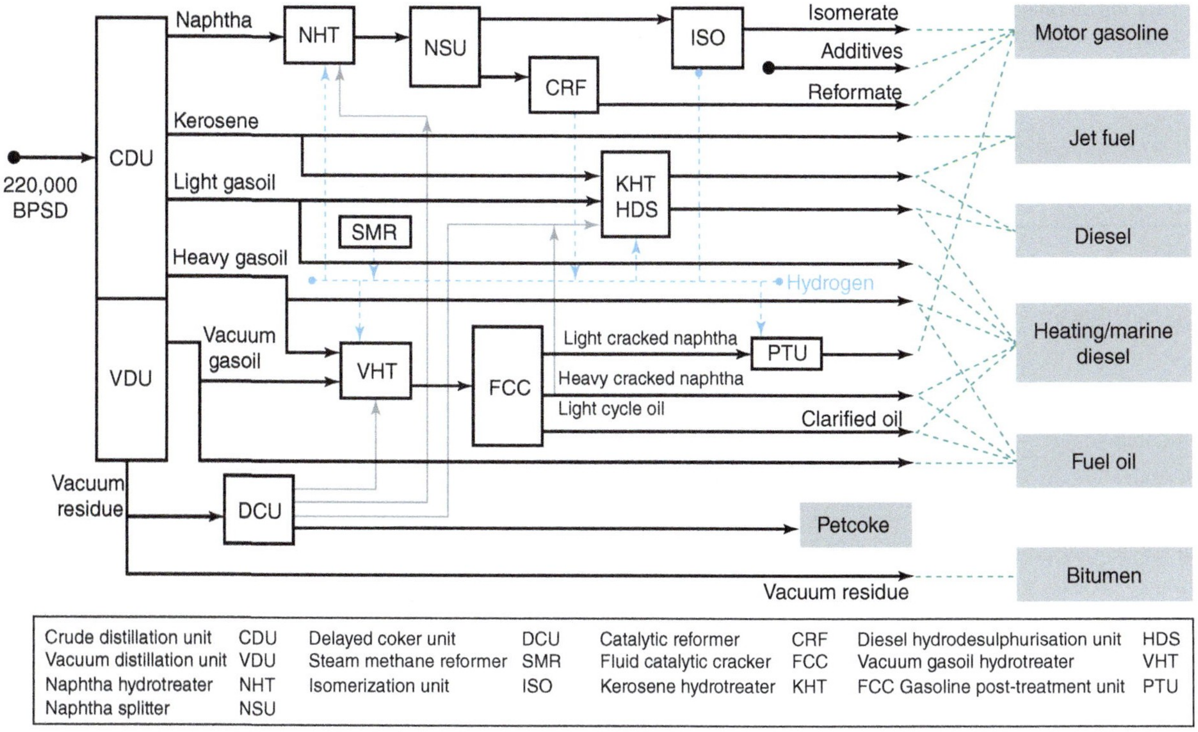

Figure 3.2 Simplified refinery block flow diagram. Source: Sunny et al./Frontiers Media/CC BY 4.0.

It then means that quantifying potential emission reductions from the refining sector and the entire oil supply chain is crucial to meeting climate targets. This work, therefore, supports strategies earlier proposed by Jing et al. (2020), which include the following key settings:

1. Naphtha catalytic reformer only takes straight-run naphtha.
2. Fluid catalytic cracker (FCC) post-hydrotreater is selected as the default hydrotreating option for FCC.
3. Hydrotreater hydrogen consumption rates are adopted and calibrated from commercial datasets (Wood Mackenzie 2020).
4. Electricity emissions factor is based on the International Energy Agency database (Emissions Factors, IEA 2019), such that each country has its own emissions factor and breakdown of contributions from CO_2, CH_4 and N_2O.
5. Amine CO_2 removal is selected for hydrogen purification at the steam-methane reformer.
6. Lower heating value (LHV) is employed.
7. The AR5 2013 IPCC 100-year GWP emissions factors are used.
8. Refinery fuel gas produced onsite is directly combusted as fuel.
9. Liquefied petroleum gas and petrochemical feedstock are burned onsite instead of being produced separately.
10. Heat and power cogeneration (CHP) is not considered.
11. Hydrogen requirements are assumed to be met by the onsite naphtha reformer and steam-methane reformer.

Some time ago, EPA (2010) provided summary of GHG reduction measures as described below:

3.2.1 Energy Efficiency Initiatives and Improvements

Benchmarking GHG performance and implementing energy management systems to improve energy efficiency are presented below, but not exhaustive, which contribute to efficiency improvement/GHG emission reduction of 4–17% of electricity consumption (Jing et al. 2020).

- improve process monitoring and control systems,
- use high-efficiency motors,
- use variable speed drives,
- optimise compressed air systems,
- implement lighting system efficiency improvements.

3.2.2 Systems Approach to Steam Generation

Analyse steam needs and energy recovery options, including:

- minimise steam generation at excess pressure or volume
- use turbo or steam expanders when excesses are unavoidable
- schedule boilers based on efficiency

3.2.3 Boiler Feed Water Preparation

It is suggested that a hot lime water softener be replaced with a reverse osmosis membrane treatment system to remove hardness and reduce alkalinity of boiler feed. It has efficiency improvement/GHG emission reduction of 70–90% reduction in blowdown steam loss; up to 10% reduction in GHG emissions (EPA 2010).

3.2.4 Leak Detection and Repair

Use organic vapour analyser or optical sensing technologies to identify leaks in natural gas lines, fuel gas lines and other lines with high methane concentrations and repair the leaks as soon as possible. The efficiency improvement/GHG emission reduction is 80–90% of leak emissions; <0.1% refinery-wide.

3.2.5 Carbon Capture

Being a post-combustion technology at end-of-pipe measures, it is noted that petroleum refineries emit CO_2 from a number of different processes, and the exhaust stacks for these emission points are numerous and scattered across the facility. The consideration of CO_2 capture and control at a refinery would likely be limited to the larger CO_2-emitting stacks, such as the Fluid Catalytic Cracking Unit (FCCU), the fluid coking unit, the hydrogen plant and large boilers or process heaters.

3.2.6 Oxy-Combustion

With a more detail discussion presented by EPA (2010), oxy-combustion being a process of burning a fuel in the presence of pure or nearly pure oxygen instead of air and fuel requirements are reduced (Barker et al. 2009). The process uses an air separation unit to remove the nitrogen component from air. The oxygen-rich stream is then fed to the combustion unit so the resulting exhaust gas contains a higher concentration of CO_2, as much as 80%. A portion of the exhaust stream is discharged to a CO_2 separation, purification and compression facility. The higher concentration of CO_2 in the flue gas directly impacts size of the adsorber (or other separation technique) and the power requirements for CO_2 compression.

3.3 Measure and Comparison of Carbon Capture Methods

It has often been acknowledged that rapid industrialisation and urbanisation have increased the demand for both energy and mobility services across the world with accompanying increases in GHG emissions. A case study reported by Adlen and Hepburn (2019) presents a combination of post-combustion CCS with a CO_2 capture rate well above 90% and other mitigation measures such as fuel substitution and emission offsets needed to transition towards carbon-neutral refinery operations. These technologies can be configured to minimise environmental burden shifting, cost-effective energy access, emissions from the combustion of refinery products, flaring and integration of the sinusoidal projection (equation below) to present a growth opportunity in a circular carbon economy (see energy transition Eq. 1.3 of Chapter 1) as follows:

$$T(t) = T_0 + \eta[c(t) - 277.04] + A \sin\left[\frac{2\pi}{\tau}(t + \phi)\right]$$

Some kind of economic motivation, together with some degree of climate mitigation potential may involve CO_2 utilisation. The removal and long-term storage of atmospheric CO_2 and its utilisation to make valuable products might offset some of the costs of climate change mitigation. On the contrary, even though CO_2 utilisation is successful does not necessarily mean it would be beneficial for the climate because the associated problems may include direct CO_2 emissions and other GHG emissions from direct and indirect land use. Thus, Table 3.2 presents comparison of the potential CO_2

Table 3.2 CO_2 capture and utilisation as potential large-scale and low-cost methods for business and environmental decisions (adapted from Adlen and Hepburn 2019).

CO_2 emission Sink/:Uses	Process description	Quantity captured/year in 2050 (Gt CO_2)	Cost per ton CO_2 (US$)	Energy Transition ≥277.04 ppmv; ≥0.500 °C/CO_2 ≥0.500 °C/Yr
CO_2 Chemicals: Use as fertilizer & polymers for durable products in buildings or cars,	Reducing CO_2 to its constituent components using catalysts and using chemical reactions to build products, such as methanol, urea	0.3–0.6	80–300	Effective
CO_2 fuels: Synfuels, syngas for transport infrastructure	Combining hydrogen with CO_2 to produce hydrocarbon fuels, including methanol	1–4.2	670	Not effective, reduce CO_2
Microalgae: Fuels and high-value chemicals	Microalgae to fix CO_2 at high efficiencies and then processing the biomass to make products,	0.2–0.9	230–920	Effective
Concrete building materials: Accelerating global urbanization	Cure cement or in the manufacture of aggregates	0.1–1.4	30–70	Effective
CO_2-enhanced oil recovery (EOR): Oil and CO_2 recovery from the well	Injecting CO_2 into oil wells can increase the production of oil	0.1–1.8	60–40	Effective
Bioenergy CO_2 capture: Carbon capture and storage	Captures CO_2 by growing trees, produces electricity through bioenergy and sequesters the resulting emissions	0.5–5	60–160	Effective
Forestry: Displace cement use	Timber is an economically valuable product that stores CO_2 in buildings	1.5	40–10	Not effective, reduce CO_2
Soil carbon sequestration: Land management techniques	Soil carbon sequestration can not only store CO_2 in the soil but also enhance agricultural yields.	0.9–1.9	90–20	Not effective, reduce CO_2
Biochar: Biochar improves agricultural soils and increases crop yields by 10%	Biochar is pyrolyzed biomass and plant material that has been burnt at high temperatures under low oxygen levels.	0.2–1	65	Effective

concentration gradients as discussed in Chapter 1 (Table 1.2) and cost of different CO_2 utilisation pathways to determine process effectiveness on energy transition.

The table shows that CO_2 utilisation could generate significant flows of CO_2 in 2050 and that some mitigation methods can be expected to be profitable. Measurement of energy transition by method of CO_2 concentration gradients proves that some pathways can effectively mitigate energy transition to solving climate change while others are not so effective. For example, CO_2 chemicals, microalgae, concrete building materials, CO_2-enhanced oil recovery (EOR) and bioenergy CO_2 capture are effective in mitigating climate change. While CO_2 fuels, forestry and soil carbon sequestration are yet to prove their effectiveness in mitigating climate change.

Although modular construction is a well-established technology in many industries, its uses and benefits in the refining and petrochemical industries have not been fully developed (Chemex Modular 2016). As opposed to traditional applications, modular refinery provides equal standards and more freedom and flexibility to a plant construction. The perceived advantages could be reduced equipment capital expenditures, labour expenses, simplified project timetables, reduced emissions to prevent or control climate change, etc. It is, therefore, expedient to delve into modular refinery design and calculations in the next section.

3.4 Benefits of Modular Refinery Design and Construction

Workflows for plant design and construction are created by repeating processes and procedures, making it easier to examine and discover potential hazards and pollutant emissions as well as chances to improve methods and timelines (Fremont 2021). Fremont considered the following three benefits of modular construction as follows:

3.4.1 Controlled Environment

Modular construction allows better control over temperature and weather variables. As a result, mitigation hazards, GHG emission, etc., should be dynamic targets. An operational petroleum refining environment as discussed in the preceding sections poses numerous options to apply hazard controls. Another advantage of modularised refinery construction is that weather and climate change have great impact on worker safety and project schedules.

3.4.2 Continuous Improvement

A controlled working environment encourages continued improvement, which can be seen as the most important safety benefit of a modular refinery operation. Since the workflows are repeating processes and procedures, to examine and discover potential hazards as well as chances to improve the applied methods and timelines are easily handled. It would be common practice that workers in petrochemical and oil and gas refineries are skilled in their work environments and are, therefore, familiar with the danger zones, while professional modular design and operations would limit the risk of accidents in machinery, equipment or other thongs. A permanent operation also instils a profound safety culture that begins on a worker's first day on the job and continues to grow on a long career in the construction (Fremont 2021).

3.4.3 Safety a Top Priority

The common practice as stated by Occupational Safety and health Administration (OSHA) Act, "Fatal Four", which includes falls, being struck by an object, electrocutions and getting caught-in or caught-between incidents, are the primary causes of occupational deaths in the private construction industry and OSHA. The refining and petrochemical sectors, therefore, could reduce accidents and fatalities by incorporating more modular components into refinery projects and general processes (Fremont 2021).

3.5 A Case Study: Design Report of Modular Refinery at Wits University

(Reprinted with permission from Mr. Michael Landgrebe of the Wits University)

The following report details the overall design of a modular refinery undertaken by final-year students of the University of the Witwatersrand, Johannesburg in South Africa as a classroom submission to the author, who was the lecturer and coordinator of the design course. The design modular refinery is capable of processing 4000 barrels of crude oil a day.

Literature Survey, Mass and Energy Balances, Preliminary Sizing of Equipment, Economic Evaluation, Entrepreneurial Analysis, Detailed Design of Fluid Catalytic Cracking Unit, Distillation Column Detailed Design, Process Control and instrumentation, Hazop, Plant Startup and Shutdown and Conclusion as presented below.

In the design, oil processing is investigated due to an increase in oil prices globally and the demand for petroleum-based products. The proposed plant utilises an atmospheric distillation column, vacuum distillation column and a FCC unit in order to yield a low-grade gasoline through prioritising the production of light straight-run (LSR) and naphthas. This is achieved by the manipulation of the Watson K factor in the heavy gas oil (HGO) fraction recovered from the distillation unit, with operating conditions altered to optimise the paraffinic content of HGO sent to the FCC. The FCC itself is designed similarly to the Universal Oil Products (UOP) model, with an annular catalyst stripper supported by the riser and a single-stage regenerator detached from the riser unit. A comprehensive investigation on the current market, profitability of the venture, unit design, process control and safety of the plant is conducted. This report places emphasis on the detailed design of the atmospheric distillation and FCC units as well as their correct material selections and structural design. Safety factors are detailed in this report and proposed control systems effectively regulate the plant with the appropriate considerations for hazardous operating levels. Correct startup and shutdown of the plant are also discussed with reference to the use of the process controllers to speed up the process, also accounting for unplanned interruptions to operations. Additionally, the selection of oil used is considered. Assuming the plant would be operated near Richard's Bay, Bonny Light crude oil could effectively be imported. This has low salt and favourable metal content and allows fairly simple pretreatment of the crude before distillation via two desalters in series that facilitate normal conditions in distillation while reducing corrosion in the plant. As the plant is a modular refinery, it is designed with the intent of further expanding operations into processing of consumer-grade vehicle gasoline after the refinery generates sufficient revenue to warrant expansion. The material balance for the system and simulation of major equipment units is performed in Aspen HYSYS, with specific dimensions of units being individually calculated using Microsoft Excel Spreadsheets. In order to perform pressure–volume relations, the Peng-Robinson equation of state is used. Simulation results are positive, with a small enough error to result in a reliable source of data. Preliminary costing performance reveals that the project is profitable even without expansion, giving a US$20 million return 10 years after initiating construction. The costing method used is subject to a wide margin of error, however, and uses exchange rates and the chemical engineering plant index (CEPCI) cost index in order to compare prices from one year and country to another, and a more detailed profitability analysis is required. Business models for the form of ownership conclude that this venture should be run as a private company, and methods of capital acquisition proposed comprise a combination of private and government grant-based funding. Overall the proposed design is sub-optimal due to an error in the design process in which the regenerator temperature has been set too low and will require a correction on the simulation and regenerator diameter calculations to ensure this design is indeed physically and economically viable.

3.5.1 Literature Survey of a Modular Refinery

A high degree of literature exists that explains the various concepts and principles that govern the many aspects associated with modular refinery. Despite the diverse themes that might be covered in literature, this survey will be limited to reporting the following aspects of modular refinery; process technology, cost estimation and market data. The objective of this literature survey is to establish the importance of modular refining to the petroleum industry and establish an avenue where a new contribution can be made a majority of this survey will be focused on critically analysing the different modular refinery plants and their respective units and identifying which direction will be embarked on as far as responding to the research question. The literature survey will be limited to publications that do not date too far back and website sources will be minimally used as reference sources because of their vulnerability to updates and shutdowns.

South Africa is one of the countries that depend on importing oil to satisfy the national demand for petroleum products. South Africa has crude oil refineries such as National Petroleum Refiners of South Africa (Natref), Chevron South Africa, Shell & BP South African Petroleum Refineries (Sapref), Sasol Refinery and Mossel Bay GTL refinery, but the demand is still greater than the supply. South Africa needs additional refineries to add to the supply capacity despite the country's 513 000 b/d production rate. In the coming years, the demand for refinery products gasoline, jet fuel and diesel will increase leaving a broader gap between supply and demand of crude oil products in the country (Mantshantsha 2013). An increase of 8, 3 and 6% in demand is reported for diesel, gasoline and jet fuel, respectively, every single year (PetroSA 2009). South Africa needs to import approximately 8000 barrels/day of finished products to meet the demand. According to PetroSA, the quantity of imported petroleum will grow to 23 000 barrels/day by 2020 if no more refineries are built (Mantshantsha 2013).

Advantages of Modular Refineries over Conventional Refineries

i. Modular refineries can be built near the location of the crude oil so as to minimise crude logistics and distribution costs. The transportation costs associated with moving crude oil from the oil field to regions of consumption are sufficiently high even for distant refineries in oil-exporting countries to consider using imported oil to be more economic (Cenam Energy Partners 2013).
ii. If light sweet crude oil is available for modular refinery, only primary distillation capacity is required under the right conditions to have significant competitive advantages. This unit needs less capital, needs less time to implement projects and adapts faster to demand changes in fast-developing regions (Cenam Energy Partners 2013).
iii. Another advantage of modular refineries is their ease and speed of construction. The constituents of the refinery are manufactured in modules and these are transported to the desired area and combined together. A 30000 barrels/day modular refining unit producing high-octane unleaded gasoline, LPG, diesel, kerosene and fuel oil can be installed in 18 months (Cenam Energy Partners 2013).
iv. The modular refinery can have only two operators to restart from cold start and have the plant in operation in hours. Modular refineries give the opportunity for rapid production of primary fuels for consumption and feed to other petrochemical industries. Modular refineries are also suitable for limited land space for refinery project (Atumah 2015).

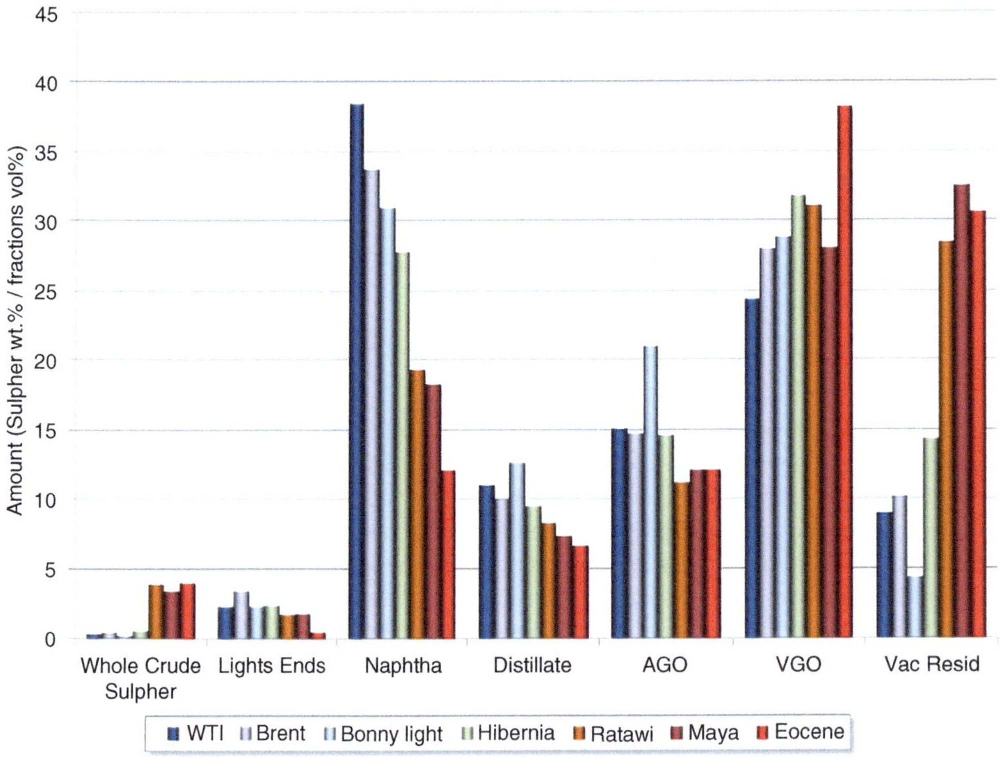

Figure 3.3 Variation in content from various crude oils. Source: Adapted from Jechura.

3.5.1.1 Crude Oil

Crude oil is processed in refineries to produce usable petroleum products. Common components of crude oil include hydrocarbons, metals such as iron, nickel, vanadium, various salts and sulphur. Crude oil is defined as either sweet or sour depending on its sulphur content – less than 0.5 wt.% is considered sweet whereas any more sulphur in the crude oil makes it sour. The compositions of crude oil vastly vary across the world; therefore, different processes are required to refine different types of crude oil; i.e. a single refining process cannot effectively convert a myriad of crude oils into a final petroleum fuel product (Yaws 2008). Figure 3.3 shows an example of a variation in crude oil composition from various sources:

Crude oil is also defined as either light or heavy. A light crude oil has a higher specific gravity and, therefore, a lower API°. This type of crude oil can be pumped and refined with a significant amount of ease compared to heavy crude oil thus making it more expensive. Heavy crude oil is cheaper because it usually contains high amounts of sulphur, nickel, vanadium and wax causing the refining process to be more complex. The heavy crude oil is broken down, for example, by cracking, to produce high-value products, increasing the cost of the process. Therefore, a decision needs to be made in terms of choosing a cheaper feedstock and having an expensive process or using an expensive feedstock and having a relatively cheaper process. Since the design of a refining process is specific to the crude oil feedstock, it is important to know the crude oil quality. A presentation by staff at Natref in Sasolburg summarised crude oil quality by the following factors:

- Acidity
- Sulphur content
- Nickel and vanadium content
- Total iron content
- Concarbon
- Mercaptan concentration
- Nitrogen content

The presence of sulphur in the crude oil is detrimental to the process equipment in the refining process, to the environment and to a vehicle's engine if there is a high sulphur content in the fuel product. Naturally, there is a cost involved in removing the sulphur from the crude oil; therefore, sour crude oils are less expensive than sweet crudes (The Cheaper Petrol Party 2016). Environmental restrictions have been implemented so as to ensure that the final sulphur content in the product fuel is low enough that it does not cause adverse effects on the equipment that will use it. Since not all of the sulphur can be removed, the acidity of the content inside process vessels can be increased so that the combination of acid and sulphur creates a layer of iron sulphide on the surface of the process equipment, protecting them from damage as well as corrosion.

The metals, salts and nitrogen in the crude oil are responsible for damaging process equipment and for poisoning catalysts. Therefore, they need to be removed in the initial stages of the refining process. Worldwide, refineries have adopted a desalting process in order to deal with the above components in the crude oil.

3.5.2 Process Operations Used within Modular Refining

3.5.2.1 Pretreatment

3.5.2.2 Desalter

The desalter is one of the mostly widely used pretreatment techniques in the crude oil refining industry. Crude oil contains a high amount of undesirable material when it reaches oil refineries via pipeline transportation. The impurities in raw crude oil are classified into two classes, i.e. oleophobic and oleophilic impurities. Oleophilic impurities are soluble in oil and are sulphur compounds, organometallic compounds, Ni, V, Fe and As. Oleophobic impurities are repelled by oil and are treated by a desalter. These include salt in the form of chloride mainly and impurities of Na, K, Ca and Mg, sediments such as sand, mud, iron oxide and water (Jones 2016). The undesirable nature of the aforementioned material stems from the negative impact that it cause downstream in the refinery. Heat exchangers experience fouling, furnaces and distillation columns experience corrosion because they operate at high temperatures that are favourable to salt hydrolysis that produces hydrochloric acid according to the following reactions (Kolmetz 2014);

1. $2NaCl + H_2O \rightarrow 2HCl + Na_2O$
2. $MgCl_2 + H_2O \rightarrow 2HCl + MgO$

The metals from the salts in the crude oil are known to cause the deactivation and sintering of catalysts downstream (Kolmetz 2014).

The operation of a desalter can be broken down into three stages. The heating, mixing and separation stages (Jones 2016). The raw crude oil is first heated up in a train of heat exchangers to 120–150 °C (New Logic Research 2003). The fresh wash water must be in the temperature range of 93–121 °C. The heated crude oil and fresh wash water converge at a mixing valve with a high differential pressure drop of 5–20 psi. The water and oil are preheated and mixed in a ratio of 1 : 20 (Kolmetz 2014). The intensive mixing of water and oil forms an emulsion with most of the oleophobic impurities being in the water phase. This emulsion is sent to a settling tank where the emulsion is broken down into an overflow and underflow (Jones 2016). The overflow mainly consists of treated crude oil and the underflow is mainly waste wash water. The desalter crude oil exits the desalter at a temperature of 88 °C. The emulsion can be disintegrated using electrostatic precipitation and/or chemical demulsifiers which can be injected into the wash water and raw crude oil streams prior to the mixing. Electrostatic precipitation uses high voltage charges to make the suspended water molecules and impurities coalesce and settle at the bottom of the pressurised vessel (Kolmetz 2014). Chemical demulsifiers or surfactants are only used when there is a large amount of suspended solids (New Logic Research 2003). It is common that demulsifiers are added at the ratio of 0.005 lb. to 0.01 lb./barrel of crude oil (Kolmetz 2014). The solids that settle from the crude oil tend to accumulate at the bottom of the desalter tank, therefore, periodic cleaning or washing is required to remove these solids. This is achieved by the installation of mud washing systems. These systems recycle a portion of the desalter underflow so that it agitates the solids that have accumulated and they are carried into the effluent water (New Logic Research 2003). The primary objective of utilising a desalter is to remove impurities in raw crude oil while minimising the quantity of oil in the water effluent and water in the overflow stream.

The desalter can be operated as a single-stage or multi-stage process. The decision on which process is suitable for operation is mainly based on how much of the undesired material should be removed to reach a certain target. The aim of many refineries is to have treated crude oil with a salt concentration of 5 ppb. Using a single-stage desalter to perform such a function is not possible. The two-stage desalter is then necessary (Jones 2016). A single-stage desalter can remove up to 95% of salt while the two-stage desalter can have an efficiency of 99% salt removal (Gary et al. 2007). A third desalter is not cost-effective, it would be better to put on corrosion inhibitors inside the vulnerable equipment. (Jones 2016)

Figure 3.4 is an illustration of a single-stage desalter. The raw crude oil undergoes desalting in one settling tank and the overflow is sent directly to a charge furnace to be heated for the next stage of refining. The red arrows in the diagram

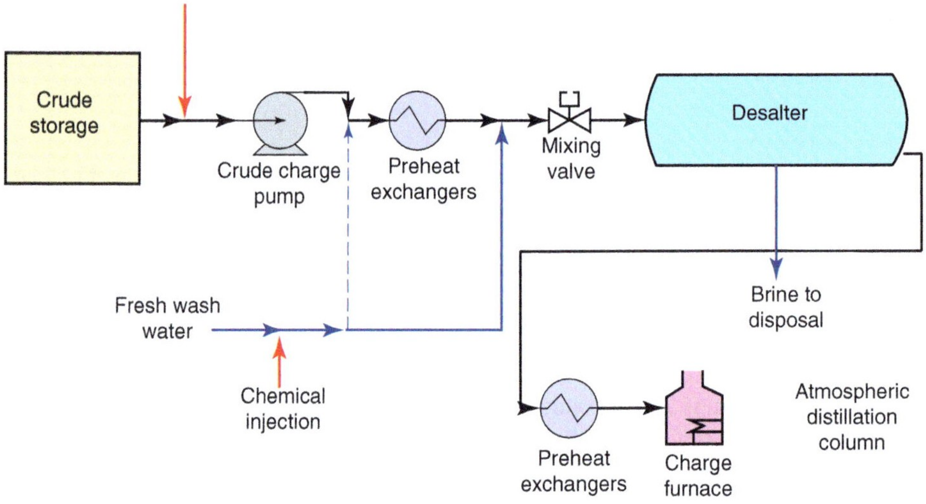

Figure 3.4 Single stage desalter. Source: Adapted from Kolmetz.

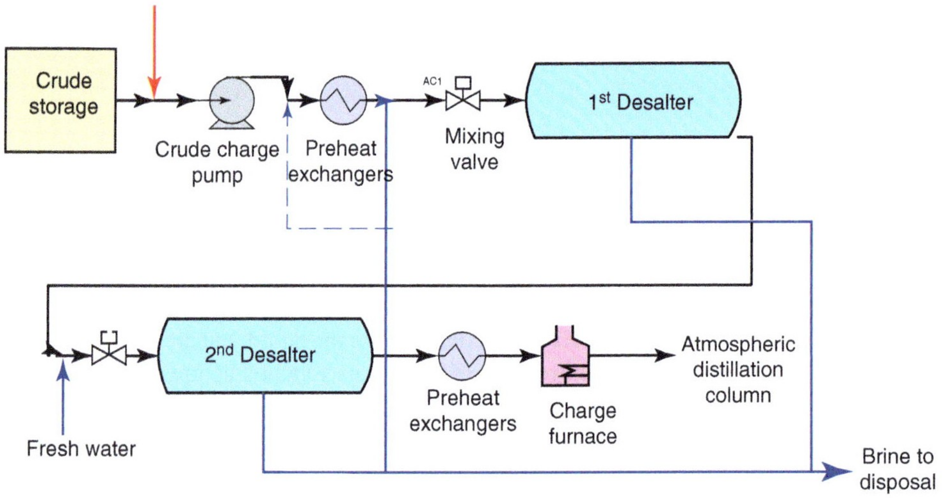

Figure 3.5 Two stage desalter. Source: Adapted from Kolmetz.

indicate the points where chemical demulsifiers are added to the system. The dotted blue line shows the possible route of the wash water if it comes into the system at a different temperature other than the required temperature.

Figure 3.5 depicts a two-stage desalter. This type of desalter has two desalter tanks for thorough removal of impurities. The crude oil that exits the first desalter tank is sent to another mixing valve where fresh water wash is added. The emulsion formed in this mixing valve enters the second desalter tank. After demulsifying, the wash water is sent for disposal or purification and some of it is recycled to the first mixing valve.

The process variables that affect the performance of a desalter are the quality of the crude, the feed rate of the crude oil, the temperature, electrical field intensity, wash water quality and feed rate, emulsifier feed rate, water level control, demulsifier feed rate and the temperature, viscosity and density of the raw crude oil. The heating stage of desalting serves to increase the temperature of the oil thereby increasing the viscosity and decreasing the density of the oil. The required viscosity of the oil is 5–15 c-Stoke. The decrease in density is favourable to the efficiency of the desalting process based on Stoke's law. Stoke's law is also used in sizing desalter tanks to allow the settling and separation of water and oil (New Logic Research 2003). It is represented mathematically by the following Eq. (3.1);

$$v = \frac{g(SG_{water} - SG_{oil})d^2}{14\mu_{oil}} \tag{3.1}$$

where V is the settling velocity of the settling particles; g is the acceleration due to gravity; SG_{water} is the specific gravity of water; SG_{oil} is the specific gravity of oil; d is the diameter of the settling particle; μ is the viscosity of the oil.

It can be observed that one of the primary variables in this equation is the difference in density of a particle and the fluid. The decrease in density of oil will increase the density difference thereby increasing the particles settling velocity (Jones 2016).

3.5.2.3 Refining Process

3.5.2.3.1 Topping Column The model chosen for the topping column is derived from various industrial models that focus on FCC for post treatment in order to yield a higher amount of gasoline and perform profitably at industrial scale. Overall it can be seen that no product extracted is classified as a pure enough substance to be sold as quality fuel (Agency for Toxic Substances & Disease Registry 2016), instead as the plant being designed is a modular refinery, post treatment for various product streams may be performed. Overall the modular refinery being designed focuses on a high yield of gasoline via cracking.

The feed to the column crude oil that has been passed through the furnace and heat exchange network so as to achieve a temperature of approximately 360 – 370 °C. Steam is also introduced via sidestream returns so as to lower the amount of products of the volatile substances such as sulphur dioxide that were not removed during pretreatment (Treybal 1981). This water is separated from the top product after being passed through a condenser and then again in a separate unit that separates the organic product from water.

Product streams have been selected with post treatment in mind, and thus precise temperatures are not highly important. Ranges for efficient operation to obtain products of acceptable purity for post treatment have been selected for preliminary study before the unit is optimised in simulation as follows based on the research of Popoola et al. (2013).

- Top product of LPG, LSR: 30 – 100 °C
- Mid- to heavy-weight naphtha: 80 – 200 °C
- Kerosene: 170 – 280 °C
- Light gas oil: 220 – 320 °C
- Heavy gas oil: 290 – 350 °C
- Residue: 350 – 390 °C

As expected, the top product has a reflux back to the distillation column. Additionally, for each side stream other than the bottom residue, there is a stream returning to the distillation column. This is due to the issue that arises when a fraction is extracted from a tray. When the liquid is removed, there are still volatile fractions within the liquid that are being extracted at lower temperature sub streams on higher plates. In order to remove some of these more volatile fractions, Treybal (1981) details that the withdrawn product from the topping column is sent to a stripping unit which uses steam to extract components with a greater volatility than the majority of component being extracted. This gaseous mixture of steam and volatile organics such as naphtha or kerosene is returned to the distillation column, on a plate above the original liquid extract.

The presence of pump-around side streams serves to remove heat and improve the mixing of streams within the column by removing liquid products from one plate and adding the removed liquid to a higher plate. Gary et al. (2007) describe that this also serves to reduce reflux ratio and will also increase the yield of gas oils that will be used in cracking (Figure 3.6).

3.5.2.3.2 Gas Oil Extraction The HGO phase is extracted to facilitate the cracking of heavy fuel into gasoline via cracking. However, not all gas oil can be used in the procedure of cracking. Lighter fuels with aromatics or not sufficiently paraffinic will lead to excess coking of the catalyst used in cracking and will simply coke the catalyst (Sadeghbeigi 2012). This notably increases operational costs through a larger regenerator being required and catalyst replacement costs. Additionally, these non-paraffinic molecules could have been used for another useful purpose, such as being sold to another plant or being converted in another refinery module to diesel. The method of determining which exact temperature range will be performed via simulations using iterations to see which range and operating conditions via internal reflux, with the determining factor of paraffinic nature being determined by the UOP K factor. Sadeghbeigi (2012) outlines the method of calculating this factor as follows:

$$K_{\text{UOP}} = \frac{(\text{CABP} + 460)^{\frac{1}{3}}}{\text{SG}} \tag{3.2}$$

where CABP (cubic average boiling point) of the mixture is in Fahrenheit; and SG represents the specific gravity of the mixture at 60° Fahrenheit. The cubic average boiling point is calculated by an equation that sums the vapour fraction of a species multiplied by the boiling point of that species:

$$\text{CABP} = \sum_{i=1}^{n} \left(f_{v_i} * T_{\text{boiling}_i}^{\frac{1}{3}} \right)^3 \tag{3.3}$$

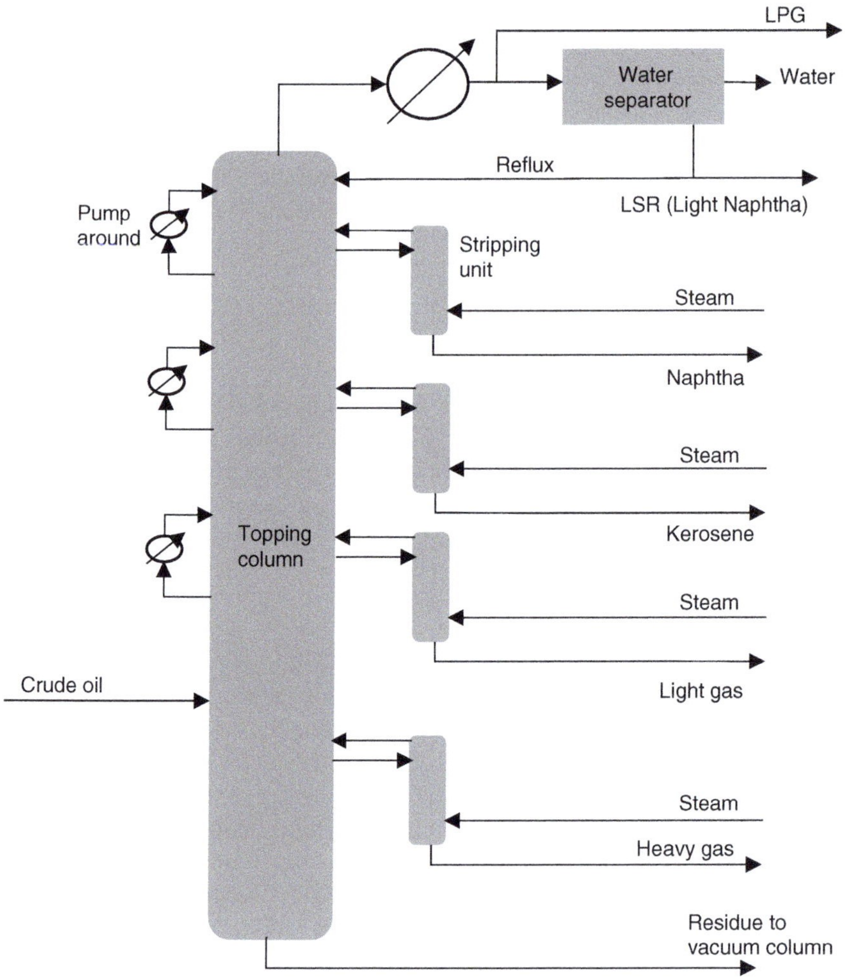

Figure 3.6 Operational setup of the topping Column.

This method is an approximation, notably due to the specific gravity being set to a value regardless of the temperature range. However, the *K* factor has proven useful in determining paraffinic nature in literature where similar operating temperature ranges and side streams were investigated.

An alternative method of obtaining the *K* factor is the Watson method, which uses the average boiling point of the mixture and the specific gravity at 60° Fahrenheit and is detailed by Gary et al. (2007).

However, even with a method of determining what operating conditions and temperature for the HGO withdrawal, a significant amount of components that could be cracked via FCC are withdrawn in the residue. The residue is thus sent to a vacuum column to extract any further gas oils in order to feed these to the cracking unit. The bottom product of the vacuum column is a residue of increased carbon number on average, but this is still useful.

This residue from the vacuum column could be sent to an additional coking unit, where the last amounts of gas oil that can be feasibly obtained are extracted via the process of delayed coking. This could be introduced at a later stage, after income is made via the plant and expansion is permitted.

3.5.2.3.3 Vacuum Distillation Vacuum distillation is a step in the refining process in which heavier gas oils trapped within residue after atmospheric distillation are recovered. It may not be possible to remove these heavier gas oils from the crude oil within the atmospheric distillation unit as the temperatures needed for the separation of these components are higher than what the components are able to withstand. According to Leffler (1985), the use of these high temperatures within the atmospheric distillation unit would lead to thermal cracking of the gas oils as well as coke formation within the fractionation column. The US Energy Information Administration (2012) states that the occurrence of these processes within the distillation column would impede the overall distillation process. This would drastically reduce separation efficiency.

Information from Fahim et al. (2009), the US Energy Information Administration (2012) and Gray et al. (2007) state that in order to prevent thermal cracking of the residue, operating temperatures should ideally be in the region of 380 °C as cracking of the components begins to occur at temperatures exceeding 415 °C. A distillation process occurring under vacuum conditions, therefore, allows for the separation of the heavier gas oil components out of the atmospheric residue due to a decrease in the vaporisation temperatures of the valuable components. Vacuum columns operate at pressures of between 10 mmHg and 100 mmHg (US Energy Information Administration 2012) (Encyclopaedia Britannica 2016).

There are multiple models of vacuum distillation that are used within the field of oil refining. These may be classified into three main categories (ATSDR, 2014 and 2015):

- Vacuum flash distillation
- Vacuum fractioning
- Vacuum re-distillation

According to Ceric (2012), vacuum flash distillation is most suited for the production of feedstock for downstream processing in the form of FCC and hydrocracking. This is because the process of vacuum flash distillation produces product fractions of gas oils that do not have a high degree of sharpness. Vacuum flash distillation is able to separate the atmospheric distillation residue into the following fractions:

- Vent gas
- Light vacuum gas oils
- Heavy vacuum gas oils
- Vacuum residue

The uncondensed vent gases are removed through steam ejectors during the process of creating vacuum conditions within the column. The light and heavy vacuum gas oils (HVGO) are removed as side fractions from the column and remixed in preparation for cracking processes. As the light and HGO fractions are remixed, it is unnecessary to achieve a high degree of separation between the two fractions (Ceric 2012) (Figure 3.7).

Vacuum fractionation is a more complex process than vacuum flash distillation. Vacuum fractioning is able to produce higher quality distillate products, as well as being able to split the atmospheric residue in a larger number of fractions. As stated by Ceric (2012), vacuum fractionation may be used to produce good-quality oil distillates which are used as the feedstock for the production of oil-based products. Vacuum fractionation is able to separate the residue from the atmospheric distillation unit into eight fractions:

(i) vents gases, (ii) vacuum oil, (iii) VD-1, (iv) VD-2, (v) VD-3, (vi) VD-4, (vii) VD-5 and (viii) vacuum residue

As with vacuum flash distillation, the uncondensed vent gases are removed from steam ejectors. The vacuum gas oils may be hydrotreated for the production of diesel or may be used as feedstock for a cracking process for the production of gasoline. The product fractions VD-1 to VD-4 are of high enough quality to be used for the production of base oils and waxes, while VD-5 may be cracked for the processing of fuel oils (Ceric 2012).

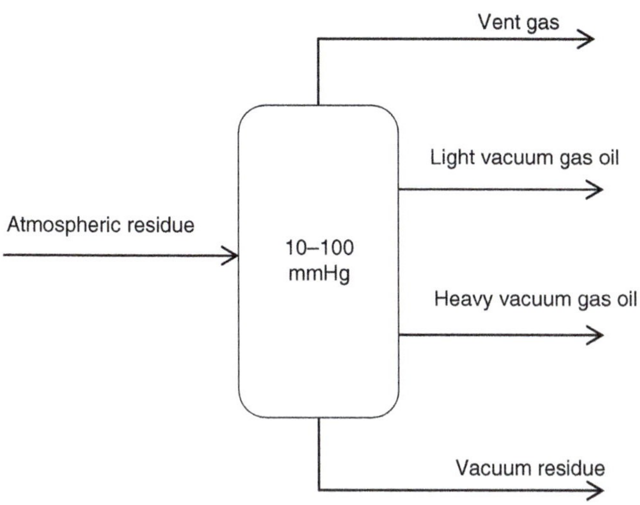

Figure 3.7 Simplified block flow diagram of vacuum flash desalination.

Vacuum re-distillation is the combination of the use of both vacuum flash distillation and vacuum fractioning. Vacuum re-distillation is still mainly used for the production of base oils, as with vacuum fractioning, however, vacuum re-distillation is able to provide higher yields for the VD-1 to VD-4 fractions (Ceric 2012).

3.5.2.3.4 Cracking In order to obtain light hydrocarbons from long-chained, a process called cracking is used whereby the carbon–carbon bonds of the long-chained hydrocarbons are broken yielding smaller chained hydrocarbons. There is a variety of cracking processes that produce usable fuels such as FCC, hydrocracking, thermal cracking, isomerisation and steam cracking. Although there are various cracking processes in use at refineries globally, each of the processes holds one thing in common: the rate of cracking and the products from each of the cracking processes are sensitive to the temperature of the cracking process as well as the type of catalyst being used.

Fluid Catalytic Cracking In the past, thermal cracking was used to convert long-chained hydrocarbons into a gasoline product; however, it has now been almost fully replaced by (fluid) catalytic cracking. This is because FCC is able to produce a significantly higher gasoline yield with a higher octane rating and liquefied petroleum gas (LPG) (Gary et al. 2007). Therefore, when looking for a process that has a high gasoline product, FCC is the preferred method. Catalytic cracking has a high yield of branched and cyclic alkanes. In addition to this, the olefins that are produced as by-products from the FCC process have monetary value, thus making this form of cracking profitable.

The use of a catalyst allows for lower temperatures to be used for the cracking process. Although there are various types of catalysts that can be used for various end products, the catalyst usually looks like granules of sand and is made to become fluid by steam. This gives the catalyst the ability to transfer and recirculate between the reactor and regenerator vessels (US Energy Information Administration 2012). Among the various types of catalysts used in FCC processes, there are catalysts such as Grace Davison Nektor ULCC and Grace Davison (Figure 3.8).

Equilibrium Catalyst and SAPREF Equilibrium Catalyst.

The FCC process consists of three stages: reaction, regeneration and separation. To summarise the process, the feed is initially reacted with the catalyst and then separated into different product streams such as gasoline and LPG through fractionation, and the catalyst is regenerated by removing the coke and is then recirculated back to the reactor. The feed to an FCC unit is the HGO from the crude and vacuum distillation columns at an approximate temperature of 370 °C and atmospheric pressure. The reactor operates at a temperature of approximately 510 °C and a pressure of 1.7 bar whereas the regenerator operates at a temperature of approximately 710 °C and 2.4 bar (Gary et al. 2007). This high operating temperature of the regenerator ensures that almost all the coke that forms on the catalyst is burned off, allowing for effective regeneration. It also serves to provide sufficient heat to vaporise the feed and provide heat to the endothermic catalytic reactions.

Commonly FCC units have two configuration types: stacked or side by side. According to the Midwest Research Institute, a large-scale refining plant should use a side-by-side configuration whereas a small-scale or modular plant should use a stacked configuration. It states that major licensors agree that: "Stacked designs require less plot space and less structural steel, thus lowering costs". However, this economic benefit is lost once the capacity of the plant is 20 000 barrels per day or higher. Therefore, with regards to a modular refinery, a stacked configuration for the FCC unit is advisable. It should be noted that the configuration of the unit does not change the major equipment pieces of the FCC unit – both still require a riser (where the feed and catalyst meet), the reactor and the regenerator. The difference between the two configurations is

Figure 3.8 Illustration of zeolite used in the FCC process (Sadeghbeigi 2012).

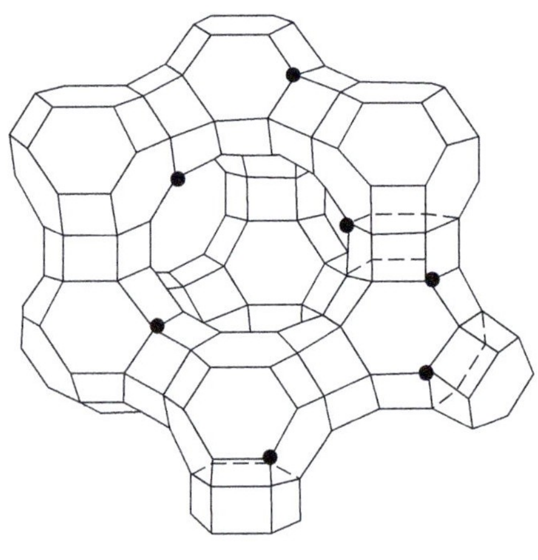

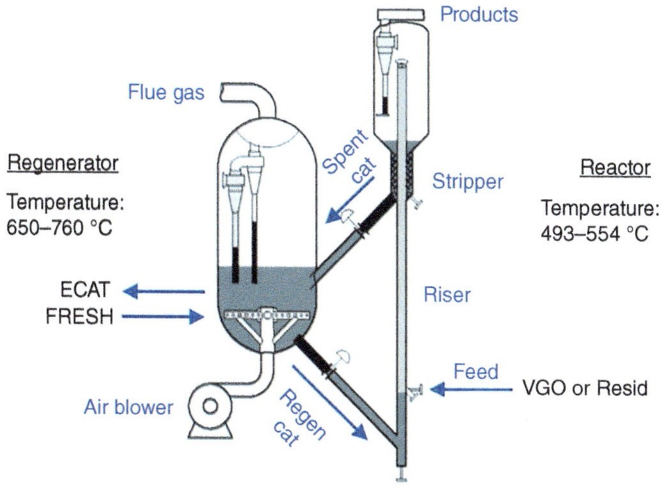

Figure 3.9 Fluid catalytic cracking unit which includes the reactor, riser and regenerator. Source: Weckhuysen and Vogt/Royal Society of Chemistry.

that for the stacked configuration, the riser is in place of the process vessels and the reactor is on top of the regenerator (Figure 3.9).

Isomerisation Isomerisation is used in the refining industry to rearrange the structure of hydrocarbons into more useful isomers so as to enhance the octane rating of the fuels. Straight-chained paraffins are converted into branched-chained paraffins – normal paraffins are converted into isoparaffins (Watanabe et al. 2008). An isomerisation unit uses the LPG products of a crude distillation unit as its feed. This process also uses catalysts, which differ depending on whether the operating temperature is high or low. For lower temperatures, an aluminium chloride catalyst is used, whereas a metal catalyst (such as platinum) is used for higher temperatures. For example, in an isomerisation unit operating at 300 K, the vaporised reactants are passed over a solid aluminium chloride catalyst (The Essential Chemical Industry 2014). Szostak (2006) states that an isomerisation process typically operates at 13.8 bar, thus pressure resistant material needs to be used for the process vessel.

Once the isomerisation reaction has taken place, separation of the products is conducted either via the use of distillation or through a molecular sieve. When using a molecular sieve, the straight-chained hydrocarbons are passed through the sieve and recycled to the isomerisation unit, and the branched chained hydrocarbons, which get trapped by the sieve, are later collected and can be converted to bigger chains for petrol.

Steam Cracking Steam cracking is used in the petroleum refining industry to break down saturated hydrocarbons into smaller hydrocarbons which are usually unsaturated. This process yields a high amount of alkenes. The feed for a steam cracking unit is usually naphtha (C_5–C_{10} hydrocarbons), gas oil, and residues from primary distillation. Liquid naphtha and gas oil are preheated so that they become vaporised. They are then mixed with steam and are allowed to react in a tubular reactor at an operating temperature ranging between 1050–1150 K for a very short time – typical steam cracking furnaces have a residence time of a few milliseconds so that the production of coke and undesired hydrocarbons is alleviated (Jukic 2013). To achieve this, the vaporised gases are quenched in order to halt the reaction.

For petroleum refining steam cracking processes, it is advised that operating temperatures of a lower severity should be used so that C_4 hydrocarbons and liquid products are favoured. This additionally minimises the chances of coke deposits forming on the reactor walls, ensuring that the efficiency of the reactor is not adversely affected. Although the form of cracking is able to produce a high yield of usable products, due to inevitable coke formation the furnace inside the tubular reactor requires decoking every few months, thus it will need to be shutdown periodically. The decoking process occurs by allowing a steam–air mixture through furnace coils in order to convert the coke into carbon monoxide and dioxide (Jukic 2013) (Figure 3.10).

Thermal Cracking In industry thermal cracking is one of the oldest techniques utilised in cracking processes. It plays an important role in petroleum refining and is used to process crude oil and process other petroleum products and these products are made available to the commercial market. Thermal cracking is exposing hydrocarbons to extreme heat in which

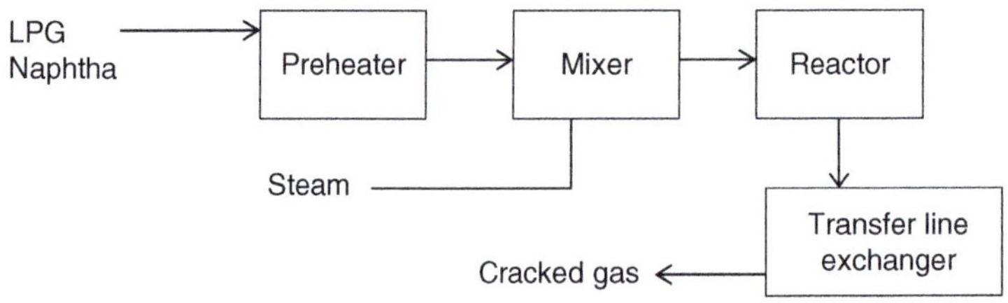

Figure 3.10 Block flow diagram of a typical steam cracking process.

the temperature ranges from 450 to 750 °C (wiseGEEK 2016). The extreme heat breaks bonds and reduces the molecular weight of the substance being cracked. As a result, free radicals at the sites where molecular bonds are broken are used to give rise to new chemical compounds through reactions like polymerisation. In the thermal cracking process, many valuable compounds are achieved making it an integral part of petroleum refining. In order to obtain satisfactory cracking which can be manipulated to desired rate, there are three important process variables to consider which are temperature, pressure and residence time. The conditions of the cracking are dependent on the feedstock and typically the feedstock for thermal cracking is heavy hydrocarbon molecules derived from the atmospheric or vacuum distillation of crude.

For many years, thermal cracking has been used in petroleum refining but in modern refining catalytic cracking is a preference over thermal cracking. Thermal cracking requires high temperatures and pressures to operate so the operating costs rise significantly and have an adverse effect on the revenue (wiseGEEK 2016). Thermal cracking has a high conversion on short-chain hydrocarbons and can also be carried long-chain hydrocarbons. One of the main reasons catalytic cracking is selected over thermal cracking is because catalytic cracking yields gasoline with a higher octane rating and also catalytic cracking produces by-product gases that are more olefinic and have more economic significance than those products produced in thermal cracking (wiseGEEK 2016). Thermal cracking still has major applications in modern refineries which are visbreaking and thermal gasoil production.

Visbreaking Visbreaking is a process of reducing the viscosity of a residue significantly which reduces the diluent requirements and the amount of fuel oil produced in a refinery (Gumilar 2009). Appropriate pre-heat is supplied to the feed before the feed enters the furnace. The feed is then sent to the furnace for heating at cracking temperatures which is in the range of 450–460 °C. Majority of the cracking takes place in the soaker (reaction chamber). To prevent excessive coke formation the products from the soaker are quenched to stop the cracking reaction. The products from the soaker then enter the fractionator which has a temperature in the range of 300–400 °C. The products that are achieved from the fractionator after separation are gas, gasoline, kerosene, gas oil and residue. The residue has a lower viscosity than that of the feed which reduces diluent requirements. The cracking temperature used is in the range of 440–450 °C at a pressure range of 5–10 barg in the soaker. The fractionator can be operated at 2–5 barg, depending on factors like furnace constraints, condenser constraints and fuel cost (Gumilar 2009) (Figure 3.11).

Figure 3.11 Configuration for the visbreaking unit. Source: SET Laboratories/http://www.setlab.com/resources/refining/thermal-cracking/#1498503315834-22c2e8e4-0376/last accessed May 17, 2023.

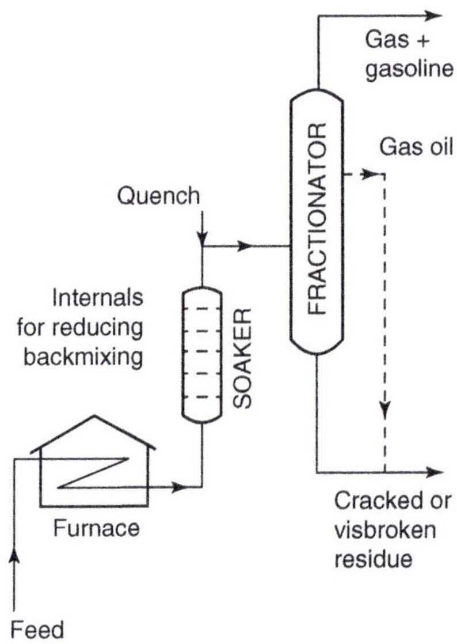

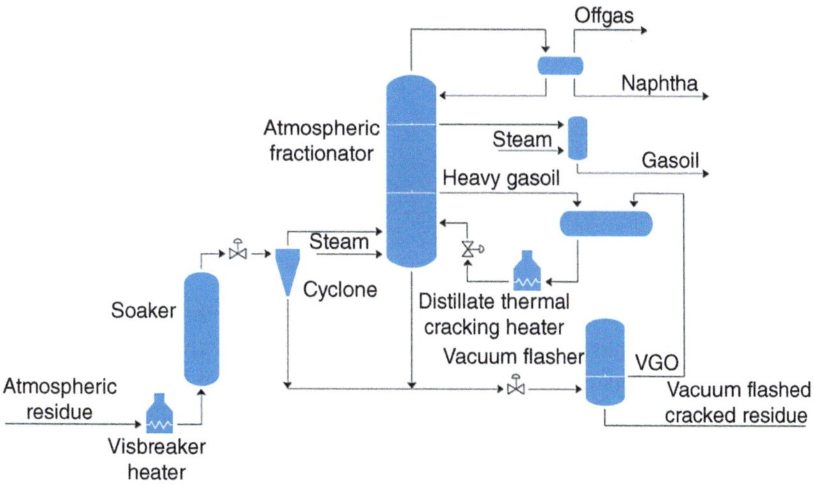

Figure 3.12 Example of a thermal gasoil process. Source: Chicago Bridge and Iron Company/https://studylib.net/doc/8197657/shell-thermal-gasoil/last accessed May 17, 2023.

Thermal Gas Oil Production The major application of thermal cracking is not just to reduce the viscosity of the feedstock but also to generate and optimize the recovery amount of gas oil. The layout is very similar to the visbreaking configuration unit. Once the residue is visbroken, it is sent through a vacuum-flash to recover heavy distillates which are sent back to the thermal cracking stage along with heavy distillates recovered from the fractionator unit. These recovered heavy distillates are sent to a second furnace which employs severe cracking conditions. This furnace operates at a temperature of 500 °C and a pressure range of 20–25 barg. The severity of the operating conditions in the first stage is highly dependent on the feedstock. If the feedstock has small molecular sizes then more severe operating conditions are employed to crack the molecules (Gumilar 2009) (Figure 3.12).

3.5.2.3.5 Coking Coking is a unit operation in a refinery that upgrades the bottoms from atmospheric and vacuum distillation columns into products of higher value. This process also produces petroleum coke. Three types of coking processes exist, namely, fluid coking, delayed coking and flexicoking. Both processes thermally crack the bottoms of the vacuum and atmospheric distillation and leave a residue of coke. Petroleum coke can be sold as fuel-grade petroleum coke or undergo additional heating to produce anode-grade petroleum coke. It also has application and usage in the electric and industrial sectors. In 2011, 132 million barrels of petroleum coke was supplied by the refining industry and most of it was used as fuel (US Energy Information Administration 2013).

Delayed Coking This type of coking is the most predominantly used, and it is relatively inexpensive (Jechura 2016). The feed to the delay coking unit is the residuum of vacuum distillation, but atmospheric distillation residuum is occasionally used. A heater, coke drum vessels, fractionator and downstream vapour processing vessels are the constituents of the process (Jechura 2016). The residuum is thermally cracked to convert it to liquid and gas product streams leaving behind petroleum coke. The term ‘delayed’ is used to identify the process because the furnace used reaches thermal cracking temperatures of 485–505 °C. The coking of the residuum is then ‘delayed’ until it reaches the coke drums downstream because of the short residence time within the furnace tubes (Ellis and Paul 1998).

The fresh feed to the process is transported to the fractionator also called combined tower which has reservoirs in the bottom where the feed combines with the condensed overhead products from the coke drums. This mixture is used as the feed to the furnace that heats it up the cracking temperatures. The heated material goes to a switch valve that permits the feed to go to either of the coke drums (Ellis and Paul 1998). One of the drums is online and accumulating with coke while the other is offline being stripped by steam and being cooled. The coke in it is removed and the drum is warmed up after having the pressure checked. A drum cycle time of 16 hours is the typical time used by delayed cokers. This is the time required to get an off-line coke drum online again (Jechura 2016).

Flexicoking This process utilises a reactor, heater and gasifier to coke the bottoms vacuum fractionation unit (ExxonMobil 2016). The residuum is first heated and sprayed into the reactor that contains hot fluidized coke. The heater and the reactor cycle the fluidized coke at such a rate that the reactor temperature remains constant at 540 °C. Cracked vapour products form in the reactor which has a scrubber attached at the top of it. The products are passed in the baffles found in the scrubber.

This is done so as to separate coke particles from the vapour products. Cool oil is then added to the scrubber to wash the accumulated coke particles from the scrubber back to the reactor. The vapour products that leave the scrubber are returned back to the fractionation tower (Stark 1998).

During the aforementioned proceedings, the coke in the reactor cools down. As it does this, it settles at the base of the reactor and is removed with steam in a baffled section at the reactor base. This is done to ensure only coking exits and no other components in the reactor. The coke flows to the heater where it is heated to 540 °C. This heated coke goes to a gasifier. This is a fluidized bed where the coke will react with air and steam producing a variety of fuel gases. The fuel gases are passed through a scrubber and thereafter to the fractioning unit (Stark 1998).

Flexicoking has a lower capital investment as a result of key factors such as an integrated simple steam/air gasification and the absence of feed furnaces. The layout of a modern flexicoking unit occupies land 20% less than a basic delayed coking unit. Flexicoking has a conversion of 95% of the coke generated (ExxonMobil 2016).

Fluid Coking Fluid coking is similar to flexicoking but with the absence of the gasifier. Vacuum residuum of about 565 °C enters a reactor overhead. It is then thermally cracked to lighter products and coke. Lighter overhead vapours exit the reactor through a scrubber and flow to the fractionation unit. Coke is cycled between a heater or burner and the reactor. This is a means of transferring heat from the heater to the reactor.

As similar to flexicoking, fluid coking produces C_4 reactor gases, full range of liquid products such as naphtha and light gas oil (LGO) and coke which have used as alkylation feed, mogas blending, FCC feed and power generation, respectively (Stark 1998).

3.5.2.3.6 Post Treatment

Hydrotreating The ultimate goal of hydrotreating is to remove contaminants sulphur, nitrogen and metals which results in a cleaner product that will comply with environmental regulations. By removing the contaminants a more pure product is generated which could be sold or it could be sent for further processing. Hydrodesulphurisation is one of the most common hydrotreatment methods employed in the processing industry.

Hydrodesulphurisation is a catalytic chemical process that is very common and commercially applied technique used to remove sulphur to meet environmental standards and drop it to tolerable amounts that the catalyst can handle this aids in prevent catalyst poisoning. The feedstock is combined with hydrogen-rich makeup gas and the recycled gas from the reactor and this mixture is heated in the furnace. The mixture enters reactor where the nitrogen and sulphur compounds are converted to ammonia and hydrogen sulphide. The reactor has a catalyst loaded in it and operates at a temperature range of 300–380 °C and a pressure ranging from 10 to 20 bar for naphtha and 30–45 bar for gas oil (Techno Manage 2016). The reaction products are sent to a liquid/gas separation process and the hydrogen-rich gas is recycled back to the feedstock stream. The hydrogen sulphide is removed from the gas rich in it and the clean gas is utilised as fuel. Finally, the liquid stream is sent to the stripping column, where hydrogen sulphide and other impurities are removed (Techno Manage 2016).

Failure to remove sulphur leads to short catalyst life span, decreased catalyst activity and reduced product output. Hydrodesulphurisation is employed commercially when naphtha is the feedstock, and the sulphur content has to be reduced to amount of 1 ppm to ensure that the platinum catalyst is not poisoned. Another application of hydrodesulphurisation is to remove sulphur compounds from kerosene and gas oil to make them suitable blending components (Techno Manage 2016) (Figure 3.13).

3.5.2.3.7 Similar Operations in Industry The modular refinery setup is widely used within the crude oil processing industry, especially for start-up processes. However, the majority of these operations are small-scale private processing plants, therefore, in-depth literature about these operations is not widely available.

As costing information of modular refineries is not available, the identified construction costs of conventional refineries may be used within rule-of-thumb cost estimates given in Section 1.6 for preliminary costing calculations (Ibsen 2006).

Cost Estimation Due to the early stages of design at this point, finding a cost estimate of the plant with reasonable error margin is challenging. While rule-of-thumb estimates are available, the majority of these estimates are for fully operational refineries, not modular refineries that only having a select few units. A method that can in fact be used is the cost curve estimate method. This method takes a plant comprised of similar process units and products and allows for a cost estimate to be performed with the only variable used in calculation being the barrels per day of crude oil that is processed.

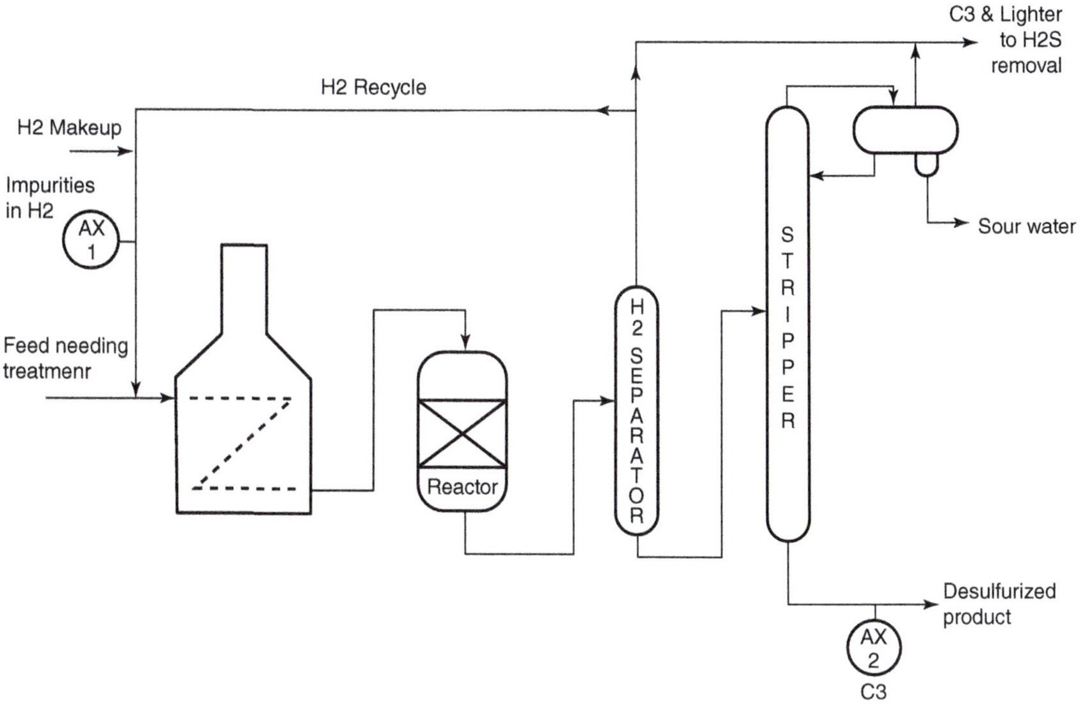

Figure 3.13 Hydrodesulphurisation process flow diagram. Source: Emerson/https://www.emerson.com/documents/automation/application-note-improving-refinery-hydrotreater-hds-unit-performance-process-gas-chromatographs-rosemount-en-71800.pdf/last accessed May 17, 2023.

The method that will be used is known as the lang factor cost estimate. The form that will be used is cited in Gary et al. (2007) as having an exponent of 0.6 and is used due to the application being a petrochemical refinery.

$$\frac{\text{Cost of plant A}}{\text{Cost of plant B}} = \frac{\text{BPD of A}}{(\text{BPD of B})^{0.6}} \tag{3.4}$$

A study by Wolf (2013) was conducted on determining the present viability of the Lang estimate by using data from various companies in order to determine the cost difference estimated by the Lang method and the actual amount of the plant cost. The study found that across various plants, petrochemical plants were found to have an average of an 8% difference between the Lang estimate and the actual cost. Thus it can be seen that for a preliminary estimate, this method works well and can provide a good indication of the cost range.

However, an issue arises in that petrochemical plant data is difficult to obtain in South Africa. Thus, taking a cost from a foreign country is required in the estimate, and this results in numerous issues arising. Labour laws, material costs, geographic and climatic factors and environmental regulations are among many factors that differ when obtaining data from international plants of similar unit composition.

Thus, it is evident that the 8% margin will almost certainly not apply when a plant cost estimate is determined via the Lang factor for international comparisons. However, the factor will still be useful when a reasonable error range is considered for the preliminary estimate, such as 50% stated by Wolf (2013), in determining where to begin looking for funding and allows for decisions on adding or removing units from the proposed modular refining based on costs.

Another issue that arises with cost estimates is that plant cost data is mostly available from different time periods, perhaps many years before proposed plant construction. Due to inflation, this results in the need for a manner of comparing costing data from different time periods. The solution to this is the Cost Index. Cost indices take into account a wide variety of factors as discussed in Turton et al. (2013), and the data used to determine the index for a year is collected from a vast quantity of plants. While cost indices are an estimate, errors can be reduced by comparing costs from as recent a source as possible. Cost indices utilise a simple linear relationship as stated by Turton et al. (2013).

$$\text{Cost B} = (\text{Cost A}) * \left(\frac{\text{Index for time B}}{\text{Index for time A}}\right) \tag{3.5}$$

There are multiple cost index options available, such as Marshall & Swift, CEPCI and Nelson-Farrar Refinery index. Issues arise once again due to geography; an example being the Nelson-Farrar index is used mainly in the United States (Gary et al. 2007), with its data majorly coming from US refineries. The index that will be used in this design is the CEPCI Index.

Process Selection While conducting the literature review, various (modular) crude oil refining processes were encountered. Generally, there are three types of modular refineries: a topping plant, a cracking refinery and a coking refinery. A cracking refinery operation was selected as the process as a high yield of gasoline is intended to be the main product. However, before any of the crude oil can be processed metals, salts and other contaminants need to be removed. To accomplish this, a desalter system will be used to remove metals and salts.

As this is a small-scale modular refinery, adding additional pretreatment units for nitrogen, sulphur, etc. will not be financially viable as the economic break-even point of the plant will be extended multiple years into the future. Thus to avoid more pretreatment units and the costs associated with them, a sweet crude oil will be used as the feedstock – it will cost more; however, this will initially be more cost-effective. A fractional distillation column will be used to separate the different fractions of the crude oil. This will yield various fuel products such as LPG, naphtha, kerosene, LGO and HGO. Some of these will need to be separated further and some will require cracking to produce more of the final product. In order to separate the residue exiting the crude distillation unit, the residue will be passed through a vacuum distillation column from which the light vacuum gas oil (LVGO) and HVGO products will be mixed with the HGO product from the crude distillation unit. They will then be sent to a FCC unit, where they will be cracked in order to obtain more of the desired gasoline product.

Various cracking methods have been reviewed and it was found that an FCC unit would produce the highest yield of gasoline; therefore, it has been selected as the reactor unit. A sweet crude oil with a moderate API° ($\approx$30) will be used so that the process is able to maximise the heavy and light (vacuum) gas oil production which in turn increases the feed to the FCC unit. As some of the process units will be operated at high temperatures, four furnaces will be used within the system along with a boiler for producing steam and a heat exchanger network to minimise utilities. The products from each process unit that will not be further processed into end products will be sold to specific markets in industry in order to acquire profits. For example, kerosene will be sold to manufactures of jet fuel whereas LPG and light naphtha will be sold to industries with paraffin demand. The fuel oil produced will be burned in the furnaces to provide heat and energy for some of the process units such as the distillation columns and reactor. It is worth a mention that the flexibility of modular refineries, i.e. the ability to add more units or components as time lapses, is the main reason why other process units are currently rejected. To reach the break-even point and be profitable within fewer years, the current process is adopted. At a later stage as operation occurs, more units will be added and the intricate details of such will be placed in the conclusion of the main report.

3.5.3 Mass and Energy Balances

The simulation of the mass and energy balance was performed using Aspen HYSYS® version 8.4. This was done so as to minimise the simplifications in methods and permitted the use of the Peng-Robinson equation of state. The property method used was the Peng-Robinson equation of state (Figure 3.14).

This was chosen as the majority of significant components dealt with are non-polar and non-electrolyte. In addition, the use of simulation software allowed interaction parameters to be accounted for, which could not feasibly be done in Microsoft Excel with the time given for the mass balance due to the extensive nature of rigorous calculation methods to incorporate physical dimensions that could not otherwise be accounted for using shortcut methods. Furthermore, due to the vast number and types of components present in each temperature class, any calculation in Excel will involve some degree of approximation. This is particularly significant for the fractional distillation, vacuum and cracking units, where selective reactions and vapour pressure separations are present. HYSYS is also proficient in the handling of petroleum assays and was a major factor in choosing Aspen HYSYS over Aspen Plus.

The majority of this process is Light naphthas (LSR), and thus our process is designed accordingly. The major units are the desalter, topping column, vacuum column and FCC and separation column for its products. The crude oil to be selected is considered this. The crude selected to be used is Bonny Light oil from total, obtained from Nigeria. Nickel content was selected to be less than 5 ppm, at 3 ppm Low sulphur content, pour point is sufficiently high and most significantly has an API of 35.09 which meets the requirement of 35API for obtaining a majority petroleum product. The desalter preceding the distillation column reduced salt content in the oil that would otherwise cause a reduction in operating span of the column and its various processes, notably the condenser and reboiling units.

3.5.3.1 Oil Characterisation

Crude oil is a very complete mixture of a wide range of molecules, comprising of over a million different components. The complexity of the makeup of the oil makes it impossible to model every single component individually and monitor their interactions with each other (Al-Assady 2009). Therefore, the concept of pseudo-components was used to model the

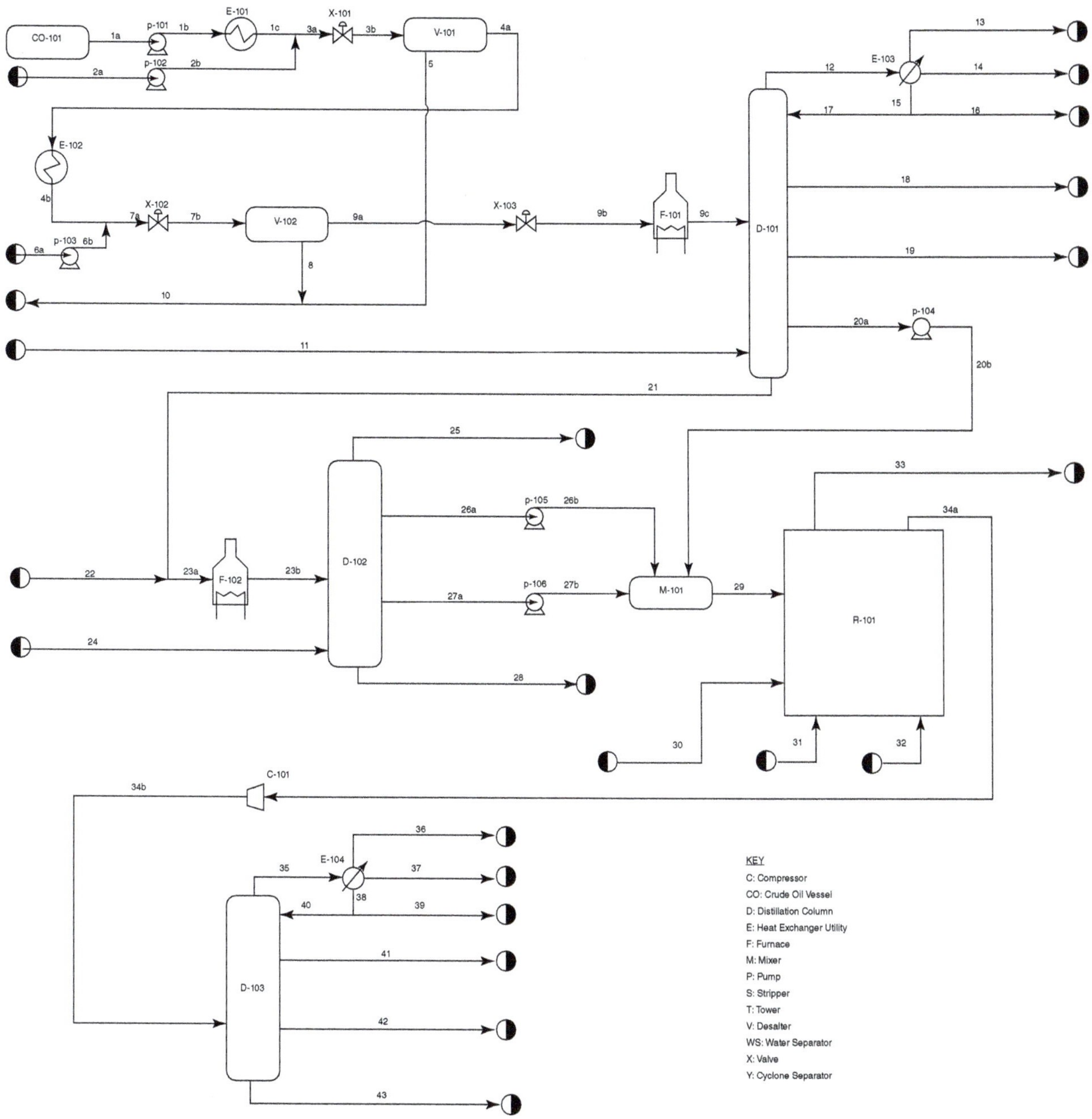

Figure 3.14 Proposed Process Flow Diagram for System.

behaviour of the crude oil. Using the pseudo-component method the light hydrocarbons (C1–C6) within the oil are identified as individual components while the heavier hydrocarbons are placed into component groups with similar properties. The oil assay is essential in the determination of stream properties of these pseudo-components. The oil assay gives the true boiling point (TBP) and °API curves for the crude oil. These curves are then used to separate the oil components into temperature classes in which aggregate boiling point and density properties may be used. The boiling point and °API properties of each of the pseudo-components may be used to determine further properties of the pseudo-components (Uppaluri 2010). The correlations between the average boiling point and °API may be used to determine the characterisation factor (K) of the pseudo-components, the characterisation factor may then be used to determine information that is essential for the performance of mass and energy balance calculations, such as molecular weight and component enthalpy using empirical relation, which have previously been determined.

3.5.3.2 Equipment Simulation

3.5.3.2.1 Topping Column The topping column is simulated using the refluxed absorber column in the HYSYS model palette. The motivation for using this column is that the ideal model to be used, the petroleum distillation column, incorporates and requires a bottoms reboiler. The process being designed uses a light oil with a focus on obtaining light fractions, thus reboiling the residue is undesired from an energy consumption point of view, and also serves to reboil some undesired products such as aromatics which could end up being withdrawn by the HGO side stream. The consequence of this would be a lower Watson K factor in the HGO and increased coking in the later FCC unit. Hence, the refluxed absorber is used to model the topping column so that steam could instead be used to heat the bottom of the column while avoiding reboiling undesired fractions.

Cut points were selected so as to obtain the products of LPG, LSR, kerosene, LGO, HGO and bottom residues. Notably, a side stream of naphtha was excluded from the final selection for a few reasons. Most significant of these were that the flow rate of the draw stream was very low in simulation, and thus economically did not warrant a withdrawal. Another reason is the purpose of the plant. The end products desired are gasoline, kerosene and diesel fuels in an impure state. LSR is most suited for the production of gasoline, and additionally, an FCC cracker is present to convert HGO into naphtha for gasoline production.

While cut points were obtained from literature, as detailed in the literature review, the particular cut of HGO was determined by simulating various temperature ranges at which HGO was withdrawn, with the variables of volumetric flow rate and Watson K factor being observed. HYSYS automatically computes the K factor via the Watson method, which uses the average boiling point of the mixture and the specific gravity at 60° Fahrenheit and is detailed in Gary et al. (2007). Ideally, a Watson K factor of 12 demonstrates a highly paraffinic feed that will possess a high cracking efficiency with little coke production in the FCC. However, a K factor of 12 could not be attained with a flow rate high enough to justify the costs associated with the FCC unit. A mixture can be cracked so long as its K factor is above 11, where a K factor of 11 indicates a primarily aromatic feed which Sadeghbeigi (2012) explains will result in high operating costs in the FCC due to the excessive coking that occurs resulting in catalyst replacement costs offsetting gasoline outputs. Additionally attempting to crack a fuel with a low K factor will simply result in much of the fuel combusting, losing product and coking catalyst. Thus, a trade-off was made and a cut point of 320–380 °C was selected, with a Watson K value for the HGO stream exiting the stripping unit being 11.65, yielding a flow rate of gasoline from the FCC of 2.03 m^3/hr, roughly 30% of the LSR flow rate from the distillation unit.

The physical column is designed to have 40 plates so as to promote adequate separation. The trays are 2.5 m in diameter and are spaced at 0.5 m. This results in the column being approximately 25 m in height, which is acceptable. The feed is introduced above the second last plate, and high-temperature superheated steam is introduced at the bottom of the column so as to provide heat to facilitate vapour flow in the column to allow distillation to commence in absence of a bottoms reboiler. The fractions exiting the top of the column flow to a partial condenser with a temperature of 72.21 °C so as to withdraw the LPG as a vapour stream, while facilitating a water separation so as to remove water from the liquid LSR stream.

Side draws from the topping column are detailed in Figure 3.15 the side draws of Kerosene, LGO and HGO were passed through strippers with 3 plates each, and a reboiler for the bottoms. The purpose of this is to remove the more volatile components in a stream that belong in higher cut points and return these streams back to the column via the top of the stripping column. The rising vapours from the reboilers aid in removing lighter fractions from a drawn stream. The boil-up ratios in these side-stripping units are not constant across the units. The kerosene stripping unit reboiler has a boil-up ratio of 0.75, while LGO and HGO are passed through a reboiler with a boil-up ratio of 0.4. The reasoning for this is due to kerosene having a large temperature range that its draw encompasses due to the absence of a naphtha withdrawal, but more significantly the boiling point of the kerosene mixture is lower than LGO or HGO, thus consuming less energy for a

Side stripper summary

	# Stages	Liq draw stage	Vap return stage	Outlet flow [m³/h]	Reboiler duty [kJ/h]
SS1	3	4_Main TS	3_Main TS	5.514	1.0373e+006
SS2	3	22_Main TS	21_Main TS	3.542	3.5356e+005
SS3	3	37_Main TS	36_Main TS	3.021	4.1042e+005

Figure 3.15 Table of side draw locations in the column.

Specifications

	Specified value	Current value	Wt. error	Active	Estimate	Current
Reflux Ratio	<empty>	3.939	<empty>	☐	☑	☐
Distillate Rate	<empty>	7.259	<empty>	☐	☑	☐
Reflux Rate	<empty>	32.52	<empty>	☐	☑	☐
Vap Prod Rate	0.9970 m3/h	0.9967	-0.0004	☑	☑	☑
Btms Prod Rate	8.022 m3/h	6.220	-0.2246	☐	☑	☐
Liquid Flow	19.27 m3/h	19.27	-0.0000	☑	☑	☑
Temperature	325.0 C	325.0	-0.0000	☑	☑	☑
SS1 Prod Flow	5.514 m3/h	5.514	-0.0000	☑	☑	☑
SS1 BoilUp Ratio	0.7500	0.7500	0.0000	☑	☑	☑
SS2 Prod Flow	3.542 m3/h	3.542	0.0000	☑	☑	☑
SS2 BoilUp Ratio	0.4000	0.4000	0.0000	☑	☑	☑
SS3 Prod Flow	3.021 m3/h	3.021	-0.0000	☑	☑	☑
SS3 BoilUp Ratio	0.4000	0.4000	0.0000	☑	☑	☑
PA_1_TRet(Pa)	80.00 C	80.01	0.0000	☑	☑	☑
PA_1_Rate(Pa)	2.500 m3/h	2.500	-0.0000	☑	☑	☑
PA_2_Rate(Pa)	2.000 m3/h	2.000	-0.0000	☑	☑	☑
PA_2_TRet(Pa)	80.00 C	80.00	-0.0000	☑	☑	☑

Figure 3.16 Illustration of the various specifications required to model the topping unit.

higher boil-up ratio with the added benefit of promoting mass transfer in the column by passing more lighter component vapours into the column. In order to model the side stripping units in HYSYS, a specification regarding the draw rates of each of the products is required. The draw rates can be calculated using the pseudo-component cut points and summing the flow rates to yield each product stream's flow rate, with the exception of the water draws at the top of the column. This method was used in agreement with the mass balance techniques as presented by Uppaluri (2010).

Aside from the side draw rates, which were required, other specifications are required to model the topping column, these are shown as inputted in HYSYS in Figure 3.16. The specification titled "Liquid Flow Rate" is the flow rate of liquid inside the column where the feed is introduced. This is calculated as detailed in Gary et al. (2007), where an over flash of 10% was taken, and the draw rate values calculated were used to obtain a flow rate of 19.27 cubic meters per hour.

Pumparounds are also present in the design of the topping column, with two being used. The pumparounds designed both move liquid from one area of the column, and have a heat exchanger present to cool down the fluid being transported. The pumparounds have the purpose of both facilitating mixing within the column, and cooling the column in various sections. Additionally, Gary et al. (2007) state that pumparounds also serve to reduce the reflux ratio, hence improving the product yield. The pumparounds were designed by specifying a flow rate which was selected based on the column flow rates at that point and specifying the cooling target. Pumps are assumed to be centrifugal with an efficiency of 75%. The details of the pumparound stream draw and returns are detailed in Figure 3.17.

While simulating the column, preliminary optimisation was performed to obtain lower heat duties and flow rates and temperatures such as for the steam feed at the bottom of the column so as to save costs. The feed temperature, which was originally set to 380 °C has been increased to 405.1 °C via the furnace. This is acceptable as the feed must remain below approximately 415 °C to avoid cracking within the column. Other minor optimisations were considered such as the boil-up ratios and steam flow rates. The system was forced to converge with a small error in the energy balance (<100 W) via an energy reconciliation stream which was added together with the feed stream with the purpose of ensuring the system converges. Further corrections to the system were applied by altering various aspects of the system such as the steam inlet rate so as to make the error to the energy balance negligible.

Liquid pump around summary

	Draw stage	Return stage	Flow [m³/h]	Duty [kJ/h]	Draw T [C]	Return T [C]	Expor
PA_1	23__Main TS	5__Main TS	2.500	-1.155e+006	300.8	80.01	☐
PA_2	1__Main TS	5__Main TS	2.000	-2.424e+005	147.9	80.00	☐

Figure 3.17 Details of pumparound circuits present in atmospheric topping unit.

3.5.3.3 Vacuum Distillation

A vacuum flash distillation unit was used in order to provide separation of the useful components (LVGO and HVGO) out of the bottoms residue from the atmospheric topping column. In complying with information obtained from Fahim et al. (2009), a heater is used to increase the temperature of the oil feed to the vacuum column to between the range of 380–400 °C, to allow for the separation needed. An additional stream of steam was also added to the oil feed prior to the heating as this is able to reduce the amount of coking that occurs during the heating process (Leffler 1985). The vacuum distillation column was simulated using the absorber column from the HYSYS model palette. This model was chosen as it operates without the use of a top condenser and or bottoms reboiler. Therefore, the absorber column is able to match the physical design of the vacuum flash column specified. The vacuum column is designed to function without the use of a top condenser with reflux to the nature of the feed. As the feed to the column contains the heavier cuts of the oil, the majority of the top product from the column consists of the very light flue gas components that had remained trapped within the residue after the topping process. Therefore, it would be unnecessary and highly energy intensive to reflux this flue back into the column. As the vacuum column operates without a condenser, pumparounds are used to provide cooling within the column and to induce the temperature gradient needed to facilitate the separation of the oil fractions.

The column was specified to operate with 4 cut points; flue gas which exits through the top of the column, vacuum residue which exits through the bottom of the column, as well as two side draws of LVGO and HVGO. The two side products drawn from the column are remixed and used as feed for FCC, therefore, a high degree of separation between these two cuts was not required. From this, the absorber column model was specified to operate using 14 stages, with 2 side draws for the LVGO and HVGO. The amount of stages used provided the degree of separation of the fractions within the column that was required while adequately removing the undesired heavier cuts through the residue stream. The side draws were placed at stage 2 and stage 9 as the desired cut fractions were concentrated the most between these stages. Two further streams were outputted from the column, a flue gas stream from the top of the column for the removal of the light ends, as well as a bottom product for the removal of the vacuum residue containing the heaviest fractions. The flow rates of each of the four streams from the column were specified as:

- LVGO – 1.547 m^3/h
- HVGO – 3.686 m^3/h

The draw rates were calculated using the desired pseudo-component cut points and summing the flow rates to yield each product stream's flow rate. This technique was used, as it provided the draw rates needed for idealised separation and is in accordance with the mass balance methods as described by Uppaluri (2010).

The top and bottom stage pressure were specified as 6.666 kPa and 8.266 kPa, as these values correspond to pressure ranges which were found by Ceric (2012). The low pressures allow for lower operating temperatures to be used, preventing the thermal cracking of the components within the column. The two pumparounds used within the column were also specified; the first drawing from the LVGO stage (stage 2) and the second drawing from the HVGO stage (stage 9). The pumparound draw stages was specified as such within the physical design of the column the side fractions are drawn through split draws from the pumparounds. As the first pumparound was drawn from stage 2, the only option would be to specify the return as at stage 1, as this is the only available stage above the draw for the return to be placed. The draw rate of pumparound one was specified to be 3.029 m^3/h (Figure 3.18).

This value was calculated by using the draw-to-feed ratio which was found to work within a previous case study by Jechura (2014), as a basis and performing iterations around that operating point until the column operation was optimised. The return temperature of this pumparound was specified at 29.44 °C. This temperature was chosen as it provided the cooling necessary to achieve the desired temperature profile within the column. The second pumparound was specified to have a draw rate of 8.022m^3/h. Again this value was calculated using the draw-to-feed ratio found within a previous case study (Jechura 2014) as a basis and performing iteration around this operating point. The specification for the return

Liquid pump around summary

	Draw stage	Return stage	Flow [m³/h]	Duty [kJ/h]	Draw T [C]	Return T [C]	Expor
TopStagePA	2_TS-1	1_TS-1	3.209	-1.050e+006	196.1	29.44	☐
PA_1	9_TS-1	2_TS-1	8.022	-1.632e+006	297.8	214.5	☐

Figure 3.18 Summary of pumparound specifications.

Specifications

	Specified value	Current value	Wt. error	Active	Estimate	Current
TopStagePA_Rate(Pa)	3.209 m3/h	3.209	-0.0000	☑	☑	☑
TopStagePA_TRet(Pa) - 2	29.44 C	29.44	-0.0000	☑	☑	☑
TopStagePA_Duty(Pa)	<empty>	-1.050e+006	<empty>	☐	☑	☐
LVGO Rate	1.547 m3/h	1.547	-0.0000	☑	☑	☑
HVGO Rate	3.686 m3/h	3.686	0.0000	☑	☑	☑
PA_1_Rate(Pa)	8.022 m3/h	8.022	0.0000	☑	☑	☑
PA_1_Dt(Pa)	83.33 C	83.33	-0.0000	☑	☑	☑
Temperature	65.56 C	86.62	0.0421	☐	☑	☐
Temperature - 2	371.1 C	371.1	-0.0000	☑	☑	☑

Figure 3.19 Summary of various specifications for the vacuum column.

stage and the return temperature were varied until the optimum convergence of the column was reached. From this, the optimum specification for the return stage and temperature were found to be stage 2 and 214.5 °C, respectively.

Additionally, superheated steam was used to provide additional heating by being fed through the bottom stage of the column. This stream was also able to act as a stripping agent to aid separation. This stripping steam was then removed through the flue gas stream at the top of the column. An energy reconciliation stream was also added as an input to the column, as this allows the HYSYS solver to converge in a quicker more stable fashion (Figure 3.19).

Once, the operating specification of the vacuum column was working and optimised the vacuum feed temperature was altered from its initial specified value of 390 °C in order to minimise the value of the energy recon stream which was fed into the column. It was found that this value was optimised with the feed having a temperature of 391.5 °C.

3.5.3.4 Fluid Catalytic Cracker

In order to maximise the amount of gasoline (naphtha) produced by the refinery, a FCC unit was used in a stacked configuration. According to the stacked configuration of the FCC system, the reactor was placed above the regenerator. The reactor was fed material by means of a riser with a height of 30 m and a diameter of 0.8 m. The regenerator vessel houses the stripper with a height of 8 m and a diameter of 3 m (annulus diameter = 1.3 m) as well as a one-stage cyclone with an inlet height of 15 m, an inlet diameter of 2.3 m and an outlet diameter of 1.3 m. The feed to the cracker comprised of HGO, LVGO and HVGO, as well as ZSM-5 catalyst so as to promote the naphthenic reactions within the reactor. According to Sadeghbeigi (2012), the following reactions take place inside the cracking unit:

- $R\text{-}CH_2\text{-}CH_2\text{-}CH_2\text{-}CH_3 + H^+$ (proton attack) $\rightarrow R\text{-}C^+H\text{-}CH_2\text{-}CH_2\text{-}CH_3 + H_2$ (a)
- $R\text{-}CH{=}CH\text{-}CH_2\text{-}CH_2\text{-}CH_3 + H^+$ (a proton at Bronsted site) $\rightarrow R\text{-}C^+H\text{-}CH_2\text{-}CH_2\text{-}CH_2\text{-}CH_3$ (b)
- $R\text{-}CH_2\text{-}CH_2\text{-}CH_2\text{-}CH_3$ (removal of H_2 at Lewis site) $\rightarrow R\text{-}C^+H\text{-}CH_2\text{-}CH_2\text{-}CH_3$ I)

The reactor operates at a temperature of 540.5 °C and a pressure of 179.1 kPa which promotes the production of olefins and naphthenic compounds. The 0.038 wt.% rare-earth present in the ZSM-5 catalyst protects acid sites, thus promoting the hydrogen transfer reaction that produces the required compounds (Sadeghbeigi 2012), with the additional benefit of negating the further cracking of gasoline once it has been produced. Under this scheme, a cracking conversion of 79% was achieved, with the reactor product stream containing 45 wt.% naphtha. Characterisation factors are used to indicate the aromaticity and paraffinicity of crude oils (Gary et al. 2007), one of them being the Watson characterisation factor K_W. As the K factor for this refinery's feed was 11.56, the crude used is classified as naphthenic. This agrees with the results obtained from the HYSYS simulation as a significant amount of naphtha is produced.

However, some of the disadvantages of operating at these conditions are that the gasoline has a lower octane rating and that the percentage of aromatics in the gasoline are higher which recently has been believed to cause adverse health effects. The simulation yielded the motor octane number (MON) of the gasoline which was 84.17 and the research octane number (RON) which was 96.24. The octane number of the gasoline produced by the FCC unit is found by averaging the RON and MON values, thus it contains an octane rating of 90.2 which is well within the EPA range of 86–94 (National Institute for Petroleum and Energy Research 1992). Due to the high temperature and low average residence time of 3.four seconds, firstly, coke formation occurred – the reactor product contained a 5.92 wt.% of coke – and this could reduce the activity of the catalyst over time, thus an air blower was used to remove the coke entrained within the catalyst particles in the one-stage regeneration vessel and secondly, a stripper was used to separate the catalyst particles from the products so that they may be reused. It is important to note that the carbon residue percentages in the feed to the reactor (riser) were minimal – it contained a Conradson carbon residue (CCR) of 1.2% and a Ramsbottom carbon residue (RCR) of 0.96% – indicating that there would be a small amount of coke formation due to the thermal conditions.

A complete combustion scheme was used in the HYSYS simulation to separate the catalyst from the coke in the regeneration process. The total combustion of air provided the system with additional energy, thus making it possible to efficiently operate the regenerator at 599.8 °C – which is less than the regenerator operating temperature of 710 °C found in literature (Gary et al. 2007). The use of total combustion additionally allowed the generator to reduce the overall coke yield and in conjunction with the air blower, removed the coke with the combustion products as flue gas. A negligible amount of CO was produced from the combustion and instead, CO_2 was produced. This further facilitated the complete combustion as the (excess) oxygen present could effectively be used to combust the air instead of reacting with CO to form CO_2. The ZSM-5 catalyst averages at approximately 1 mm thus catalyst loss by entrainment within flue gas can be minimised, ensuring that the initial amount can be regenerated and reused for the full 5-year lifespan of the catalyst. This is able to serve the additional benefit of reducing the load on the cyclone so that the flue gas and nitrogen can more effectively be removed from the regeneration vessel. Furthermore, having a cyclone with an inlet height of 15 m ensures that the air blower can only blow gases into the cyclone, once again minimising the possibility of catalyst loss through the flue gas exit stream.

The metal content within the feed is at an acceptable fraction (in terms of the criteria outlined by Sadeghbeigi 2012), thus adverse effects on the catalyst and reaction kinetics due to the presence of metals are minimised. Figure 3.20 shows the fractions of metals in the feed:

Additionally, Figure 3.21 shows the minimal effects due to coking on the catalyst due to the operating conditions specified for the FCC system:

3.5.3.5 FCC Fractionator

The FCC fractionator was modelled as a reflux absorber column in HYSYS. This is because the fractions within the oil from the FCC closely represented the oil fractions within the feed crude. Therefore, it was determined that the separation process could be run under atmospheric conditions. Additionally, as the product from the FCC was already at a high temperature, further heating would not be required, therefore, simplifying the atmospheric distillation process. The FCC product was

Nickel Content [ppmwt]	2.438
Vanadium Content [ppmwt]	0.3146
Iron Content [ppmwt]	4.411
Copper Content [ppmwt]	0.0000
Sodium Content [ppmwt]	0.0000

Figure 3.20 Fractions of metals in the FCC unit feed.

Reactor Section

Activity on Pathways to C Lump	1.374
Activity on Pathways to G Lump	0.5857
Activity on Pathways to L Lump	1.615
Metals Coke Activity	9.141e-005
Catalyst Deactivation Factor	0.7987
Catalyst Surface Area Parameter	−10.72
Effluent per Mass of Catalyst into Stripper	66.22
Stripper Parameter	2.347
H to C Ratio for Coke	0.8494
Coke Burn Activity	15.30
CO Heterogeneous Burn Activity	−1.002
CO Homogeneous Burn Activity	11.00
Heat of Cracking Parameter	1.630e-002
Kinetic Coke Activity Factor	8.206e-002

Figure 3.21 Coking parameters for FCC reactor.

specified to be fed into the bottom of the column, as it was at a high temperature allowing for heat to be provided for the separation process. As the feed to the FCC did not contain high amounts of heavier fractions the number of stages for the column could be reduced when compared to the distillation column used for the original crude feed.

The FCC fractionation column was specified to operate having five cut points; LPG, naphtha, kerosene, cycle oil and residue. It was decided that the cycle oil would be used within process integration to contribute to the heating requirements of the plant operation. Therefore, it was unnecessary to separate the cycle oil cut further into light cycle oil and heavy cycle oil, as these would not be sold as products. From this, the FCC fractionator was specified to operate with 30 stages, with a top and bottom operating pressure of 255.5 kPa and 135.8 kPa. These operating pressures were used, as they provided the lowest duty across the condenser. At the top of the column, a 3-phase condenser was used. This allowed for the separation of the LPG gas while condensing the water and the top oil product from the column for reflux. The 3-phase condenser also allowed for the further separation of the water and the naphtha product. The draw rates from the column were specified as follows:

- LPG – 4.920 m^3/h
- Naphtha – 2.207 m^3/h
- Kerosene – 2.760 m^3/h
- Cycle Oil – 0.4880 m^3/h
- Residue – 0.2310 m^3/h

As mentioned above, the draw rates were calculated using the desired pseudo-component cut points and summing the flow rates to yield each product stream's flow rate. This technique was used, as it provided the draw rates needed for idealised separation and is in accordance with the mass balance methods as presented by Uppaluri (2010).

An additional specification for the fractionator was needed by HYSYS in order to calculate the operational results of the column. Therefore, iterations were performed using the reflux ratio until the optimal convergence was reached. From this, it was found that the column would operate optimally when a reflux ratio of 10 was specified.

3.5.3.5.1 Comment on Products The design focuses on a high output of gasoline, with other desired products being kerosene and LGO that can be sold to another party to manufacture diesel. The term gasoline is used loosely, as the quality of the products from the plant is not of sufficient quality to be actually used as vehicle fuel. The plant designed is a modular refinery, and gasoline purification units such as an isomerisation unit to lower Reid Vapour Pressure or an alkylation unit to increase octane unit can be added at a later date.

LSR from the topping column and naphtha from the FCC unit both are used to make gasoline, with this being the target product, it is essential that the biggest output from the process is the naphthas. The LSR stream from the topping column and the FCC naphthas both have similar fractions as expected, with the FCC naphtha being slightly skewed towards the components of higher boiling points. The flow rates of these are 7.259 m^3/hr and 2.027 m^3/hr. These combined are significantly higher than any other product outputted from the process, with kerosene having a draw rate of 5.514 m^3/hr from topping and 0.448 m^3/hr from cracking. LGO of acceptable quality was obtained from the topping column at a flow rate of 3.021 m^3/hr while the Gas oil stream from the FCC is too impure to have any commercial use.

As many furnaces and heat exchangers are present in the process, it seems plausible that the impure cycle oil from the cracking unit, with a flow rate of 0.448 m^3/hr at 160 °C could be used to provide heat to various heat exchangers and be combusted in a furnace to reduce operating costs of electricity for heating, cooling and steam generation. Other streams that can be used for this purpose are the high-temperature residues from the Vacuum Distillation unit and the FCC Unit.

3.5.3.5.2 Comment on Overall Mass Balance As shown in the mass balance table in Figure 3.22, the overall mass balance error is −0.008%. This error is a result of the flow of water in the topping column and FCC separation units being modelled by approximation across the condensers. This error is due to the condenser in the refluxed absorber functioning as a liquid–liquid extractor while also removing gas in the form of LPG. The error is due to calculation inaccuracies associated with using the Peng-Robinson equation of state when a high flow rate of water is present. Water is a polar mixture and is thus not intended to be modelled by the Peng-Robinson equation. Hence, an approximation occurs in the condenser unit and some of the water remains unaccounted for over the topping column. However, as the error over the topping column is 0.02% between inlet and outlet, the model is assumed to be sufficiently accurate in simulating the distillation column. The error in the FCC separator is −0.0575%, and the reasoning for the error being present is the same as for the topping column. This is because both columns used the same absorption column model and condenser method, and both had steam feeds introduced instead of a bottom reboiler.

Overall mass balance					
In	Out		Error	Fractional Error	Percentage Error
39849.18	39852.38		−3.198691623	−8.027E-05	−0.008026995

ADU balance					
In	Out		Error	Frac	Percent
24025.44	24020.6		4.838192811	0.000201378	0.020137789

VDU balance					
In	Out		Error	Frac	Percent
6369.773	6369.773		−8.18545E-12	−1.28505E-15	−1.28505E-13

FCC balance					
In	Out		Error	Frac	Percent
20888.35	20888.35		−8.14496E-07	−3.89928E-11	−3.89928E-09

FCC frac balance					
In	Out		Error	Frac	Percent
13971.24	13979.28		−8.03688362	−0.000575245	−0.057524466

Figure 3.22 Overall and major unit mass balance error.

In addition to this, error is also due to the energy reconciliation streams present in the topping column, with a final stable value of 642 kJ/hr. being added to the model in the simulation. These streams have already been discussed in the topping column section of the report, and could not be totally eliminated due to equations of state (Peng-Robinson) being used to model the system cannot perfectly predict fluid properties as they occur in actual process units. Thus, oscillations do occur at a low level throughout the plant, which is why the error can never be perfectly 0 in any simulation using equations of state to model properties.

Overall, the mass balance across inputs and outputs for the entire process has a low enough error to indicate that the results of the simulation conducted are of a high enough accuracy to be used to physically model the plant.

3.5.4 Preliminary Sizing of Equipment

Preliminary sizing of equipment is performed to gauge whether or not the proposed plant will be profitable or not. More detailed analysis of individual units, namely, the topping column and FCC being sized is discussed later in this report.

3.5.4.1 Desalters

The desalters used in this process were sized by modelling them as decanters, as both units of operation are settling tanks that are used to separate an emulsion into two phases using the density difference between the liquids. The following assumptions were made in the process of designing the desalters

(1) The velocity of the dispersed phase is greater than the velocity of the continuous phase.
(2) The diameter of the dispersed phase particle was taken as 150 μm as advised by Towler and Sinnot (2013).
(3) The height of the desalters is twice the diameter of the respective vessels.

Sizing a decanter requires one to know the settling velocity of the droplets of the dispersed phase. The settling velocity of the droplets was calculated using Stoke's law, mathematically expressed in Eq. (3.6) as reported by Towler and Sinnot (2013)

$$U_d = \frac{d^2 g(\rho_d - \rho_c)}{18\mu_{\text{oil}}} \tag{3.6}$$

The dispersed and continuous phases in this case are the crude oil and water, respectively. The settling velocity of the dispersed phase particles of the respective desalters was then used to calculate the interfacial area that exists between the phases. The mathematically expression of the volumetric flow rate of the continuous phase was used to compute the interfacial area. This is shown in Eq. (3.7) as reported by Sinnot (2005):

$$L_c = \frac{U_c}{A_i} \tag{3.7}$$

where L_c is the volumetric flow rate of the continuous phase, U_c is the velocity of the continuous phase and A_i is the interfacial area.

Since the process of designing a decanter is based on the assumption that the velocity of the dispersed phase is greater than the velocity of the continuous phase, Uc in Eq. (3.7) is substituted with Ud. After the interfacial area was calculated, the radius of the vessel was computed using Eq. (3.8) as reported by Sinnot (2005):

$$r = \sqrt{\frac{A_i}{\pi}} \tag{3.8}$$

The diameter of the vessel is the easily calculated from the radius. The height of the vessel was taken as twice the diameter as Sinnot (2005) reported this as a reasonable value.

The diameter and height of V-101 were calculated to be 2.44 m and 20 m, respectively. V-102 was sized to have a diameter of 0.972 m and height of 12.25 m.

3.5.4.1.1 *Heat Exchangers*

The area of the heat exchangers used in this process was calculated using Eq. (3.9) as reported by Sinnot (2005):

$$\mathrm{Q} = \mathrm{UA}\Delta T_m \tag{3.9}$$

where Q is the heat duty of the heat exchanger, W; U is the overall heat transfer coefficient, $\mathrm{W/m^2\,°C}$; A is the area of the heat exchanger, m^2, $\Delta_m{}^T$ is the mean temperature difference, °C.

3.5.4.1.2 *Assumptions*

The overall heat transfer coefficient was estimated to be $900\,\mathrm{W/m^2\,°C}$ as reported by Sinnot (2005).

E-101 was sized to have an area of $6.06\,\mathrm{m}^2$ with a duty of 5 188 000 kJ/hr. With a duty of 2 353 000 kJ/hr., E-102 was sized to have an area of $3.73\,\mathrm{m}^2$.

3.5.4.1.3 *Furnaces*

The construction of a furnace is divided into two sections, namely, the radiant and convection section. These sections are named after the type of heat transfer that occurs from the combustion of a fuel to the particular stream that needs to be heated. The preliminary sizing of furnaces focuses mostly on the radiant section (Towler and Sinnot 2013). The method of Lobo and Evans as reported by Kern 1950 was used to find the tube surface area and the number of tubes used. The following specifications and assumptions were made to fulfil the sizing of the furnaces:

(1) Methane was specified as the source of the fuel.
(2) The efficiency of the furnace was assumed to be 75% as advised by Sinnot (2005).
(3) The steam to atomise the fuel is specified at 0.3 kg/kg of oil.
(4) The average flux in the radiant section was assumed at $30\,\mathrm{kW/m^2}$.
(5) The overall exchange factor F was assumed to be at 0.57.
(6) The tube surface temperature was assumed to be the temperature of the inlet stream.
(7) The ratio of the total hourly heat transfer to the cold surface and the product of the equivalent cold plane surfaces ($\Sigma Q/\sigma AcpF$) is twice the average flux.

The calculation to find the number of tubes and the tube surface area in the first furnace of the process commenced with the computation of the required heat released by the fuel during combustion Eq. (3.10) was used.

$$Q_F = \frac{\text{Total duty of furnace}}{e} \tag{3.10}$$

The heat released was used to calculate fuel flow rate required using Eq. (3.11) as reported by Kern (1950).

$$\text{Fuel required} = \frac{\mathrm{Q_F}}{\text{Energy value of methane}} \tag{3.11}$$

The energy value of methane is 55 530 kJ/kg.

The combustion process occurs in the presence of air. The amount of air that is required was calculated using the air–fuel ratio. Eq. (3.12) was used to calculate the flow rate of air required.

$$\text{Air Required} = (\text{Fuel Required})(\text{Air} - \text{fuel ratio}) \tag{3.12}$$

There is a certain amount of heat energy that is released when air is burning and this contributes to the total radiant duty of the furnace. To find out how much energy is released by the air Eq. (3.13) was used.

$$Q_A = (\text{Air Required})(\text{Heating value of Air}) \tag{3.13}$$

To take into account the non-ideality of heat transfer from combustion to the material in the inlet stream a fraction of the heat liberated from the combustion of the fuel is subtracted from the sum of Q_F and Q_A. This is the heat loss to the furnace walls. Towler and Sinnot (2013) show that this can be estimated to be 2% of the heat liberated by the fuel. Some of the heat released leaves the radiant section with the flue gas produced during combustion. Therefore, the total radiant duty is illustrated in Eq. (3.14).

$$Q = Q_F + Q_A - Q_W - Q_F \tag{3.14}$$

One of the requirements for finding Q_G is the temperature of the flue gas. Kern (1950) illustrates a graph that was used to find this temperature. To read the flue gas temperature, ($\Sigma Q/\sigma AcpF$) was computed using one of the assumptions listed above, and T_s was taken as the temperature of the inlet stream. Eq. (3.15) was then used to find Q_G

$$Q_G = C_{p_G}(\text{Fuel Required} + \text{Air Required} + \text{Atomizing steam}) \tag{3.15}$$

C_{PG} is the heat capacity of the flue gas at $T_{G.}$

The outside diameter of the tubes and the exposed tube length were used to find the tube surface area. Towler and Sinnot (2013) reported that an exposed tube length of 40 ft. can be used and that an outside diameter of 75–150 mm is typically used. For F-101 and F-102, the outside tube diameter was chosen as 150 mm and 250 mm, respectively, to get a reasonable tube area. The exposed tube length was taken as 40 ft. Eq. (3.16) is the mathematically formulation of the tube surface area:

$$\text{A} = \pi(\text{tube outside diameter})(\text{exposed tube length}) \tag{3.16}$$

where A is the tube surface area, m^2.

Finding the estimate of the number of tubes to be used in the furnace was done using Eq. (3.17):

$$\text{N}_t = \frac{Q}{A(\text{average flux})} \tag{3.17}$$

The tube surface area of F-101 was found to be 5.77 m^2 with 52 tubes. F-102 was sized to have 9.63 of tube surface area and 171 tubes.

3.5.4.1.4 Topping Column Two major dimensions that are calculated to preliminary size the fractionation column are the height and diameter of the column. The method used was adapted by Sinnot (2005). The following assumptions were made:

(1) The plate spacing is 0.5 m as advised by Sinnot (2005).
(2) The maximum vapour rate was assumed to be the maximum vapour flow rate in the column.

The diameter was calculated using the maximum allowable vapour velocity which was found using Eq. (3.18):

$$U_v = \left(-0.171l_t^2 + 0.27l_t - 0.047\right)\sqrt{\frac{(\rho_l - \rho_v)}{\rho_v}} \tag{3.18}$$

The height of the fractionation column is the product of the number of plates in the column and the plate spacing as shown in Eq. (3.19).

$$\text{Column Height} = (l_t)(N_t) \tag{3.19}$$

The diameter of the column was found to be 2.44 m with 40 plates; therefore, the height of the column should be 20 m.

3.5.4.1.5 Vacuum Column The vacuum column was sized as a packed column. Certain assumptions and estimates had to be made to complete the sizing of this column. The following assumptions and estimates were made;

(1) Raschig rings were used as the type of packing.
(2) A packing size of 50 mm.
(3) A pressure drop of 8 mm of water/m of packing across the column.

The vacuum column was designed using Onda's method which requires the computation of the flow factor Flv and a term denoted as K_4 to find the appropriate dimensions of the column. Eq. (3.20) shows the mathematically representation of the F_{lv}

$$F_{lv} = \frac{L_w^*}{V_w^*}\sqrt{\frac{\rho_v}{\rho_l}} \tag{3.20}$$

The flow factor was then used along with the pressure drop assumption to find the term K4. This term can also be expressed in another way, as shown in Eq. (3.21):

$$K_4 = \frac{13.1\left(V_w^*\right)^2 F_p\left(\frac{\mu_L}{\rho_L}\right)^{0.1}}{\rho_v(\rho_L - \rho_v)} \tag{3.21}$$

where V_w^* is gas mass flow rate per column cross-sectional area

Eq. (3.21) was rearranged to make V_w^* subject of the equation and calculated.

3.5.4.1.6 Pumps The sizing parameter of a pump is the power. The sizing of the pumps in the process was modelled according to (Biegler et al. 1997) which gives the power of the pump. The power of the pumps was modelled according to the following equation:

$$W = \frac{\mu(P2 - P1)}{\rho \eta_m \eta_p} \tag{3.22}$$

The pressures of the inlet and outlet streams that go into the pump can easily be found from the mass and energy balances while the density of the stream can be found on the HYSYS simulation. To achieve results from Eq. (3.22) two assumptions needed to be made the first assumption is the motor efficiency is 0.9 and the second assumption is the pump efficiency is 0.5 (Table 3.3). After applying these assumptions to Eq. (3.22), the following results were obtained and depicted in Table 3.4

The power that was projected from Eq. (3.22) is close to values obtained in the ASPEN HYSYS simulation with slight deviations. The deviations could be a result of assuming different efficiencies and the fact that the ASPEN HYSYS models the pump according to an adiabatic efficiency of 75%.

3.5.4.1.7 Compressors Only one compressor was used in the process and the parameter which needed to be accounted for was the compressors power and was modelled according to (Biegler et al. 1997). To achieve the theoretical power 3 assumptions must be made an ideal, isentropic and adiabatic expansion takes place.

Table 3.3 Power projected for the compressor C-101.

C-101	
Molar flow (gmole/s)	130,75
Gamma (Cp/Cv)	1114
Pressure IN (kPa)	172
Pressure OUT (kPa)	255,5
Temperature IN (K)	813,15
Gas constant (J/gmol.K)	8314
Theoretical work (J/s)	356978,82
Actual work (J/s)	496200,56
In kW	496,20056

Table 3.4 Projected power of the pumps in the process.

	P-101	P-102	P-103	P-104	P-105	P-106	P-A01	P-A02
Pressure IN (kPa)	101,325	101,325	101,325	223,337	6,789	7,651	213,437	197,88
Pressure Out (kPa)	844,607	844,607	844,607	300,037	300	300	250,708	250,708
Motor efficiency	0,9	0,9	0,9	0,9	0,9	0,9	0,9	0,9
Pump efficiency	0,5	0,5	0,5	0,5	0,5	0,5	0,5	0,5
Molar flow (kgmole/hr)	122,139	62,527	62,526	9,28	3,683	8,093	9,298	12,51
Density (kgmole/m^3)	4,58	53,53	53,53	2,102	2,057	1,739	2,799	5,376
Power (kJ/hr)	44048,38438	1929,35192	1929,32106	752,485	1166,635	3023,424	275,1346	273,1805
Power (kW)	12,23566233	0,53593109	0,53592252	0,20902	0,324065	0,83984	0,076426	0,075883

The theoretical power of the compressor was modelled according to the following equation:

$$W_{\text{theoretical}} = \mu\frac{\gamma}{\gamma-1}RT_1\left[\left(\frac{P2}{P1}\right)^{\frac{\gamma-1}{\gamma}} - 1\right] \tag{3.23}$$

The actual work can be calculated from the following equation:

$$W = 1\,W_{\text{theoretical}} \tag{3.24}$$

3.5.4.1.8 Mixer There is only one mixer (M-101) in the process which will be modelled as a vessel according to (Biegler et al. 1997). The sizing parameters that are needed from these vessels are volume, height and diameter. The retention time will be assumed as five minutes (Table 3.5).

The volume of the vessels can be achieved through the equation:

$$V = 2\frac{\mu\tau}{\rho} \tag{3.25}$$

where τ: retention time (hr). To achieve the diameter of the vessel the length-to-diameter ratio is assumed to be $4\left(\frac{L}{D} = 4\right)$, therefore, the diameter can be calculated through the equation:

$$D = \left(\frac{V}{\pi}\right)^{\frac{1}{3}} \tag{3.26}$$

3.5.5 Economic Evaluation

A preliminary costing method is discussed here that is expected to have a significant error margin. The reasoning for this is that performing the detailed design to obtain the exact dimensions required for costing and obtaining quotations for fixed

Table 3.5 Design parameters for mixer (M-101).

M-101	
L/D	4
Flow rate of liquid (kgmole/hr)	21.05
Liquid density (kgmole/m^3)	1.962
Retention time (hr)	0.0833333
Volume (m^3)	1.7881414
Diameter (m)	0.8287381
Height (m)	3.3149525
Pressure (kPa)	300
Temperature (°C)	313.5

equipment would have been unfeasible at this sections point in time. A more accurate and intensive analysis of the plant economics needs to be considered prior to plant construction or physical planning.

3.5.5.1 Cost of Equipment

Although there are various methods to determine preliminary cost estimates for chemical plants, this report will focus on the use of the module costing method to determine the fixed capital investment (FCI) required to purchase the equipment. This is done by calculating the bare modulus costs at base and non-base conditions. The data constants used by Turton et al. (2008) are from the year 2001, thus inflation needs to be taken into account. The CEPCI index makes it possible to determine the value of something relative to a base year:

$$\text{Cost of A at 2016} = \text{Cost of A at 2001} \times \frac{\text{CEPCI at 2016}}{\text{CEPCI at 2001}} \tag{3.27}$$

The CEPCI value for 2001 is 394.3, whereas the CEPCI value for 2015 (October) is 547.4

$$\log_{10}(Cp^0) = K_1 + K_2 \log_{10}(x) + K_3 (\log 10(x))^2 \tag{3.28}$$

where the K values are constants and x is the equipment parameter (area, volume, shaft power) which are tabulated below (Table 3.6):

In order to account for pressure, a pressure factor F_p is used. The F_p for pumps and heat exchangers is calculated differently from pressure vessels:

For pumps and heat exchangers:

$$\log_{10}(F_p) = C_1 + C_2 \log_{10}(P) + C_3 (\log 10(P))^2 \tag{3.29}$$

where the C values are constants specific to the equipment type and P is the pressure in bar (Table 3.7).

Table 3.6 Cost data for equipment (Turton et al. 2008).

Equipment type	Equipment description	K_1	K_2	K_3	Capacity	Units	Min size	Max size
Heat exchangers	Fixed tube sheet	43 247	−0,303	0,1634	Area	m^2	10	1000
	Floating head	48 306	−0,8509	0,3187	Area	m^2	10	1000
	Double pipe	33 444	0,2745	−0,0472	Area	m^2	1	10
Process vessels	Horizontal	35 565	0,3776	0,0905	Volume	m^3	0,1	628
	Vertical	34 974	0,4485	0,1074	Volume	m^3	0,3	520
Pumps	Centrifugal	33 892	0,0536	0,1538	Shaft Power	kW	1	300
Towers	Trays and packed	34 974	0,4485	0,1074	Volume	m^3	0,3	520
Trays	Sieve	29 949	0,4465	0,3961	Area	m^2	0,07	12,3

Table 3.7 Pressure factors for pumps and heat exchangers (Turton et al. 2008).

Equipment type	Equipment description	C_1	C_2	C_3	P (bar)
Heat exchangers	Fixed tube sheet (Both shell & tube)	0	0	0	$P < 5$
		0,03881	−0,11 272	0,08183	$5 < P < 140$
	Floating head	0	0	0	$P < 5$
		0,03881	−0,11 272	0,08183	$5 < P < 140$
	Double pipe	0	0	0	$P < 40$
		0,6072	−0,912	0,3327	$40 < P < 100$
		131 467	−126 574	30 705	$100 < P < 300$
Pumps	Centrifugal	0	0	0	$P < 10$
		−0,3935	0,3957	−0,00226	$10 < P < 100$

Table 3.8 Material factor constants for various equipment (Turton et al. 2008).

Equipment type	Equipment description	Material of construction	F_m	B1	B2
Heat exchangers	Fixed tube sheet	Carbon steel	1,39	1,63	1,66
	Floating head	Carbon steel	1,39	1,63	1,66
	Double pipe	Carbon steel	1	1,74	1,55
Process vessels	Horizontal	Carbon steel	1	1,49	1,52
	Vertical	Carbon steel	1	2,25	1,82
Pumps	Centrifugal	Carbon steel	1,55	1,89	1,35

For pressure vessels:

$$Fp = (((P + 1)D)/(2(850 - 0{,}6(P + 1))) + 0{,}00315)/0{,}0063 \tag{3.30}$$

where P is pressure in units of bar and D is diameter in units of m (Table 3.8).

With this information, the equipment bare modulus cost at base and non-base conditions can be found for each piece of equipment using the following equations. At base conditions, the F values are equal to 1. The bare modulus cost at non-base conditions is:

$$C_{\mathrm{BM}} = Cp^0(\mathrm{B1} + \mathrm{B2}F_mF_p) \text{ for pressure vessels} \tag{3.31}$$

$$C_{\mathrm{BM}} = Cp^0\mathrm{NF}_{\mathrm{BM}}F_q \text{ for trays} \tag{3.32}$$

The bare modulus cost at base conditions is:

$$C_{\mathrm{BM}} = Cp^0(\mathrm{B1} + \mathrm{B2}) \text{ for pressure vessels} \tag{3.33}$$

$$C_{\mathrm{BM}} = \mathrm{Cp^0N} \text{ for trays} \tag{3.34}$$

Cost, C_{BM} and $C_{\mathrm{BM}}{}^0$ data can be read from any standard equipment costing for heat exchanger. The bare modulus cost at non-base condition is used to obtain the total modulus cost by summing all the individual bare modulus costs of each piece of equipment and multiplying by 1.18. The total modulus cost for the refinery was estimated to be US$12884878.

$$C_{\mathrm{TM}} = 1{,}18 \sum C_{\mathrm{BM}} \tag{3.35}$$

The sum of total modulus cost and the half the summed bare modulus cost of each piece of equipment at base conditions yields the grass roots cost. The grass roots cost of the refinery was estimated to be US$16276371. As the grass roots cost includes the total modulus cost, it will always be higher.

$$C_{\mathrm{GR}} = C_{\mathrm{TM}} + 0{,}5 \sum C_{\mathrm{BM}}{}^0 \tag{3.36}$$

Once, all these variables have been obtained, the FCI required for the refinery can be obtained. Upon completing the equipment costing section of the economic evaluation, an estimate of US$17536639 was determined for the FCI for the refinery.

$$\text{Fixed Capital Investment (FCI)} = \text{Land Cost} + \mathrm{C_{GR}} \tag{3.37}$$

3.5.5.2 Cost of Manufacturing

The total cost of manufacturing represents the yearly costs associated general, fixed and direct manufacturing costs, where direct manufacturing costs are comprised of costs related to raw materials, waste treatment, operating labour and utilities. A basis of 8000 plant operating hours a year was used for the yearly operational period of the refinery. A total cost of US$56254666 per annum is required to manufacture and run the refinery.

3.5.5.2.1 Operating Labour Cost (C_{OL}) This cost was calculated by first obtaining the number of particulate (P) and non-particulate ($\mathrm{N_{np}}$) operations in order to determine the $\mathrm{N_{OL}}$ which subsequently yields the number of operators for the refinery. A $\mathrm{N_{OL}}$ of 6.63 was calculated for this modular refinery using Eq. (3.38):

$$\mathrm{N_{OL}} = (6{,}29 + 31{,}7\mathrm{P}^2 + 0{,}23\mathrm{N_{np}})^{0{,}5} \tag{3.38}$$

Labourers are to work 5 shifts a week. There will be three separate shifts every working day, each lasting for eight hours. Under this scheme, 28 operating labourers will be needed at the refinery. An hourly wage of US$12 will be paid to each labourer which amounts to an annual salary of US$22685. Thus the operating labour cost will be US$635175 per annum.

3.5.5.2.2 Utilities Cost (C_{UT}) There are two criteria used to evaluate utility costs: flow and duty. As energy (Q) streams were used to cool condensers, the duty-based utility was calculated. An electricity cost of 0.06$/kWh (Turton et al. 2008) was used to cost the annual electrical usage by utilities. The net electrical cost was calculated as US$193714 per annum.

3.5.5.2.3 Raw Materials (C_{RM}) For the refining process, 3 raw materials are required, namely: the ZSM-5 catalyst, water and the crude oil. The hourly flow rates of each of these materials were converted to a yearly flow rate and then costed accordingly. The catalyst cost was found to be US$34 per kg; the cost of water per m^3 was found to be US$0.01; and the cost of crude oil feed was found to be US$26.33 per barrel in 2016. Thus, the cost of raw materials is US$39608644 per annum.

3.5.5.2.4 Waste Treatment (C_{WT}) The catalytic cracking unit operation produces flue gas after complete combustion, which contains a significant amount of CO_2. As such, a carbon tax needs to be paid yearly at a rate of US$10 per tonne of CO_2 emitted. The mass flow rate of emitted CO_2 was calculated as 1540.13 kg/hr. thus a pollution abatement cost of US$135492 needs to be paid per annum.

3.5.5.2.5 Manufacturing Costs The cost of manufacture (COM) is calculated by Eq. (3.39). The individual costs calculated previously are used to calculate the COM – for this refinery, the COM was estimated as US$55767842 per annum. The COM without depreciation, COM_d, is calculated using Eq. (3.40). The COM_d value for this refinery was estimated to be US$54014178 per annum (Table 3.9).

$$\text{COM (\$/year)} = 0{,}280\text{FCI} + 2{,}73C_{OL} + 1{,}23\,(C_{UT} + C_{WT} + C_{RM}) \tag{3.39}$$

$$\text{COM}_d\text{ (\$/year)} = 0{,}180\text{FCI} + 2{,}73C_{OL} + 1{,}23(C_{UT} + C_{WT} + C_{RM}) \tag{3.40}$$

3.5.5.2.6 Revenue Along with the main gasoline (naphtha) product, several other valuable materials were sold in order to obtain revenue. The Oil Price Information Service (OPIS) International Feedstocks Intelligence Report of February 2016 was used to obtain prices for the product materials. Due to the prices being reported in cents per US gallon, a volume conversion factor of 264.17 was used as there are 264.17 US gallons in 1 m^3.

3.5.5.2.7 Depreciation The economic value of a chemical plant along with its equipment shall diminish as time proceeds. Equipment in a plant has an expected lifespan and over time must be replaced. When the chemical plant shuts down it can only expect a fraction of its total value when sold (Turton et al. 2008) (Table 3.10).

The Modified Accelerated Cost Recovery System (MARCS) method of depreciation was used to determine the depreciation. This method uses the straight line method given by Eq. (3.41) and double declining balance method given by Eq. (3.42).

Table 3.9 Summary of yearly costs of manufacturing and operation of the refinery.

Parameter	Value (US$/year)
Operating Labour	635 175
Utilities	193 714
Raw Materials	39 608 644
Waste Treatment	135 492
COM	55 767 842
COM_d	54 014 179
Direct Manu. Costs	43 665 696
Fixed Manu. Costs	3 395 859
General Manu. Costs	9 193 111
Total Costs	56 254 666

Table 3.10 Income from selling valuable materials produced from the refining process.

Stream	Product	Quantity (m^3/hr)	Quantity (m^3/year)	Unit Price (US$/$m^3$)	Income (US$/year)
13	LPG	0.10	8730.38	340.00	2 968 328.95
16	Naphtha	7.26	63 592.68	325.00	20 667 620.22
18	Kerosene	5.51	48 301.20	262.03	12 656 511.01
19	LGO	3.54	31 026.76	213.95	6 638 287.54
36	LPG	4.96	43 401.77	340.00	14 756 601.38
39	Naphtha	2.03	17 758.00	325.00	5 771 350.14
41	Kerosene	0.45	3925.16	262.03	1 028 522.65
42	Other	14.13	123 741.95	197.73	24 467 916.92
				Total	**88 955 138.81**

The straight line method is given by:

$$d_k^{SL} = \frac{\text{FCI}_L - S}{n} \tag{3.41}$$

where: FCI_L = fixed capital investment excluding land cost; S = salvage value; n = life of the equipment. The double declining method is given by:

$$d_k^{DDB} = \frac{2}{n}\left[\text{FCI}_L - \sum d_j\right] \tag{3.42}$$

where: FCI_L = fixed capital investment; n = life of the equipment; $[FCI_L - \sum d_j]$ = book value

The double declining method is used first then it changes when the straight line method has a higher depreciation for that specific year. The straight line method is then used for the remaining years. A half year convention is used in this method of depreciation (Turton et al. 2008) (Table 3.11).

3.5.5.2.8 Profitability Analysis A profitability analysis is undertaken for a prospective project to investigate its ability to generate profit in the future. Most shareholders are interested in a particular project because of its potential to make profit; therefore, this part of the economic evaluation is essential. A profitability analysis was conducted for this particular project and the results are illustrated in the below sections.

3.5.5.2.9 Discounted Payback Period (DPBP) The DPBP is the length of time it takes the project to generate or recover back the fixed capital cost after startup. The DPBP of this particular project is approximately two years – a DPBP of 1.57 years was calculated. This can be viewed as a positive point because two years is a relatively short period of time and as a rule of thumb, the project with the shortest DPBP is the most desirables as stated by Turton et al. (2008).

Table 3.11 Depreciation over the five year period using the MACRS method.

Year	Yearly depreciation (US$) (double declining method)	Yearly depreciation (US$) (straight line method)
Half of year 1	3 255 274	
2	5 208 439	2 893 577
3	3 125 063	2 232 188
4	1 875 038	1 875 038
5	1 125 023	1 875 038
Half of year 6		937 519
	Total Depreciation	16 276 371

3.5.5.2.10 Net Present Value The net present value of a project can be defined as the cumulative discounted cash position at the end of a project. It is the sum of the cash inflow and cash outflow of a project discounted to time 0. The net present value was computed to be US$19751197. A positive net present value is a good indication of profitability. The calculation of the NPV of the project commenced with the computation of its cash flow for each year from startup until the end of the project. Eq. (3.43) was used to perform this calculation as shown by Turton et al. (2008):

$$\text{Cash flow} = (R - \text{COM}_d - d_k)(1 - t) + d_k \tag{3.43}$$

where: R is the revenue of the project for the respective year, COM_d is the cost of manufacturing without depreciation of the project for the respective year; d_k is the depreciation of the project for the respective year; t is the taxation rate which is specified as 0.29.

The cash flows for each year of the project are then discounted to the present year. This was done using Eq. (3.5.4.18) as shown by Turton et al. (2008):

$$F = P(1 + \text{i})^{\text{n}} \tag{3.44}$$

where, F is the non-discounted cash flow; P is the discounted cash flow; i is the interest hurdle rate which is specified as 0.15; n is the number of years from the present time (Figure 3.23)

3.5.5.2.11 Present Value Ratio (PVR) The PVR is a ratio of the present value of all the positive cash flow and present value of all the negative cash flow. It serves as a comparison between the positive and negative cash flows of the project throughout the duration of the project. The PVR of this project was found to be 1.51. A PVR greater than 1 is an indication of profitability; therefore, this process has good potential to generate profit (Turton et al. 2008).

3.5.6 Entrepreneurial Analysis

Crude oil is one of the most prominent natural resources in use within the modern world due to its high calorific value and wide range of uses. Crude oil is made up of a multitude of different hydrocarbon components. The refining of crude oil allows for these components to be separated into numerous different fractions, each of which having various different chemical and physical properties. These variations in the products that may be produced through the processing of oil allow for the products to have a wide range of usage, ranging from vehicular fuels to cosmetic oils.

The wide variety of uses of products from oil refining has created an increasing demand for the petroleum products. From this knowledge, a modular refinery was proposed which would be able to generate petroleum-based products for sale. A preliminary design of the refinery was carried out, in which the refining process was laid out and a mass balance was completed to identify the production rates, which could be achieved. During the preliminary design stage, an economic evaluation was also performed to ensure that the proposed plant was viable in a monetary sense, ensuring that the refining process was able to achieve an economic gain in terms of the value of process inputs and outputs.

Although the viability of the process design and economics was determined, it is still imperative to perform an entrepreneurial evaluation on the proposed crude oil facility. This evaluation will aid in the identification of a market gap, which may be present within the petroleum industry and may be taken advantage of. The analysis will also aid in the identification of the legal viability of the proposed refinery and in the identification of past experiences in oil refining from which crucial lessons in business operations may be learned. This report presents an evaluation of a proposed 4000 barrel per day modular refinery.

3.5.6.1 Downstream Opportunities

The gas and oil industry may be divided into three sectors; upstream, midstream and downstream. The upstream sector encompasses the exploration of potential oil reserves as well as the extraction of the oil. The midstream sector generally

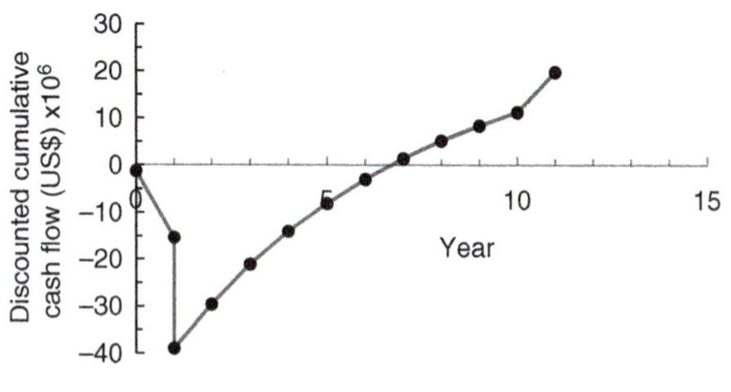

Figure 3.23 The cumulative cash flow over the project life showing the NPV at the end point.

accounts for the transport and storage of the crude oil, as well as the marketing of the raw oil for refining. The downstream petroleum sector refers to the processing of crude oil after it has been extracted and transported; this includes both the direct refining of the oil, as well as the marketing of the petrochemical products (SAPIA 2016). The scope of the downstream sector, therefore, includes elements from both the industrial and consumer markets.

At present the majority of South Africa's resource reserves are coal based. Due to the large availability of coal in South Africa, it is currently being used to supply 77% of the country's energy needs (South African Department of Energy 2016). Oil is not traditionally seen as having a large potential for extraction, however, concerns over depleting coal reserves have led to increased emphasis on the exploration of oil reserves to determine the capacity which may be extracted (Havemann 2014). It has been suggested that South Africa has the potential of up to 9 billion barrels of available offshore oil reserves. If viable reserves are found they could be able to provide for up to 80% of current oil imports, with estimated extraction rates of up to 370 000 barrels per day (Carnie 2014).

Although South Africa currently does not have a high crude oil production rate, there is still a high demand for petroleum-based products. To meet this demand crude oil from various parts of the world is imported and refined locally. South Africa has various operational crude oil refineries; Natref, Chevron South Africa, Sapref, Sasol Refinery and Mossel Bay GTL refinery. South Africa currently has the capacity to refine 513 000 barrels per day of oil, however, the demand for petroleum products is still greater than the supply. In the coming years, the demand for refinery products gasoline, jet fuel and diesel will increase leaving a broader gap between supply and demand of crude oil products in the country (Mantshantsha 2013). An increase of 8, 3 and 6% in demand is reported for diesel, gasoline and jet fuel, respectively, every single year (PetroSA 2009). South Africa needs to import approximately 8000 barrels per day of finished products to meet the demand. According to PetroSA, the quantity of imported petroleum will grow to 23 000 barrels/day by 2020 if no more refineries are built (Mantshantsha 2013). Therefore, there is a large potential for the development of small-scale operations in the downstream petroleum sector.

3.5.6.2 Design Production Rates

From the preliminary design of the modular crude oil refinery, the following overall results were achieved. The processing of 4000 barrels per day of raw oil resulted in the following production rates of the refinery products in Table 5.1. Not all products produced from the refining process may be sold directly into the market as usable products; however, they may be sold onto further petroleum processing plants in which they may be converted into final products for consumer sales. The production rates achieved will help in the contribution towards satisfying the gap in the supply and demand for the final products. An evaluation of the sales market indicated that US$69705094 of revenue may be generated per annum from the sale of the products produced at the end of the proposed refining process.

3.5.6.2.1 Product Applications

Gasoline Gasoline is currently the most used commercial vehicle fuel, making it an essential good in society. Gasoline is a highly volatile mixture, and its use is mainly as vehicular fuel, with some application in chemical production as a solvent.

Gasoline is obtained through the LSR from atmospheric distillation and naphtha is produced in the through cracking. More treatment is required; however, before this blended fuel is of sufficient quality to be used in vehicle fuel. This quality is determined in terms of properties such as octane number, vapour pressure and antiknock properties. The fuel mixture typically has butane added from the isomerisation process in order to reduce the vapour pressure of the mixture, described by Gary et al. (2007). Vapour pressure is limited to certain levels of the Reid Vapour Pressure that vary with ambient temperature. The vapour pressure is regulated so as to avoid vaporisation of fuels leading to volatile organic compounds (VOC) emissions. Alkylation is one of the processes that are used to increase octane number, which reflects the antiknock ability of the fuel. The octane number is also increased via additives such as MTBE or ethanol (Masters and Wendell 2014). Knocking is caused by early combustion of the fuel within an engine.

Gasoline combustion releases harmful emissions into the environment, making it a potent pollutant. Carbon monoxide, carbon dioxide, sulphur dioxide, VOCs, nitrous oxides and particulate matter are some of these pollutants. Vehicles have universally adopted catalytic converters and many have abandoned leaded fuel additives, however, environmental concerns are still significant. Carbon monoxide is a toxin that prevents the haemoglobin from transporting oxygen about the body and can be fatal with high levels of exposure. In long term, low levels of exposure numerous side effects are possible, ranging from heart issues to neurological disorders, detailed in Masters and Wendell (2014), 380–381). Most carbon monoxide is now reacted in the catalytic converter to form carbon dioxide. Carbon dioxide emissions contribute towards the greenhouse effect, nitrous oxides can cause respiratory disorders such as asthma and sulphur dioxide results in acid rain which can cause agricultural damage. Hence, the use of gasoline as a fuel results in significant environmental emissions that cause harm to people and the environment.

The price of gasoline (as of 17 April 2016) in the South was US$0.66/L (Engen) for 93 Unleaded petrol, higher than that of inland diesel 0.005% Sulphur content, at US$0.57/L (Engen). Diesel is another competitive vehicular fuel that is discussed at a later stage.

Diesel Diesel is obtained from a refinery via the LGO fraction which undergoes hydrocracking. Similarly to petroleum, diesel is used as transport fuel to fuel commercial vehicles; however, diesel also has significant uses as a fuel in other functions such as locomotives, industrial machines and is particularly favoured as a fuel for heavy vehicles such as trucks.

Diesel is less volatile than petrol, allowing for greater compression ratios. This results in a greater fuel efficiency. As mentioned in the previous paragraph, diesel also currently is cheaper than petrol per litre, which combined with the greater efficiency can result in consumer cost savings.

Environmentally diesel has the benefit of lower carbon monoxide emissions than gasoline, but as the engine operates at higher temperatures from the compression ratio, a significant amount of nitrous oxides which negatively affect the respiratory system are produced. Vehicles using diesel fuels additionally emit harmful particulate matter which is detailed in Masters and Wendell (2014, 410) as a potential carcinogen.

Sulphur-Based Products Much of the refining process focuses on the removal of sulphur from the fuels to prevent SOx emissions. Thus the process of hydrotreating and operations such as the Claus process yield large quantities of solid sulphur (Gary et al. 2007). This is a solid and can thus be collected and transported to chemical plants which convert elemental sulphur into useful products. Sulphur is highly flammable in its elemental state.

Chung et al. (2013, pp. 518–524) document that elemental sulphur uses are more limited than sulphur in compound form. This is due to the differing properties of elemental sulphur from sulphur bound, such as the inability to dissolve elemental sulphur in water. Methods are available to convert this elemental product into a more useful form.

A major use of elemental sulphur is the production of sulphuric acid via the contact process. Elemental sulphur is a feasible source material in the contact process and is stated in Ashar and Golwalkar (2013) as being capable of large-scale producing 98:5% sulphuric acid. In this process, solid sulphur is combusted to form sulphur dioxide which undergoes a series of reactions with the end product being sulphuric acid. Sulphuric acid has widespread uses throughout industry. In petrochemical refining, sulphuric acid is used as a catalyst in the alkylation process which increases gasoline's octane number, as detailed by Gary et al. (2007). Sulphuric acid is used as a reagent in many chemical processes with end uses relating to a vast array of products such as cleaning agents, pharmaceutical products and batteries. Sulphuric acid is a corrosive substance (Turton 1998).

Residues Residues from the refining process can be sold as products as there are various uses for these heavy compounds. Residue, especially that which has been further treated by a vacuum distillation unit, is fit for use in the production of tar. This has been investigated by Satou et al. (1998). Bitumen is used in the construction of roads and is thus an integral aspect of infrastructure.

Additional residue uses are fuel uses, with specific types of residue being able to function as fuel for ships while others may be used to fuel furnaces in processes (Gary et al. 2007).

LPG The top product of the fractional distillation column undergoes gas treatment to remove methane and ethanol so as to leave a propane and butane mixture. This gas mixture is Liquefied Petroleum Gas. LPG has extensive uses in the consumer market where it is commonly used to fire gas heaters and used as a cooking fuel at cheap cost. This has made LPG extensively used in communities without access to an electricity grid and in low-temperature countries.

LPG has also seen some use in vehicles as a fuel (Masters and Wendell 2014).

Much of the cost incurred with LPG purchasing is associated with high costs of the steel casing and transportation. This is as a result of LPG being an extremely volatile substance (Afrox 2016).

Kerosene Kerosene is the primary fuel used in aviation. According to Gary et al. (2007), commercial jet fuel can be made by a refinery when the kerosene fraction undergoes hydrocracker treatment. However, fuel requirements are stringent and the actual compositions with regards to aromatics and naphthas must meet certain requirements. Blending the kerosene product with naphthas is also performed to satisfy commercially acceptable criteria (ASTDR report).

With a growing number of flights, and virtually no other competing fuel base other than synthetic jet fuels which are not commercially available, demand for this product is high.

Lubricating Oils Various different types of lubricant are from the fractional distillation of crude oil, and see extensive use both industrially and commercially. These lubricating oils take on forms ranging from viscous liquids to wax. The purpose of lubricant is to prevent structural damage resulting from metal abrasions that arise from structural motion or friction. Udonne (2010) detailed the added usefulness of removing heat from lubricated areas.

Lubricating oils are used extensively industrially to prevent wear on machines with moving parts. This covers static operations in a factory and in transportation. Similarly, consumers use grease application on rudimentary mechanical systems such as door hinges and locks. Motor oil presents a great demand for lubricating oils, as they function as a means of avoiding mechanical wear by reducing the coefficient of friction between abrasive parts. Motor oil also serves to transport the extensive amounts of waste heat produced in the engine that was not converted to work to other areas where the heat can be released without affecting the vehicle's operation. Finally, the oil has the added benefit of physically transporting dirt and other solids such as metal that was scraped off during abrasion away from critical engine parts, thus avoiding a build-up of solids that would eventually lead to the engine malfunctioning. Another consumer use is pharmaceuticals. Petroleum jelly is a wax-like substance derived from heavier lubricating oils and is used medicinally to soothe chafes common amongst sportsmen who undertake activities such as running or cycling. Dermatological uses are also present with petroleum jelly.

3.5.6.2.2 Required Technologies The conversion of crude oil into usable products may be separated into three stages; pretreatment, oil refining and post treatment. There are various different technologies that may be used depending on the products which are desired.

Pre Treatment Desalting is the most commonly used pretreatment within the oil refining process. Desalting allows for the removal of inorganic salts, suspended solids and water-soluble trace metals which may be found in the raw crude oil. It is important for these components to be removed as they will have adverse effects on the processing operations and equipment if present. There are two types of technologies that are used for desalting; chemical desalters and electrostatic precipitators. Each of these methods has economic costs and benefits. Electrostatic precipitators have negligible amounts of raw materials that need to be added during separation; however, the use of high levels of electrical power increases the utility costs. Chemical desalters on the other hand have a lower utility costs; however, they have a high demand for the addition of chemical demulsifiers during processing, increasing the raw material costs.

It was decided that the pretreatment of the raw crude oil feed was to be carried out by using electrostatic desalters. This was used as it would reduce the raw materials needed by the process allowing for decreased material costs. Two desalter units were used as this allowed for increased effectiveness in the removal of unwanted components. The cost of the two desalter units was found to be US$4129 and US$2949.

Refining Crude oil refining mainly involves the separation of various components in the crude oil allowing for a greater degree of usage for the individual fractions. The refining process may, however, also include various processes in which unusable fractions are converted into more useful and valuable cuts.

Atmospheric distillation is generally taken as the basis for the refining process. An atmospheric topping column allows for the separation of the raw crude oil feed into more valuable fractions such as LPG, naphtha, kerosene, LGO, HGO and heavier residue products (Treybal 1981).

An atmospheric distillation column was used for the topping process in which the oil fractions were separated. An additional atmospheric distillation column was also sized for use in treating the cracker product. The costs of these topping columns and the FCC fractionation column were found to be US$187667 and US$53619, respectively.

Vacuum distillation is used to allow for the further separation of components that could not feasibly be separated by atmospheric distillation. This stage uses the atmospheric residue as a feed into a packed distillation column under vacuum conditions. This allows for the further separations of the products into LVGO, HVGO and vacuum residue (Gray et al. 2007).

A vacuum distillation column was used for the separation of the fractions within the atmospheric residue. This column allowed for aiding in the increased conversion of the oil into the desired products. The cost of the vacuum distillation column was found to be R200 000.

Cracking is a process in which the heavier cuts produced from the crude oil fractionation are converted into more valuable lighter cuts. There are 2 prominent techniques for the cracking of oil fractions; thermal cracking and catalytic cracking. Thermal cracking makes use of high temperatures and pressures to break longer, heavier hydrocarbons into shorter molecules. Thermal cracking is not always seen as economically viable as the high temperatures and pressures used to create large utility costs which may offset any economic value gained through the conversion of the oil fractions (wiseGEEK 2016). Catalytic cracking operates by catalytically converting the longer, heavier hydrocarbon molecules into shorter more valuable fractions. The use of catalysts allows for lower operating temperatures which decrease the processing costs associated with the cracking process. Catalytic cracking is also able to convert a higher percentage of the feed into more valuable naphtha product, allowing for a higher degree of economic gain (Gray et al. 2007).

A FCC was used for the process of cracking the heavier oil fractions into lighter more valuable products. A catalytic cracker was used as it allowed for reduced utility costs over the use of thermal cracking. The cracking process allowed for the increased extraction of more valuable products from the crude oil increasing the profits obtained. The cost of the FCC reactor unit was found to be US$107239 with the catalysts costing US$1072386 per annum.

Post Treatment Post treatment units are used within the oil refining process to convert the oil fractions produced into a product in a more usable form. This usually involves the removal of contaminants or the conversion of the components to a more useful form.

Hydrotreating is used to remove contaminants within the fractionation products to allow for them to be compliant with economic and environmental regulations. Hydrodesulphurisation is the most common hydrotreating process incorporated into the refining process (Techno Manage 2016). Isomerisation is the process in which the hydrocarbon molecules within an oil fraction are rearranged into a more valuable form, ideally with a higher octane rating. This process is usually carried out catalytically allowing for lower temperatures and pressures to be used (Watanabe et al. 2008).

No post-treatment units were used in the production process in order to reduce initial capital and operational costs of the process. As the proposed refinery is modular in nature the addition of post treatment into the process may be introduced to allow for the conversion of the final products into products that are directly marketable, such as gasoline and diesel, increasing the value of the materials. However, it was determined that allowing preliminary processes to occur before the introduction of these units will create an initial cash flow into the production which may be used to contribute to the capital costs associated with product post treatment, increasing the overall profitability.

3.5.6.3 International and South African Experience in Modular Refining

3.5.6.3.1 Past Experiences Majority of the refineries constructed in Africa are between medium to large scale size this can be accounted for by the time that these refineries were built and the technology that was available at this time Atumah 2015. A total of 42 refineries in Africa with a capacity of 3 217 600 barrels per day, the major refineries are situated in Egypt, Algeria, Libya, South Africa and Nigeria. There are 9 refineries (774 900 bpd) in Egypt, Algeria with 5 refineries (303 700 bpd), Libya with 5 refineries (380 000 bpd), South Africa with 4 refineries (545 000 bpd) and Nigeria with 3 refineries (445 000 bpd) according to (Atumah 2015). It is imperative to keep up with the ever-growing demand of oil and this could be made possible by operating a modular refinery with the benefits of a quick startup and expansion can be made without the production being shutdown (Atumah 2015).

A modular refinery has been constructed in Papua New Guinea with the aid of InterCorp Oil (Ogj.com 2001). The plant was constructed in later part of 2002. The primary objectives of Papua New Guinea were to eliminate the need for dependency on foreign petroleum. The products which are generated from the plant were marketed to Papua New Guinea and also to near export markets. During the time of construction, there were no environmental issues encountered and could proceed with ease. The federal government of Nigeria has approved the establishment of modular refineries since the issue of constructing 18 refineries in 2002 was unsuccessful (Atumah 2015). Since the startup for a modular refinery is relatively short and has a low-cost investment this could encourage and be an incentive for private investors to engage and be a solution to this predicament and this will also reduce the theft of crude oil (Oil Review Africa 2012). President Muhammadu Buhari has issued 65 Nigerian companies licences to construct modular refineries with the primary concern of reducing the large import costs by improving the quality and capacity of refining to satisfy the local demand market (Atumah 2015). Due to the advantages that modular refineries have many problems encountered with the production of oil can be eliminated.

Finally, this year the 116 million US dollar modular refinery for integrated oil and gas is ready for action (Bhutta 2013). The location of the modular refinery is in Nigeria, Tomaro Island Port. The plant capacity is 20 000 bpd and has already received preliminary approval. The generated products include diesel, kerosene and export-quality aviation fuel and fuel oil along with the promise of great employment for the youth.

3.5.6.3.2 Critical Lessons In the past, it can be noted that modular refineries were not profitable which could be attributed to the low level of investment from private investors. The demand for products in the past could also have been another factor that resulted in modular refineries not being profitable. The current demand for petroleum products will now make modular refineries a very lucrative business for private investors to engage in. The feedstock of the crude largely affects the economics of the refinery. Another factor that must be addressed is the location and logistics, ideally have the refinery in places with existing farm tanks and similar facilities. Modular refineries can be satisfactory quantities thereby eliminating foreign petroleum.

3.5.6.4 Health and Safety

While conducting an entrepreneurial assessment regarding the type of modular refinery to construct and its feasibility, it is important to address health, safety and environmental issues that may be possible, and find technological solutions to help alleviate these issues or any other issues that may arise. This is especially important as it can determine whether or not a license to construct the refinery is granted – this still holds true when considering the expansion of a section of an already operational modular refinery as hefty fines and lawsuits can be faced if environmental standards and safety procedures are not honoured, as they may adversely affect human and ecological life. It is equally important to promote health and safety to instil responsibility in workers and contractors once technological and equipment safety standards and procedures have been considered.

3.5.6.4.1 Responsible Health and Safety The most basic protection that workers and individuals at the refinery can receive is collectively known as personal protective equipment (PPE)which includes hard hats, fire retardant clothing, safety eyewear, etc. It is essential that anyone in the vicinity of the process vessels at the modular refinery wears PPE, as it can save their lives. Tesoro (2016) states that its employees receive special training in order to ensure that they are safe and have sufficient knowledge with regards to responding to dangerous or hazardous situations. Some of the training procedures include (Tesoro 2016):

- Emergency response tactics
- Firefighting procedures
- Evacuation and alert procedures
- Safe handling and use of refinery chemicals
- General safe working procedures

Additionally, Tesoro (2016) identifies that its environmental, health and safety (EHS) policy ensures optimal personal, process and transport safety – thus this company is a good example to follow when assessing all-around procedures for best practice. The United Kingdom Petroleum Industry Association (2012) similarly emphasises the importance of health and safety policies and procedures. Adopting their strategy of monthly meetings where best practices are reiterated and information is shared regarding safety in the operation of new technologies would be beneficial to the modular refinery's safety awareness for individuals (Table 3.12).

3.5.6.5 Environmental Standards and Procedures

3.5.6.5.1 Environmental Laws and Regulations Currently, environmental protection laws are in effect, thus any processing plant needs to ensure that land, water sources and the atmosphere are not harmed via hazardous waste and emissions (U.S. Environmental Protection Agency 1996). The Air Quality Act of 2004 (Act No. 39) states that an atmospheric emissions licence needs to be applied for from the licencing authority in the vicinity of the processing plant (Government Gazette 2005). Relative to other refinery air emission laws, South Africa has quite lenient laws – for example, according to the Groundwork Organisation (2002), South African refineries are allowed to emit 82 tonnes of SO_2 per day whereas other

Table 3.12 Production Rate of Refinery Products.

Product	Quantity (BPD)
LPG	150.43
Naphtha	1095.74
Kerosene	832.26
LGO	534.61
LPG	747.84
Naphtha	305.98
Kerosene	67.63
Other	2132.15

Table 3.13 National ambient air quality standards for SO_2 (Government Gazette 2009).

Averaging period	Concentration	Frequency of exceedance	Compliance date
	National Ambient Air Quality Standards for Sulphur Dioxide (SO_2)		
10 minutes	500 μg/m³ (191 ppb)	526	Immediate
1 hour	350 μg/m³ (134 ppb)	88	Immediate
24 hours	125 μg/m³ (48 ppb)	4	Immediate
1 year	50 μg/m³ (19 ppb)	0	Immediate

The reference method for the analysis of sulphur dioxide shall be ISO 6767.

Table 3.14 National ambient air quality standards for CO (Government Gazette 2009).

	National Ambient Air Quality Standards for Carbon Monoxide (CO)		
1 hour	30 mg/m³ (26 ppm)	88	Immediate
8 hour (calculated on 1 hourly averages	10 mg/m³ (8.7 ppm)	11	Immediate

The reference method for analysis of carbon monoxide shall be ISO 4224.

countries in Europe currently producing 2 tonnes per day. This is because the main law governing air pollution in South Africa, the Atmospheric Pollution Prevention Act, No. 45 of 1965 is outdated and does not conform to newer international standards (Groundwork 2002). Additionally, Groundwork (2002) states that South African refineries do not have to measure their air emissions, estimating the level of emissions is sufficient. This provides the modular refinery with leeway when it comes to air abatement because the government is not too strict in this regard. Tables 3.13 and 3.14 indicate examples of the assessment of ambient air quality against national standards for criteria pollutants (Government Gazette 2009). At a media briefing held at Parliament in Cape Town in February 2015, the National Air Quality Officer granted several companies postponements to meet minimum emission standards, until April 2020 (Fourie 2015). This presents the modular refinery executives with an opportunity to make such an application for leniency in the first few years of operation until further expansions for flue gas treatment technologies are constructed and enforced.

3.5.6.5.2 Environmental Abatement Technologies In the design of the modular refinery, the use of cyclones is presented for flue gas separation from the products and catalyst particles. The cyclones can be optimised, as a supplementary form of flue gas treatment; however, this is not very effective on its own. Once financial break-even has occurred, it is intended to use flue gas desulphurisation (FGD) technologies in order to treat the flue gas by removing the sulphur dioxide from the flue gas. Some FGD technologies have the capability to remove up to 99% of sulphur from the flue gas (UK Air Quality Organisation 2016), making it possible for the modular refinery to be well in line with minimum emission standards by 2020.

3.5.6.5.3 Flue Gas Desulphurisation Among the FGD technologies are the limestone/gypsum system, spray dry system, seawater scrubbing process, and the Wellman-Lord process (Global Environment Centre Foundation 1996). From these technologies, the limestone/gypsum system and the seawater scrubbing process seem the most feasible as they are able to yield a 90 and 99% removal of SO_2, respectively (UK Air Quality Organisation 2016). Beychok (2012) states that the high % removal of SO_2 using the seawater scrubbing process is due to the use of additives; without using additives, the %removal of SO2 from the flue gas is reduced to 95%. Due to the very high removal of SO_2, this type of wet FGD technology is most commonly used in the United States (Beychok 2012) and will thus be employed when it is time to expand the modular refinery after the breakeven point, as it meets the standards for environmental and health and safety standards – air quality at the refinery will no longer be hindered once this technology is in place.

3.5.6.5.4 The Seawater Scrubbing Process The seawater scrubbing process is so effective because it is able to use the natural alkalinity of seawater to absorb acidic gases (The Institute of Energy 2013). If seawater is used for cooling, the same source

Figure 3.24 The Hanbal Masstech's SO_2 scrubber, which was patented by the Hanbal Masstech Ltd. located in Gimhae in Korea. It is a compact, flexible square-shaped FGD unit creating the smallest possible footprint. Source: Long et al./MDPI/CC BY 4.0.

of seawater can be used to absorb SO_2 from the flue gas in an absorption unit. The product from the absorption unit is then treated using a water treatment facility using another water feed and air to oxidise the absorbed SO_2 and convert it into sulphate ions (The Institute of Energy 2013). This method is able to collectively treat the flue gas and the resultant wastewater from the process, thus the water used to treat the flue gas can be discharged with the other water effluent streams in the process. One of the several recently commercialised advanced and intensified seawater FGD systems, which simultaneously satisfy regulations, high efficiency, and low space requirements is the Hanbal Masstech's SO_2 scrubber (Figure 3.24). Literature also states that the cost associated with a large wet FGD technology that is able to achieve 94% SO_2 removal is \$75 000 000 meaning that for the proposed modular refinery design (4000 barrels per day), the price will be significantly lower, making this an attractive investment.

3.5.6.5.5 The Claus Process The National Renewable Energy Laboratory (2006) suggests the use of the Claus process for sulphur recovery and desulphurisation. It states that for flue gas emissions of less than 1 t per day, a common method of sulphur removal is to react the sulphur with materials such as liquid hexahydrotriazines or zinc oxide. The Claus process is able to remove 97% of SO_2 by using multiple stages (National Renewable Energy Laboratory 2006). Although this method of sulphur removal is very high, it will not be used as it requires reactors that use catalytic reactions which suggests the cost of such a process is relatively high.

3.5.6.6 Waste and Residue Management

Various different types of waste will be produced from different process units in the modular refinery designed in the previous section. The desalter will produce waste water which contains salts and metals. The vacuum distillation column will produce residue in the form of tar (refinery tar), sludge and light cycle oil which is considered a waste product in this process. The FCC unit will produce flue gas that needs to be desulphurised if the suggested sweet crude oil is not used as well as spent catalyst. Finally, the fractionation column subsequent to the FCC unit will produce fuel oil as a waste. Contaminated fuel products are also considered as waste products; however, it will be assumed that none of the saleable products are contaminated (Rankish 2014).

There are two ways to achieve optimal waste management: one method is the traditional implementation of waste management systems and the other is the integrated EHS management system. The latter system encompasses aspects of the previous EHS section discussed above and, therefore, it seems to be a more fitting system to use when considering a viable waste management scheme for the modular refinery (Figure 3.25).

Figure 3.25 Summary of key elements in waste management (IPIECA 2014).

As residue and waste adversely affect the health of human beings and the environment, it is necessary to find effective methods of managing them. In addition to health and safety, waste and residue minimisation allows for better resource utilisation as less by-products are formed and more volume is available for desired materials within process vessels (IPIECA 2014). For the proposed modular refinery, it is intended to make use of some of the by-products as energy sources. For example, the fuel oil and some of the light cycle oil will be combusted to provide energy for process units that require high temperatures, such as the catalytic reactor and the regenerator. The residue from the topping column will be upgraded using the vacuum distillation column. From an entrepreneurial perspective, more needs to be done in order to effectively manage waste and residue – managing waste and residue after production is not enough, it is necessary to find ways to minimise waste and residue. This is especially important when there is little to no infrastructure for correct waste disposal. The IPIECA guidebook (2014) suggests that the guidebook should be used for waste management schemes in countries that have relaxed emission and waste management laws – this is ideal for the proposed modular refinery as South Africa currently does not have extensive regulations compared to other countries such as the UK (Groundwork 2002). This is where the philosophy of reduce, reuse or recycle plays an important role. In addition to these three aspects of waste management, treatment and disposal are also measures that can be taken to deal with waste.

The waste management techniques listed above can be applied to the modular refinery in various ways. However, some techniques may negatively affect the yield of the final product while reducing waste. Reducing waste can be accomplished by finding alternative feed materials or changing operating conditions to improve selectivity of the desired products – a crude oil with a lower API can be used in order to minimise the production of residue. Waste can be reused as a raw material for a different process so that its quantity is reduced – a good example of this is the combustion of tar and fuel oil to provide heat and energy for other process units. Recycling of wastes can be achieved retreating waste in order to reobtain the original raw materials. Finally, if the above proves to be to complicate, thus being uneconomical, the waste can be disposed to a secondary location. The IPIECA guidebook (2014) states that waste can be stored for a short term so that its quantity accumulates until there is enough of the waste to be treated, disposed or recycled in an economic manner; it classifies short term to be less than or equal to a year.

Waste can be treated physically, chemically or biologically (IPIECA 2014): physical methods include drying beds, hydrocyclones and centrifuges; chemical methods include solidification, stabilisation and encapsulation; and biological methods are bioremediation and composting. There are also different types of disposal techniques: among others, incineration and landfill. The latter is a cause for concern in its own right as toxic chemicals and bacteria can cultivate over time and lead to more adverse health effects, for the environment and those living close to the landfill site. For our modular refinery, waste from the desalter can be minimised by using demulsifiers and electrostatic precipitation in order to minimise the amount of sludge formed – this has the additional benefit of improving separation. The IPIECA guidebook (2014) additionally recommends the use of shear mixing to maximise the use of water in the desalter and the use of low-pressure water to reduce turbulence which in turn prevents sludge formation. Fuel products and residue that are considered to be waste will be burned for fuel in order to minimise utility costs for process units operating at high temperatures. With regards to spent catalyst waste, the IPIECA guidebook (2014) states that catalyst waste can be reduced but by sacrificing the yield of the desired product. This can be done by using base–metal catalysts instead of noble-metal catalysts as the former can withstand changes in process feedstock and process operating conditions.

3.5.6.6.1 Effects of Petroleum Regulations In South Africa, the Department of Energy has the task of regulating the petroleum industry. The downstream petroleum industry is regulated in accordance with the provisions of the Petroleum Products Act, 1977 (Act No. 120 of 1977) as amended by the Petroleum Products Amendment Act of 2003 and the Petroleum Products Amendment Act of 2005 (Petroleum Products Act). The National Energy Regulator of South Africa (NERSA) is also been given responsibility for regulating the petroleum pipelines industry (Dentons 2015).

News24 (2012) reports that petrol price regulation is done by the government for the purpose of ensuring that petroleum products are available as widely as possible to the users. It further states that the reason for needing the regulation is that international fuel prices fluctuate on a daily basis. The price of petroleum products is made up of two main components, i.e. the international and the domestic sections. South Africa has no control over the international section, commonly known as the Basic Fuel Price, as this is mostly made up of Free-On Board. Free-On Board or otherwise known as the international market petroleum spot prices is the price quoted by the numerous places where South Africa obtains it is fuel from. The Basic Fuel Price also consists of freight costs, which are the cost associated with shipping of the fuel, and the insurance placed on the valuable goods transported. Cargo dues to the Port Authority, a governmental or semi-governmental body managing the ports, are also taken into account along with storage costs once the fuels reach shore. The domestic segment of the price firstly includes the transport cost for transporting the fuel from the coastal port to the depot to be stored and secondly, the cost associated with delivering the fuel to the service stations. Department of Energy (2016) reports that NERSA regulates the prices of pipeline gas and petroleum for the purpose of reducing the monopoly in the energy sector, increasing competition and economic growth within the sector.

The major two major advantages of a modular refinery that in essence attract private investors are the low investment cost and promised period in which the revenue will be brought in. These two advantages would attract any private investors but the current system of price regulation of petroleum plays an imperative role in their decision of participating in an investment of a modular refinery. If the proposed the production price of which the petroleum is generated is larger than the regulated price this will indicate that the plant is running a loss and will not be economical feasible to the private investor, therefore, a private investors will not engage in such an investment. Selling the petroleum products below production costs will not encourage any private investors to acknowledge modular refineries and this affects the operation of the modular refinery as there is no source of capital to construct the modular refinery. However, the situation can be reversed at when the petroleum product prices are sold higher than production costs and the regulated prices at this point many private investors will be attracted making modular refineries a popular investment in South Africa, therefore, increasing the local supply of petrol reducing the need of foreign petroleum products. So the regulated petroleum prices directly dictate whether a private investor will engage and, therefore, operation of the modular plant. By the pricing being regulated, this gives each and every petroleum refinery equal opportunity, therefore, this ensures that even smaller refineries can be given opportunities such as the modular refinery.

If the current system of regulated petroleum prices is removed then the petroleum price can be dictated by the modular refineries and other petrol-producing refineries. The operation of modular refineries can be adversely affected if other petroleum refineries are selling the petroleum products at a lower price than the modular refinery is selling it at. Consumers will value good quality petroleum products at low prices. The costs of the plants will now dictate the price of the petroleum and the prices of the petroleum from the modular refinery will be in direct competition with other petroleum-producing refineries. The competitiveness between modular refineries and other refineries which could result in elimination of modular refineries if other refineries are favoured more. Private investors will also want low capital investments with high revenues making a modular refinery a great target as they will now be able to manipulate the prices of the petroleum product and seek optimum profits this shall improve the operability of modular refineries.

3.5.6.7 South African Laws and Regulations

Department of Energy (2016) states that the petroleum pipelines industry is regulated by the NERSA which is also the regulatory body of the country's electricity and piped gas industry. NERSA regulates all three industries using various laws that are applicable to each industry. The Petroleum Products Amendment Act of 2003 and petroleum price regulation along with regulation related to petroleum product specifications are used for the petroleum industry. These legislations generally apply to any establishments that will be dealing with petroleum products. The Petroleum Products Amendment Act of 2003 mostly focuses on the acquisition of a licence that permits an entity to carry out certain activities. Petroleum price regulation in South Africa focuses on setting a price that protects consumers and which petroleum products are legal to produce in South Africa. The regulation related to petroleum product standards states all the grades of fuels that should be produced in South Africa.

3.5.6.7.1 Petroleum Products Amendment Act (Act No. 58, 2003) Department of Energy (2004) reports that Section 2A 1(a) of The Petroleum Products Amendment Act (Act No. 58, 2003) states that a person, as defined by the act, may not manufacture petroleum products without a manufacturing licence issued by the controller of petroleum products. If a situation occurs whereby a person engages in an activity that goes against this section of the Act, the controller of petroleum products must inform the person to stop such activity from that moment onwards by written notice. The controller may permit the person to continue with such activity pending an application and the issuing of a licence. If the application is not successful then the person must be notified by the controller to cease such activity. If it is applicable, the person will also be notified to rectify any state of affairs that occurred during the period in the notice.

Section 2A (4) stipulates that in the event of the application of a manufacturing licence the owner of the particular property must be the individual to do the application and the individual performing the application must have written permission from the owner. Section 2A (6) states that a licenced manufacturer must only be involved in sales with only licensed wholesaler or licensed retailer except for matters involving the export of petroleum products. Section 2B (1) stipulates that the controller of petroleum products must be the one issuing out licences. Section 2B (3) that a licence that was issued by the controller to the person shall remain valid as long as the person follows all the conditions of the licence and the licensed activity stays a going concern. According to Section 2B (5), the controller may issue out a temporary licence upon application which will be invalid after a period of six months. The person who obtains a manufacturing licence is bound to sell petroleum products that produced from coal, natural gas or vegetable matter and must also produce petroleum products that are up to par with the prescribed and relevant specifications and standards. This is according to section 2B (3) (a) and (b), respectively.

3.5.6.8 Acquisition of Manufacturing Licence

According to South African Government (2016) a person must complete the prescribed application form (form DME 399). After being informed by the controller of petroleum products of the success of your application, the person will be required to R10 000 if they were not operating before the Petroleum Product Amendment Act, No. 58, 2003 was promulgated and R 1000 if otherwise.

3.5.6.8.1 Petroleum Specifications and Standards Through the Petroleum Product Act, 1977 (Act No. 120 of 1977), the Minister of Minerals and Energy made Regulations Regarding Petroleum Specifications and Standards in 2006. Section 3 (1) of the Regulations shows the grades of petrol that can be sold to an end-consumer. Section 3 (2) stipulates that metal-free unleaded petrol grades produced must comply with the South African National Standard for Unleaded petrol, SANS 1598 and it further shows other requirements for permitted metal-free petrol. Section 3 (4) states the same but for metal-containing unleaded petrol. The allowed grades of diesel that can be sold to an end user are stipulated in Section 5. Section 6 (1) of the Regulations states that selling petroleum products that do not comply with these regulations for use is not allowed and Section 6 (2) further mentions that a licensed manufacturer cannot produce petrol and diesel grade for use that does not meet the standards of the Regulations. Section 7 stipulates that inspection and testing of petroleum products will be performed if ever there is a contravention of the Regulations. Section 7 states the labelling specifications of petroleum products, respectively.

3.5.6.8.2 Legislations as Applied to Modular Refinery South Africa has regulatory legislation in place concerning the petroleum industry and these have to be fitted and applied to the relevant entity one is establishing. There is no legislation concerning modular refineries in particular but there are legislations that affect the construction and operation of such refineries. The regulatory body of the petrochemical industry along with the electricity and piped gas industry is the NERSA, therefore, all modular refineries have to comply and oblige to the requirements of this body.

The Petroleum Products Amendment Act of 2003 is an act that needs to be followed whilst operating a modular refinery. This act applies to petrochemical establishments that wish to act as wholesalers, retailers and manufacturers of petroleum products. The main operation of a modular refinery is to produce petroleum products so modular refineries have to comply with the relevant requirements found in this Act. Provided the entity constructing the modular refinery is of the opinion that the time in which the application is being processed is lengthy, it can apply for a temporary licence in the meantime, which will be valid for a period of six months. The entity owning the modular refinery must completely fill out form DME 399 and submit it to the regulator along with other information as requested by the energy regulator. After being notified of the success of the application, the modular refinery will have to pay out US$536.

The regulation of prices performed by NERSA does have an impact on modular refineries as some of the prices of the products that will be manufactured and sold are regulated, namely, LPG and petrol. The other legislation that modular refineries will have to adhere to is the regulations regarding petroleum product specifications and standards. Complying with this means that modular refineries can only produce and sell a certain grade of petroleum products, respectively, that are in accordance with the Regulations.

3.5.6.8.3 Business Model Funding the project will start with selecting a form of ownership that will facilitate its ability to obtain sufficient funding. As the funding required to build the refinery is estimated at US$19949508, the form of ownership of the project needs to comply with relevant requirements of the various funding institutions in South Africa, which will enable it to obtain the funding.

3.5.6.8.4 Form of Ownership There are various forms of ownership in South Africa. A few can be immediately ruled out. The following are discussed with reference to the South African Companies Act of 2008. Sole ownership is nearly impossible for a refinery due to the vast amounts of capital required to construct the plant and begin operations. In addition to this the lack of limited liability means that should the plant fail, the owner is responsible for all debts incurred, which can amount to enormous figures. The same applies to a partnership where limited liability is not present. Close corporations cannot be registered since the New Companies Act was established, and these too lacked capital raising ability by being limited to 10 partners. Thus the most effective forms of ownership would be narrowed down to a private company, and a public company.

The following are a few of the differences between a public and a private company:

- The name of a private company ends with "(Pty) Ltd"; that of a public company ends with 'Ltd'.
- There is at least on shareholder of a private company; public companies require at least seven.
- The transferability of shares is restricted for a private company; but there is free transferability of shares for a public company. This means that public companies can raise funds from the public through listing on a stock exchange
- A private company need not lodge financial statements with the Companies and Intellectual Property Commission (CIPC 2016); whereas a public company must appoint an auditor and make financial statements available for review
- Voting rights in a private company are flexible while voting rights in a public company are proportional to the number of shares the voter holds.

From the above, it can be assumed that public funding would be inefficient as the project is still at its initial start and would require much time to convince the public to invest. Another factor in deciding this is that the modular refinery designed in its early stages relies on secondary refineries that process the products further. This dependency leads to potentially volatile profits and reduces financial stability.

Thus a private company would be more suited to the initial stages of the project, raising capital to construct and begin plant operations. At a later stage, the company could go public and raise further capital so as to expand the modular refinery so as to facility higher level production, notably inclusion of alkylation and isomerisation units to output vehicle quality gasoline that could be directly sold as a product.

3.5.6.8.5 Capital Acquisition Due to the degree being studied by our group members being Chemical Engineering, our experience with business affairs and particularly the procedure for raising capital is heavily limited. Thus, in addition to personal research, an accounting firm was informally consulted on methods of funding available to large-scale industrial projects. Acknowledgements to Mr. I. Black, who is registered with SAPIA for his assistance in providing some insight into these matters.

Funds can be raised from numerous sources such as loans from financial institutions, grants from government institutions, investors or securities. In South Africa funding options are available for large-scale industrial projects so as to reduce the risk of investment for the owners and increase the chance of success of the refinery. A suitable method of funding would be to partially fund the project by a governmental grant, and partially via private investment. In South Africa, investments are not only made on a risk versus reward basis but also on an equity scale (Industrial Development Corporation 2014). This means that businesses that have a high level of representation of previously disadvantaged employees in ownership, management and general staff will have a vastly greater chance of obtaining funding from investors than a business that does not focus on employment equity (Ojambo and Bakhsh 2013).

The South African government has an initiative to empower the black population in South Africa, and a means of doing so is by allowing access to specialised funding through government subsidiaries such as the Industrial Development Corporation, the Department of Trade and Industry and the National Empowerment Fund. Numerous loans at preferential interest rates can be made from these institutions as well as certain Grants pertaining to the Industrial industry. However, this form of accessing funding can only be obtained by the company's shareholders consisting of certain percentage of Black ownership.

After ensuring the required levels of representation in shareholding, the government institutions should be willing to grant access to funding a majority portion of the initial capital required to fund the project. The funding will be accessible over a period of time as the project reaches certain progression phases.

The funding from the government institutions will, however, only be granted if the existing shareholders contribute a certain percent of the capital required. This is where other sources of funding can be investigated. The fact that the funding from the government institutions such as the Industrial Development Corporation and Department of Trade and Industry is subject to the requirement of partial from the shareholders, it would be more difficult to obtain funding from the local banks as the National Credit Act regulates lending and restricts the banks to lend money if income cannot be proven (Arde 2015).

Issuing securities could also be difficult as in negotiations the shareholders would need to convince the new potential holders of the securities that their investment will be secure. Thus, possibly the preferential capital acquisition method would be to approach venture capital companies to obtain the remaining portion of capital required. Venture capital companies invest based on opportunities presented and the risk attached to them (Investopedia 2003). These companies would structure a deal that could be achieved with all relevant parties benefitting from their investment. In certain circumstances, the venture capital company might request a portion of the shareholding.

3.5.6.9 SWOT Analysis

A strengths, weakness, opportunities and threats (SWOT) analysis is a technique that assesses a project's SWOT. This method is used to make decisions regarding the best route for an initiative and to offer solutions to problems (Community Toolbox 2015). Strengths as the positive aspects of the business that are completely under the control of the business. These are the positive attributes that the business chooses to use for its greater good. Weaknesses are explained as factors that are in the control of the business but these still prevent the corporation from operating at its optimum. Weaknesses increase the competitive edge of the business's competitors. Both strengths and weaknesses are regarded as internal factors. Opportunities are positive aspects that are out of the control of the business but still have a positive impact on it. These are possibilities that the business can exploit for it to do well. Threats are defined as factors that put the business at a risk. Although the business has no control over them, it can devise a plan to combat or avoid them. Opportunities and threats are regarded as external factors.

A SWOT analysis was performed on the construction and operation of a modular refinery in South Africa. The results from the analysis are illustrated below:

3.5.6.9.1 Strengths

1. Modular refineries require lower capital costs to construct compared to conventional oil refinery (Energy Mix Report 2014).
2. The speed of construction is fast and with low degree of difficult.
3. Modular refineries parts can be mounted together to produce an output suitable for the demand present.
4. Less land is required to construct a modular refinery.
5. Modular refineries have a reduced demand for skilled labour, this reduces the cost and time for training employees and makes the employment process quick and fast (Energy Mix Report 2014).

3.5.6.9.2 Weaknesses

1. Modular refineries have less capacity than conventional refineries.

3.5.6.9.3 Opportunities

1. Regulation of prices is an opportunity for modular refineries. As mentioned in the literature section, the price of regulated fuels consists of freight costs, insurance and cargo duties. None of these costs are incurred by a modular refinery as these fuels are produced by the refinery itself. So there is a large gain in profit as the only costs incurred by a modular refinery in the fuel price are those of domestic shipping.

2. Government subsidies will be positive cash flow into the business and, therefore, a positive external contributor.
3. There are few refineries in South Africa and some of the fuels have to be imported to meet the South African demand. Therefore, constructing a modular refinery will serve to supply this demand. (Energy Mix Report 2014)
4. At the current moment there is relatively low oil price and this means there is less amount money that has to be spent on the acquisition of the crude oil.
5. Diesel is currently in high demand by Eskom as it is needed to operate the turbines in some of the power stations. The plants use about 57 000 l of diesel an hour per turbine. Diesel-powered trucks are also contributors to South Africa's growing demand for diesel (Vanderschuren et al. 2008).

3.5.6.9.4 Threats

1. South Africa has no price regulation on kerosene. This is a threat to modular refineries as conventional refineries with larger capacity can sell their products at a lower price and still make a reasonable profit while attracting more of the demand whilst modular refineries risk operating at a loss if they sell at low prices.

3.5.6.9.5 Entrepreneurial Analysis Crude oil is one of the most prominent natural resources in use within the modern world due to its high calorific value and wide range of uses. From the entrepreneurial evaluation conducted insight into the market and legal viability was gained aiding in the determination of the overall business potential of the proposed modular refinery.

A variety of opportunities are present in the downstream petroleum sector, with regard to crude oil refining and product marketing. The technology required by these downstream applications is currently readily available, with costs estimated to be within a viable range for construction. There also currently exists a market gap in South Africa with respect to the supply and demand of petroleum industry which may be taken advantage of. The product rates, of raw untreated products, achieved through the refining of 4000 BPD of crude oil will be able to satisfy this market gap while generating a significant amount in sales revenue. As the plant is modular in design future post-treatment units may be installed to increase the revenue generated through product sales.

It is important to identify health, safety and environmental issues which may occur during operations of the refinery process, as this may affect the procurement of construction licensing. As Tesoro (Tesoro 2011) has a high safety record, the safety model imposed by this company will be used as a basis for plant safety. To protect against safety issues during plant operations PPE equipment will be imposed on all employees, with regular safety meetings being imposed.

The environmental laws imposed on oil refineries in South Africa are outdated and, therefore, very lenient when compared to the rest of the world, therefore, as engineers we are obligated to provide our own higher standards for emission control. Upon the financial breakeven point of operations a seawater scrubbing process will be used for FGD will be used for the treatment of effluent flue gas, due to the cheap and efficient nature of operations. This effluent treatment will allow for the modular refinery to achieve globally acceptable emission levels.

As a high degree of waste and residue will be produced during the refining process it is vital to manage this waste correctly. An integrated EHS waste management system will be applied as it encompasses the health and safety requirements discussed above. The IPIECA guidebook will also be used as it compensates for the lack of adequate waste management laws in South Africa, encouraging the philosophy of the reduction, recuse and recycling of waste produced.

NERSA regulates the pricing of petroleum products in order to increase completion and economic growth. Regulated fuel pricing may have various effects on the operation of modular refineries as the set prices will directly affect the profitability of the operations. The national deregulation of petroleum pricing might be catastrophic to small-scale operations such as modular refining as larger convention refining will be able to produce cheaper products due to the economics of scale. However, the current likelihood of deregulation occurring is very slim as the government seeks to protect the countries petroleum industry.

Modular refineries are regulated by NERSA; therefore, the refinery is obligated to operate within the restrictions set out by this body. The Petroleum Products Amendment Act of 2003 applies to petrochemical establishments that wish to act as wholesalers, retailers and manufacturers of petroleum products, therefore, the modular refinery will need to operate abiding to the standards set out. Further regulations are also in place that set the grades to which petroleum products need to meet in order for them to be sold. These regulations will aid in defining the products that are produced during the refining process; however, they will also affect the economics of the refining process due to pricing regulations.

The ownership of the modular refinery will be taken in the form of a private company. This will allow for a high number of partners sharing the ownership, while still allowing for funds to be granted for startup and operations. This form of ownership will be best suited for the initial stages during times when financial stability is not guaranteed. A large portion of the funding for the initial stages of the project may be gained through the application for government grants, with the remaining capital being generated through investments with venture capital companies. This will be alleviated the pressure of private capital being needed for the refinery startup while also contributing to an increased chance of operational success.

From this entrepreneurial analysis, it may be seen that there is a market demand for modular refining in South Africa. It may also be seen that South African operational and environmental laws and regulations are suited for the encouragement of the processing of natural resources. Therefore, there exists a high potential for success in the construction and operation of a modular refining plant.

3.5.7 Detailed Design of Fluid Catalytic Cracking Unit

The FCC design is designed similar to the UOP model, with an annular catalyst stripper supported by the riser and a single stage, instead of two-stage regenerator, detached from the riser unit. This is chosen over a stack design such as the Kellogg design (Gary et al. 2007) due to the weight of catalyst in the regenerator compensating for lower operating temperatures. and the calculated optimal regenerator diameter to minimise entrainment being 6.6 m in internal diameter. Overall having a stacked design would add extra weight onto an already heavily laden regenerator and thus would increase construction costs through having to use more expensive and thicker materials of construction as well as increasing the costs of supporting the structure (Reuters 2015). Internal cyclones are used to circulate catalyst as the calculated cyclone sizes are acceptably small, and additionally as the regenerator is optimised for minimal entrainment. Discussed in this section are the sizing and optimisation of the regenerator, and the physical designs of the regenerator and riser units.

Basis Results and Simulation

The original mass balance was performed using a combination of Microsoft Office Excel 2013, and simulation using Aspen HYSYS. HYSYS was primarily used to accurately simulate the reaction kinetics for coking, cracking and regeneration without needing to make inaccurate assumptions via a lump conversion or a linear reaction constant equation. This resulted in there being no need for external kinetics equations, as basis values that applied to the kinetics have been carried over from HYSYS. The simulations were performed using the Peng-Robinson equation of state, chosen specifically for the system due to hydrocarbons mostly consisting of non-polar non-electrolyte components. Some of the base values used in this assignment were optimised for yielding a high conversion of feedstock and minimising coke on the catalyst. Almost all extensive values regarding equipment sizing is manually calculated, using target ranges given in literature, notably the FCC Handbook (Sadeghbeigi 2012), Coulson and Richardson Volume 6 (Sinnot 2005) and Perry's Chemical Engineering Handbook 8th edition (Perry and Green 2008). The optimal values directly taken from the simulation that is used as a basis for sizing equipment and performing calculations on entrainment or pressure drop are the loadings of catalyst within the stripper and regenerator, and the catalyst circulation rate. Stream flow rates and compositions are taken from the mass balance. The reason for directly using the simulated catalyst loadings is due to an issue with the regenerator that was encountered after the mass balance and simulations had been performed. This is further elaborated in the regenerator section of this report.

3.5.7.1 Regenerator

3.5.7.1.1 Preface to Detailed Design of Catalyst Regenerator The sizing of the regenerator was the final unit I designed. Unfortunately, I only noticed a week from the due date for this part that the regenerator design temperature of 600 °C is outdated and results in a high degree of catalyst coking due to incomplete combustion occurring. Ideally, I would have designed the reactor to operate above 680 °C in order to facilitate complete combustion and reduce the catalyst loading, however, editing this part of the mass balance would take excessively long due to its significance on all unit dimensions. Thus I have decided to design the reactor operating at a temperature of 600 °C, and compensate for this by increasing the regenerator catalyst loading, as well as the catalyst circulation rate. Overall after having performed all calculations for dimensions and completed the layout of the unit, I am satisfied that the unit has been well designed by my standards and I had little further problems with this issue.

3.5.7.1.2 Introduction to Sizing Method The method used to determine the dimensions of the regenerator centred about the bed height of the catalyst. A catalyst loading of 170 t is sought (Obtained from HYSYS Simulation), while catalyst entrainment rate is minimised. A pressure drop of 70 kPa was fixed across the bed during the mass balance, and this serves to eliminate a degree of freedom in the calculations. Hence the major calculations involved for determining the bed dimensions are iterating diameter monitoring the amount of catalyst entrained, and determining the maximum bed height for various regenerator diameters by selecting the height of the bed for a pressure drop of 70 kPa. Both results are compared graphically and the ideal bed dimensions are determined accordingly.

3.5.7.1.3 Entrainment As the average diameter of the particles is determined from the particle size distribution as 75 μm, which is less than 100 μm, the Zenz and Weil equations can be used:

$$K_{i_\infty} = 1.26 * 10^7(\rho_f U_{ps})\left(\frac{U_{ps}^2}{gd_{avg}\rho_p^2}\right)^{1.88} \text{ if } \left(\frac{U_{ps}^2}{gd_{avg}\rho_p^2}\right) < 3 * 10^{-4} \tag{3.45}$$

An additional requirement for use of this equation is that the superficial velocity of fluidized solids is less than 1.2 m/s. There is no direct way to determine U_{ps}. Rather, performing a force balance on particles can be used to calculate the terminal velocity of the bed, which can then be used via empirical correlation to determine U_{ps}. Assuming spherical particles, a force balance on fluidized particles yields

$$\frac{\pi d_{avg}^3 g}{6}(\rho_p - \rho_f) - \frac{C_D\rho_f U_T^2}{2}\left(\frac{\pi d_{avg}}{4}\right) = 0 \tag{3.46}$$

The drag coefficient is unknown but can be determined by assuming the Stokes viscosity relations apply, allowing the terminal velocity to be calculated as $U_T = 0.3496$ m/s. Checking the Reynolds number for this velocity indicates that the postulated relationship is correct.

The following relationships are derived by Rhodes (2008).

The Richard and Zaki relationship relates the particle velocity to the terminal velocity

$$U_p = U_T\varepsilon^n \tag{3.47}$$

where n is determined via the Khan-Richardson equation through use of the Archimedes number. These relationships are straightforward empirical equations for which all data is present. These calculations are shown in the FCC calculations spreadsheet. As the Khan-Richardson equation is non-linear, the solver in Excel is used to determine n by iterating n until an error of less than 10^{-8} is attained. The values of n vary slightly per iteration of bed diameter, with an average of $n = 2.856$.

The Richard and Zaki equation applies to batch reactors, for a continuous system the flow rate of the air moving through the bed is accounted for:

$$U_p = U_T\varepsilon^n + \frac{Q}{A} \tag{3.48}$$

As $U_{ps} = \frac{U_p}{(1-\varepsilon)}$, the superficial solids velocity can be determined:

$$U_{ps} = U_T\varepsilon^n(1-\varepsilon) + \frac{Q}{A(1-\varepsilon)} \tag{3.49}$$

Iterating this calculation for various diameters including and above 1 m yields expected results, with a maximum value of 0.45 m/s. Hence, velocities are less than 1.2 m/s, and the Zenz and Weil equation can be used. The application of Zenz and Weil is simulated using a batch method, where the entrainment is divided by the simulation time to mimic a continuous system. The reason for this is that directly modelling a continuous system via iteration results in another nonlinear equation that requires solver. This is to be avoided as the resulting Excel iterative file becomes inflexible to future change, and the error margin increases each time a numerical method is used. Overall the entrainment rate versus diameter of bed graph is obtained and illustrated in Figure 3.26.

In order to perform the Zenz and Weil calculations for entrainment, a particle size distribution is required. As none could be obtained for the specified catalyst, a generic volumetric particle size distribution discussed in (Sadeghbeigi 2012) is utilised. Assuming a particle density of 2100 kg/m^3, it is possible to simulate a particle size distribution and perform the necessary calculations.

It is noteworthy that the entrainment rate is significantly lower than the catalyst circulation rate. Ideally, a diameter where the marginal reduction of catalyst entrainment becomes minimal, approximately at 10 m diameter. However, this is economically unsupported as the increasing diameter serves to increase the height, thickness, weight and thus material costs for supports, refractory materials and the thickness of the vessel. From Figure 3.26, it can be seen that catalyst entrainment rates are quite low for diameters above 3 m. As with the riser, a series of cyclone units is assumed, with the gas stream from one cyclone flowing to the inlet of the next. Assuming cyclone efficiency of 99.995%, the catalyst recovery is calculated for successive cyclones, the results being tabulated in Table 3.15.

From Table 3.15, it can be seen that multiple cyclones significantly reduce catalyst losses at low diameters; however, the marginal benefit reduces as the diameter increases. It is thus possible to utilise a single cyclone, thus reducing equipment and maintenance costs. Cyclones in the regenerator are exposed to high levels of erosion corrosion, both inside and outside

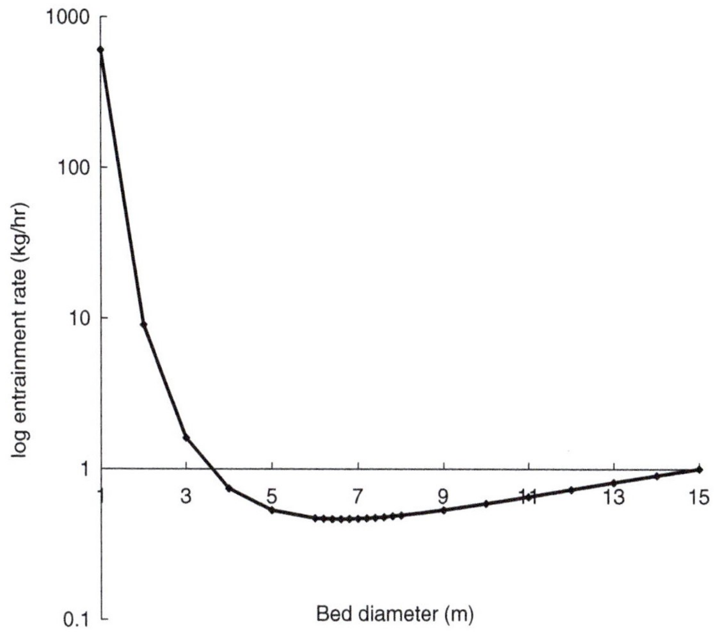

Figure 3.26 Semi-log graph indicating the entrainment rate for a range of diameters for the fluidized bed.

Table 3.15 Effect of different cyclone arrangements on catalyst loss for 99.995% efficient cyclones. Derived from calculations performing the Zenz and Weil method.

Bed Diameter (m)	Entrainment Rate (kg/hr)	Single cyclone Catalyst Lost to flue (kg/hr) × 10^{-5}	Two Stage Cyclone Catalyst Lost to Flue (kg/hr) × 10^{-9}	Two Single Stage Cyclones Catalyst Lost to Flue (kg/hr) × 10^{-5}	Two Multi-Stage Cyclones Catalyst Lost to Flue (kg/hr) × 10^{-10}
1	605.30	3026.52	1513.26	1513.26	7566.30
2	9.08	45.39	22.69	22.69	113.47
3	1.62	8.07	4.03	4.03	2.02
4	0.75	3.73	1.86	1.86	9.33
5	0.53	2.67	1.34	1.34	6.68
6	0.48	2.37	1.19	1.19	5.93
7	0.47	2.35	1.18	1.18	5.88
8	0.49	2.47	1.24	1.24	6.19
9	0.54	2.68	1.34	1.34	6.71
10	0.59	2.99	1.48	1.48	7.39
11	0.66	3.29	1.64	1.64	8.22
12	0.73	3.66	1.83	1.83	9.15
13	0.82	4.08	2.04	2.04	10.20
14	0.91	4.54	2.27	2.27	11.35
15	1.01	5.04	2.52	2.52	12.60

the unit, and thus having fewer cyclones reduces costs for refractory linings and corrosion allowance in the structure of the cyclone. Investigating the relationship between pressure drop and the diameter of the vessel will aid in determining the optimal diameter for the vessel, and hence the arrangement and number of cyclones in the regenerator. The pressure drop across the fluidized bed is calculated through the use of the Ergun equation (Rhodes 2008):

$$-\frac{dP}{dh} = \frac{150\mu U_f(1-\varepsilon)^2}{d_{avg}^2\varepsilon^3} + \frac{1.75\rho_f U_f^2(1-\varepsilon)}{d_{avg}\varepsilon^3} \tag{3.50}$$

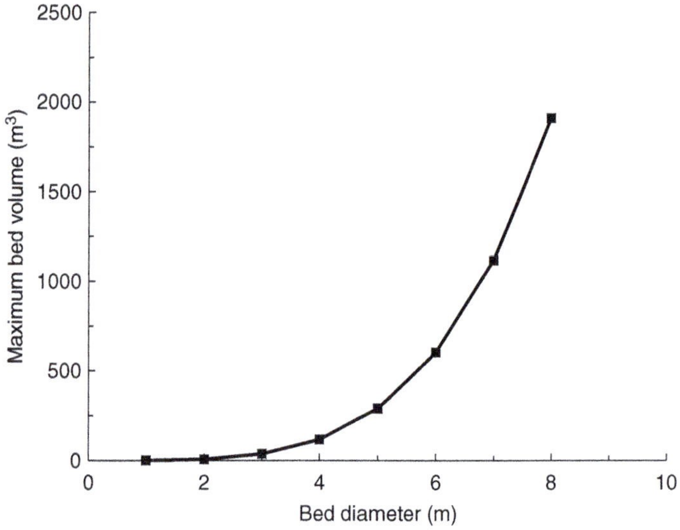

Figure 3.27 Graph illustrating the maximum bed volume for a 70 kPa pressure drop for a series of bed diameters.

For packed beds, the pressure drop across the bed increases with gas velocity until the incipient velocity is reached. After this point, the pressure drop remains approximately constant. The incipient velocity is calculated to be 0.0047138 m/s, and is exceeded for all iterations of a diameter less than 19 m. The fluid velocity is calculated by dividing the volumetric flow rate of air from the gas distributor by the cross-sectional area of the bed. The maximum pressure drop is set to 70 kPa, and Figure 3.27 illustrates the maximum bed height for this pressure drop for various diameters.

In order to compensate for a reduced regenerator temperature, a high catalyst loading is required. The simulation in HYSYS recommends a catalyst loading of 170 100 in order to reduce the catalyst coke to 0.2%. Using the assumed bulk density of 500 kg/m^3, the required volume of the reactor bed is calculated to be 340.2 m^3. Figure 3.31 indicate that a bed diameter of 6.6 m with the minimum entrainment rate would be ideal for the system. Even if this is only a local minimum, higher diameters than this will result in excessive material costs, and mass transfer phenomena issues due to the fact that mixing becomes an increasingly significant issue as the diameter of the regenerator increases. The result is mass transfer becoming a limiting factor and the regenerator bed functioning improperly. A smaller diameter vessel would facilitate the use of a basic aeration device at the bottom of the regenerator and would reduce the need for a complex catalyst injection mechanical design, allowing simple ski inlets to be used with the catalyst standpipe from the riser stripping unit. The first diameter that allows the required volume to be attained is 5.1 m. This is not selected as the actual bed diameter, however, as this would require an undesirable bed height of 17 m, and contingency must be made in the event that bulk density is slightly higher than the unit is designed for. Hence, a value of 6.6 m is chosen as the bed diameter, with a height of 9.94 m selected, as the diameter of 6.6 m minimises the objective function of entrainment rate versus diameter and thus optimises the system. The pressure drop across the bed for these dimensions is 21.75 kPa, a rather low-pressure drop for a fluidized bed operating above atmospheric conditions, which is favourable as the flow rate of fluidizing medium is reduced.

At this point, the manual calculations can be compared to the simulation results from Aspen HYSYS. As the mass balance was performed in HYSYS, certain dimensions are available with regards to the sizing of the regenerator. The simulation yielded a recommended bed diameter of 7.6 m, and a bed height of 7.5 m. These results are not used to design the regenerator, as having a larger diameter makes aeration and adequate mass transfer of oxygen to catalyst-coked sites more difficult. The optimal value chosen is that which provides the minimum entrainment. In either case, the manual results obtained are similar to that of the simulation, reaffirming that calculations regarding entrainment are not necessarily incorrect, as HYSYS may be optimizing for a lower pressure drop. Rather, a more reliable indication of whether or not the entrainment calculations are correct is a check on the superficial velocity of entrained particles by comparing the values obtained manually and via simulation. The HYSYS simulation calculates a velocity of 0.04388 m/s, whereas the manual calculation provides a superficial velocity of 0.04838 m/s. The results are sufficiently similar, and the difference is most likely due to the assumption of spherical particles and use of Stokes law rather than exact data. At this point, it can be seen that the designed regenerator is very likely to be an accurate representation of a real regenerator as manual and simulated calculations performed in different manners yield similar results.

3.5.7.1.4 Regenerator Auxiliary Equipment The resultant entrainment rate is 0.46806 kg/hr. Using a single cyclone reduces the catalyst loss to flue to 2.34×10^{-5} kg/hr. Being significantly lower than the catalyst loading or the circulation rate, it is unnecessary to add further cyclones. The cyclone is sized according to the FCC Handbook's recommended inlet and outlet velocities for certain cyclone layouts, with the inner dimensions being sized according to high-efficiency cyclone correlations detailed in Utikar et al. (2010). The cyclone calculations are performed in the FCC spreadsheet, and dimensions are illustrated in the main FCC diagram in Figure 3.33. As one cyclone is used with an outer diameter of 1.1 m, no enlargement of the upper compartment of the regenerator above the bed is needed. The cyclone inlet is located at a height of 14 m, whereas the HYSYS simulation yielded a recommended height of 15 m, indicating that cyclones have likely been set up correctly. The vapour phase from the cyclone is directly removed from the regenerator as the flue gas and undergoes further treatment to ensure environmental regulations are met with regards to carbon monoxide and other emissions. Pressure is controlled in the vessel via the accumulation of gases that are not removed by the cyclone to yield a vessel pressure of 241 kPa. An outlet port is constructed in order to directly purge any accumulation to prevent damage to the vessel as a safety measure. This port is only opened when the control actuator indicates that the design pressure has been exceeded.

Catalyst is circulated at a rate of 241902.3 kg/hr, and is added and removed from the regenerator via standpipes. These standpipes operate through use of slide valves which along with aeration every 2 m to generate the required pressure drop. This is necessary as the pressure of the Riser terminator is lower than that of the regenerator, meaning that catalyst flow will not occur naturally. Velocity in the pipe is limited to avoid excessive abrasion due to the largely solid particle flow. In order to simplify calculations, the bulk density of 500 kg/m^3 is assumed constant from the catalyst bed to the stripper loadings throughout the standpipe. While this is not an exact representation of the density inside the standpipe, the pressure drop is limited to 70 kPa, and hence the actual bulk density of catalyst particles will not greatly vary from this approximation. For a catalyst flux of $835.5\frac{\text{kg}}{\text{m}^2\cdot\text{s}}$ and a velocity of 1.67 m/s, a standpipe of 0.32 m internal diameter is selected. This correlates with the recommended values in the FCC Handbook for standpipe design. The standpipe of catalyst flowing from the riser stripper to the regenerator is 69 kPa, the difference between the two units. This correlates with the simulation's recommended pressure drop and with the desired range for pressure drop over a slide valve in the FCC Handbook. The pressure drop across the standpipe leading from the regenerator to the bottom of the riser is obtained from Aspen HYSYS as 98.78 kPa, which exceeds the recommended maximum. Instead, a pressure drop of 70 kPa is set. Due to the high flow rates of solid particles, erosion corrosion is a major issue, causing metal to be worn down quickly, potentially causing a temporary halt to operations. In order to avoid this, highly erosion-resistant refractory linings are added. These are detailed later in the report. A final consideration at this level is physical layout related, the standpipes need to be at a minimum angle of 45° to permit effective operation.

The air supplied to the regenerator cannot simply be fed through a single port from the bottom of the vessel due to the large diameter of the bed (Sadeghbeigi 2012). Insufficient oxygen diffuses through catalyst particles to coked regions and thus limits the rate at which catalyst regenerates and overall results in the catalyst particles having a higher coke content than designed for. Thus a device is constructed that passes air through the bottom of the regenerator and is then distributed among four main branched arms at 90° that contain 16 nozzles that eject the air at 30 m/s. The physical design with dimensions for the air distributor is illustrated in Figure 3.27. 5270 m^3/hr of air at atmospheric conditions is required for the design, and calculations of dimensions were based off of this value. Ideally, a quarter of this flow rate is sent into each major branch, which then ejects the steam through small 15.6 mm diameter nozzles 150 mm in length. Unlike other equipment in the process, the refractory applied is on the exterior. This will further be detailed in the section discussing the thickness and mechanical factors of each item of equipment (Figure 3.28).

3.5.7.2 Riser Reactor

The riser design follows a similar design as the UOP FCC reactor, an example of which is located in Sadeghbeigi (2012), in which a stripping unit is designed around the annular region of the riser. As the riser outer diameter is 0.972 m, the catalyst stripper can be designed without being heavier than the riser can structurally support. This is done to save material costs and the need for additional external support. The riser reactor layout is completely illustrated with all significant dimensions in Figure 3.31.

3.5.7.2.1 Riser The riser was specified during the simulation to be 30 m tall, with a 0.8 m diameter. This is selected with regards to the velocity of fluid and the residence time in the riser. A target residence time of 3 seconds is desired, having low residence times will result in feedstock not cracking; whereas a residence time significantly exceeding 3 s will result in over-cracking (Sadeghbeigi 2012), where desired products are broken down into lighter, less desired products. The ultimately desired product is naphthas as the purpose of this plant is to produce gasoline feedstock. The riser product flows

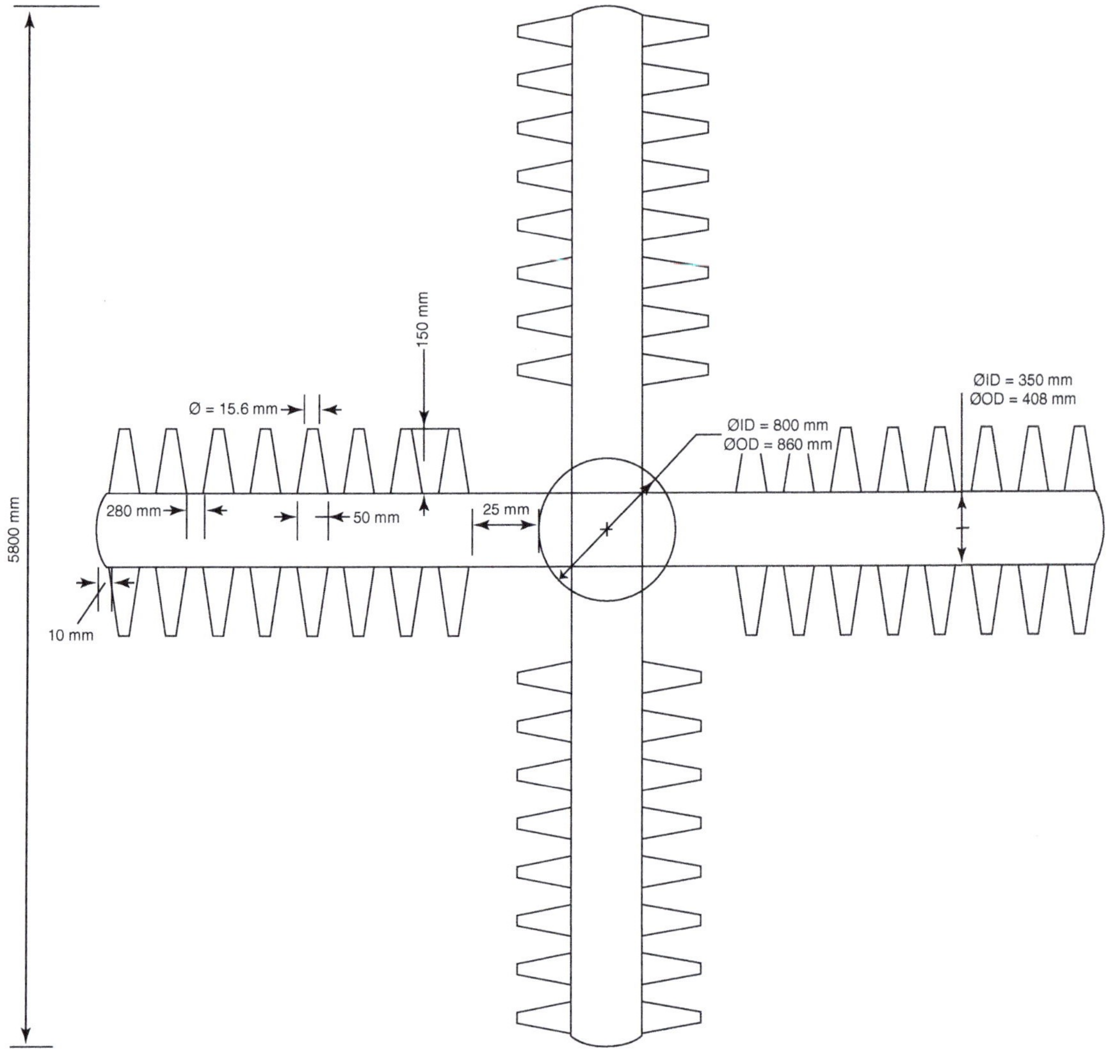

Figure 3.28 Top-down view of design drawing of regenerator air distributor.

with a volumetric flow rate of 4.7158 m^3/s. Using the area and height to calculate the residence time off of this flow rate, a residence time of 3.198 s. This is not exact, as the volumetric flow rate will change in the riser as reaction occurs. The HYSYS simulation gives the riser residence time as a slightly higher value of 3.362 s, close to the target without creating an undesirably high catalyst flux through the riser, causing more abrasion and erosion on the riser. This allows the riser to have a longer lifespan while allowing a thinner refractory and reduced downtimes for maintenance. Similarly, the velocity of fluid in the riser is calculated near a minimal allowed value of 9.382 m/s. As the riser is the primary support for the reactor side of the FCC, preserving its operation is priority, and having a slightly lower selectivity of products is acceptable.

The feed for the riser uses a wye design for catalyst being fed into the riser at the bottom of the vessel, where the catalyst is entrained by blaster steam at 200 °C, 1000 kPa with a flow rate of 1220.4 m^3/hr. Located 4 m above the catalyst entry into the riser, the feed oil is injected into the riser along with steam, weighting 3% of the feedstock mass flow rate. These targets are selected based off of the guidelines given for feed nozzle design in the FCC Handbook (Sadeghbeigi 2012). The nozzle is sized to permit an injection rate of 60 m/s into the entrained catalyst and blaster steam mixture and occurs under a pressure drop of 400 kPa. As the flow rate of oil is low due to this process being a modular refinery, only one oil feed into the riser is needed. The exact design of the riser feedstock nozzle can be examined in the diagram in Figure 3.31.

One area in which the riser designed does significantly deviate from literature is the catalyst-to-oil ratio. This ratio is very high, at 31.38. This result was determined by Aspen HYSYS simulation, and accounts for the lower-than-desired regenerator temperature which is discussed in the Regenerator section preface. Because the regenerator temperature is low, more coke is present on the catalyst, lowering activity and thus more catalyst is needed to ensure a favourable conversion in the reactor.

Another reason for the high catalyst circulation rate is that no recycle is present in this system, hence cycle oils need to be removed with a higher conversion (Sadeghbeigi 2012).

3.5.7.2.2 Terminator The riser product is then passed through two cyclones in series with each other, sized similarly to the regenerator cyclone via the FCC Handbook velocity recommendations, and the swift correlations for a high-efficiency cyclone. For a 214902.3 kg/hr, catalyst circulation rate, and assuming a single cyclone operating with a 99.995% efficiency, a single cyclone results in a 10.745 kg/hr catalyst loss. This is a substantial amount, vastly more than the regenerator loss and would result in a significantly high operating cost. Adding a secondary cyclone results in a $5.37 \cdot 10^{-4}$ kg/hr, a vastly more favourable result. The actual cyclone operation becomes more complex, however, as the riser product is fed directly into the primary cyclone. The issue with this is that the riser exit velocity is 9.382 m/s, significantly less than the recommended minimum of 18 m/s for a primary cyclone (Sadeghbeigi 2012). In order to eliminate this issue, a contraction is added to the riser pipe before entry into the cyclone. The contraction internal diameter is 0.556 m to yield an entrance velocity into the primary cyclone of 19.45 m/s, which meets the FCC Handbooks guidelines on cyclone ranges. The physical dimensions of the cyclones are given in the layout in Figure 3.39. The product gas having passed through both cyclones is extracted from the top of the riser terminator, and fed directly into a fractionation column which separates the product gas into various fractions.

3.5.7.2.3 Stripper After the catalyst particles are separated via the cyclones, these particles are heavily coked and cannot be directly returned to the regenerator. These catalyst particles are first stripped with steam and endure a long residence time in the riser terminator to promote further cracking of volatile components bound to the catalyst which leave the vessel via a pressure-controlling port and are fed into the same separation column as the product gases of the riser. The riser terminator is designed for cost reasons to house the cyclones internally. The cyclones are rather large, thus requiring a terminator with a 5 m internal diameter and a 6 m height. A conical section 0.3 m high at an angle of 30° to the horizontal connects the riser terminator to the stripper. Diplegs from the cyclones transport catalyst to the surface of the stripper bed. The simulation in HYSYS determines that a 3765 kg catalyst loading is required, however, this is not the only consideration when designing the stripper. A minimum residence time of 60 s for the catalyst particles is required, and mass transfer between the steam and particles needs to be accounted for. As the stripping process is located inside the annular region between the risers outside diameter and the internal diameter of the stripper, the area is given by $\frac{\pi}{4}(d_{strip_i} - d_{riser_o})$. Setting a stripper internal diameter as 1.55 m, with the outer diameter of the riser being 0.972 m allows for a catalyst flux of $58.69\frac{\text{kg}}{\text{m}^2\cdot\text{s}}$. This falls within the required range of fluxes the FCC Handbook recommends. During the mass balance a stripping steam flow rate of 854.8 kg/hr was calculated, at 200 °C, 1000 kPa. This is feasible as a ratio of $0.0035\frac{\text{kg Steam}}{\text{kg Catalyst}}$ is obtained, within the recommended range (Sadeghbeigi 2012), when divided by the catalyst circulation rate.

Having determined the area of the annulus region, the required stripper height can be determined. Using the 3765 kg catalyst loading, the minimum height of the stripper is calculated as 6.56 m. However, this does not account for the required residence time. The minimum height of the stripper for a 60s residence time is as follows:

$$H_{\text{min}} = \frac{(\text{Catalyst Circulation Rate})\tau}{\rho A} = 7\text{ m} \tag{3.51}$$

While 60s is the minimum required residence time to crack the majority of compounds still bound to the catalyst, ideally a higher residence time is desired to ensure a minimal flue gas production and hence a loss of product via the regenerator discard. In addition to this, adequate mixing of the catalyst particles and the stripping steam is required. In order to achieve this, six KFBE (Köln-Frechen-Benzelrather Eisenbahn) annular structural packings are loaded into the stripper. These are specified to be 10 mm thick and spaced at 1.2 m. These packings are rather expensive, but offer a superior mass transfer and lifespan when compared to disk and doughnut annular fittings, and are also less prone to fouling or erosion than tray baffles. The overall residence time will be higher than 60 s, and thus the minimal height of the stripper being 7 m is designed.

The pressure drop is monitored to ensure a target of 20 kPa is not heavily exceeded, As the stripper operates with loadings in a packed bed, the Ergun equation is used to calculate pressure drop; assuming spherical particles packed with a bulk density of 500 kg/m^3 (Figure 3.29).

3.5.8 Material Selection and Refractories

The most significant form of material damage in the unit is erosion corrosion, however, other factors are present that aid in selecting the ideal metal to be used to construct various units.

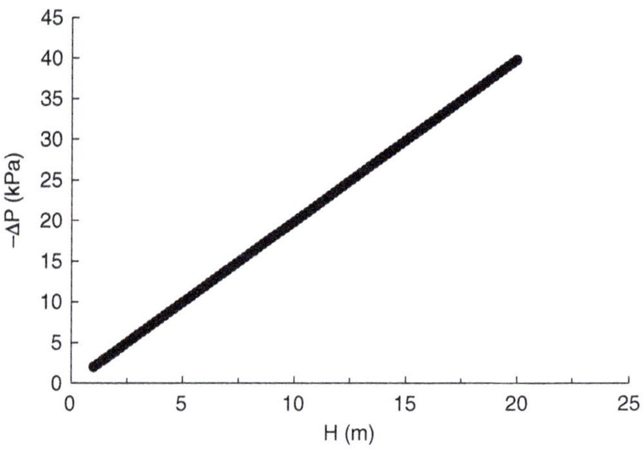

Figure 3.29 Graph of pressure drop versus stripper height for the packed catalyst loadings.

3.5.8.1.1 Corrosion Factors As mentioned, this is the primary concern. A very high catalyst circulation rate is present and is almost always entrained in a gaseous medium flowing at high velocities. This wears down metals at a significantly higher rate than any other form of corrosion considered for the unit, and cannot be overcome simply through good material selection. Rather, refractory linings are utilised with these refractories being chosen for resistance to erosion as the primary selection factor. The additional benefit of using refractory linings is it is possible to reduce the temperature the wall of a vessel or piece of equipment is exposed to. The thickness of the refractory is determined based on the erosion resistance required, and if the equipment is being designed for a hot wall or cold wall layout. In a cold wall layout, the refractory causes the metal support to be subjected to lower temperatures. Table 6.2 provides a list of equipment and refractories used, the general recommendations of which often used the recommendations and general classifications used in the FCC Handbook (Sadeghbeigi 2012). For thick walls of standard shape, vee anchors are applied. In cyclones hex mesh is used for strength with thin layers and odd shapes, exclusively used with the cyclones (Table 3.16).

Table 3.16 Refractory Details for each item of equipment, reference for layout format to (Sadeghbeigi 2012).

Equipment	Refractory approach	Refractory class	Exact refractory	Refractory thickness (mm)	Bulk density (kg/m^3)	Anchor used
Riser	Cold wall	Extreme erosion resistant castable	ATCHEM 85 VC castable	125	3043.4	Vee
Riser terminator	Cold wall	Medium weight insulating castable	Kaolite 2500-HS castable	75	1320	Vee
Riser stripper	Cold wall	Medium weight insulating castable	Kaolite 2500-HS castable	75	1320	Vee
Riser primary cyclone	Hot wall	Extreme erosion resistant castable	ATCHEM 85 VC castable	25	3043.4	Hex mesh
Riser primary cyclone	Hot wall	Extreme erosion resistant castable	ATCHEM 85 VC castable	15	3043.4	Hex mesh
Regenerator cyclone	Double sided refractory, hot wall	Extreme erosion resistant castable	ATCHEM 85 VC castable	2x25	3043.4	Hex mesh
Regenerator	Cold wall	Moderate density & erosion castable	Rescocast 15 castable	100	1968	Vee
Regenerator air distributor	Hot wall (Exterior refractory)	Extreme erosion resistant castable	ATCHEM 85 VC castable	25	3043.4	Hex cells
Catalyst standpipes	Cold wall	Extreme erosion resistant castable	ATCHEM 85 VC castable	125	3043.4	Hex cells

3.5.8.1.2 Graphitisation Above 455 °C, decarburisation occurs as carbon in the steel interacts to form graphite. This significantly weakens the material structure and can lead to failure (Perry and Green, 2008). All process units and fittings operate at a temperature above this temperature, hence high carbon steels are undesirable for hot wall equipment.

3.5.8.1.3 Intergranular Corrosion Austenitic stainless steels are vulnerable to intergranular corrosion between 425 and 815 °C due to chromium carbides being insoluble in this range and precipitating, causing sensitisation. The riser system as well as the regenerator operate in this range, hence rendering hot wall stainless steel undesirable as a construction material.

3.5.8.1.4 Hydrogen Induced Corrosion Unlike the distillation unit, lower concentrations of sulphides are present as much of these are removed in the vacuum residue and other fractions. However, hydrogen sulphide is present to some extent, and steam is heated to high temperatures and can contribute to hydrogen induced corrosion. This form of corrosion attacks alloys of titanium, zirconium, niobium or vanadium heavily. Hence, alloys of these components are avoided for both hot and cold wall designs, as alternate substitutes are present and more suitable to the system (Perry and Green 2008).

3.5.8.1.5 Stress Corrosion Cracking This occurs due to H_2S within the temperature range of 60 to 200 °C. This is not an issue with any of the process units in the FCC.

3.5.8.1.6 Crevice and Pitting Corrosion Few chloride ions are present at this stage of the process, and amounts of chromium, nickel and molybdenum serve to eliminate any significant corrosions of this type (Popov 2015).

3.5.8.1.7 Overall Material Selection Aided by Perry's Chemical Engineering Handbook (Perry and Green 2008) and the mechanical design sections indicating the required strength of the metal, based off of the corrosion factors the material of construction can be determined. As refractories are used for each piece of equipment, it permits metals to be used that would otherwise be unsuitable due to thermal conditions. However, this approach does rely on the refractories being adequately maintained. While the materials are designed mechanically to resist the worst-case scenario of temperature and erosive components, these factors will rapidly disrupt the structural integrity of the metal and require the refractory to be re-applied again.

After examining the mentioned criteria, two general classes of materials are chosen: carbon steel and low alloy steel. Low alloy steels are particularly applicable to cold wall items, and offer an increased design stress, at greater cost. Two general metals are selected with Perry's Chemical Engineering Handbook as a reference. ASTM 53B, a carbon manganese steel, while initially selected as the material to construct the riser terminator and stripper, immediately failed the required design stress tests and was ruled out as a suitable material. ASTM A387, consisting of 2.25% chromium and 1% molybdenum is selected as a suitable material for the riser terminator, stripper and catalyst standpipes. This, however, possesses inadequate design stress for use with the riser or regenerator without having an unnecessarily high thickness, being economically inefficient. Hence, a stronger alloy is used, A335 consists of 9% chromium and 1% molybdenum, and possesses higher tensile, yield and design stresses. This is only used with cold wall designs due to the increased susceptibility to intergranular corrosion (Perry and Green 2008).

3.5.8.2 Mechanical Stress and Thickness Calculations

3.5.8.2.1 Flanges Due to the high solids circulation rate, all input and output flanges on any unit are forged ring flanges (Coulson et al. 2005).

3.5.8.2.2 Design Stress Having discussed the design stress, it is necessary to discuss the method of obtaining the value of this factor. While design stress values are available for broad categories such as carbon steels or low alloy steels, these are only sufficiently accurate for preliminary, not detailed design. Design stresses are difficult to obtain without performing actual tests, due to the diverse region of operating conditions. There are experimental correlations available, but are not always available for all metals under all conditions. The Larson-Miller correlation is used in this report to obtain the design stress under exact temperatures for certain metals. These calculations are performed in the detailed mechanical calculations for each material calculation in Excel Spreadsheet and are explicitly shown in the regenerator calculations. The Larson-Miller equation used is as follows (Prager et al. 2014):

$$\text{Parameter} = \frac{(T + 273)(C_{LM} + \log(t_r))}{1000} \tag{3.52}$$

In this equation, C_{LM} is a constant that varies depending on the metal used. In all cases of application, this constant is supplied by the reference material. t_r Represents the time in hours for which the plant operates. Assuming operation occurs for 10 years with minimal downtime, 87200 hr is set as the total operation time. Once the parameter is obtained, it is graphically used to determine the design stress for the particular metal. The references consulted to obtain the graphs required are Canale et al. (2008) and Prager et al. (2014). The Prager resource uses a graph-based off of imperial units, and the needed conversions are performed.

3.5.8.2.3 Thickness of Equipment Metal Longitudinal and circumferential stress calculations can be easily performed for varying thicknesses of the vessel, but these are not suitable bases for the design stress. The design stress takes into account the weight and wind factors that simple stress calculations do not. The first check is whether or not the proposed material exceeds the longitudinal and circumferential stresses for a reasonable thickness. The actual longitudinal stress will increase, often drastically as with the riser, for tall, heavy columns.

The way calculations for thickness are performed in a similar manner to how Coulson and Richardson vol. 6 (Sinnot 2005) approach the topic. An arbitrary thickness – 1 mm is chosen and iterations of 1 mm increments are applied. A rigorous calculation method is then manually performed with the end goal of calculating the Overall Longitudinal and tensile stresses. Describing in detail the exact calculations that are performed in the stress calculations Excel files would not be an informative exercise, as these calculations are covered in literature, notably Coulson et al. (2005), from which the format is applied. Rather, the general scope of the calculations is detailed here. Taking the example of the regenerator, using the thickness for the iteration, the weight of the metal casing without the refractory is calculated. This figure is then summed with the weight of catalyst loadings and the refractory weight. A total of 5% weight is added to this figure in order to account for the weight attributed to fittings such as nozzles and flanges. This obtained figure is used as the total vessel weight for successive calculations. Maximum wind speeds of 50 m/s are accounted for, a figure that is reasonable even when experiencing coastal gale force winds. This is done as the plant would ideally be situated in the Richard's Bay area where crude is imported. After the wind loading is calculated, the bending moment is obtained and the overall bending stress is obtained. Finally, the actual downwind longitudinal stress is obtained by subtracting the dead weight stress and the bending stress from the originally calculated longitudinal stress. The upwind longitudinal stress is obtained by subtracting the dead weight stress from the original longitudinal stress, while the bending moment is added to this figure. Performing multiple iterations, the first thickness to operate with stresses below the design stress is chosen as the design thickness. A 2 mm corrosion allowance is added onto this figure to ensure an adequate operating lifespan. This process is applied to the riser, terminator and stripper and regenerator, and the results Table 3.17 are also shown in Figure 3.31.

3.5.8.2.4 Riser Supports Two types of supports were investigated: saddle and skirt supports. Saddle supports are generally used for smaller or horizontal vessels. As the riser is 30 m high, this type of support is totally unfeasible. Skirt supports make use of a bolt ring and base anchor design, and the column rests on the bolts rather than the concrete directly. This allows for adjustments to the bolts to compensate for a gradient, allowing the column to be balanced perfectly, resting completely perpendicular to the horizontal.

An iterative approach similar to the method discussed in Coulson et al. (2005) is used to size the bolts. With a skirt diameter of 986 mm and a ring diameter of 1180 mm calculated through determining the base ring diameter and compressive load. A 50 mm overshoot of the skirt by the base plate is assumed in order to fit an existing bolt model for more accurate sizing dimensions. The number of bolts used is iterated to yield a target bolt diameter of roughly 30 mm, sufficiently higher than the lower limit of 25 mm while leaving a corrosion allowance. A total of 16 bolts are the closest whole number of units, yielding a diameter of 31.66 mm. Bolts are spaced at 193 mm.

Table 3.17 Sample of data from mechanical stress calculations, and the design thickness for major equipment units.

Equipment	Design stress (N/mm^2)	Maximum longitudinal stress (N/mm^2)	Vessel thickness (mm)	Refractory ID (mm)	Steel ID (mm)	Steel OD (mm)
Riser	100.07	72	11	800	950	972
Riser terminator	69.3	33.05	6	5000	5250	5262
Catalyst stripper	69.3	33.05	6	1550	1700	1712
Regenerator	70	49.12	8	6600	6800	6816

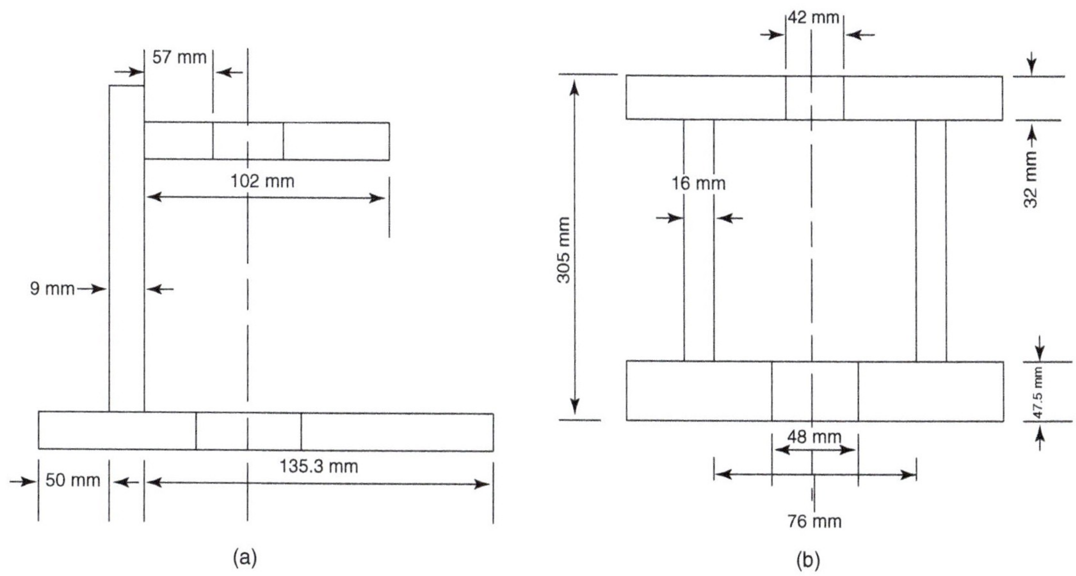

Figure 3.30 (a) Side view of skirt support bolt flange. (b) Front view of skirt support bolt flange.

The dimensions of the flange ring are then determined by using the model given by Sinnot (2005) to acquire an existing bolt size. The size recommended for the nearest bolt diameter is M36. This is chosen as no bolt exactly corresponds to 787 mm^2 bolt area and M36 has a bolt area of 817 mm^2. Overall, the design for the bolt flange ring and base is illustrated in Figure 3.30a,b.

Credits once again to Sinnot (2005) for this common bolt sizing table and dimensions.

Vessels were closed via ellipsoid heads. The reasoning for using this simple form of head is because low pressures are present. A joint factor of 1 is assumed. The thicknesses of these heads are illustrated in Figure 3.31 and are calculated for each major process vessel in an FCC calculations spreadsheet.

3.5.8.3 Distillation Column Detailed Design

Overall, a 40-plate sieve plate column is designed. In designing the column factors such as flooding, weeping and channelling are used to size the diameter of the column and the dimensions of the plates. These calculations were performed by iterating based off of the profiles of temperatures and velocities on the plates in the column.

3.5.8.3.1 Simulation Due to the sheer number of components present in the feed oil, using assumptions and modelling the column in Microsoft Excel would detract from the accuracy and reliability of data provided. Simulating in Aspen HYSYS allows for each component to be accounted for, rather than simply using lump parameters for temperatures. Particularly for this process, this is essential as the distillation column is designed to optimize the Watson K factor for the HGO stream being fed to the FCC. HYSYS accounts for groups of aromatic and paraffinic calculations in order to output an accurate K factor. Hence, the actual simulation with regards to the mass and energy balance, as well as varying properties along plates is performed in Aspen HYSYS. More specific details, however, regarding the sizing of the distillation column, thickness and support calculations and the exact sizing of the plates are manually performed using Microsoft Excel. This is done to ensure the designed unit is suitable for the system, and factors such as cost, corrosion and lifespan can be taken into account. The simulation is modelled in Aspen HYSYS using the Peng-Robinson equation of state. This property method applies to non-polar non-electrolytic fluids. The assumption of non-electrolytic nature is made as the feedstock undergoes a desalting process and any effect of residual ions in the oil on electrolytic nature is insignificant. The column was simulated using a refluxed absorber, with side strippers for the outlet streams. The side strippers each utilize kettle reboilers to heat up and return the more volatile components in the side stream to the column. Three side strippers are present for the Kerosene, LGO and HGO streams, and these were covered more extensively in the mass balance report.

The residue leaving the bottom of the column is not reboiled, as the heavier fractions and sulphides present in this are undesired, and the residue is directly sent to the vacuum column for separation into vacuum gas oils and residue. This necessitates an alternate means to heat the bottom of the column. The method chosen is the passing of superheated steam at 325 °C, 255.5 kPa through the bottom of the column, thus vaporizing light fractions from the feed plate. Although lighter

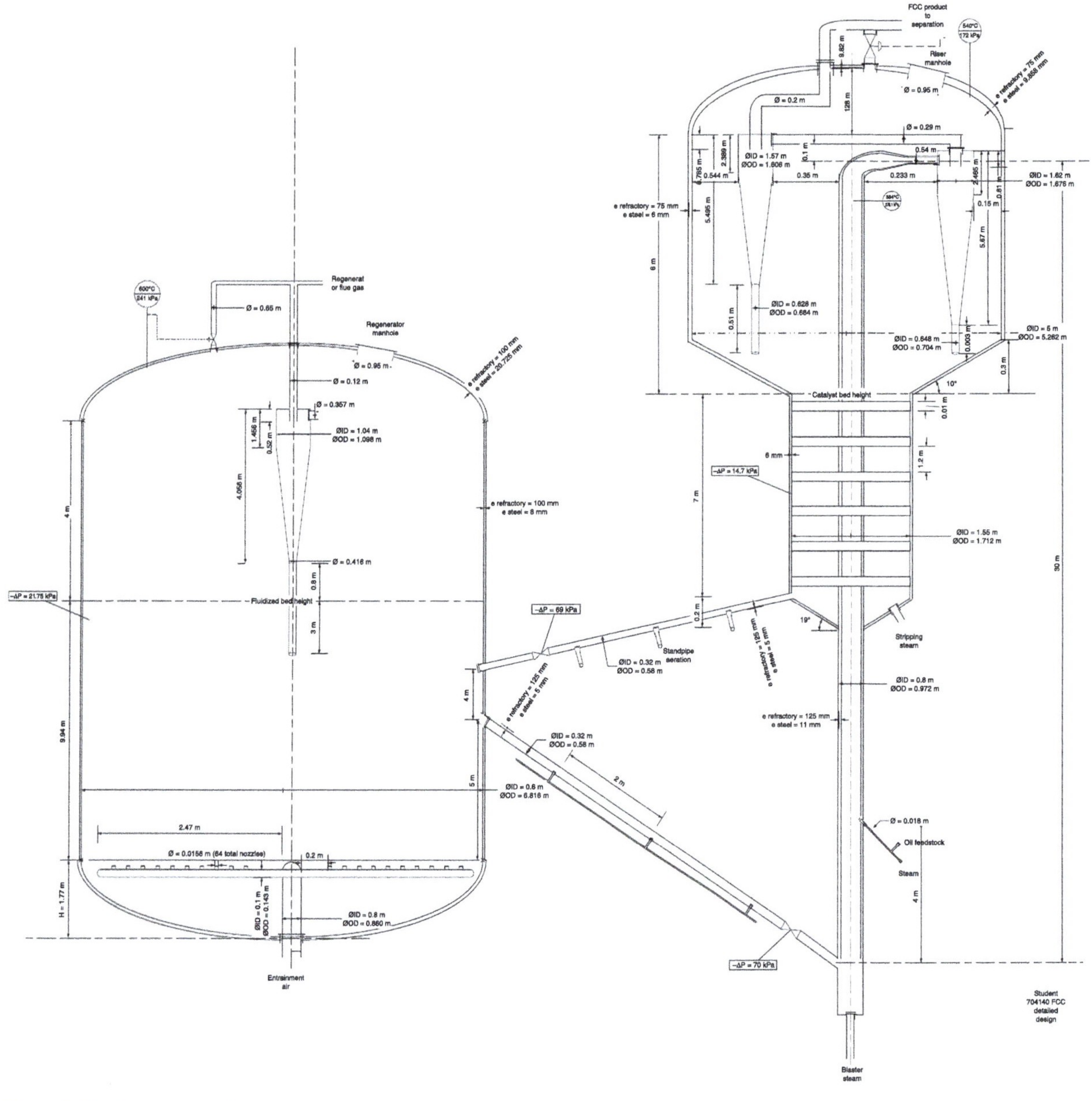

Figure 3.31 Detailed fluid catalytic cracker design.

fractions are lost to the residue, they are recovered in the vacuum column, thus the designed process feasible. The column is simulated using 40 plates with 0.5-m spacing, while the exact dimensions of the plates are covered in the section on plate calculations. Due to the steam being fed into the bottom of the column and the vast number of components present in the column, an error in the energy balance is expected. This is rectified by adding a reconciliatory energy stream to the bottom of the column during simulation and minimising the value. A final energy offset of 2 W is recorded, a sufficiently low value when considering the scope of the unit and heat transfer involved. Once, this layout was set up, sensitivity analysis was performed with the objective of minimizing the Watson K factor. Manipulated variables included reboiler boilup ratios, feed oil and steam temperature and the layout, flow rate and cooling amount for pumparounds. Due to the extensive amount of variables considered, at this level, an objective function cannot be obtained. Rather an iterative approach was tested, starting with the inlet conditions of the feed oil and steam. Due to the costs associated, a temperature slightly above the minimum required temperature is set, which also reduces the number of variables. 405 °C, 225.5 kPa is set for the feedstock. At this point now, only the pumparounds and the boilup ratios are considered for optimizing the *K* factor. Figure 3.32 is an example of how one of the side stripping units is adjusted.

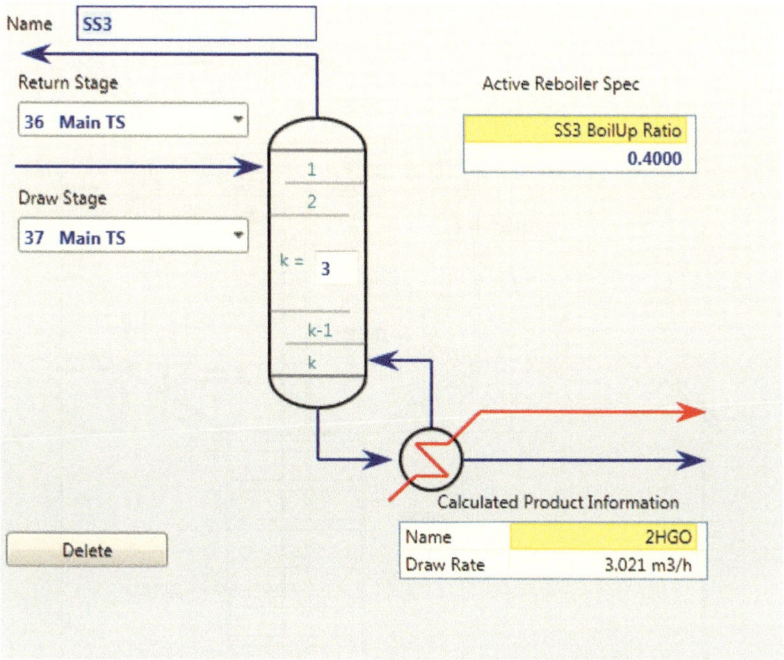

Figure 3.32 HYSYS simulation control of heavy gas oil side stripper.

As long, straight chain paraffinic components increase the K factor, the boilup ratios of the HGO and LGO side strippers are set at 0.4, whereas the Kerosene reboiler has a boilup ratio of 0.75 to optimize the amount of naphthas recovered from the top of the column. These values were set in conjunction with iterating different configurations of the pumparounds. A rather unorthodox arrangement for the pumparounds was finalised, with liquid draws from plates 1 and 23, with the liquid phases being cooled to 80 °C for both draws. Both of these pumparounds are returned to plate 5. The reasoning behind this is this setup vastly improves mixing in the column, and additionally serves to transport aromatics and volatile components away from the gas oil side draws to compensate for the reduced boilup ratio, while additionally reducing the naphthas being drawn from the kerosene stream. Kerosene is not a primary product for this refinery and compromising on its throughput is acceptable (Figure 3.33). The complete stream summary regarding the feed and draw plates can be seen in Figure 3.32 for the detailed design of the distillation column.

3.5.8.3.2 Plate Selection Bubble cap trays offer the greater flexibility with regards to the liquid-to-vapour flow rate ratios at a point in the column than sieve trays. This is particularly applicable to a modular refinery as the feedstock or throughput is likely to vary as time passes. Cheaper and lighter oils could be obtained for long-term operation and hence a point of concern with regards to plate choice is being able to operate under variable conditions. With sieve plates, should a situation arise in which the vapour rate is significantly lower than the liquid rate, weeping is liable to occur which will severely impact on the plant's operation. While valve plates are cheaper than bubble cap trays and do indeed allow for more flexible operations than sieve trays (Coulson et al. 2005), the fractionating column is at the heart of this process and it is imperative particularly before modules are added to the plant that the required fractions are obtained at reasonable flow rates.

In order to decide on which plate is to be used, the cheapest option is first tested. The calculations, performed in the distillation column Excel file calculations, utilized many graphical approaches that are dependent on the type of plate used. Using sieve plates, the results are promising, with all the major checks being performed such as residence time and pressure drop being satisfied. These checks are further elaborated on in the section detailing the method of calculation. Using a turn-down ratio of 2:1 (Perry and Green 2008), entrainment and weeping do not occur when using a sieve plate. In addition to sieve plates being the least expensive type of plate, the pressure drop is also the lowest, allowing for cheaper equipment and working costs. A final consideration is corrosion. As discussed, the environment inside the column is not severely corrosive. Plates can be constructed from type 304 stainless steel to save costs, and possess a sufficiently long lifespan.

Hence it can be seen that sieve plates are an adequate choice in the column, and as plates cost significantly less than other major equipment needs, these could be upgraded to bubble cap trays in a worst-case scenario of having to process an oil of vastly different composition.

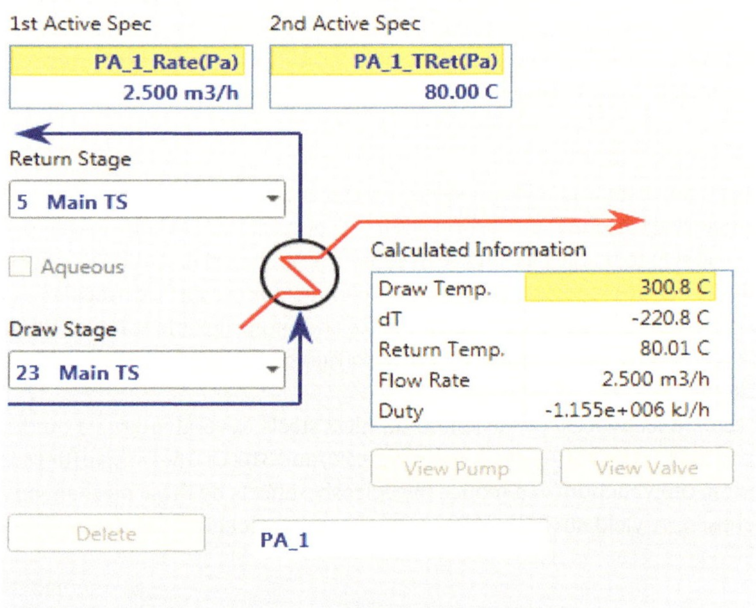

Figure 3.33 HYSYS simulation, pumparound specifications.

3.5.8.3.3 Operating Temperature and Pressure The Feed temperature is constrained to be less than 415 °C due to the presence of cracking in the column above this point. Using this condition optimisation was performed in attempt to lower the overall duties on the reboilers in the side draw stripping units. A final value of 405 °C is chosen. The bottom of the distillation column is at a pressure of 225.5 kPa and the top at 135.08 kPa.

3.5.8.4 Material Selection

Above 455 °C, carbon steels will experience graphitisation wherein carbon molecules inside a metal bond to form graphite molecules (Perry and Green 2008), reducing the structural integrity of the metal. This is, however, not a concern in the system as the maximum temperatures dealt with in the column are below 415 °C.

3.5.8.4.1 Inter-Granular Corrosion Austenitic stainless steels are susceptible to intergranular attack from 425 to 815 °C (Popov 2015) due to chromium carbides being insoluble in this range and precipitating. Fortunately, with temperatures being restricted to a maximum of 415 °C, this is not a concern for the unit and hence stainless steels are not ruled out for use.

3.5.8.4.2 Hydrogen Induced Corrosion This form of corrosion is of particular concern in material selection for the process as a whole. Although sulphides are treated before distillation via the desalting units, a small concentration of hydrogen sulphide, as well as sulphides present in other hydrocarbons do pass through the system. This is problematic as above 200 °C hydrogen induced corrosion becomes significant. The effects of this are trapping, blistering and precipitation of hydrogen that dissolves into the lattice. Furthermore, methane and hydride formation may occur, causing decarburisation and further structural weakening. Overall hydrogen induced corrosion lowers the tensile strength of the metal and increase brittleness as an example of the consequences. (Perry and Green 2008). This form of corrosion is due to sulphides and other trace poisons and affects metal alloys consisting of titanium, niobium, vanadium, tantalum and zirconium (Popov 2015). The effects of hydrogen induced corrosion can be reduced through the addition of inhibitors, but these could reduce product quality. Hence steels consisting of these metals are ruled out as a material of construction.

3.5.8.4.3 Stress Corrosion Cracking Ferric stainless steels undergo stress corrosion cracking (SCC) when exposed to H_2S, austenitic stainless steels experience SCC when exposed to heated chloride solutions, and parts of the column do fall within the 60–200 °C range in which this form of corrosion occurs. A means to increase the resistance of austenitic stainless steels to SCC is through a nickel content greater than 8%. Carbon steels, however, will not undergo SCC in this column.

3.5.8.4.4 Pitting and Crevice Corrosion Crevices and surface irregularities in the metal are likely to be present due to the height of the column, the addition of plates and the weirs that are present with a bubble-cap layout. In addition to this, chloride ions are present in the feed in trace quantities. Perry and Green (2008) stated that this environment is likely to propagate corrosion. Popov (2015) detailed that pitting and crevice corrosion is reduced on alloys of Cr, Ni, Mo. From this point, it is seen that the best choice of construction material would be austenitic stainless steel due to the presence of these metals.

3.5.8.4.5 Stress Factors Examining various carbon and austenitic stainless steels, a trend is observed that while the yield strengths of both are usually within the same range as each other, the tensile strengths of austenitic stainless steels are generally higher. Hence austenitic stainless steel is shown to be the best choice of construction material. Although carbon steel is significantly cheaper than austenitic stainless steel, due to the sheer size of the column, making surface irregularities likely, and the high temperatures involved, the stainless steel option provides a safer and more flexible solution while adhering to the original budget in which stainless steels were accounted for as the construction material. Finally, a choice has to be made regarding which exact type of austenitic stainless steel to use.

After identifying several commonly used and economically feasible austenitic stainless steels, several are ruled out for titanium, niobium and excessive chromium content. AISI types 304(Cr:18–20%,Ni:8–10.5%) and 316(Cr:16–18%,Ni:10–14%, Mo:2–3%)remain. As stated in Pitting and Corrosion, molybdenum can reduce the corrosive effects by these mechanisms. Thus AISI type 316 austenitic stainless steel is chosen. A yield strength of 248 MPa, a tensile strength of 565 MPa (Perry 25–36) 55% elongation and 149 HB hardness.

3.5.8.5 Column Diameter, Plate Spacing and Design

The method of calculating the dimensions of the plates to be used is influenced heavily by Perry's Chemical Engineering Handbook (2008) and Coulson et al. (2005). In order to obtain a functioning column, several conditions have to be met; flooding, weeping, entrainment, pressure drop, channelling, residence time, physical plate dimensions and number of holes. All manual calculations are performed in Microsoft Excel. Before discussing calculation details, a graph of liquid and vapour mass flow rates for each plate is given in Figure 3.34. The maximum and minimum flow rates are frequently used in calculations for proving or rejecting a criteria across all plates based on a maximum or minimum velocity. The column profile is rather odd, with an unexpected shape. This is, however, due to the arrangement of pumparounds and side strippers.

In this approach, there are five major variables that are altered to achieve a working plate distribution, namely, plate diameter, weir height, hole diameter, plate thickness and plate spacing. Although in the method used the diameter is obtained via empirical correlation, and the plate spacing is set based off of the diameter, the remaining variables are iterated until all required conditions are flagged as true. Firstly, the diameter of the column is determined. Fair's correlation for entrainment and flooding is graphically utilised to determine the diameter and fluid velocity while accounting for flooding. The column is designed to operate with a maximum of 90% flooding (Perry and Green 2008). As a trial, the velocity is set to 80% to the velocity obtained from Fair's graphical method so as to be able to perform further calculations, notably the net area which calculates the actual velocity of liquid on a plate. It is expected that the actual flooding percentage will go up when the final calculation results are obtained, but as long as the value is below 90% and at least 80%, the selected column diameter is acceptable. It is worth noting at this point that flooding calculations are based off of the maximum velocity in the column,

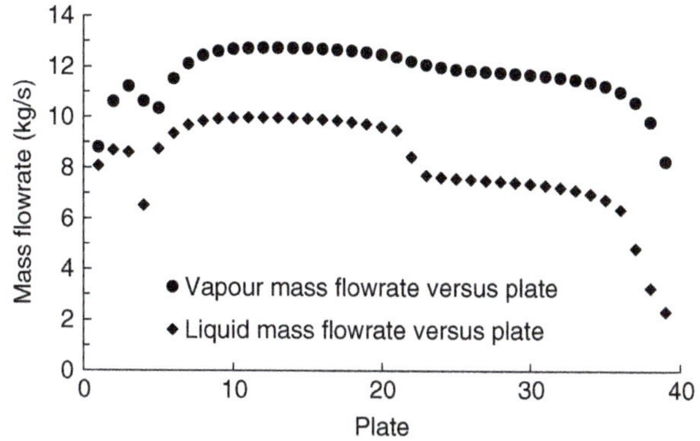

Figure 3.34 Column mass profiles for liquid and vapour flow.

Table 3.18 Initial iterations versus final values to satisfy all criteria for the sizing of plates.

Controlled variable	Initial condition from literature (mm)	Final condition after all tests successful (mm)
Plate diameter	–	1951.3
Weir height	50	40
Hole diameter	10	8
Plate thickness	5	5
Plate spacing	500	500

that is if the column does not experience flooding at the plate with the highest velocity of liquid, flooding will not occur elsewhere in the column. Table 3.18 illustrates the changes that occurred with the values of controlled variables as opposed to the values recommended by literature.

After obtaining a diameter, it is possible to decide the number of passes that the plate design should accommodate. Sinnot (2005) performs this by utilising a graphical method devised by Huang and Hodson, determining the flow arrangement via the diameter and flow rate of liquid in the column. Using the maximum liquid volumetric flow rate of 0.0150 m^3/s and a diameter of 1.95 m, it can be seen that the flow arrangement is a single pass one. Similarly, all liquid flow rates on each plate are checked on the Huang and Hodson graph (Sinnot 2005) to ensure that the single pass arrangement is valid throughout the column. Hence, single pass plates are used throughout the column.

Having access to the net diameter also permits flooding to be examined. Dividing the gas volumetric flow rate by the net area yields the actual velocity of vapour for the maximum liquid flow plate. This is performed at the maximum liquid flow rate as if flooding does not occur at this plate, flooding will not occur at any other plate. Dividing the actual vapour velocity at this plate by the vapour velocity obtained by use of Fair's flooding correlations yields an 87.4% flood. This is higher than the original trial of 80%, but is less than 90% and is thus an acceptable amount. This 87.4% flood is used together with the flooding parameters obtained when obtaining the vapour velocities for each plate in order to determine the fractional entrainment. Fair's entrainment correlation graph given in Perry and Green, (2008) indicates that for a sieve column at 87% flooding, a fractional entrainment of $\psi = 0.063$ is obtained. Sinnot (2005) stated that in general a fractional entrainment value of less than 0.1 should be targeted. Hence, it can be seen that the column diameter of 1.952 m holds and more in-depth calculations can be performed in order to determine plate dimensions (Figure 3.35).

As sieve trays are used, a maximum liquid turn-down ratio of 2:1 is present (Perry and Green, 2008). The turn-down ratio is used to determine the weir crest height, the graph of which for each plate is illustrated in Figure 3.36. Sinnott uses the Eduljee Weeping correlation graph in order to obtain a constant from the sum of the minimum weir crest with the weir height, which is then used to determine the minimum velocity of vapour passing through holes in a tray (Coulson et al. 2005)

$$u_{v,\min} = \frac{K - 0.9(25.4 - d_{\text{hole}})}{\sqrt{\rho_v}} \tag{3.53}$$

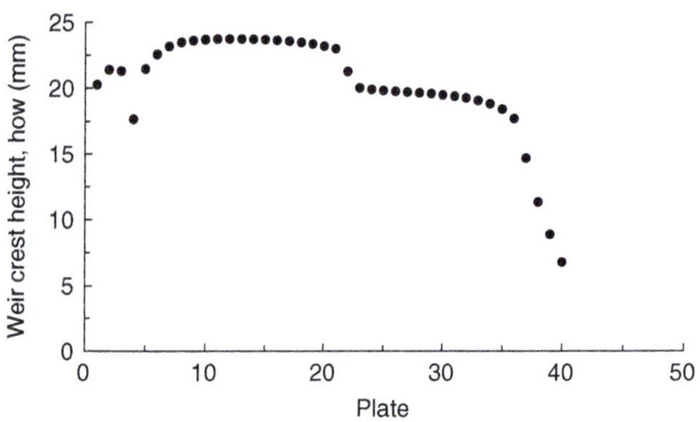

Figure 3.35 Weir crest height measured for each plate.

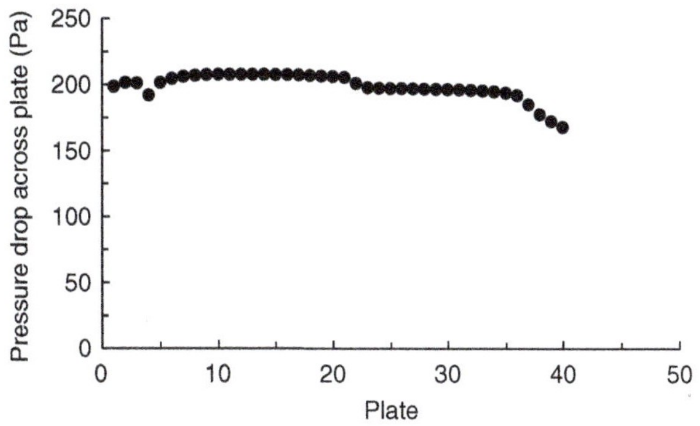

Figure 3.36 Pressure drop over individual plates calculated from total liquid head.

The velocity of vapour at any point in the column cannot be lower than this minimum allowable vapour velocity. Should this occur, weeping will ensue wherein liquid begins to fall through holes in the tray (Perry and Green, 2008), due to the rising vapours being unable to generate the force required to prevent this. After performing calculations, the theoretical minimum velocity allowed before weeping occurs is determined to be 5.638 m/s. Dividing the minimum vapour flow rate by the net area calculated by subtracting the downcomer area from the diameter of the column yields the actual minimum vapour velocity in the column, being 5.932 m/s. This passes the check of being greater than the minimum allowable velocity. This value is achieved through iterating primarily the tray hole diameter to increase the velocity through the holes of the plate.

After confirming that weeping does not occur with the proposed plate layout, the residence time of liquid backed up in the plate downcomers can be examined. Calculating the residence time requires the backup in the downcomer to be known. This in turn requires the pressure drop across the plate to be known. The method used to calculate these quantities follows the graphical method of obtaining a discharge coefficient detailed in Sinnot (2005). The calculations performed differ from the method prescribed in literature in that the pressure drop is calculated for each plate. This is performed so as to obtain the overall pressure drop for the column. As the pressure drop is given as a liquid head, using the liquid density and the gravitational acceleration constant, the pressure drop in pascals is obtained. The simulation performed accounts for a maximum pressure drop of 62.05 kPa. In adding the liquid heads across all 40 plates, a total pressure drop of 35.3 kPa is obtained. This value is well below the maximum pressure drop and thus indicates that calculations thus far are correct, however, the phenomenon of channelling has not been considered at this point. Channelling causes vapour to unevenly pass through the trays, resulting in reduced mass transfer and high entrainment (Perry and Green 2008). This is present in systems with a low-pressure drop across a plate. However, as the maximum vessel pressure is set to the low value of 225.5 kPa, and the maximum liquid flow rate is lower than 50 m^3/hr, the effects of channelling are assumed to be insignificant.

From the total pressure drop across a plate, the downcomer backup of liquid is obtained by summing the weir, weir crest, downcomer loss and total plate height. Practically, this value cannot exceed the physical limitation of plate spacing and height of the weir, hence, a check is performed to ensure the downcomer backup is an acceptable figure. Sinnott further recommends that backup height is less than half the sum of the distance between plates and the weir heights. This check is satisfied, with the downcomer backup being 207.8 mm high, with the check being 270 mm. Finally, the residence time is calculated via:

$$\tau_{\text{plate}} = \frac{h_b \rho_L A_{\text{downcomer}}}{\text{Liquid Turn Down Rate}} \tag{3.54}$$

The plate with the highest velocity possesses the lowest residence time of approximately seven seconds. This is above Coulson et al. (2005) recommended 3-second minimum residence time, and all plates satisfy this criteria. This ensures that adequate mass transfer occurs at each plate, thus facilitating the removal of volatile components from heavier fractions.

Overall, the plate sizing regarding weir height, hole diameter, plate thickness and spacing are indicated in Table 3.18, while all constraints are listed in Table 3.19.

3.5.8.5.1 Physical Plate Design As the diameter of the column exceeds 1.5 m, the calming zone width is set to 100 mm, and the support ring width for the plate is set to 50 mm. From graphical method, the perforated angle is 88°. The weir length of 1331 mm is obtained during the flooding and weeping calculations. Dividing the total area occupied by holes by the area of a single hole yields the number of perforations, with there being 4938 holes of 8 mm diameter located in the

Table 3.19 Constraints and actual design values for plates.

Variable	Value	Logical operator	Constraint	Value
Flooding	87.4%	>		90%
Fractional entrainment	0.063	<		0.1
Max liquid flow actual minimum capour velocity (m/s)	5.932	>	Theoretical minimum vapour velocity (m/s)	5.638
Total pressure drop (kPa)	35.33	<	Simulation pressure drop (kPa)	62.05
Plate backup height (mm)	207.8	<	0.5 (Plate space + weir height)	270
Fractional entrainment	0.063	<		0.1

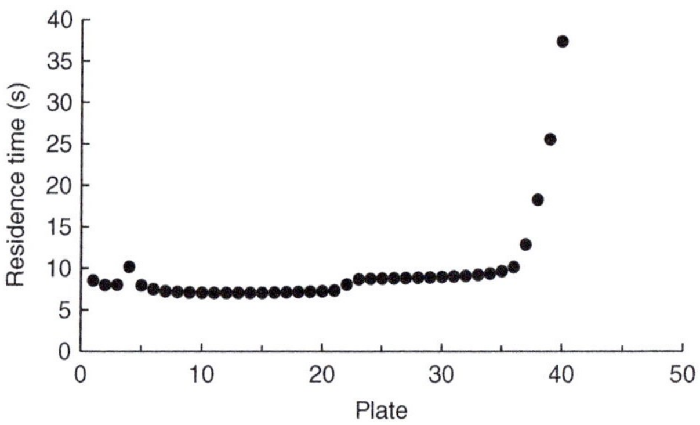

Figure 3.37 Graph of the residence times for liquid for each plate in the column.

perforated area. The hole diameter is determined from the flooding and weeping calculations performed and is constrained to be greater than 5 mm, at which Perry and Green, (2008) state blockages and scaling become increasingly prevalent. Figure 3.37 illustrates the design for the plates of the column. The central area indicates the perforated zone wherein the 4938 holes are located.

In addition to designing the plates, the actual setup of how the plates fit in the column and the downcomers and tray arrangements are designed. Table 3.20 tabulates data from the calculations performed in excel with reference to the sizing of plates and downcomers inside the column, while Figures 3.38 and 3.39 provides a visual two-dimensional representation of the designed arrangement.

Table 3.20 Calculation results summary for the detailed design of the downcomer and plate layout.

Dimension	Symbol	Value
Plate diameter	D_c	1951.26 mm
Weir height	h_w	40 mm
Max weir crest	h_{ow}	23.73 mm
Min weir crest	h_{ow}	9.06 mm
Weir length	I_w	1330.76 mm
Loss under downcomer	h_{dc}	5.84 mm
Downcomer backup	h_b	207.7840 mm
Apron height	h_{ap}	30 mm
Number of holes		4938
Distance between holes		22.8 mm
Perforation diameter		8 mm

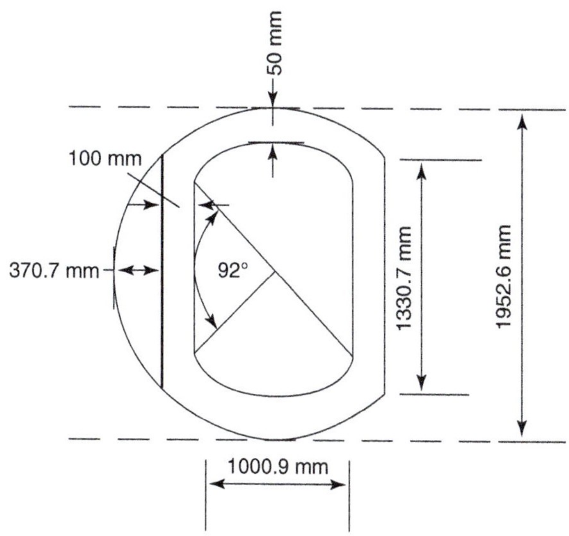

Figure 3.38 Design of plates in the distillation column, credits to Coulson et al. (2005) for the basis of the diagram used to illustrate calculated plate values.

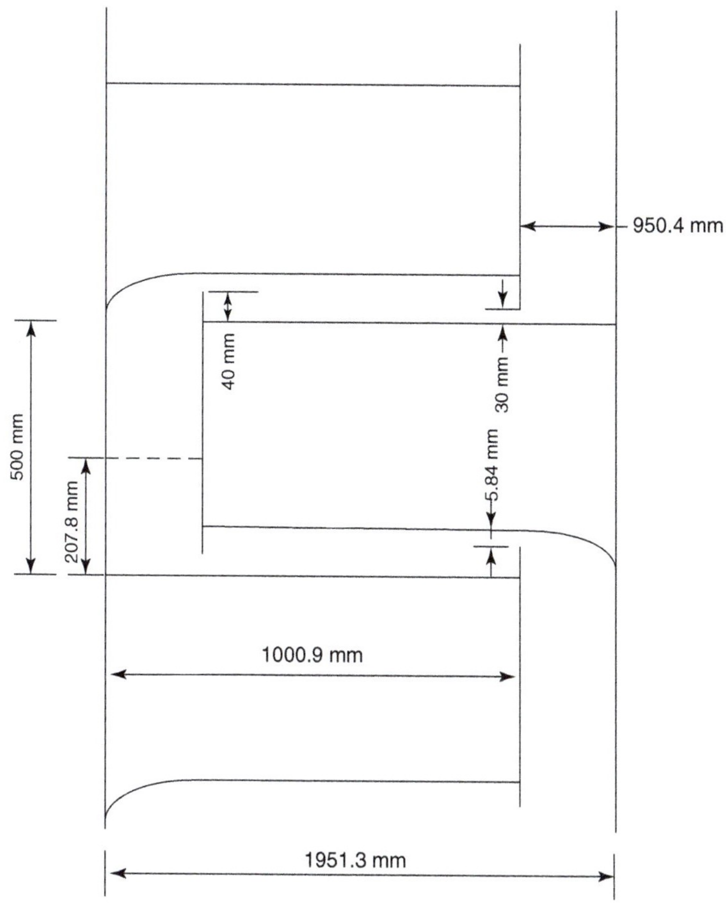

Figure 3.39 Proposed layout of plates and downcomers in the column. Note that the curved areas indicate the top of the tray, with the thickness of the tray being visible.

3.5.8.5.2 Column Thickness Calculations Due to the vast number of equations associated with the calculations, finding a thickness for the vessel via a direct method is impractical. Instead, the thickness of the vessel is iterated 1 mm at a time, and calculations are performed on each iteration. In order to obtain a design stress to use as the basis of comparison for results, the design pressure and temperature must be specified. Due to the high flow rate of steam into the system and the fact that the column is designed to operate at a low pressure, a 10% inflation of the maximum vessel pressure is chosen to yield the design pressure, 248.05 kPa. The temperature inside the vessel has less room for error, as the maximum temperature in the vessel for the current simulation is 369.6 °C, and higher temperatures than approximately 415 °C begin to risk cracking of molecules within the column. Hence, a 5% increase in maximum temperature is chosen as the design temperature, 400 °C.

In order to fully obtain the height of the vessel, the top and bottom heads of the vessel must be accommodated. As the vessel operates at a low pressure, ellipsoidal heads may be used. The thickness is calculated from the formula given in Coulson et al. (2005) for ellipsoidal heads and assuming a joint factor of 1 as 3 mm. However, the actual thickness of the vessel as will be seen is higher than 3 mm, thus the ellipsoidal head is assumed to be the same thickness as the column itself. For a 2:1 axis ellipsoid, a total vessel height for a 40-plate column with spacing of 0.5 m and tray thickness of 0.04 m, is calculated as 21.2 m.

As the vessel operates at a relatively high temperature with a thin-walled vessel, insulation is required in order to prevent excess energy costs associated with heat loss. The use of refractories is pointless as, unlike the FCC, high-speed solid particles are not present and erosion is not a risk factor in the column. Hence a lightweight insulant is chosen by Coulson et al. (2005) as Glass wool, which possesses a density of 130 kg/m^3. The insulation thickness is set to 50 mm.

Mechanical calculations follow literature due to the standard nature of the design and are extensively covered in Sinnot (2005). A brief summary of the calculations to provide an idea of the scope will now be discussed. The total weight of the vessel is calculated through adding factors such as the steel vessel weight, the weight of the liquids inside the column at any given time and plate and insulation weights. After adding 5% to this to account for feed nozzles and manholes, the wind loading and bending moment are obtained. After performing several manipulations such as obtaining the dead weight stress on the vessel, the upwind and downwind stresses are obtained. Taking the absolute values of these, the results are checked to locate the first thickness that yields both absolute stresses lower than the design stress. This thickness has 2 mm added to it for corrosion allowance, and thus the final design thickness of the steel of the column is determined as 6 mm.

3.5.8.6 Distillation Column Supports

As the column is vertical and above 20 m high, a base ring and bolt design is selected as the means of anchoring the column. Calculations regarding skirts and bolts have been performed where the basis values for vessel weight are obtained directly from the column thickness calculations, notably the final 6 *mm* thick steel vessel. 316 stainless steel is used as the material for constructing bolts and skirts, allowing a design stress of 175 N/mm^2. A 4 × 4 m concrete base 6 m deep is constructed to serve as a base for the anchoring. The calculation approach involves determining the required skirt thickness, determining the base ring width, calculating a minimum required bolt diameter and finally finding a standard bolt size to use with the supports. The final fully designed bolt flanges are illustrated in Figures 3.40a and b. The primary assumption made with

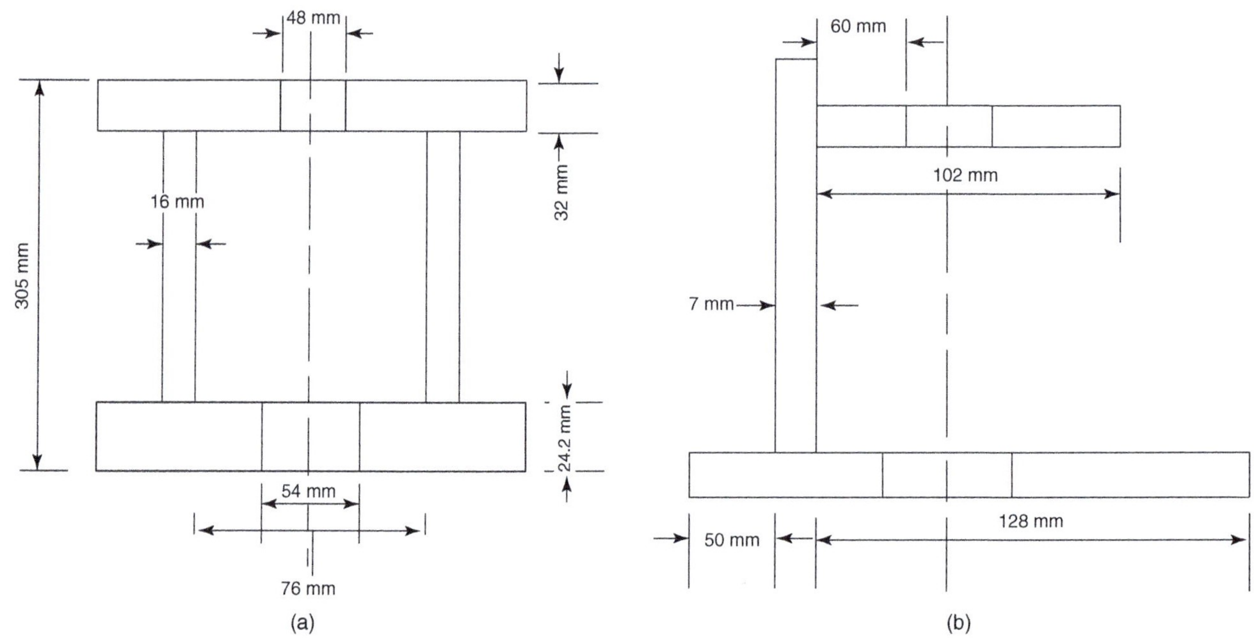

Figure 3.40 (a) Front view of the base ring and skirt support for bolt anchoring. Reference to Coulson et al. (2005) for providing bolt sizing method. (b) Side view of the base ring and skirt support for bolt anchoring.

regards to the skirt of the anchor, is that the joint factor is 1 and the skirt is at 90° to the base. As with the vessel thickness calculations, the skirt thickness is iterated, with each iterated thickness being subject to a variety of equations regarding mechanical calculations in order to determine whether or not the proposed thickness is suitable. With reference to bolt calculations performed by Sinnot (2005), the skirt is designed by calculating the bending stress, skirt dead weight, tensile stress and compressive stress. Tests are applied to the skirt and a 2 mm corrosion allowance is allocated. The bending stress for the skirt is calculated per iteration via the bending stress and moment for the column, along with Young's Modulus is externally obtained as a standard value for type 316 Stainless steel. Similarly, the dead weight stress is calculated. Once these figures are obtained for the iteration, the tensile and compressive stresses can be calculated:

$$\text{Tensile Stress} = \text{Bending Stress} - \text{Dead Weight Stress} \tag{3.55}$$

$$\text{Compressive Stress} = \text{Bending Stress} + \text{Dead Weight Stress} \tag{3.56}$$

If the bending stress is less than the design stress, and the compressive stress is less than a factor considering the maximum designed stress load on the concrete base, the skirt can be used. Adding 2 mm for corrosion and a final skirt thickness of 7 mm is selected. This is a rather high figure but is justifiable from calculations carried out as the compressive load check fails until a 5 mm thickness, before corrosion allowance, is iterated. For a skirt thickness of 7 mm, a skirt diameter of 1975 mm and a bolt ring diameter of 2042 mm is obtained. Bolts in sets of 4 are then iterated under the condition that all bolts must be above 25 mm in diameter. A final bolt count of 8 is obtained, and a standard bolt size is appropriately chosen. M42 bolts are chosen from the standard bolt sizing table given by Sinnot (2005) and result in a bearing pressure on the concrete base of 2.08 N/mm^2, which is lower than the design maximum base pressure of 4 N/mm^2. The bolts are spaced at 775.7 mm (Figure 3.41).

3.5.9 Process Control and Instrumentation

Maintaining control over process variables is vital in the operation of each processing unit, as well as for the modular refinery as a whole. The control of process variables allows for control over the mass and energy that flows through the system. This is important, as it directly impacts the performance of the process. The control of the mass and energy flows within the system is also important in allowing the refining process to operate under stable, stead state conditions. Additionally, process control measures need also be put in place to insure the safe operation of process units in order to prevent major incidents from occurring during the life cycle of the plant.

It is also vital to ensure process safety in during day-to-day operations, as well as during process startup and shutdown. In order to achieve this a set of guidelines are created, in which operational hazards are accounted for, which may be followed in order to achieve incident-free operations. These guidelines are developed in the form of a HAZOP analysis and start-up, shut-down procedures.

The process in which the process control, HAZOP analysis and start-up/shut-down procedures will be developed consists of the following section; pretreatment in which salts and other *z* are removed from the crude oil through the use of desalters, atmospheric distillation in which the crude oil is separated into component fraction, vacuum distillation in which the residue from the atmospheric distillation is further separated to remove valuable fraction, and finally, a FCC which is used to upgrade the heavier components of the crude oil into lighter more valuable gasoline products.

3.5.9.1 Control Philosophy

In order to achieve complete control over all process variables four different control mechanisms will be used across the refining process; flow control, level control, temperature control and pressure control. The use of these controllers will allow for the measurement, either direct or indirect, of all process variable if used in the correct configurations.

Control systems may be used in two different configurations; feedback and feedforward. Feedback control is the most widely used method for controlling process variables, operating by measuring the output of a process and then taking corrective action by altering the input to restore, or maintain the output variable at the desired set-point. Feedforward control operates by measuring the input of a process and using predictive action to maintain the process output at the desired set-point. Feedback control tries to minimise the difference between the set point (desired output) and the actual result. Thus the feedback system is generally used as it is able to offer reliable control through direct measurement. For this reason, a feedback control system will be used throughout the plant, with the exception of flow control. This is as when it is imperative that there are no deviations from the desired output result a feedforward control system is best. The feedforward control system offers the advantage of immediate disturbance correction, whereas the use of a feedback system within flow control would lead to disturbance propagation.

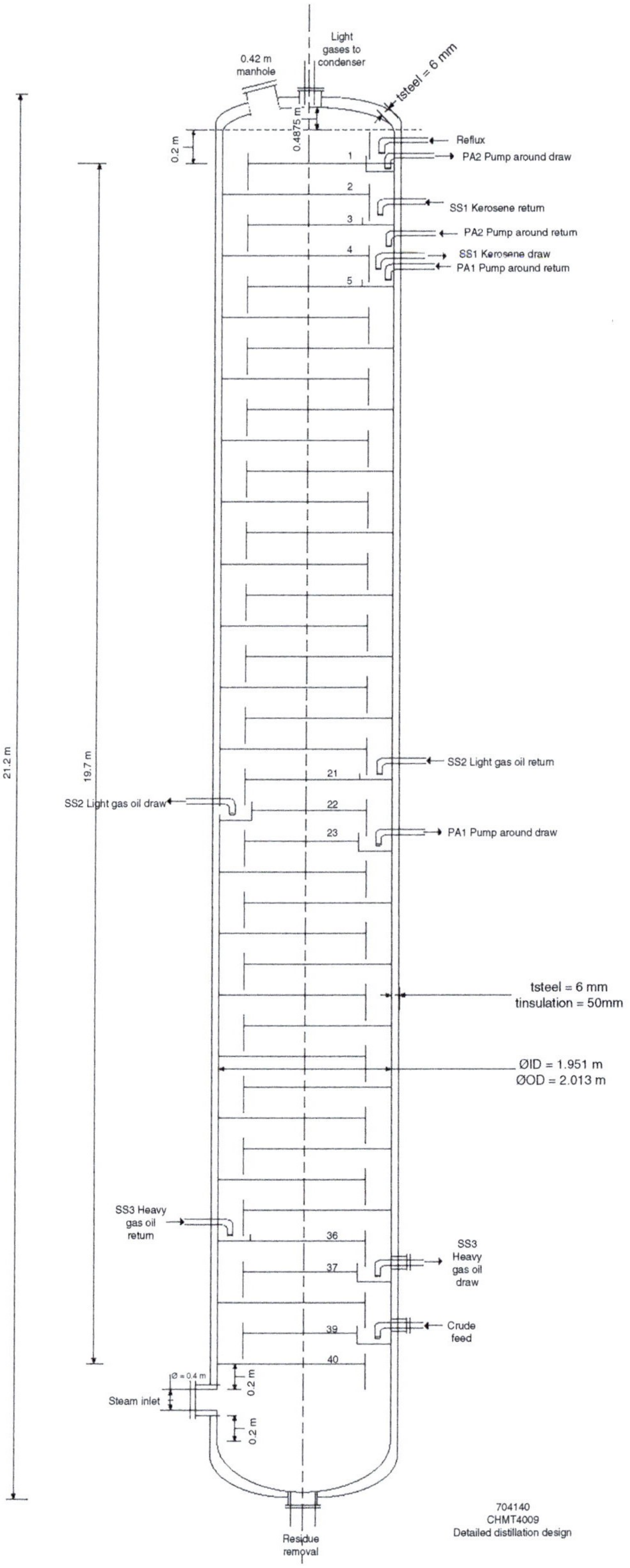

Figure 3.41 Distillation Column Design.

3.5.9.2 Variable Measurement

3.5.9.2.1 Pressure The measurement of pressure is a vital part of the plant operation as it allows for both direct variable (operating pressure) and indirect variable (level, flow) measurement. A benefit of this is that pressure measurement is a fairly simple process. Bourdon tubes will be used as they allow for the conversion of a pressure differential into a mechanical motion which may be measured. For the majority of the units in the refining process, it will be sufficient to measure the gauge operating pressure with respect to atmospheric pressure, however, for the vacuum distillation unit the absolute pressure will be measured as the vessel operates below atmospheric pressure.

3.5.9.2.2 Flow Measurement The flow of material through a pipe will be measured using a differential pressure device. These devices operate on the principle, in which a restriction is created within the pipe line of the stream. The restriction creates a pressure drop in the stream. The size of the pressure drop may then be used to calculate the rate at which the fluid flows.

3.5.9.2.3 Level The measurement of liquid level within a vessel may be very difficult depending on the operating conditions as well as the fluid mediums within the vessel. To attain a reliable measurement a differential pressure method will be employed. Simplified, this method involves the measurement of the differential pressure between the vapour phase and the bottom of the liquid phase. This pressure difference may then be used to determine the liquid head. This method may encounter difficulties when the vessel is operating at high pressure as the signal of the pressure differential may be dwarfed by the operating pressure. This, however, should not be an issue within the scope of operation of the designed modular refinery as the majority of operations occur at relatively lower pressures (1–2 bar. a).

3.5.9.2.4 Temperature The temperature of a system may be measured in many different ways, ranging from traditional mercury thermometers, thermostats, thermocouples and resistive temperature devices (RTD). For the use in the designed modular refinery, a RTD setup will be used. These devices work on the principle in which the electrical resistance of a pure metal, such as platinum, varies with changing temperature. The measurement of the electrical resistance would, therefore, allow for the determination of temperature. As it is required to use highly purified metals within the measurement device RTDs may be relatively expensive, however, they are able to offer a much higher response time increasing the degree of control that may be achieved over the process.

3.5.9.2.5 Safety To ensure safe operation of the process vessels, pressure safety valves will be installed. These valves will be designed to automatically open when the internal pressure of the vessel reaches the limits of the design pressure. The valves will then automatically close once the pressure has been restored back to the acceptable operating range. The use of self-governing valves for safety will be beneficial, as they will remain in operation in case of the failure of the rest of the control system.

3.5.9.3 Control Schemes

3.5.9.3.1 Desalters Pressure Sinnot (2005) stated that pressure controls are mostly used in processes that involve vapour and gases. In the pretreatment of the modular refinery, the crude oil is in its liquid state. However, IPS (2000) recommends that, as a minimum, desalters must be equipped with pressure transmitters and indicators and also a pressure safety valve. Therefore, pressure transmitters and indicators will be fitted on the desalters to measure and show the desalter pressure. If any case the pressure gets too high than the required operating pressure, the desalter vessel may rupture and this could lead to serious injuries and equipment damage. The pressure safety valve will, therefore, be used to reduce the desalter pressure when it gets too high. The desired desalter pressure is 758 kPa and the indicator should illustrate such for sound operation. The pressure around the pump pumping crude oil will be controlled using a pressure controller and stabilised at a value slightly above 758 kPa (Sinnot 2005; IPS 2000).

3.5.9.3.2 Temperature The monitoring and controlling of temperature of the desalters will be done by a controller that manipulates the temperature of the incoming crude oil to maintain the desired temperature. The controller will correct disturbances that could result from malfunctioning heat exchangers or even a drastic change in ambient weather conditions. The controller will manipulate the amount of steam that is used in the heat exchangers hearing the oil to bring about the required temperature. The set point that will be used is the desalter operating temperature of 88 °C. The valve will be placed on the pipeline of heat exchanger steam. When the desalter is at a lower temperature, the controller will signal the valve to increase the amount of steam used to heat up the crude oil. When the desalter temperature is higher, then it will signal that the steam flow rate be reduced to decrease the crude oil temperature.

3.5.9.3.3 Flow The flow of the process water, which is mixed with the crude oil before entering the desalter, is controlled so that the correct amount of water is used in the desalting process. If an incorrect amount of water is mixed with the oil it could result in less salts and contaminants removed or even flooding of the vessels. This can be prevented by installing flow controls around the process water pipeline. The controller will be automated to take the water flow rate measurements from a flow transmitter and compare it to a set point and physically control the water flow by an actuator valve. If the flow rate is more than the desired value, the valve will reduce the flow by closing appropriately and the opposite will occur if the flow is less than required. The set point of process water will be 22 528 kg/hr. as required.

3.5.9.3.4 Level Sinnot (2005) stated a level controller is necessary where an interface between 2 phases exits. The level controller serves as a way to keep the interface at the desired level (Sinnot 2005). The level of the interface between the water and oil inside the desalters must, therefore, be monitored and controlled. A level control loop will be installed on the desalters to carry out this function. The controller will be automated to take signals of the measurements of the height of the interface inside the vessel by a level transmitter. An actuator valve will be fitted on the pipeline of the effluent water to act as the final control element. If the height of the interface decreases, then the valve will close appropriately to reduce the flow rate of the effluent process water from the desalter, this will in turn restore the level of the interface. The vice versa will occur when the opposite happens in order to maintain the required interface height.

3.5.9.4 Atmospheric Distillation

3.5.9.4.1 Pressure The atmospheric distillation column operates at maximum pressures that are lower than 3 bar, and a design pressure was selected when determining the vessel thickness to accommodate moderately higher pressures. This allows less stringent control over pressure inside the vessel from a safety standpoint but is still controller via a proportional integral (PI) controller to avoid altering the equilibrium conditions within the vessel significantly. Pressure is controlled by the withdrawal of gaseous top product, and an emergency manual venting valve is present at the top of the column. The PI controller utilises a set point of 165 kPa as the target of control and varies the positioning of the outlet gas control valve to regulate vessel top pressure. Pressure changes inside the vessel are significant with changes in crude oil feed and stripping steam introduced at the bottom of the column. In order to ensure that stripping steam input reflects the feed of oil, a ratio controller compares the flow rates through both feeds and adjusts them to enforce a desired crude feed-to-steam ratio.

3.5.9.4.2 Temperature Temperature in the column is controlled by pre-heating the crude oil feed in a furnace prior to distillation, and superheated steam is added to the bottom of the column to compensate for the absence of a residue reboiler. The feed oil is controlled by a PI controller that targets a set point of 405 °C by controlling the flow rate of oxygen into the furnace that heats the crude oil. The design temperature does not permit a much higher temperature, as the oil risks initiating cracking inside the column at temperatures exceeding 415 °C. Hence, once the crude oil has been heated to its set point, any further temperature increases required in the column are supplied by increasing the flow rate of superheated steam into the bottom of the column. Temperature inside the column is regulated via the use of pumparounds and returns via side stream reboiled vapours. The objective during design was to ensure a favourable Watson K factor for the HGO stream, and thus an unconventional layout of pumparounds is present that serves to improve mixing and cool streams within the column.

3.5.9.4.3 Flow As described in the section on pressure, a ratio controller is used to ensure the stripping steam introduced into the bottom of the column changes with a varying crude oil inlet. This permits variable feedstock and reduces the need for manual control or reprograming of the PI controllers. This technique is applied to the product flow rates on the side streams of the unit, linking these flow rates to that of the crude feed by a specified ratio. Flow of utility streams is regulated based on temperature control PI controllers that provide the required amounts of heating or cooling required by altering the flow rate of either water or superheated steam.

3.5.9.4.4 Level The flow rate of residue leaving the bottom of the column for the vacuum distillation unit is controlled via a level controller to ensure that flooding does not occur. In the event of flooding occurring an emergency drainage system is provided to manually purge the distillation column as well as side strippers should it be necessary. Level in the main column is controlled by altering the flow rate of residue leaving the bottom of the column and is controlled in the side strippers by reducing the flow rate of each side draw so as to respond to changes in the system quicker. Ratio controllers serve to reduce the chances of flooding occurring if the inlet flow rate increases. This also facilitates the automatic startup of the column by allowing the residue to only be removed once the maximum level is reached, resulting in the equilibrium of the vessel being attained faster and reducing the time it takes to achieve steady operation.

3.5.9.5 Vacuum Distillation

3.5.9.5.1 Pressure The vacuum conditions within the distillation column are created through the use of steam ejectors which entrains the flue gas stream exiting the top of the column. The pressure in the vacuum distillation column will be controlled by manipulating the steam flow rate through the vacuum ejector system. Increasing the steam flow rate will decrease the pressure while decreasing the flow rate will increase the pressure. The pressure indicator will be placed at the top of the vessel with a set point of 6.66 kPa absolute.

3.5.9.5.2 Temperature The temperature in the distillation column will be controlled on two fronts. The preheat furnace will control the temperature of the feed into the vacuum column which will in-turn control the bottom temperature of the column. Under the designed process conditions the set point for the column feed temperature will be 390 °C. The temperature profile of the column will be controlled by manipulating the temperature of each of the pumparound returns entering the column, by altering the rate at which cooling water flows through the heat exchangers.

3.5.9.5.3 Flow The flow of the column withdrawals would need to be controlled to be stable at the steady state values. These flows will be controlled by measuring the flow ratio between the column feed and the side withdrawals. This would allow for corrections to be made in case the feed flow into the column varies due to operational demands. The stripping steam feed flow rate into the bottom of the column will also be controlled by using the flow ratio with the oil fed. The top and bottom withdrawals of the column (flue gas and vacuum residue, respectively) will not, however, be controlled in this manner. This is because the flow rate of bottom product will be used to control the liquid level in the column. The flue gas flow rate will, therefore, not have to be controlled, as the mass balance dictates that direct control of 5 of the 6 streams flowing into and out of the column will allow for indirect control over the 6th and final stream.

3.5.9.5.4 Level It is important to control the liquid level within the column in order to prevent flooding of the column, or liquid back-flow through the vapour inlets. The liquid level in the column will be controlled by manipulating the flow of the bottom column product (vacuum residue). The liquid level in the column should be kept below the height of the stripping steam inlet while allowing for a high enough residence time to insure effective separation of the light components from the heavier components being removed.

3.5.9.6 Fluid Catalytic Cracker

The FCC unit is used to convert gas oils fed from the mixing unit into gasoline products. For this to occur efficiently the temperatures and pressures of operation in each section of the FCC need to be controlled. Additionally, the flow rates of streams providing feed materials need to be controlled. For the regenerator, level control is used to ensure that catalyst over loading does not occur. Air and fluidising steam inlets require ratio controls as they are dependent on the flow rate of other streams – air flow rate depends on the amount of catalyst and steam flow rate depends on the flow rate of the feed gas oil. Thus the control schemes used in the P&ID of the FCC unit include feed-forward control and ratio control. The riser is the unit where the cracking reaction occurs. The regenerator ensures that the used catalyst particles can be regenerated through decoking and can, therefore, be used again. Thus it is necessary to ensure that temperature, pressure, level and flow rates are monitored and controlled effectively for smooth operation.

3.5.9.6.1 Pressure The pressure of the riser is controlled by controlling the pressure of the feed stream that contains the gas oil mixture. It is required that the FCC riser operates at a pressure of 1.79 bar and since there will be a pressure drop across the dispersion nozzle at the feed, a slightly higher pressure will be used as the set point of the pressure controller. The intended pressure control is achieved using a pressure control valve.

3.5.9.6.2 Temperature It is required to operate the riser of the FCC at a temperature of 540.5 °C because as the riser is adiabatic, it is expected that the temperature will decrease along the length of the reactor as the reactions taking place are endothermic, thus this temperature reduction is compensated using a higher feed temperature. In order to achieve and control the temperature of the riser and its contents, the flow rate of heated catalyst (From the regenerator) entering the riser is controlled using the temperature control valve as shown in the FCC PID. The amount of air entering the regenerator is used to control the temperature of the regenerator, which needs to be operated at a temperature of 590 °C. Since the regeneration process is comprised of combustion to decoke the catalyst, which is an exothermic process, the amount of air entering the regenerator can be used to control the temperature of the regenerator as the air can absorb some of the heat or used as energy required for combustion. This is also done using a temperature control valve.

3.5.9.6.3 Flow Effective flow rate control of the streams in the FCC unit is extremely important to ensure effective functionality of the process. This is why multiple flow rate and ratio controls have been implemented in various sections of the FCC as seen in the FCC PID. The flow rate of the fluidising steam is controlled using the ratio controller. Sadeghbeigi (2012) suggested steam to gas oil feed ratio of 1/3 though a simulation yielded an effective ratio of 0.685 which can be maintained using the ratio controller (Sadeghbeigi 2012). The flow rate of the FCC product stream is also controlled using a controller that monitors the flow rate of the products exiting the riser. This is to ensure that if there are possibly high flow rates, they can be reduced to prevent flooding or over-loading of subsequent process equipment.

3.5.9.6.4 Level The regenerator will require level control so as to ensure that there is enough catalyst within the regenerator and to ensure that the loading of the regenerator does not become critical. This is achieved by controlling the flow rate of the stream containing spent catalyst (from the reactor vessel) using a flow control valve.

3.5.9.6.5 Mode of Control The controller logic (mode of control) will be chosen to allow for the most rapid response to changes and disturbances without causing instability in the system. To allow the control system to operate in a stable way the mode of control which is able to limit response oscillation will be chosen.

Throughout the refinery, various automated control systems are present that aid in regulating flow throughout the various systems, and ensuring that certain limits are not exceeded. Additionally, the automated controls assist during plant startup and shutdown. Due to the large flow rates and the number of components being controlled, having a finely tuned system is impractical and is also unnecessary as a high degree of precision is not needed due to the nature of units such as the distillation column separating based on temperatures rather than exact components. However, basic on–off controllers are not suitable. Although these controllers perform well with certain large volume units (Bolton 2004), due to the presence of a dead band in which no control action is taken, which greatly restricts the response time and can lead to safety issues by the controller not responding fast enough to a rising temperature or pressure, for example. The use of proportional-only controllers eliminates this issue by constantly tracking and correcting changes in set point. These controllers have the fastest response time of any controller, but suffer from setpoint offset (Bequette 2003). This results in the controller being unable to exactly achieve the desired set point. While this is not much of an issue for many units, temperature effects are prominent in the distillation unit. The furnace is required to heat the crude feed to 405 °C but temperatures higher than 415 °C will result in cracking initiating inside the distillation unit. Hence, temperature does need to be accurately controlled and fluctuations or deviations from the set point.

This leaves the use of PI and PID controllers as possible viable solutions. These controllers possess integrators that force the controller to completely remove any set point error by increasing control action through accumulating the error. The issue with this is that while the set point will eventually be reached, the time that it takes to achieve this is significantly longer than control via a proportional-only controller would be (Bolton 2004). PID controllers reduce this through use of a differentiator which significantly shortens the control time such that Proportional only times are approached. This is performed by acting on the gradient of the disturbances rather than the disturbances themselves, allowing for rapid control times. Conversely, this results in even slight disturbances in conditions being amplified, and controllers react to otherwise benign conditions aggressively. This can cause damage to valves but also results in oscillating control. This is highly undesirable. Thus while PID controllers effectively eliminate integral windup, they are not suited for large imprecise systems with large quantities of flow being dealt with in which disturbances are inevitable. The time that it takes to reach a steady state can be reduced through use of manual control in certain sections when starting up or making alterations to upstream flow conditions, making PI controllers a viable option that consistently provides adequate control action while remaining largely unaffected by disturbances in the process. The final choice of control action is thus provided by PI controllers.

3.5.9.6.6 Results of Calculations The results of the implementation of the control mechanisms were used in the creation of P&ID diagrams. The diagrams are separated into individual sections allowing for an increased level of detail as well

Table 3.21 P&ID instrumentation codes.

Code	Instrument
FT	Flow transmitter
FI	Flow indicator
FCV	Flow control valve
FFC	Flow ratio controller
FFV	Flow ratio valve
PT	Pressure transmitter
PI	Pressure indicator
PCV	Pressure control valve
PSV	Pressure safety valve
PdT	Pressure differential transmitter
PSV	Pressure differential indicator
TT	Temperature transmitter
TI	Temperature indicator
TCV	Temperature control valve
LT	Level transmitter
LI	Level indicator
LCV	Level control valve

Table 3.22a P&ID line numbering section codes.

Code	Section
DS1	Pretreatment section 1 (Desalting)
DS2	Pretreatment section 2 (Desalting)
ADU	Atmospheric distillation Section
VDU	Vacuum distillation section
FCC	Fluid catalytic cracking section

as increased ease of understanding. Each control loop will comprise three elements; a transmitter which allows for the measuring of the process variable, an indicator which converts the measurement into a control signal and a control valve which alters the manipulated variable. All three elements will work together to maintain the process variable at the desired setpoint. The following key will be used in the P&ID diagram (Tables 3.21 and 3.22a):

The line numbering of the P&ID will follow the format of:

###-$$$-@@@

Where:

– Section Code

$$$ – Line Number

@@@ – Pipe Diameter (mm)

The section code will follow the key:

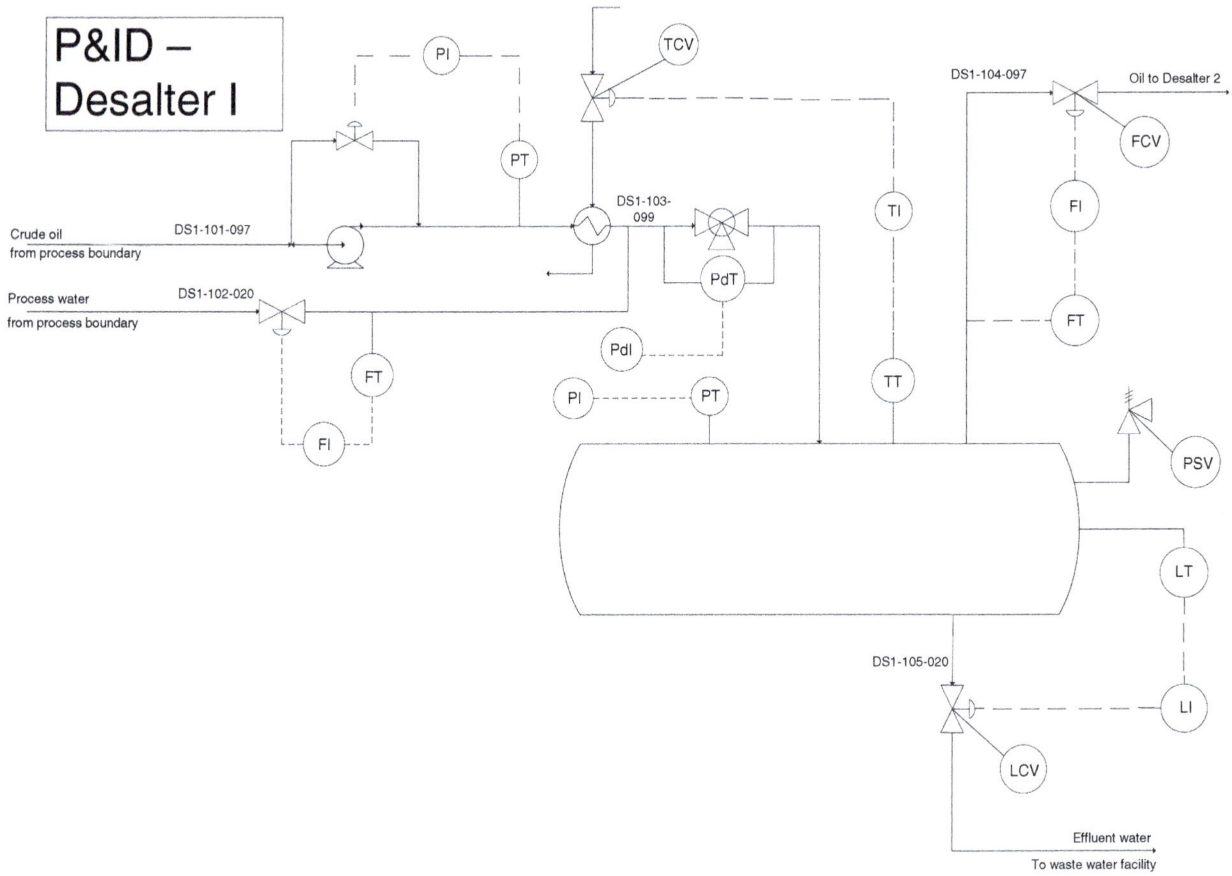
P&ID –
Desalter I
PI
TCV
DS1-104-097
Oil to Desalter 2
PT
FCV
Crude oil
from process boundary
DS1-101-097
DS1-103-099
TI
FI
Process water
from process boundary
DS1-102-020
PdT
FT
PdI
FT
PI
PT
TT
FI
PSV
LT
DS1-105-020
LI
LCV
Effluent water
To waste water facility

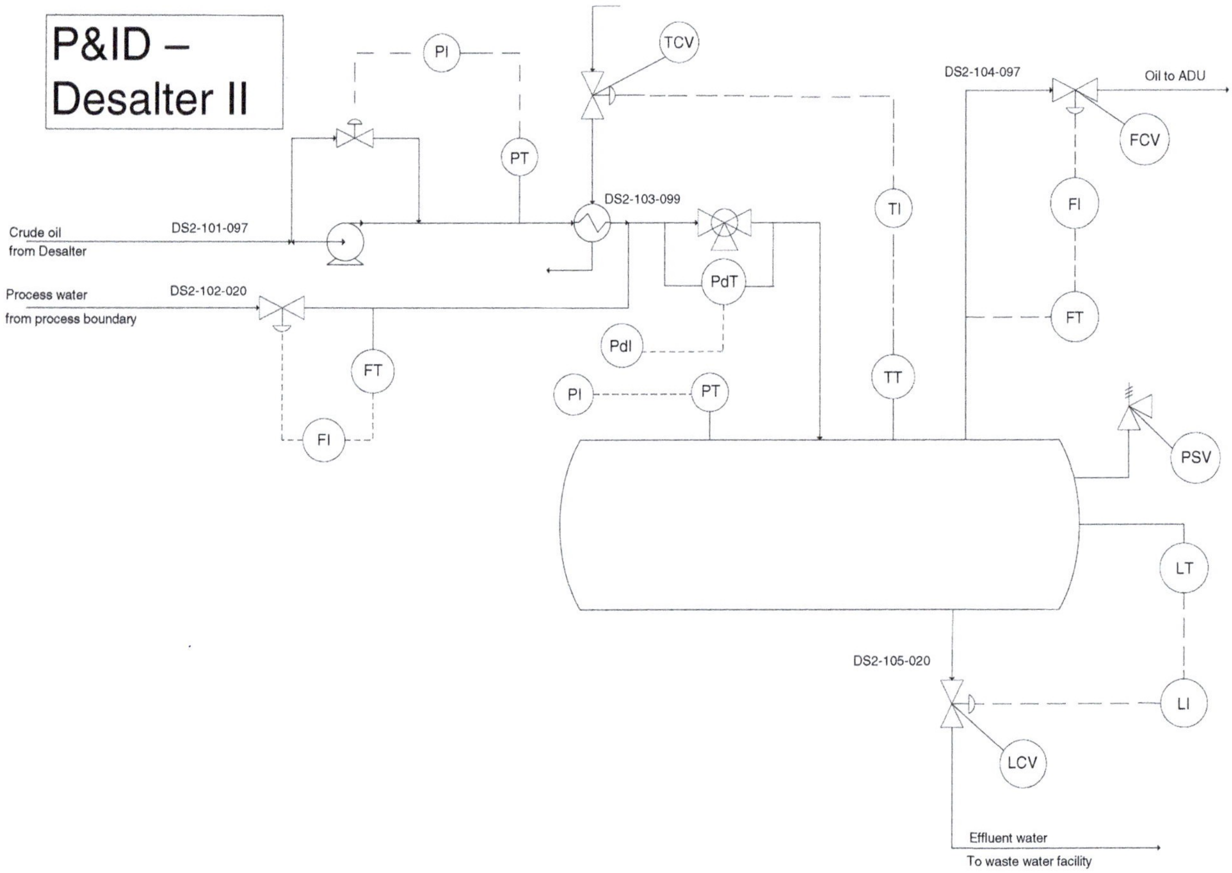
P&ID –
Desalter II
PI
TCV
DS2-104-097
Oil to ADU
PT
FCV
DS2-103-099
Crude oil
from Desalter
DS2-101-097
TI
FI
Process water
from process boundary
DS2-102-020
PdT
FT
PdI
FT
PI
PT
TT
FI
PSV
LT
DS2-105-020
LI
LCV
Effluent water
To waste water facility

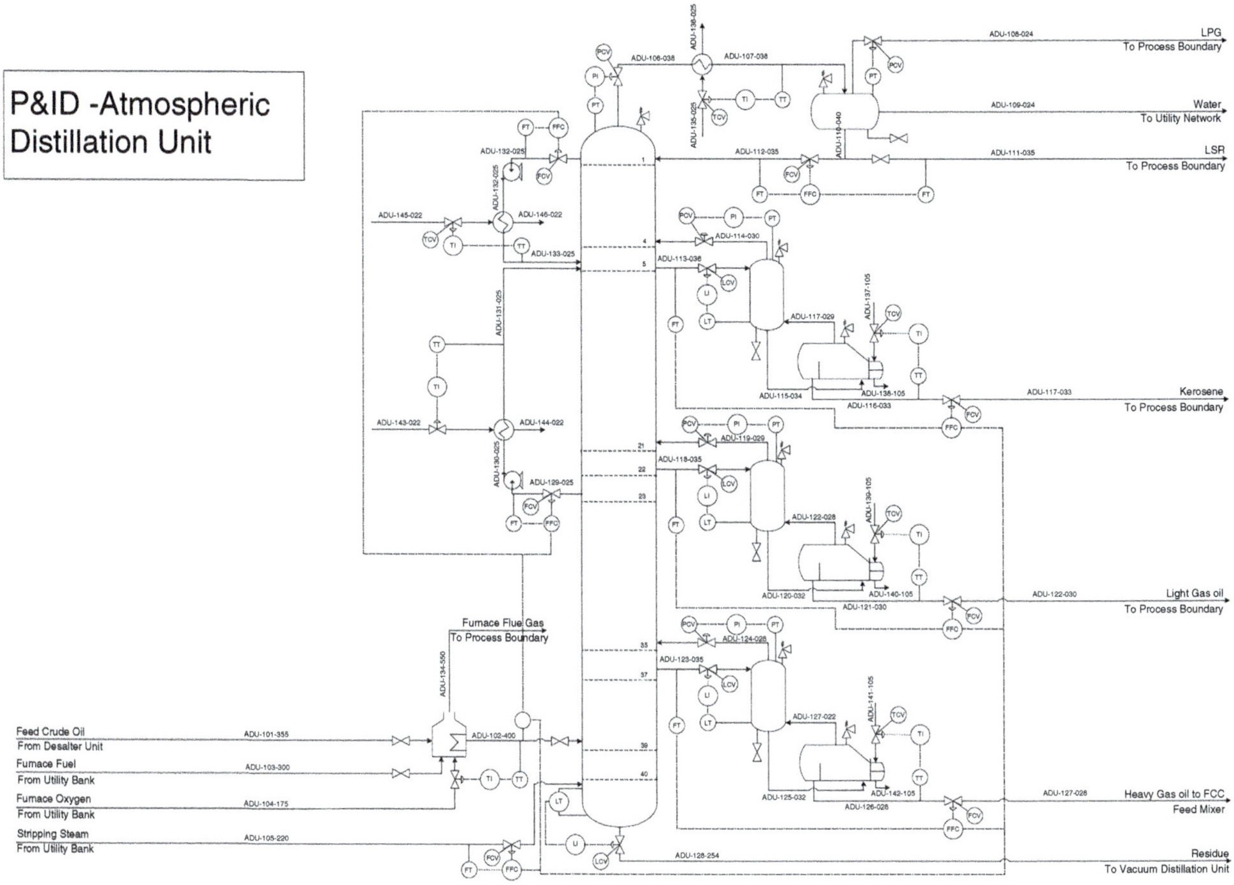
P&ID -Atmospheric Distillation Unit
LPG
To Process Boundary
Water
To Utility Network
LSR
To Process Boundary
Kerosene
To Process Boundary
Light Gas oil
To Process Boundary
Heavy Gas oil to FCC
Feed Mixer
Residue
To Vacuum Distillation Unit
Furnace Flue Gas
To Process Boundary
Feed Crude Oil
From Desalter Unit
Furnace Fuel
From Utility Bank
Furnace Oxygen
From Utility Bank
Stripping Steam
From Utility Bank

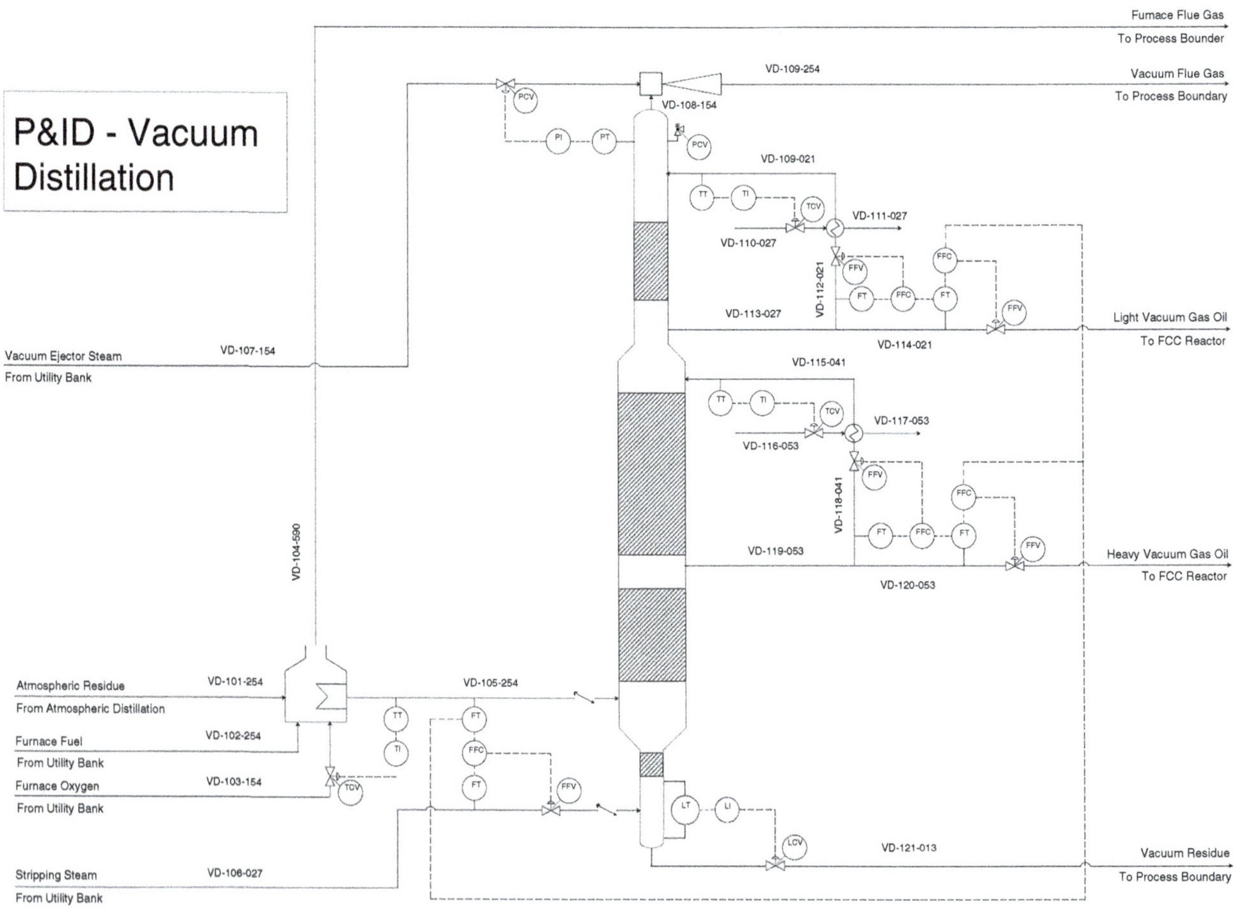
P&ID - Vacuum Distillation
Furnace Flue Gas
To Process Bounder
VD-109-254
Vacuum Flue Gas
To Process Boundary
VD-108-154
VD-109-021
VD-111-027
VD-110-027
VD-112-021
VD-113-027
VD-114-021
Light Vacuum Gas Oil
To FCC Reactor
Vacuum Ejector Steam
VD-107-154
From Utility Bank
VD-115-041
VD-117-053
VD-116-053
VD-118-041
VD-119-053
Heavy Vacuum Gas Oil
To FCC Reactor
VD-120-053
VD-104-590
Atmospheric Residue
VD-101-254
From Atmospheric Distillation
VD-105-254
Furnace Fuel
VD-102-254
From Utility Bank
Furnace Oxygen
VD-103-154
From Utility Bank
VD-121-013
Vacuum Residue
To Process Boundary
Stripping Steam
VD-106-027
From Utility Bank

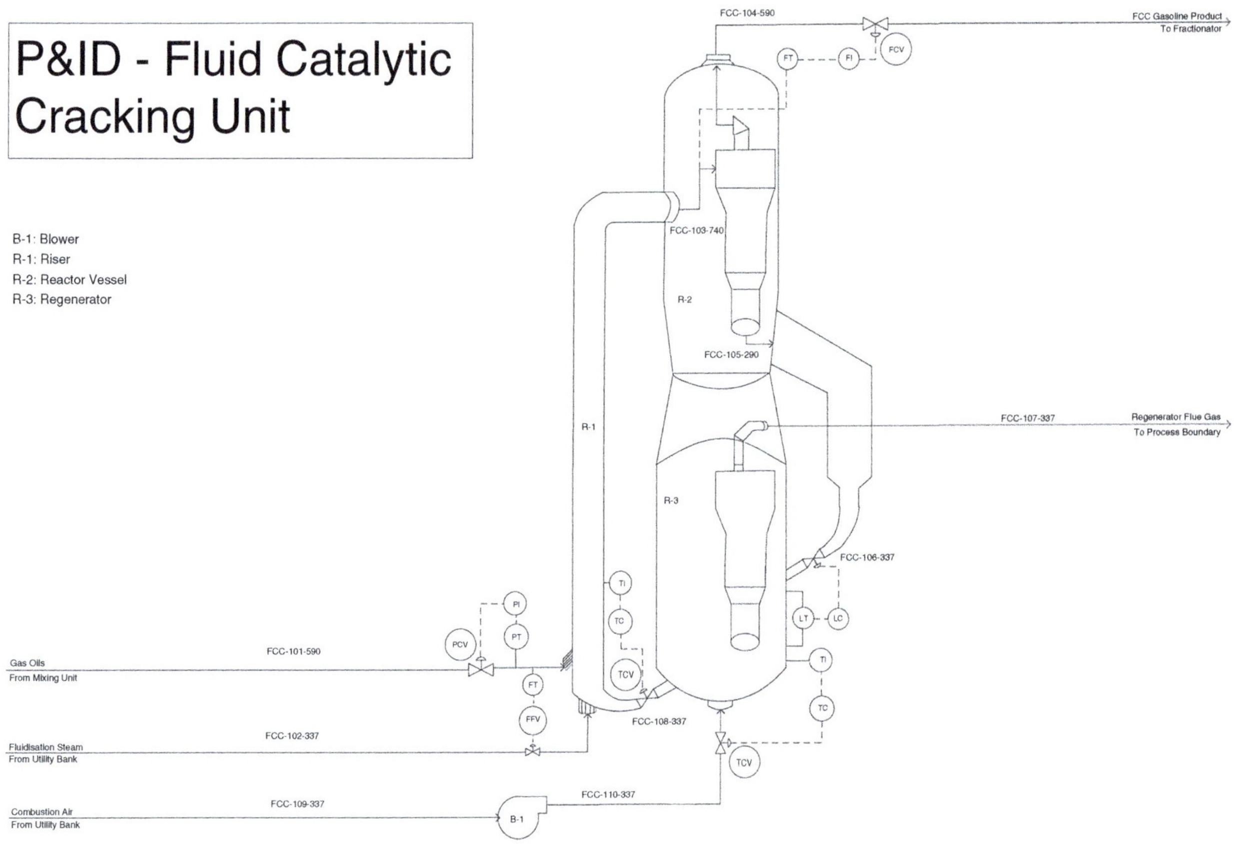

3.5.10 HAZOP

A hazard and operability (HAZOP) study is considered to be a structured and systematic analysis of a planned or process which is existing or an operation that serves to identify and evaluate problems that may present risks to personnel or equipment, or inhibit an efficient operation (Rausand 2005). The HAZOP approach is the most utilised method of identifying and evaluating hazards in the chemical industry over other known methods which include the what-if analysis, error analysis and failure mode analysis. A HAZOP study typically makes use of the piping and instrument diagrams (P& ID) alternatively plant model is sufficient and this serves as a guide for an analysis of every section and component of a process (Product Graphics 2016). The study allows for an in-depth analysis of safety issues during the design phase and allows for control actions to be placed in action for safer and more efficient operation in the process.

For the HAZOP analysis, a team was employed to make use of the P&IDs in which they examine the impact of potential changes to parameters which include flow, temperature, pressure and level. The HAZOP study is a systematic, step-by-step approach to brainstorming possible deviations; determining the likelihood of the deviation (is there a realistic cause); evaluating existing protections; and estimating the resulting impact and potential catastrophic deviation (Product Graphics 2016). The potential causes of failure are identified from the process design and this results in seeking out existing safeguards and protection systems which must be analysed to see if these safeguards and protective systems are able to accommodate these potential failures (Product Graphics 2016). When a hazard condition is identified a recommendation should be made for process or system modifications, or further study by a specialist may be required (Product Graphics 2016).

Table 3.22b HAZOP guide word definitions.

Guide Word	Meaning
Low/No	Process variable is significantly lower than operating range or has stopped completely
High	Process variable is significantly higher than operating range
Reverse	Process variable is moving in the opposite direction to operating conditions

A HAZOP analysis was completed for each of the four major units within the modular refinery process; desalters, atmospheric distillation column, vacuum distillation column and FCC. The results of the HAZOP analysis will be presented in a tabular form, with guidewords being used to indicate the type of operational variation that the process parameters may undergo. The guide words used and their definitions may be found in Table 3.22b

Vessel Purpose
Desalter
Remove salts and contaminants from crude oil

Parameter	Guide word	Cause	Consequences	Action
Line	**DS1-102-020**			
Intention	Transport mixture of oil and water to desalter			
Flow	Low/no	Mixing valve gets stuck	Level in desalter decreases	Fit low-level alarm
	High	Failure of flow control system in process water line	Flooding of desalter	Fit high-level alarm
	Reverse	Desalter pressure increases. Desalter flooding.	Pipe rupture due to the accumulation of oil and water	Fit non-return valve
Temperature	High	Failure of temperature control system	Possibly vaporise some fraction of the crude oil, this will increase pressure in system	Fit high temp warning alarm
	Low	Heat exchanger failure	Coalescence rate will be decreased, settling rate will, therefore, be decreased	Fit low temp warning alarm
Line	**DS1-103-099**			
Intention	Transport oil to the second desalter			
Flow	High	Failure of flow control system	Flooding on the next desalter	Fit high-level alarm
Pressure	High	Pressure safety valve gets stuck	Damage to heat exchanger equipment downstream	Have a vent line connected from desalter to sewer
Line	**DS1-104-097**			
Intention	Transport effluent process water from desalter			
Flow	Low/No	Failure of flow controller	Pump for pipeline wash system will have no water, therefore, solids accumulate in desalter	Fit low flow rate alarm as warning
	High	Failure of flow controller	Level in desalter decreases	Fit a high-flow rate alarm as warning

Vessel Purpose
Topping Column (Atmospheric Distillation Unit)
Fractionate crude oil and extract products based on temperature classes

Parameter	Guide word	Cause	Consequences	Action
Line	**ADU-102-400**			
Intention	Transport Crude Oil to the Distillation Unit			
Flow	High	Excessive crude oil feed to system or disturbances stemming from desalting unit irregularities and feed control actuator not adequately responding	Liquid flow inside the column exceeds design limitations, resulting in flooding. Results in inefficient mass transfer in system and quality of products compromised	Manually close inlet valve if possible, purge contents of the vessel through bottom of the column. Seal all side stream valves to temporarily halt operations so as to avoid further product loss until column flooding ceases to occur
	Low/No	Desalter operations halted, flow in pipes restricted	Fractionation results in undesired products or simply ceases completely	Close all outlet valves and monitor level in the column. Should level meter exceed maximum design specifications, open purge valve at the bottom of the column to return liquid level to desired value. Use of ratio controllers to prevent product quality being hampered
Pressure	High	Back mixing	Vessel risks rupture, higher pressures increase risk of flooding	Open the manual control valve at the top of the unit to equalise pressure inside the vessel. Use a PID controller that automatically controls the pressure inside the vessel. Use of 1-way valves to
	Low	–	Column operates at near atmospheric conditions, no effect	–
Temperature	High	Furnace oxygen flow rate	Cracking can occur inside the column, leading to losses of product	PI controller used to regulate furnace oxygen with a setpoint of 405 °C
Line	**ADU-106-038**			
Intention	Vapour extract from column top			
Flow	High	Pressure valve not responding accurately, inlet steam or crude oil flow rate too high	Flooding, disruption of equilibrium and loss of product quality	Stringent control of inlet flow rates and appropriate pressure valve control
	Low/No	Flooding in column, pipe fouling or inlet flow disruption. Possibly pressure control failure	In the case of pressure control, vessel pressure may rise and result in an explosion. Else causes product losses	Emergency pressure release valve manually activated or automated
	Reverse	Flooding	Flooding aggravated	Fit non-return valve
Pressure	High	Vessel pressure too high	Equilibrium inside vessel affected, possible safety risk of explosion	Regulate inlet flow rates, increase flow rate of outlet gas to reach steady state, in emergency open safety release valve
	Low	Low inlet steam flow rate	Column flooding or weeping may result, no serious risks	Use of ratio controllers to ensure sufficient steam is always present regardless of crude flow rate
Temperature	High	Vessel temperatures too high	No effect	–
	Low		No effect	

Line	**ADU-113118123**			
Intention	Kerosene, Light gas oil and heavy gas oil draws			
Flow	High	Increased vapour fraction of column feed, inefficient control scheme for regulating withdrawals	No serious effects	No Action
	Low/No	Decreased vapour fraction of column feed, inefficient control scheme for regulating withdrawals	No Hazardous Consequence	Ensure draw valves are operating efficiently to avoid pressure buildup inside vessel
Pressure	High	Inlet steam flow rate too high	Pressure irregularities in stripping column, may result in hazardously high pressures	Pressure control system established for side strippers
	Low	Failure of steam ejector pressure control	No Consequence as vessel designed to handle complete vacuum	No action
Temperature	High	Column temperature profile deviant. Column not at steady state	No hazardous effects	Use of ratio controllers to ensure adequate steam flow rate into
	Low	Column temperature profile deviant. Column not at steady state	No hazardous effects	Use of ratio controllers to ensure adequate steam flow rate into
Line	**ADU-128-254**			
Intention	Residue transport to vacuum distillation unit			
Flow	High	Inlet crude flow rate too high, steam ratio controller not operating effectively Flooding may be occurring	No Hazardous Consequences	No Action
	Low/No	Level controller not operating correctly, possible pipe blockage	Flooding, will lead to halted production	Fit alarm to alert to low levels of flow, emergency liquid drain manual valve can be used
Pressure	High	Steam inlet flow rate may be too high	High corrosion of residue carrying pipes, may eventually rupture	Proper mechanical design of pipes, steam inlet flow rate regulated by ratio controller
	Low	Steam inlet flow rate too low	Possible low flow rates, and column likely to flood	Effective regulation of steam, manual drainage of column possible via emergency valve
Level	High	Column flooding	Halted production	Use of level controllers, alarms and emergency drainage valves
	Low	Column level too low, level controllers may be too stringent and flow rate of residue may be too high	No hazardous effects	Better tuning of level controller
Temperature	High	Inlet crude or steam flow rate too high	Pipe damage, possible residue cracking	Alarms to cease inlet flow rates so as to avoid loss of oil, better control over inlet temperatures
	Low	Inlet crude or steam flow rate too low	No hazardous effects	–

Line	**ADU-105-220**			
Intention	Bottom of Column stripping steam			
Flow	High	Failure of flow control system	Excessively high pressure vessel, without removing pressure due to instrument failure could lead to explosions	Signal alarm for excess flow rate. Use of a ratio controller to tailor steam flow rate to suit the crude inlet flow rate. Emergency release valve at the top of the column can be opened if needed
	Low/No	Failure of flow control system. Pipeline blockage	Vessel flooding	Check ratio controller and install sensor alarm to indicate steam flow rate is too low
Pressure	High	Steam boiler outlet pressure too high	Excessively high-pressure vessel, without removing pressure due to instrument failure could lead to explosions	Use of a pressure control valve after steam exits the boiler to drop steam pressure to desired levels, alternatively reduce steam flow rate into column to compensate for the higher pressure
	Low	Steam boiler outlet pressure too low	Flooding in vessel likely to occur	Halt operations until steam reboiler issue resolved
Temperature	High	Steam boiler outlet temperature too high	Vessel contents may become too hot, resulting in cracking occurring inside the column	Reduce steam inlet flow rate into column flow rate to compensate for the increased temperature
	Low	Steam boiler outlet temperature too low	Vessel contents may not sufficiently vaporise, resulting in flooding	If feasible, increase steam flow rate to compensate for reduced temperature; if not, cease operations until steam boiler issue

Vessel Purpose

Vacuum Column

Separate out valuable products from atmospheric residue

Parameter	Guide word	Cause	Consequences	Action
Line	**VD-105-254**			
Intention	Transport heated atmospheric residue to the vacuum column			
Flow	High	High output from the ADU. Flow control valve not responding	Flooding within the column, leading to process failure	Purge vessel through bottom of the column
	Low/No	Failure in ADU system. Flow control valve no working. Pipeline Blockage	VDU system operations halted	Shutdown the flow of all VDU utility streams
	Reverse	Vessel Pressure Rise. Vessel Flooding above level of pipeline inlet	Build-up of residue within the ADU system. Pressure build-up within the pipe may lead to rupture	Fit non-return valve. Shut off manual flow valve
Temperature	High	Failure of furnace control	Excessive cracking and coking, leading to wear and potential blockage of the VDU system	Fit high-temperature warning alarm
	Low	Failure of furnace control	Decreased viscosity of the atmospheric residue, leading to potential blockage of the VDU system	Fit low-temperature warning alarm

Line	**VD-106-027**			
Intention	Transport stripping steam to the column			
Flow	High	Failure of flow control system	Increased Vapour rate in the packed column may lead to flooding conditions. Liquid build-up may result in mechanical failure of the vessel due to increased loading	Fit alarm for high flow rate warning
	Low/No	Failure of flow control system. Upstream Utility supply issue. Pipeline Blockage	No Hazardous Consequence	No action
	Reverse	Vessel flooding above level of the pipeline inlet	Residue components may create a blockage within to pipeline	Fit non-return valve
Line	**VD-108-154**			
Intention	Transport flue gas from the column to the steam ejectors			
Flow	High	Increased vapour fraction of column feed	No hazardous consequence	No action
	Low/No	Decreased vapour fraction of column feed	No hazardous consequence	No action
Pressure	High	Failure of steam ejector pressure control system. Blockage of all vessel outlets	No hazardous consequences for failure of steam ejectors. Pressures above 1 atm may lead to vessel rupture	Fit automatic pressure safety valve
	Low	Failure of steam ejector pressure control	No consequence as vessel designed to handle complete vacuum	No action
Line	**VD-113-027**			
Intention	Transport LVGO product out of the column			
Flow	High	Failure of flow control system	No hazardous consequences	No action
	Low/no	Failure of flow control system. Pipeline blockage	Build-up of liquid within the column will lead to vessel flooding	Fit alarm to alert to low levels of flow
Line	**VD-119-053**			
Intention	Transport HVGO product out of the column			
Flow	High	Failure of flow control system	No hazardous consequences	No action
	Low/No	Failure of flow control system. Pipeline blockage	Build-up of liquid within the column will lead to vessel flooding	Fit alarm to alert to low levels of flow

Line	**VD-121-013**			
Intention	Transport vacuum residue out of the column			
Flow	High	Failure of flow control system	No hazardous consequences	No action
	Low/No	Failure of flow control system. Blockage of pipeline due. Low temperature will lead to increased fluid viscosity hindering flow	Vessel flooding above the level of the stripping steam and VDU feed pipelines will lead to blockages and possible reversal of flow within these lines. May lead to equipment damage or line rupture	Fit alarm to warn of high level
	Reverse	Increased pressure in downstream processes	Vessel flooding above the level of the stripping steam and VDU feed pipelines will lead to blockages and possible reversal of flow within these lines. May lead to equipment damage or line rupture	Fit alarm to warn of high level
Level	High	Increased throughput through the vessel. Failure of level control system	Vessel flooding above the level of the stripping steam and VDU feed pipelines will lead to blockages and possible reversal of flow within these lines. May lead to equipment damage or line rupture	Fit alarm to warn of high level
	Low	Decreased throughput through the column	No hazardous consequences	No action

Vessel Purpose

Fluid Catalytic Cracker

Convert feed gas oil material into gasoline product

Parameter	Guide word	Cause	Consequences	Action
Line	**FCC-101-590**			
Intention	Transport gas oil mixture to FCC riser			
Pressure	High	Build-up of gas oils within the pipeline; blockage within the pipe	Pipeline rupture	Shut off the valve controlling flow from the mixing unit
Line	**FCC-102-337**			
Intention	Transport catalyst fluidizing steam to the FCC riser			
Flow	High	Failure of flow control system resulting in no valve response	FCC riser becomes flooded with steam, blocking feed and catalyst inlets. This could in turn lead to the entire FCC unit failing as all vessels and pipes will be over-loaded leading to mechanical failure	Fit alarm for high flow rate warning and use an automated switch to block off steam flow if the flow rate is too high
	Low/No	Pipeline is blocked or there is insufficient supply of steam from the utility source	Not a potential hazard although riser functionality will be negatively affected	No action

Line	**FCC-103-740**			
Intention	Transport content within FCC riser to the reactor vessel			
Flow	High	Flow rates coming into FCC riser are too high; once reactions have occurred, exit velocities are too high	No hazard	No action
	Low/No	There is a blockage within the riser or the streams feeding material to the riser	No hazard	No action
Line	**FCC-104-590**			
Intention	Transport FCC products to the fractionator column			
Flow	High	Velocity of products from cyclone too high or the control system is not working effectively	Subsequent process vessels could become overloaded	Use an alarm to notify workers of the issue and manually reduce the flow rate using a valve
	Low/No	Cyclone not working efficiently	Not a potential hazard but a waste of time and resources because little or no product	No action
Line	**FCC-105-337**			
Intention	Transport catalyst particles to stripping unit			
Flow	High	Previous build-up of catalyst in reactor vessel causing surge of catalyst coming out of the bottoms of the cyclone in the reactor vessel	No hazard	No action
	Low/No	Cyclone not working effectively or is blocked so catalysts exiting in product stream	No hazard but if solids present in products streams then fractionator will be damaged	No action
Line	**FCC-106-337**			
Intention	Transport material from stripping unit to regenerator through standpipe			
Flow	High	Failure of level control system	Too much material will be fed into the regenerator vessel which will result in excessive loading which could make the system fail	Use an alarm to notify workers of this control system failure and use a manual valve to control the flow
	Low/No	There is a blockage in the pipeline or in previous equipment; control system failure	Standpipe might fatigue; previous flow area will become flooded	Stop the process and check if there is no blockage; if it is due to control system failure, use a manual valve to control flow
Level	High	Too much material is being supplied to regenerator or insufficient flue gas removal	Regenerator will become over-loaded	Use an alarm to notify workers when the level is too high; manually reduce flow rate of substances into regenerator
	Low	Low level of catalyst	No hazard	No action

Line	**FCC-107-337**			
Intention	Transport flue gas and coke particles from the regenerator vessel to waste treatment			
Flow	High	A lot of air is being fed into the regenerator	No hazard	No action
	Low/No	Cyclone not effectively removing flue gas from regenerator; pipeline is blocked	Flue gas will either build up in regenerator or cyclone and cause mechanical failure if too much or will cause pipeline to rupture	Use an alarm to notify workers so they can see what the problem is and resolve it
Line	**FCC-108-337**			
Intention	Transport regenerated catalyst to FCC riser			
Flow	High	Too much catalyst has been fed initially or material other than catalyst is present	No hazard	No action
	Low/No	Pipeline/standpipe is either blocked or the control system has failed	A pipeline blockage would cause the catalyst particles to remain in the regenerator, increasing pressure and loading. This could result in mechanical failure	Use an alarm to notify workers of the issue and use an external manual valve to fix the flow rate
	Reverse	Pressure is too high at the feed point or within the FCC riser	Gas oils and steam flow into regenerator instead of the FCC riser and thus over-load the regenerator	Use a manual valve to immediately shut the flow so that adverse consequences can be avoided
Temperature	High	Either the temperature of the catalyst or steam being fed to riser is too high	Unfavourable catalyst conditions which in turn could lead to catalyst deactivation or reduction in desired product; could harm workers and damage equipment	Use an alarm to notify workers when the temperature deviates too much; use a manual valve to decrease catalyst flow into FCC riser
	Low	Not enough catalyst, steam or FCC riser is no longer adiabatic	No hazard	No action
Line	**FCC-109-337**			
Intention	Transport air from utility bank to blower			
Flow	High	Too much air is being fed to the blower from the utility bank	Blower might become over-loaded leading to mechanical failure and injury of workers	Set an alarm to notify workers of high flow rates; use manual valve to decrease flow rate
	Low/No	Pipeline is blocked or there is little to no supply from the utility bank	No hazard	No action

Line	FCC-110-337			
Intention	Transport hot air from blower to regenerator			
Flow	High	Too much air is being fed to the blower	Regenerator might become flooded leading to inadequate functionality or even vessel damage	Use manual valve to decrease the flow rate of air going to regenerator
	Low/No	the pipeline could be blocked or there is little to no air exiting the blower	No hazard	No action
Temperature	High	Air temperature not enough to maintain regenerated catalyst temperature; control system malfunction	Coking could occur within the regenerator leading to cyclone overloading	Use an alarm to notify workers of high temperatures; increase air blower feed to regenerator
	Low	Too much air is being fed to the regenerator; control system malfunction	No hazard	No action

3.5.11 Plant Startup and Shutdown

3.5.11.1 Plant Startup

3.5.11.1.1 General Plant Startup Plant start-up procedures are required for the commencement of process activity for both the initial commissioning of the plant, as well as the restarting of process activities after maintenance operations. The commencement of processing activities may be a very hazardous process as well as being sensitive to slight disturbances in operating conditions. It is, therefore, essential to flow the correct procedures in for to safely attain stable operations. The start-up procedures will closely follow those as recommended by KLM Technology Group, which satisfy the required standards and codes as given by the American Petroleum Institute and the American Society of Mechanical Engineers (KLM Technology Group 2011).

Upon the completion of the construction of the modular refinery, the following steps must be taken to ensure all units have been constructed and are integrated correctly:

Operation Check Out – Ensure that all lines, valves and instrumentation have been installed correctly as prescribed by the P&ID. It is also required that all vessel nozzles and drains are check, as well as any refractory linings and vessel insulations.

Hydrostatic Testing – All vessels should be hydrostatically tested in order to ensure that they are able to withstand any mechanical stresses that would be experienced during plant operations. For vessels in which hydrostatic testing is not possible, pneumatic pressure tests should be carried out. All valves should be included in the testing to ensure proper operation. All vessels should be completely drained of water once hydrostatic testing has been completed.

Vessel Inspection – Before the vessels are bolted shut after testing the internals should be checked to ensure the proper installation of equipment.

Line Flushing – All pipelines should be cleaned of any scale build-up and flushed with air to ensure the removal of any debris, which may have accumulated during the construction process.

Final Check – Once all vessel and line checks have been completed, a final inspection should be carried out to ensure that all components have been connected correctly and safely. All monitoring instrumentation and alarm systems should also be checked to ensure that they are operating correctly.

Once all the preliminary checks have been completed the plant operation may be started, with the startup of each section being carried out sequentially.

3.5.11.1.2 Desalters The startup of the pre-treatment stage of the process will begin by cleaning the desalters using the process flow water. The mixing valve will be turned on to allow the water to flow to the desalters. Thereafter the water will flow to the desalters to remove any.

3.5.11.1.3 Atmospheric Distillation Due to the size of the column and all auxiliary equipment such as the side strippers and pumparounds, startup of the column is expected to take a significant period of time. The process requires constant monitoring and adjustment. Process controllers can be used to assist with reaching steady state, but the undertaking must be performed in a specific order so as to avoid safety issues and equipment damage. The column can be started by initially running with total reflux. Preheated feed from the furnace is added to the column, while the heating steam is added from the bottom of the column. Initially, all side draw valves in the column are closed. Once, the pressure of the vessel reaches the operating pressure, the top valve removing the gases from the column is opened and total reflux is ensued. Controllers maintain the pressure within the column during this process, with additional feed or steam being added as needed. Once, equilibrium is reached the pumparounds can be activated. Equilibrium is assumed to occur once sensors indicate that the column profile ceases to change significantly. Once, the pumparound return temperatures indicate that the set point is reached, the side draws of the column are opened and the same level procedure is followed with the stripping vessels, with withdrawal to the reboiler being controlled via level controllers on the strippers. The inlet valves are not yet operated continuously, as no final products are being withdrawn from the system boundary. Allowing the inlet valves to remain open would cause a safety hazard via pressure and the emergency pressure release valve would be activated. The product valves are only opened once the stripping columns have attained unchanging column plate profiles, and the system is at steady state. After steady state has been attained across all units, the product valves are opened. It is at this point that the feed valve and steam inlet are operated continuously, and their exact conditions being controlled to achieve the final desired set point. When the column temperature profile across plates begins to experience only insignificant variations, the system can be fully transferred to automated control and the unit can run continuously and at steady state.

3.5.11.1.4 Vacuum Distillation At the startup of the vacuum column, inert nitrogen will be pumped through the feed line and into the column. This will allow for the feed furnace as well as the pumparound heat exchangers to be started. For the oxygen should be allowed to flow and once it has stabilised the furnace fuel may be introduced and ignited. For heat exchangers, the cooling water will be pumped through. Once, the furnace has reached the desired operating temperature, the stripping steam will start to be fed into the column until the desired flow is reached. The vacuum steam may then be fed into the steam ejectors in order to create the operating vacuum pressure within the column. At this point, the nitrogen flow may be turned off, and the flow of atmospheric residue may be fed through the heating furnace and into the vacuum distillation column

3.5.11.1.5 Fluid Catalytic Cracker The FCCU start procedure is based on the procedure developed by Chaudhuri (2011) and is given by the following description: The FCC unit can be divided into 3 components which are the riser reactor, the regenerator and the main FCC fractionator. The supply of power and utilities must firstly be ensured then the airflow to the regenerator can be added while the flue gas lines to the waste heat boiler are closed. The air is blown out into the atmosphere. The control valve performance is checked and a conduction of a leaking test is done by draining the water and when there is sufficient air pressure build-up. The catalyst can be loaded to the regenerator vessel where it is fluidised. Catalyst circulation between the reactor and regenerator is allowed to occur when there is a sufficient level of build-up. The slide valves are examined in order to analyse the overriding control action for temperature performance for temperature and differential pressure drop over the control valve. The steam can now be introduced into the riser reactor, the stripper section, upstream slide valve of the spent catalyst and the downstream slide valve of the regenerated catalyst. Fluidization is now possible and occurs between the steam and the catalyst. The gasoil feed can now be fed to the riser reactor and the cracked products can be sent to the FCC main fractionator under cold circulation conditions. As the reactor temperature is established, the cold circulation is stopped and the products from the FCC fractionator are directed to their slop tanks and re-processed until the desired conditions are achieved. The flue gas from the regenerator can now be directed to the wastewater heat boiler.

There are a few points of safety which must be noted in the start-up procedure (Chaudhuri 2011).

- The isolation of hydrocarbons from air
- The gradual rises in temperature
- Ensuring good care is provided for the refractory lining
- The differential pressure drops over the slide valves
- The maintenance in freeness in the distributors for air and steam

3.5.11.2 Plant Shutdown

Plant shut-down procedures will be laid out and followed for periodical plant maintainable and inspection. The periodic nature of these inspections will help in the prevention of the occurrence of major incidences during operation which may prevent the operation of the modular refinery for large periods of time. In carrying out the shutdown of operations, the operation of each section will be stopped sequentially.

3.5.11.2.1 Desalters The shut-down process of the pretreatment stage will begin by switching off the pump that pumps crude oil to the desalters. This is to stop the flow of oil and prevent it from entering the desalter. The heat exchangers will be switched off as it will be a waste of energy to keep them running during the shut-down process. To prevent any contaminants from making it down during the shut-off stage, the process water will be kept running to pretreat the remaining oil. Once, the flow of oil has subsided, the process water pumps will be switched off. To prevent the mixing valve from malfunctioning when the plant is operational again or getting stuck while the plant is shutdown, the mixing valves will be returned to their normal positions. All controllers will be switched off.

3.5.11.2.2 Atmospheric Distillation In order to avoid product losses, the column must be shutdown effectively. In emergencies, all inlet valves can be closed and the column can be purged by opening the top safety pressure release valves and the liquid drainage valves at the bottom of the column. The same procedure can be applied to the side stripper columns. Ideally, the inlets of the column have their flow rates reduced slowly over time, allowing the ratio controllers to effectively generate products of acceptable quality such that profits are not lost by losing the contents of the vessel. Eventually, operations will be halted when the feed oil flow rate becomes sufficiently low such that the velocity inside the column results in flooding or weeping and little product will result. Once the level controls indicate that the column is beginning to flood, all inlet valves are closed and the remains of the columns contents are distilled, with the remaining liquid mixture being drained and the pressure release valve at the top of the column being opened.

3.5.11.2.3 Vacuum Distillation For the shutdown of the vacuum distillation, the feed valve to the system should be closed, in order to ensure that no atmospheric residue enters the column. Once, the column feed has been stopped, the flow of stripping steam will be stopped. The flow of vacuum steam will also be stopped to allow the column to return the ambient pressure. All oil will then be drained out of the column (note that heat exchange vessel should still be kept during the draining process to ensure the viscosity of the oil stays low enough to pump out). Once, the oil has been drained from the column, the feed furnace as well as the pumparound heat exchangers may be turned off. Cooling water will then be passed through the column, with the aim of both return the vessel to ambient temperature, as well as wash out any residual oil trapped within the column. Once the column has reached ambient temperature, compressed air will be passed through in order to remove any volatile vapours trapped within the column, as well as dry out the column, ensuring that it is safe for maintenance work.

3.5.11.2.4 Fluid Catalytic Cracker The feed gasoil stream that is allowed to enter the riser reactor must be cut off by closing the valve. The power supply to the FCC unit should now be turned off. The airflow which is allowed to flow into the regenerator should be shutdown by closing the valve allowing the air into the regenerator. The circulation of catalyst must be stopped which goes into the riser reactor by closing the regenerated catalyst recycle valve. The spent catalyst must be allowed to flow into the regenerator from the riser and this is achieved by allowing the spent catalyst valve to be opened. The catalyst from the regenerator can be unloaded by adjusting the valve at the addition hopper. The regenerated catalyst can be unloaded by opening the exit valve for the lock catalyst hopper upstream in the regenerator. A check-up on the cyclones is required to check if the cyclones are still in good condition and have not worn out or show signs of corrosion (Jawad 2009).

3.5.11.3 Conclusions on Design

Through the completion of the process control systems, an operating system was developed that would allow for automated control over the process allowing for stable operations at the steady state conditions which were determined through the process design in previous sections. This was done through the use of feedback control loop which is able to control the variables pressure, temperature and level. Feedforward control loops were used to control system flows. Through the control of these four variables, complete control over the process is obtainable. All control loops will be governed by automated PI controls which would ensure a high level of response without leading to process fluctuations.

Through the completion of a HAZOP analysis process operating hazards were identified and steps put in place to ensure that adequate prevention measures were accounted for. The HAZOP analysis together with the process control schemes were used to draw up P&ID diagrams which may be used to illustrate the connections between vessels and pipelines within the final plant as it is contributed. The P&ID document also contains the correct placement of all valves and instrumentation and how they should be connected.

Finally, process start-up and shut-down procedures were developed. These procedures will aid in ensuring the correct steps are taken during plant commissioning as well as operational stops needed for plant maintenance. The use of the prescribed procedures will allow for the start-up and shut-down process to be performed in a safe manner while preventing any damage to personnel as well as to the process equipment.

As described in Section 3.5.7, the FCC regenerator temperature was designed to operate at 600 °C, far lower than what should occur in the twenty-first century. While the designs herein do account for this and considerations are made that increase cracking efficiency significantly, the result is a high catalyst replacement rate that ultimately lowers the profitability of the plant through increasing working costs. Additionally, an exceedingly high catalyst circulation rate is present as a result of the low regenerator temperature. A main recommendation that could be made regarding this design is redoing the simulations regarding the mass balance and design of the FCC unit and downstream FCC product separator to accommodate a regenerator operating at 700 °C. Furthermore, due to the presence of a slight error surrounding the distillation column, a different equation of state could be used in simulation to reduce errors that arose.

With regards to the FCC unit, further optimisation could additionally be performed by detailed analysis of effects such as entraining medium flow rate, the temperature of gas in the riser, cyclone sizing or regenerator temperature itself. This can be performed using Aspen software via optimisation and sensitivity tools.

Further research and planning is required with regards to the expansion of the modular refinery such that fixed expansion plans at certain periods can be defined. Ideally, the refinery could produce consumer-grade gasoline and in the process eliminate the dependency on a secondary refinery to further process the naphthas and the light straight –run (LSR). This would greatly increase the profitability and future financial security of the plant. Possible further expansion opportunities by way of increasing the throughput of crude oil also need investigating. While the plant is designed to process light crude oil, other oils can be used. A cheaper alternative can be identified which would aid in reducing raw material costs.

Furthermore, this exact process used in the report may not be the ideal solution to producing a mainly gasoline-related output. The marginal effect of a coking unit could be investigated. A far more significant impact, however, would be investigating the alternative of hydrocracking as opposed to FCC. Performing a simulation and preliminary costing will help quantify what the best plant layout is.

In summary, although the present environmental calamities and devastations, and of course pressure groups demanding immediate downgrading of use of fossil fuels for petrochemicals, feed chemicals and even power productions, care should be exercised to avoid immediate reverse of the industrial revolutions achieved many years ago. As a result, Chapter 4 below presents energy transition and power reforms from coal. Thus, since power supply is major to the social and economic development of any nation some case studies are cited in the African context because power system in South Africa is dominated by coal-fired thermal generators.

References

Adlen E. & Hepburn C. 2019. 10 Carbon Capture Methods Compared: Costs, Scalability, Permanence, Cleanness. https://energypost.eu/10-carbon-capture-methods-compared-costs-scalability-permanence-cleanness (accessed 13 February 2023).

Afrox, 2016. Liquefied Petroleum Gas. Product Reference Manual. Afrox. Available at: www.afrox.co.za/internet.global.corp.zaf/en/images/Section%205%20-%20%20%20%20%20%20%20%20LPG266_154562.pdf (accessed 29 February 2016).

Agency for Toxic Substances & Disease Registry. Substance Index for ToxProfiles. Available at http://www.atsdr.cdc.gov/toxprofiles/tp76-c4.pdf (accessed 15 April 2016).

Al-Assady, Q. (2009). Characterization of petroleum fractions. *Iraqi J. Mech. Mater. Engine.* 9 (2): Available at: http://www.iasj.net/iasj?func=fulltext&aId=64185 (accessed 16 March 2016).

Andress, D., Dean Nguyen, T., and Das, S. (2010). Low-carbon fuel standard—status and analytic issues. *Energy Policy* 38: 580–591.

Arde, A.(2015. Credit Providers Must Properly Test Affordability. Available at: www.iol.co.za/business/personal-finance/news/credit-providers-must-properly-test-affordability-1835207 (accessed 20 April 2016).

Ashar, N.G. and Golwalkar, K.R. (2013). Processes of manufacture of sulfuric acid. In: *A Practical Guide to the Manufacture of Sulfuric Acid, Oleums, and Sulfonating Agents*, 9–30. https://doi.org/10.1007/978-3-319-02042-6_2.

ATSDR, 2014. Toxicological Profiles – Fuel Oils. Agency for Toxic Substances & Disease Registry. Available at: http://www.atsdr.cdc.gov/toxprofiles/tp75-c3.pdf (accessed 27 February 2016).

ATSDR, 2015. Toxicological Profiles – Gasoline. Agency for Toxic Substances & Disease Registry. Available at: http://www.atsdr.cdc.gov/toxprofiles/tp72-c3.pdf (accessed 27 February 2016).

Atumah, S. 2015. FG Licenses 23 Modular Refineries. Vanguard. Available at: FG licenses 23 modular refineries Read more at: http://www.vanguardngr.com/2015/10/fg-licenses-23-modular-refineries (accessed 23 February 2016).

Barker, D.J., Turner, S.A., Napier-Moore, P.A. et al. (2009). CO_2 capture in the cement industry. *Energy Procedia* 1: 87–94.

Bequette, B. (2003). *Process Control*. Upper Saddle River, NJ: Prentice Hall PTR.

Beychok, M. 2012. Air Pollution & Air Quality. Flue gas desulphurisation. Available at: http://www.eoearth.org/view/article/173180 (accessed 13 April 2016).

Bhutta, Z. (2013). Oil and Gas Production: PSO to Build $600 Million Oil Refinery in K-P. The Express Tribune. Available at: http://tribune.com.pk/story/534317/oil-and-gas-production-pso-to-build-600-million-oil-refinery-in-k-p (accessed 26 February 2016).

Biegler, L., Grossmann, I., and Westerberg, A. (1997). *Systematic Methods of Chemical Process Design*. Upper Saddle River, NJ: Prentice Hall PTR.

Bolton, W. (2004). *Instrumentation and Control Systems*. Oxford: Newnes.

Bredeson, L., Quiceno-Gonzalez, R., Riera-Palou, X., and Harrison, A. (2010). Factors driving refinery CO_2 intensity, with allocation into products. *Int. J. Life Cycle Assess.* 15: 817–826.

Canale, L., Yao, X., Gu, J., and Totten, G. (2008). A historical overview of steel tempering parameters. *Int. J. Microstruct. Mater. Prop.* 3 (4/5): 474.

Carnie, T. 2014. SA to Drill for Offshore Oil. IOL. Available at: www.iol.co.za/news/politics/sa-to-drill-for-offshore-oil—zuma-1765872 (Accessed 16 April 2016).

Cenam Energy Partners, 2013. The Case for Modular Refineries. Cenam Energy Partners. Available at: www.google.co.za/url?sa=t&rct=j&q=&esrc=s&source=web&cd=2&ved=0ahUKEwjV9LboiKDLAhXM1x4KHWwIAbYQFggfMAE&url=https%3A%2F%2Fcenamenergypartners.files.wordpress.com%2F2012%2F06%2Fmodular-refineries.pdf&usg=AFQjCNHfih60YkPPW9Bi9geMPPq6386E_Q&sig2=mAztQKRNniZKj9sMIhBpRQ&bvm=bv.115339255,d.ZWU&cad=rja (accessed 24 February 2016).

Ceric, E. (2012). *Crude Oil, Processes and Products*. Sarajevo: Petrolinvest.

Chaudhuri, U. (2011). *Fundementals of Petroleum and Petrochemical Engineering*. USA: Taylor and Francis.

Cheaper Petrol Party, 2016. Oil Refinery. Available at: http://www.cheaperpetrolparty.com/Refinery.php (accessed 10 February 2016).

Chemex Modular, 2016. Modular Client Experience. Available at: http://chemexmodular.com/experience (accessed 24 February 2016).

Chicago Bridge & Iron (2012). *Shell Thermal Gasoil*, 1ste. Available at: https://studylib.net/doc/8197657/shell-thermal-gasoil (accessed 30 May 2023).

Christensen, A. and Hobbs, B. (2016). A model of state and federal biofuel policy: feasibility assessment of the California low carbon fuel standard. *Appl. Energy* 169: 799–812.

Chung, W.J. et al. (2013, 2013). The use of elemental sulfur as an alternative feedstock for polymeric materials. *Nat. Chem.* 5 (6): 518–524. https://doi.org/10.1038/nchem.1624.

Cipc.co.za. 2016. CIPC :: Compliance Obligations. Available at: www.cipc.co.za/index.php/manage-your-business/manage-your-company/private-company/compliance-obligations/ (Accessed 20 April 2016).

Community Tool Box. 2015. Section 14. SWOT Analysis: Strengths, Weaknesses, Opportunities, and Threats. Available at: http://ctb.ku.edu/en/table-of-contents/assessment/assessing-community-needs-and-resources/swot-analysis/main (accessed 15 April 2016).

Coulson, J., Richardson, J., and Sinnott, R. (2005). *Chemical Engineering*. Oxford, Frankfurt [u.a.]: Pergamon Pr.

Crude Oil Pretreament in Petroleum Refining (2007). In: *The Encyclopedia of Earth*, 1ste.

Dentons 2015 Wholesale Licensing in the Downstream Petroleum Industry in South Africa and Certain Other African Jurisdictions. Available at: http://www.dentons.com/en/insights/alerts/2015/october/22/wholesale-licensing-in-the-downstream-petroleum-industry?ParentItemID=ktrrmYwQ8YF3a5SKjru8rjQh4mofQ7sXIz2lFCnRwTHpmyWPUqCnxA==&Archive=False#?utm_source=Mondaq&utm_medium=syndication&utm_campaign=LinkedIn-integration (accessed 19 April 2016).

Department Of Energy. 2004. Petroleum Products Amendment Act, 2003. Available at: www.energy.gov.za/files/policies/act_petroleumproducts_58of2003_amendment.pdf (accessed 15 April 2016).

Ellis, P. and Paul, C. (1998). *Tutorial: Delayed Coking Fundementals*, 1ste. Great Lakes Carbon Corporation Available at: http://inside.mines.edu/~jjechura/Refining/DECOKTUT.pdf (accessed 26 February 2016).

Emerson (ed.) (2012). *Improving Refinery Hydrotreater (HDS) Unit Performance with Process Gas Chromatographs*, 1ste. Emerson Process Management Available at: http://www2.emersonprocess.com/siteadmincenter/PM%20Rosemount%20Analytical%20Documents/GC_AN_42-PGC-AN-REFINING-HYDROTREATER-HDS.pdf (accessed 26 February 2016).

Encyclopedia Britannica 2016 Petroleum Refining: Vacuum Distillation. Available at: http://global.britannica.com/technology/petroleum-refining#ref624037 (accessed 20 February 2016).

Energy Mix Report. 2014. Crude Oil Refining In Africa & The Way Forward. Available at: energymixreport.com/crude-oil-refining-in-africa-the-way-forward (accessed 15 April 2016).

Engen. 2016. Fuel Price. Available at: www.engen.co.za/home/apps/content/products_services/fuel_price/default.aspx (accessed 15 April 2016).

EPA, 2010. Available and Emerging Technologies for Reducing Greenhouse Gas Emissions from the Petroleum Refining-Industry

Essential Chemical Industry, 2014. Cracking and Related Refinery Processes. Available at: http://www.essentialchemicalindustry.org/processes/cracking-isomerisation-and-reforming.html (accessed 24 February 2016).

ExxonMobil, 2016. Resid Conversion. Available at: http://corporate.exxonmobil.com/en/company/worldwide-operations/catalysts-and-licensing/resid-conversion (accessed 27 February 2016).

Fahim, M.A., Al-Sahhaf, T.A., and Elkilani, A. (2009). *Fundamentals of Petroleum Refining*, Revisede. Elsevier.

Fourie, M. 2015. Breathing Space for Polluters – But What About Public Health? Available at: www.dailymaverick.co.za/opinionista/2015-03-09-breathing-space-for-polluters-but-what-about-public-health/#.Vxczdka8qUu (accessed 09 April 2016).

Fremont C.A. 2021. Three Key Benefits of Modular Construction. Construction Business Review of Wednesday, June 16. https://www.constructionbusinessreview.com/news/three-key-benefits-of-modular-construction-nwid-222.html#:~:text=Reduced%20equipment%20capital%20expenditures%20and%20labor%20expenses%2C%20and,construction.%20Here%20are%20three%20benefits%20of%20modular%20construction%3A (Accessed 24/02/2023)

Gala-Martın, A., Tulus, V., Dıaz, I. et al. (2021). Sustainability footprints of a renewable carbon transition for the petrochemical sector within planetary boundaries. *One Earth* 4: 565–583.

Gary, J.H., Handwerk, G.E., and Kaiser, M.J. (2007). *Petroleum Refining Technology and Economics*, 5the. London: CRC Press.

Global Environment Centre Foundation. 1996. Seawater Scrubber for SO2 Removal. Available at: http://nett21.gec.jp/AIR/data/Air-104.html (accessed 13 April 2016).

Government Gazette. 2005. NEMA Amendment Act 39. Available at: www.environment.gov.za/sites/default/files/legislations/nema_amendment_act39.pdf (accessed 13 April 2016).

Government Gazette. 2009. National Ambient Air Quality Standards. Available at: https://www.gov.za/sites/default/files/gcis_document/201409/328161210.pdf (accessed 30 May 2023).

Gray, M.R. et al. (2007). The relationship between chemical structure and reactivity of Alberta Bitumens and heavy oils. *Can. J. Chem. Eng.* 69 (4): 833–843. https://doi.org/10.1002/cjce.5450690404.

Groundwork. 2002. Oil Refineries, Your Health and the Environment: What You Need to Know. Available at: www.groundwork.org.za/factsheets/Oil%20Refineries.pdf (accessed 10 April 2016).

Gumilar, A. 2009. Thermal Cracking. Chemical Engineering Processing. Available at: http://chemeng-processing.blogspot.co.za/2009/01/thermal-cracking.html (accessed 18 February 2016).

Havemann, L. 2014. Exciting Time Ahead for SA Oil, Gas Industry. BD Live. Available at: www.bdlive.co.za/opinion/2014/02/20/exciting-times-ahead-for-sa-oil-gas-industry (Accessed 15 April 2016).

Ibsen K., 2006. *Cost Estimates of Small Modular Systems*. Fluidized Catalytic Cracking.

Emissions Factors, IEA, 2019.

IEAGHG (2017). *Understanding the Cost of Retrofitting CO2 Capture in an Integrated Oil Refinery*. 2017/TR8. August. Cheltenham: IEAGHG.

Industrial Development Corporation. 2014. Broad-based Black Economic Empowerment. Available at: www.idc.co.za/about-the-idc/bbbee.html (accessed 20 April 2016).

Investopedia. 2003. Venture Capital Definition. Available at: http://www.investopedia.com/terms/v/venturecapital.asp (accessed 20 April 2016).

IPIECA. 2014. Petroleum Refinery Waste Management and Minimisation. An IPIECA Good Practice Guide. Available at: http://www.ipieca.org/sites/default/files/publications/Petroleum_refinery_waste_management_and_minimization_2014.pdf (accessed 13 April 2016).

IPS (2000). *Engineering Standards for Process Design of Crude oil Electrostatic Desalters*. Iranium Ministry of Petroleum.

Jawad, Z. 2009. *FCC Fundamentals: Turnaround Best Practices*.

Jechura, J. 2014. Crude Tower Simulation – Hysys v8.6. Colorado School of Mines. Available at: http://inside.mines.edu/~jjechura/Refining/Crude_Tower_Simulation-HYSYS_v8.6.pdf (accessed 7 March 2016)

Jechura, J. 2015. Refinery Feedstocks & Products – Properties & Specification. Colorado School of Mines. Available at: http://inside.mines.edu/~jjechura/Refining/02_Feedstocks_&_Products.pdf (accessed 20 April 2023).

Jechura, J. 2016. Delayed Coking. Colorado School of Mines. Available at: http://inside.mines.edu/~jjechura/Refining/06_Delayed_Coking.pdf (accessed 27 February 2016).

Jing, L., El-Houjeiri, H.M., Monfort, J.C. et al. (2020). Carbon intensity of global crude oil refining and mitigation potential. *Nat. Clim. Change* 10: 526–532. https://doi.org/10.1038/s41558-020-0775-3.

Jones, D. (2016). *Crude Oil Desalting*, 1ste. Available at: http://www.crudeoildesalting.com/blog (accessed 18 February 2016).

Jukic, A. 2013. Petroleum Refining and Petrochemical Processes. University of Zagreb, Faculty of Chemical Engineering and Technology. Available at: https://www.fkit.unizg.hr/_download/repository/PRPP_2013_Steam_cracking_Olefins.pdf (accessed 15 February 2016).

Kern, D.Q. (1950). *Process Heat Transfer*. UK: McGraw-Hill International Book Company.

KLM Technology Group 2011. Start-Up Sequence and General Commissioning Procedures (Project Standards and Specifications). pp. 1–12.

Kolmetz, K. (2014). *Kolmetz Handbook of Process Equipment*, 1ste. KLM Technology Group Available at: http://kolmetz.com/pdf/EGD2/ENGINEERING-DESIGN_GUIDELINES_safety_in_process_equipment_design_rev_web.pdf (accessed 20 April 2023).

Leffler, W.L. (1985). *Petroleum Refining for the Nontechnical Person*, 2nde. Pennwell Books.

Lei, T., Guan, D., Shan, Y. et al. (2021). Adaptive CO_2 emissions mitigation strategies of global oil refineries in all age groups. *One Earth* 4: 1114–1126. https://doi.org/10.1016/j.oneear.2021.07.009.

Long, N.V.D., Lee, D.Y., Jin, K.M. et al. (2020). Advanced and intensified seawater flue gas desulfurization processes: recent developments and improvements. *Energies* 13: 5917. https://doi.org/10.3390/en13225917.

Mantshantsha, S. 2013. SA's Fuel Supply Growth. Financial Mail. Available at: www.financialmail.co.za/2013/01/24/sa-s-fuel-supply-growth (accessed 17 February 2016).

Masters, G.M. and Wendell, P.E. (2014). *Introduction to Environmental Engineering and Science*, 2014. Harlow, Essex: Pearson.

Moretti, C., Moro, A., Edwards, R. et al. (2017). Analysis of standard and innovative methods for allocating upstream and refinery GHG emissions to oil products. *Appl. Energy* 206: 372–381.

National Institute for Petroleum and Energy Research (1992). *Petroleum Product Surveys*. Motor Gasolines.

National Renewable Energy Laboratory. 2006. Equipment Design and Cost Estimation for Small Modular Biomass Systems, Synthesis Gas Cleanup, and Oxygen Separation Equipment. Task 1: Cost Estimates of Small Modular Systems. Available at: http://www.nrel.gov/docs/fy06osti/39943.pdf (accessed 13 April 2016).

New Logic Research (ed.) (2003). *Using V-Sep to Treat Desalter Effluent*, 1ste. New Logic Research Available at: http://www.vsep.com/pdf/DesalterCaseStudy.pdf (accessed 16 February 2016).

News24. 2012. Petrol Price Regulation, How & Why. Available at: m.news24.com/news24/MyNews24/Petrol-Price-regulation-How-Why-20120903 (Accessed 15 April 2016).

OGCI at Work: Commited to Climate Change (OGCI, 2018).

Ogj.com. 2001. InterOil Builds First Refinery in Papua New Guinea. Available at: http://www.ogj.com/articles/print/volume-99/issue-12/processing/interoil-builds-first-refinery-in-papua-new-guinea.html (accessed 22 August 2016).

Oil 2018 (IEA, 2018).

Oil Review Africa, 2012. Call for a Dozen Modular Refineries in Niger Delta. Available at: http://oilreviewafrica.com/downstream/downstream/call-for-dozen-modular-refineries-in-niger-delta (accessed 26 February 2016).

Ojambo, F. and Bakhsh, N. 2013. Uganda Seeks Investor to Build $2.5 Billion Oil Refinery. Bloomberg Business. Available at: http://www.bloomberg.com/news/articles/2013-10-08/uganda-seeks-investor-to-build-2-5-billion-oil-refinery (accessed 27 February 2016).

Perry, R. and Green, D. (2008). *Perry's Chemical engineers' Handbook*. New York: McGraw-Hill.

PetroSA, 2009. Annual Report 2009. PetroSA. Available at: www.petrosa.co.za/discover_petroSA/Documents/PetroSA_Annual_Report_2009.pdf (accessed 19 February 2016).

Popoola, L.T., Babagana, G., and Susu, A.A. (2013). A review of an expert system Design for Crude oil Distillation Column Using the neural networks model and process optimization and control using genetic algorithm framework. *Adv. Chem. Eng. Sci.* 3 (2): 164–170. https://doi.org/10.4236/aces.2013.32020. Available at: http://file.scirp.org/Html/7-3700155_30523.htm (accessed 25 February 2016).

Popov, B. (2015). *Corrosion Engineering*. Elsevier Science.

Prager, M., Osage, D A., Panzarella, C. H., Brown, R G. 2014. Development of A Material Databook for API STD 530. ASME 2014 Pressure Vessels & Piping Conference 20–24 July 2014. PVP2014–28538.

Product Graphics 2016. What is Hazop?

Rankish. 2014. What is Waste Management? Available at: http://www.conserve-energy-future.com/waste-management-and-waste-disposal-methods.php (accessed 13 April 2016).

Rausand, M. 2005. HAZOP. Technical report.

Reuters, 2015. Construction Costs To Delay Kuwait's Al zour Oil Refinery. Arabian Business. Available at: http://www.arabianbusiness.com/construction-costs-delay-kuwait-s-al-zour-oil-refinery-597026.html#.VtcQI_l97IV (accessed 27 February 2016).

Rhodes, M. (2008). *Introduction to Particle Technology*. Chichester: Wiley.

Sadeghbeigi, R. (2012). *Fluid Catalytic Cracking Handbook*. Oxford: Butterworth-Heinemann.

SAPIA. 2016. SAPIA – South African Petroleum Industry Association. Available at: www.sapia.org.za/Overview/South-African-fuel-industry (accessed 16 April 2016).

Satou, M., Mikuni, M., Hattori, H., Nagaishi, H., Yoshida, T. 1998. Thermal Decomposition Characteristic of Vacuum Residue in Athabasca Tar Sand Bitumen. Available at: https://web.anl.gov/PCS/acsfuel/preprint%20archive/Files/43_3_BOSTON_08-98_0426.pdf. (accessed 15 April 2016).

SET Laboratories, 2023. Schematic of Visbreaking unit. Available at: https://www.setlab.com/resources/refining/thermal-cracking/#1498503315834-22c2e8e4-0376 (accessed 30 May 2023).

Sinnot, R. (2005). *Coulson & Richardson's Chemical Engineering Design*. UK: Elsevier.

South African Department of Energy. (2016). Overview: Coal Resources. Available at: www.energy.gov.za/files/coal_frame.html (accessed 15 April 2016).

South African Government. 2016. Petroleum Manufacturing Licence. Available at: http://www.gov.za/services/petroleum/petroleum-manufacturing-licence (accessed 15 April 2016).

Stark, D. 1998. Notes on Petroleum Processing. Department of Aerospace and Mechanical Engineering University of Notre Dame. Available at: https://www3.nd.edu/~msen/Teaching/DirStudies/Petroleum.pdf (accessed 27 February 2016).

Sunny, N., Bernardi, A., Danaci, D. Bui, M. Gonzalez-Garay, A. Chachuat, B. 2022. A pathway towards net-zero emissions in oil refineries *Front. Chem. Eng.*, 4, 804163. https://doi.org/10.3389/fceng.2022.804163

Szostak, R. 2006. Isomerization in Petroleum Refining. The Encyclopedia of Earth. Available at: http://www.eoearth.org/view/article/153923/ (accessed 24 February 2016).

Techno Manage, 2016. Hydrotreating. Available at: http://technoemanage.com/Process%20Technology%20Database/Refinery/Hydrotreating/hydrotreating.htm (accessed 26 February 2016).

Tesoro, 2011. Material Safety Data Sheet Naphtha. Tesoro. Available at: http://www.collectioncare.org/MSDS/naphthamsds.pdf (accessed 27 February 2016).

Tesoro. 2016. Health & Safety. Available at: http://tsocorp.com/social-responsibility/health-safety/ (accessed 09 April 2016).

The Future of Petrochemicals, IEA, 2018.

The Institute of Energy (2013). *Combustion and Emissions Control*, 53. Elsevier.

Tollefson, J. (2018). Global industrial carbon emissions to reach all-time high in 2018. *Nature* https://doi.org/10.1038/d41586-018-07666-6.

Towler, G. and Sinnot, R.K. (2013). *Chemical Engineering Design*, 2nde. UK: Elsevier.

Treybal, R.E. (1981). *Mass-Transfer Operations*, 3rde. London: McGraw-Hill.

Turton, R. (1998). *Analysis, Synthesis, and Design of Chemical Processes*. Upper Saddle River, NJ: Prentice Hall PTR.

Turton, R., Bailie, R.C., Whiting, W.B. et al. (2008). *Analysis, Synthesis, and Design of Chemical Processes*, 3rde. Upper Saddle River, NJ: Prentice Hall PTR.

Turton, R., Bailie, R.C., Whiting, W.B. et al. (2013). *Analysis, Synthesis, and Design of Chemical Processes*, 4e. New York: Prentice Hall.

U.S. Energy Information Administration 2012. Vacuum Distillation is a Key Part of The Petroleum Refining Process. http://www.eia.gov/todayinenergy/detail.cfm?id=9130 (accessed 20 February 2016).

U.S. Energy Information Administration, 2013. Coking is a Refinery Process That Produces 19% of Finished Petroleum Product Exports. Available at: http://www.eia.gov/todayinenergy/detail.cfm?id=9731 (accessed 24 February 2016).

U.S. Environmental Protection Agency, 1996. Study Of Selected Petroleum Refining Residuals. U.S. Environmental Protection Agency. Available at: http://www3.epa.gov/epawaste/hazard/wastetypes/wasteid/petroleum/study_p3.pdf (accessed 15 February 2016).

Udonne, J.D. (2010, 2011). A comparative study of recycling of used lubrication oils using distillation, acid and activated charcoal with clay methods. *J. Pet. Gas Eng.* 2 (2): 12–19.

UK Air Quality Organisation. 2016. Air Pollution. Industrial Emissions Controls: Sulphur Dioxide. Available at: http://www.air-quality.org.uk/27.php (accessed 09 April 2016).

Uppaluri, R. (2010). *Refinery Process Design*. Indian Institute of Technology.

Utikar, R., Darmawan, N., Tade, M. et al. (2010). *Hydrodynamic Simulation of Cyclone Separators*. Computational Fluid Dynamics.

Vanderschuren, M., Jobanputra, R., and Lane, T. (2008). Potential transportation measures to reduce South Africa's dependency on crude oil. *J. Energy Southern Africa*. 19 (3): 20–29.

Watanabe, K., Chiyoda, N. and Kawakami, T. 2008. Development of New Isomerization Process for Petrochemical By-products. Gasoline Production Technology Group, Research and Development Center, Cosmo Oil Co., Ltd. Available at: www3.kfupm.edu.sa/catsymp/docs/18%20symppsium%20papers/15%20cosmo.pdf (accessed 13 February 2016).

Weckhuysen, B. and Vogt, E. 2015. Fluid Catalytic Cracking: Recent Developments on The Grand Old Lady of Zeolite Catalysis. Source: http://pubs.rsc.org/en/content/articlelanding/2015/cs/c5cs00376h#!divAbstract (accessed 20 April 2013).

wiseGEEK, 2016. What is Thermal Cracking?Available at: http://www.wisegeek.org/what-is-thermal-cracking.htm (accessed 18 February 2016).

Wolf, J.B.W. (2013). Principles of transcriptome analysis and gene expression quantification: an RNA-seq tutorial. *Mol. Ecol. Resour.* 13 (4: 559–572. https://doi.org/10.1111/1755-0998.12109.

Wood Mackenzie, 2020. Oil Refining & Marketing Analysis. https://www.woodmac.com/industry/oil-and-gas/oils-refining-ngls/ (accessed 11 February 2023).

World Energy Outlook 2018 (IEA, 2018).

World Energy Outlook, IEA, 2016.

Yaws, C. (2008). *Thermophysial Properties of Chemicals and Hydrocarbons*. William Andrew.

4

Energy Transition and Power Reforms from Coal

Power supply is major to the social and economic development of any nation being the component of the Sustainable Development Goals of United Nations. South Africa for instance is a Southern African country with population of 62 million average growth of 1.43% (Ateba et al. 2019). Since power system in South Africa is dominated by coal-fired thermal generators, the energy demands increases with increasing population and in turn more coals are burned, while in other African nations Oil & Gas are used for power generation, which exacerbates greenhouse gases emissions and environmental concerns. However, abundance of renewable energy in Africa in general calls for new energy policies that will promote renewable energy-sourced power generation and ensure a cleaner environment for South Africa and the continent as a whole. Comparison of the carbon footprint and CO_2 emission from 2009 to 2018 as presented in Table 4.1 below, carbon print and CO_2 emission have slowly reduced relative to year 2009. However further reduction is required from burning fossil fuels and coals at power stations.

Coincidentally, South Africa power sector has experienced many reforms from before and after the era of 1948–1994 leading to program that is based on competition between independent power producers (IPPs) and Eskom (a government-owned non-profit company that was established by the Electricity Act of 1922) to provide electricity to consumers (Ayamolowo et al. 2022).

In the American scenario, according to Abby Neal (2020) the coal industry is centred in the regions of Western, Appalachian and the Interior. In 2018, the top five coal-producing states accounted for 71% of the nation's coal resources including Wyoming, West Virginia, Pennsylvania, Illinois and Kentucky. While In 2019, coal production amounted to 706 MMst with a decrease of 7% from 2018. Abby Neal (2020) also reported that state, local and private-sector entities have begun to create programs to ease transitions of displaced workers into new careers and coal regions into more diverse economies as the use of coal diminishes. These are efforts to create more diverse and stable economies so that workers and their families can remain in their communities.

Meanwhile, the National Economic Transition Platform (NET Platform 2020) of the United States recommended seven policies to guide federal support for communities in the transition from coal as:

- Investments in local leadership;
- Economic development through diverse entrepreneurs and businesses;
- A broad system of workforce development pathways;
- Reclamation and restoration of former coal sites;
- Investments in physical, social and digital infrastructure;
- Protections for workers, communities and the environment during bankruptcies;
- Coordination to ensure communities have direct access to federal resources.

These strategies in the NET Platform are considered consistent with research on transition plans that identify replacing and stabilising revenue streams; planning, funding and executing environmental remediation. On the local context, the various communities see the positive outlook as key components for effective energy transition strategies (Haggerty et al. 2018). The NET Platform is also seen as an all-embracing federal transition program to support a holistic, community-driven approach to economic development and diversification (NET Platform 2020).

Measuring Climate Change to Inform Energy Transitions: Carbon Footprint Calculations, First Edition. Sunny E. Iyuke.

Table 4.1 Carbon footprint and CO_2 emission report.

Year	Metric ton (Mt) C/year	CO_2-emission (Mt)
2009	95	348.08
2010	90	329.76
2011	90	329.76
2012	91	333.424
2013	89	326.096
2014	92	337.088
2015	88	322.432
2016	88	322.432
2017	90	329.76
2018	92	337.088

Source: Adapted from Mäki.

4.1 Planning for a Just Carbon Transition

From 2 to 14 December 2018, the global climate change community convened in the heart of Poland's Silesian coal region, which brings to picture recognition of those who work in the coal industry and their efforts to address climate change in Poland. The Polish government uses the opportunity of hosting COP24 to draw attention to the need for a transition towards a resilient, low-carbon world. The central focus is on social aspects to provide decent jobs for those who will be migrating from employment in the fossil fuel activities.

The need to reconcile demands for climate action and social justice in a neighbouring France to Poland, a so-called Yellow vests Movement submitted statement demanding social justice in transition to a low-carbon economy (Elliott 2019). Even though other social movements have recognised the interrelationship of economic and climate justice for years, the Paris Agreement does recognise the role of a 'just transition' as integral to climate transformation. Where a 'just transition' is defined as check list of boxes, which nations should grapple with as they put together their long-term national climate strategies.

Guided partly by the International Labour Organisation's guidelines for a just transition towards environmentally sustainable economies and societies for all, countries around the world are shifting towards a climate-resilient and low-carbon economy. These efforts are beginning to bring about changes on how people live and work. As a result, economies that rely on the production and consumption of fossil fuels should be transforming from the current skill practice, for example, the coalmine in Silesia presented in Figure 4.1.

Figure 4.1 A coal mine in Silesia, Poland. Source: Krzysztof Kwaśny/Wikimedia commons/CC BY-SA 3.0.

As countries plan thriving low-carbon economies and forward-looking for a people-centred climate strategies, some countries are already developing promising approaches to building social equity into climate solutions. For example, Costa Rica included this cross-cutting issue directly in their decarbonisation plan to be sure no workers are left behind when moving towards net-zero emissions by 2050, the Germany's long-term strategy developed a phase-out plan for coal by 2038, while in the case of South Africa, by 2030, it is expected that 24, 100 MW of coal power plants will be de-commissioned (Visser et al. 2018) and appropriate arrangements made for the workers' transition to the renewable energy employ.

In the Paris Agreement, countries are invited to prepare long-term, low emissions development strategies in addition to the national climate commitments they pledge every five years. These strategies are a far-sighted approach to climate and development to limit warming to 1.5–2 °C by the end of the century. It is believed that long term strategies can do more than reduce emissions and help governments to prepare for workforce. In this vein, the University of the Witwatersrand, commissioned a case study project to specifically train the graduating students in the design of production plant for Dimethyl Ether (DME) from Gasification as liquid fuel, a transportable product for power generation and transportation. It is expedient to mention that the design exercises below are not meant for actual production plant but for educational and training purposes.

4.2 Coal Gasification and DME Production

Kabir and Bhattacharya (2011), describes a reactive and spontaneously combustible, dried Victorian brown coals that is used in mine-mouth power stations, which is known for low efficiency and high CO_2 emission. They emphasised that production of DME is an option that could be alternative liquid fuel for power generation and transportation. Kabir and Bhattacharya (2011) studied drying equilibrium model for coal gasification and a kinetic model for DME synthesis. The effect of various parameters on the performance of the process is evaluated, which confirms that synergy of the reactive system for direct synthesis of DME is properly achieved when CO_2 is removed from the fuel gas prior to feeding to the DME synthesis reactor. They receive a maximum yield of DME for molar ratio of H_2 to CO to be 1.41 and 0.81 at the gasifier outlet and the DME reactor inlet, respectively. Process heat integration and design of appropriate catalyst for gasification and DME synthesis can result in further improvements in the process. Initial gasification and catalytic water gas shift reaction experiments indicate favorable gas composition for the DME synthesis process.

In the simulation study of the Victorian brown coal gasification (Kabir and Bhattacharya 2011), a yield reactor and a Gibbs reactor are used, where oxygen and steam are used as gasifying agents. By keeping the ratio of C:O_2 feed to the gasifier as 1, the effect of steam to carbon feed ratio to the gasifier on gasifier affects temperature and H_2 to CO molar ratio in the product as shown in Figure 4.2. Since the syngas composition as the ratio of H_2 to CO molar ratio plays a very important role in the synthesis of DME and methanol production, increase in the flow of steam to the gasifier favors the water–gas shift reaction (WGSR) as;

$$CO + H_2O \leftrightarrow CO_2 + H_2 \tag{4.1}$$

It is known that CO_2 has negative effect on DME yield (Brown et al., 1991), even though the addition of steam increases H_2 in the product and balances H_2 and CO ratio in CO-rich coal syngas, CO_2 is produced as well. The data presented in Figure 4.2 therefore presumes that 40% of steam fed to the gasifier is reacted when steam to carbon feed ratio is kept at 1. This conversion rate gradually decreases with increasing steam. If the steam is fed in excess, DME yield is hindered since it is one of the products of methanol dehydration as shown in Eq. (4.2). Thus, a trade-off must be established between CO_2 and/or steam in syngas with H_2:CO ratio to achieve optimum DME yield.

$$2CH_3OH \leftrightarrow C_2H_6O + H_2O \tag{4.2}$$

However, introducing CO_2 separation at this stage can benefit from a carbon capture and storage (CCS) system. Thus, since DME yield is strongly dependent on the ratio of H_2 and CO ratio in the reactor feed gas, DME production increases with increasing steam feed rate and reaches a maximum for the steam to carbon ratio of approximately 1.7 as shown in Figure 4.3.

As steam is further fed into the reactor, even though the H_2:CO ratio is increased in the syngas, DME synthesis decreases because the steam begins to hinder the synergic effect by destabilising the feed ratio at which the production should be optimum. In this case, Kabir and Bhattacharya (2011) conclude that at optimum, H_2 to CO molar ratio at the gasifier outlet and DME reactor inlet is 1.48 and 0.81, respectively (Figure 4.3).

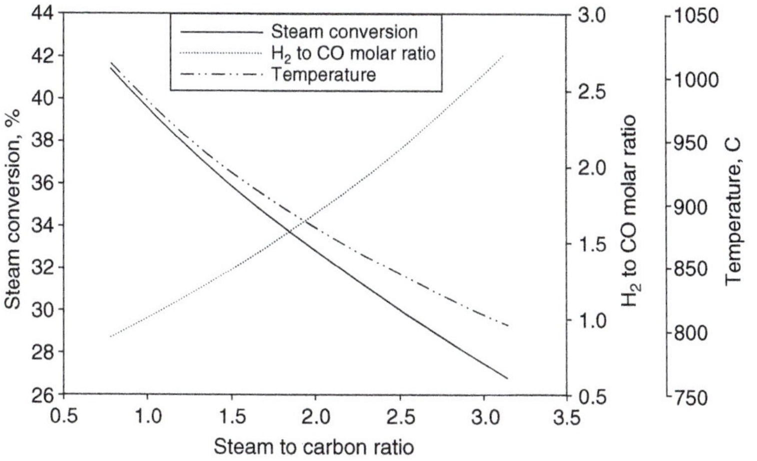

Figure 4.2 Effect of steam to carbon feed ratio on gasification parameters (Kabir and Bhattacharya 2011).

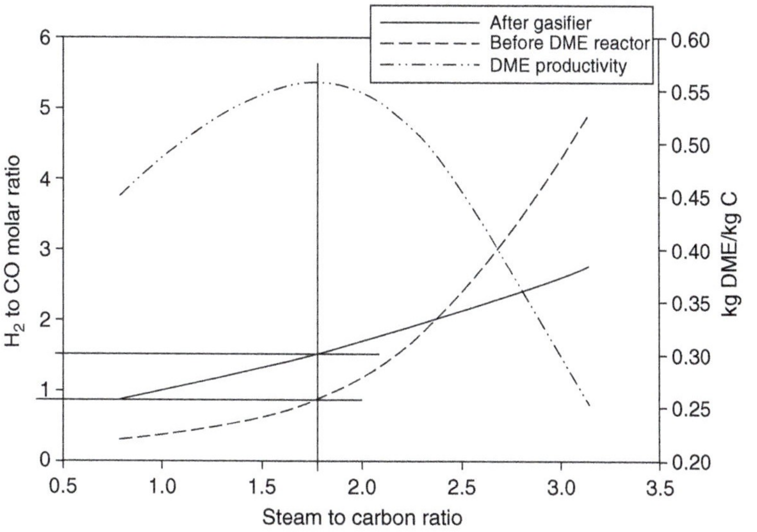

Figure 4.3 Effect of steam feed to the gasifier on H_2:CO molar ratio and DME production (Kabir and Bhattacharya 2011).

Their observation, therefore, necessitates a Balance of Plant Design for simultaneous DME and Methanol production to determine operational, safe, environmental friendly and economically feasible plant design as case study presented in the next section.

4.3 A Case Study: Balance of Plant Design for Simultaneous DME and Methanol Production, and Power Generation at Wits University

(Reprinted with permission from Darren Bouttell, Ishaam Ismail, Ketalya Reddy of the Wits University).

Balance of Plant Design for simultaneous DME and Methanol production, and Power Generation is covered in the students' final year design project report at the School of Chemical and Metallurgical Engineering at the University of the Witwatersrand, Johannesburg (Bouttell et al. 2022). The primary issue resolved is that by 2050, a formal and phased decommissioning of coal power plants is anticipated for the South African market. There would be a 35 GW electricity shortfall as a result, opening up a market for the research and use of alternative energy sources. Sasol Lurgi gasifiers may be adapted to a multiproduct facility that can produce syngas out of coal and natural gas (methane), which can be used to generate power.

The detailed design of a production process using the Integrated Gasification Combined Cycle (IGCC) is presented in the case study. A process that generates 200 MW of power, 10 000 tons of DME per year, and 5000 tons of methanol annually is planned, with the facility running for 330 days per annum. The report includes a detailed process flow diagram (PFD). This is an amalgamation of all machinery employed during manufacture. The PFD and an in-depth examination of the reactor to create DME are also described. The design chosen to move forward in the design model for the dehydration of methanol uses the NRTL-RK method, with a minimum reflux ratio of 0.42. The actual number of stages sits at 24, with a 45% column efficient. The column diameter is split into two parts with the diameter above the feed stage being 0.95 m and that below being 0.82 m. The column height sits at 12 m and the plate pressure drop for the top and bottom of the column are 173 and 157 mmHg, respectively.

Due to the thermodynamic and financial benefits, indirect synthesis for the production of DME is chosen. The conversion of 80% is the dominant industrial methodology for DME synthesis. The simplicity and inexpensive capital expenses of a fixed-bed reactor led to its selection as the reactor type. It is chosen to use the catalyst γ-Al_2O_3. γ-Al_2O_3 has a larger operational temperature range because it deactivates above 673 K. 35 tons of catalyst are calculated to achieve the needed conversion. 473–530 K would be the ideal temperature range for DME to be generated, with the lower temperature boundary yielding the largest amount. Additionally, operating the catalyst at lower temperatures is more advantageous.

The design uses cooling water that is 25 °C, with a 15 °C rise anticipated and ΔTLM is calculated to be 50.67 °C. The aspen simulation estimates the reactor's net heat duty to be around 13 MW. Therefore, the heat transfer surface area necessary to fulfil these parameters is determined to be roughly 245 m^2. Due to the feasibility of the design, a tube diameter of 40 mm is used. A rotated square-pitched arrangement is used to calculate the number of tubes to be 380. It is determined that the overall shell thickness is 7.27 mm. A cut segmental baffle arrangement has been selected for the reactor under analysis. For horizontal vessels supported at two cross sections, saddle supports work well. These supports are placed in the areas of the vessel that experience the least bending moment, which is approximately 21% from either end. Because the reactor vessel under consideration works at 10 bar, a torispherical head and closure is ideal and practical.

Grade 316 stainless steel has been used in the material construction since it resists rusting readily and is a sturdy material that requires little upkeep and repair, just as the tubes. Since the design stress for vessels made of grade 316 stainless steel at reactor operating temperature is 117.5 N/mm^2, the required thickness of the head and closure is found to be 10.54 mm after adding a 3 mm corrosion allowance. The head and closure thickness is roughly 3 mm thicker than that required by the shell. This might potentially increase the vessel's longevity and lead to long-term cost savings.

For the pipes, a 2 mm corrosion allowance is utilised. Therefore, the ideal pipe diameters for the feed and product are found to be 74 and 94 mm, respectively. The pump P-102 serves as the foundation for this design. A single-stage centrifugal pump operating at 3500 rpm is employed with a flow rate of 250 m^3/h and a head developed of approximately 150 m (calculated from the Aspen simulation). Thus, 0.263 m is determined to be the ideal pipe diameter. It is safe to presume that the pipe is 10 m long. By dividing the volumetric flowrate of the fluid by the pipe's cross-sectional area, the fluid's velocity is calculated to be 1.27 m/s.

In this plant, one compressor is employed to boost air pressure prior to combustion for the gas turbine. Compressor C-101 is responsible for raising the pressure at room-temperature air to 13 bar. Calculations resulted in an intermediate pressure of 3.6 bar. This results in an isentropic efficiency of 85% and a compression ratio of 3.6. Each stage's work is calculated to be 137.7 kJ/kg air. When isentropic efficiency is taken into consideration, the work for the entire compressor is 324 kJ/kg air instead of the original 275.5 kJ/kg air. With an air flow of 549 000 kg/h, or 153 kg/s, the compressor has a total work of 49.5 MW. Pneumatic actuators have been chosen because of their superior linear accuracy and repeatability, ability to withstand high temperatures, compliance with safety regulations due to the use of air rather than potentially dangerous fluids, light weight, minimal maintenance needs and durability of their components, making them cost-effective.

The report also looks at the column design's process control to demonstrate what safeguards would be necessary for the process. To guarantee the safe operation of the plant and the proper composition of the materials flowing, control measures pertaining to the pressure, temperature and level of the column and pipes must be maintained constant. The choice of valves, a crucial element in managing these factors, is also made. To further highlight the significance of the control and alarm systems that must be implemented in the plant, the hazard and operability (HAZOP) analysis is carried out to determine where control measures should be placed. Along with the risks of these consequences, the environmental and human effects caused by the operation of the distillation column are also evaluated. A number of laws and regulations are also discovered, and the plant must abide by them in order to avoid legal repercussions.

The Impumelelo mine in Mpumalanga is chosen as the coal mine for this venture. The Sasol mine is the deepest and is located far from any populated areas. The mine has the ability to provide Sasol Secunda Synfuels Operations with 10.5 million tpa of bituminous coal. A facility located close to the mining site would be advantageous since it would lower transportation costs and offer local populations employment possibilities. The Kusile Power Station, the Komati Power Station and the Hendria Power Station are adjacent power stations that can be used as backup power supplies in case of an emergency even though the plant produces electricity for self-sustainability. A basic plant layout that shows how the plant is to be designed to operate efficiently taking the requirements of the plant into account is presented in the report. It is anticipated that the plant would run for 330 days a year, with the remaining days being used for upkeep. After five years of operation, the project started to generate some revenue after having very little at first. At the end of the 12-year term, the net present value will be US$659517, with a net present value ratio of 5.18. It is clear from the high net resent value (NPV) ratio that the investment will generate excellent returns. Furthermore, the duration of the discount repayment is two years, which is a relatively short period of time.

4.3.1 South Africa Transits from Coal

In the case of South Africa transiting from fossil fuel dependence, a formal, staged decommissioning of coal power facilities has been planned for the South African market by 2050. As a result, there would be a 35 GW electricity shortage, creating a market opportunity for the investigation and application of alternative energy sources. A multiproduct plant that can generate electricity using coal and natural gas (methane) can be retrofitted with Sasol Lurgi gasifiers to create syngas. Along with resolving South Africa's electricity issue, this will help battle climate change by lowering greenhouse gas emissions and preventing waste from leaving the production process.

The demand for new advancements to counteract greenhouse gases is driven by the increase in emissions of various pollutants. Due to its global availability, coal has long been regarded as an energy source that may be utilised to generate electricity (Majozi, 2009). These advancements have been made possible by a variety of coal technologies, including IGCC and Underground Coal Gasification (UCCG). The IGCC configuration is distinguished by lower SO_X and NO_X emissions as well as lower CO_2 and water emissions. After going through many cleaning steps, the coal-to-synthetic-gas (syngas) process flows via a gas turbine to generate heat and electricity (Rajcoomar and Ramjeawon, 2017).

4.3.2 The Design Scope and Objectives

The problem at hand is to complete a detailed plant design in its entirety, which involves the reactor design and taking into consideration the several technical reports to choose and which distillation column fits the plant design the best. One of the objectives is to retrofit an existing gasifier to create syngas that are used in an IGCC process to generate electricity. It also aimed to construct a distillation column that will separate methanol from the waste water stream of the DME plant. It is also aimed to analyse and design the DME manufacturing plant. In order to select the best design to use moving forward, a technical review that combines a detailed design of the DME and subsequently incorporates entire plant components design to produce 200 MW of energy, 10 000 tons of DME and 5000 tons of methanol.

Thus, the scope of the problem statement includes the following aspects:

- Key components and characteristics are identified and analysed as part of a technical review to compare the specific detailed designs of each group member.
- A theoretical review on the fundamentals of reactors.
- A final PFD which would include improved designs of the distillation column and methanol dehydration reactor.
- Mass and energy balances for the balance of the plant design.
- Equipment selection and design which includes reactor design and optimisation.
- Process control, safety and environmental impact with regards to the process control and instrumentation diagram (P&ID), HAZOP and environmental assessment.
- Process economics in terms of equipment costing and profitability assessment of the design.
- Plant location and plant layout including site selection and applicable Acts important to the design.

4.3.3 Technical Review

The purpose of the technical review is to present several column specifications for usage in the distillation column to separate methanol and water. The base design is presented in Figure 4.4, which is used for the design and synthesis of a single multicomponent distillation column. The distillation column is developed as a preliminary design by each team member, who also discovered several values for the design parameters, as shown in Table 4.2.

The parameters looked at for the individual student detailed design are: (1) Thermodynamic model, (2) Minimum reflux ratio, (3) Actual number of stages, (4) Column efficiency, (5) Column diameter, (6) Column height, (7) Plate pressure drop, etc.

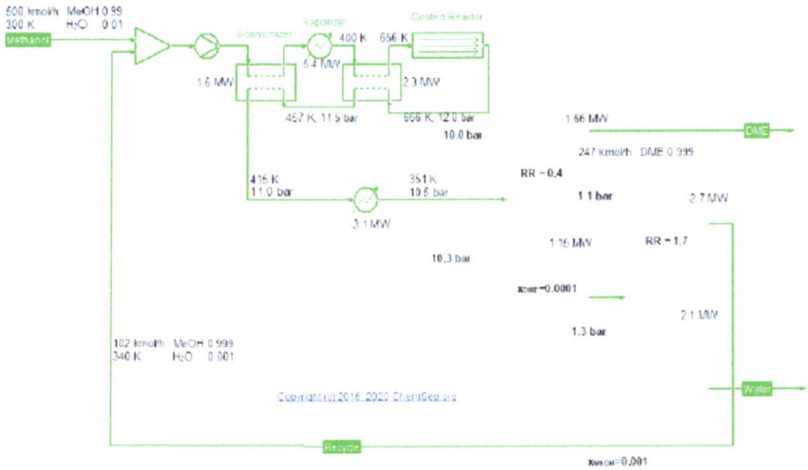

Figure 4.4 Dehydration of methanol to dimethyl ether.

Table 4.2 Table comparing the different column specifications.

Parameters Thermodynamic model	Design A NRTL-RK	B PSRK	C NRTL-RK	D NRTL	E NRTL-RK	F UNIQUAC	G UNIQUAC
Feed condition	Boiling	84% Liquid	66% Liquid	100% Liquid	100% Liquid	100% Liquid	100% Liquid
Minimum reflux ratio	0.42	0.367	0.6	0.285	0.26	0.32	0.37
Graphical correlation	Nmin versus Rmin	Erbar-Maddox	Erbar-Maddox		Ntheo versus R/Rmin		
Actual number of stages	24	27	26	19	26	21	19
Column efficiency	45%	49%	47%	43%	44.78%	46%	49.4%
Tray spacing (m)	0.5	0.5	0.4	0.7	0.45	0.5	0.6
Tray types	Single Pass Sieve	Single Pass Sieve	Single Pass Sieve	Single pass Sieve	Single pass sieve	Single pass sieve	Sieve tray
Operating pressure (bar)	10.5	10.5	10.5	10.5	9.88	10.3	10.4
Column diameter (m)	Top – 0.95 Bottom – 0.82	0.914	1.3	Top – 0.9 Bottom – 0.7	1.09	0.94	1.0669 m
Column height (m)	12	14.13	13	28	21.78		
Plate pressure drop (mmHg)	Top – 173 Bottom – 157	97.51	104.26	Top – 78 Bottom – 31	Dry – 38.48		140
Type of reboiler	Kettle	Kettle	Kettle	Kettle	kettle	Kettle	kettle
Type of condenser	Total	Total	Total	Total	Total	Total	Water-cooled

Through the process of elimination, the best-detailed design was picked from the criteria in Table 4.2. Starting with the thermodynamic model, the model that best fitted the design was the NRTL-RK which would mean that the options for the detailed design, the best fits would be designs A, C, D and E.

The column height and efficiency can be looked at together as the column efficiency indicates how well the column separates the components while the column height affects the cost of the column. What needs to be noted is that an increase in column height would be subjected not only to cost but more risk to wind damage and stress thus it is better to design a smaller column so as to prevent any damage to it. The difference of column efficiencies between the two designs is not large enough for it to be considered a deciding factor. However, some columns with a large height can be eliminated such as Design D which has the largest column height out of all the designs and the lowest column efficiency of 43% between A, C, D and E.

One of the core ideas in chemical engineering is the traditional trade-off between the number of trays and energy usage in distillation design. According to conventional knowledge, using additional stages reduces reflux ratios and reboiler tasks for a given separation (Luyben 2012). The minimum reflux ratio can be used as a design parameter as decreasing this ratio would decrease the duty of the condenser and reboiler and lead to a decrease in operation cost, which would then decrease the overall cost of the plant. Although increasing number of stages would lead to better separation, it would also lead to an increase in cost. Thus, it would be best to look at a design that has a minimum number of stages as well as the smallest minimum reflux ratio. By cancelling the other design options and moving further to look at the minimum reflux ratio, option E has the smallest ratio. However, this design parameter needs to simultaneously be taken to account with the actual number of stages, where A had the minimum actual number of stages. This means both E and A can be further evaluated to see which design would suit the plant best.

The size of the column is another element that may now be examined. In comparison to a larger column, a smaller column that meets the same design requirements is preferable since it would have reduced capital expenditures. Design E was taken out of the column possibilities because Design A had lesser dimensions than Design E. Because Design A's column has a smaller diameter, there is less risk of weeping in the rectification portion, which would reduce the column's effectiveness. Design A also chose an acceptable thermodynamic package, NRTL-RK, which accurately predicts the behaviour of the system. While Design A has a higher pressure drop than E, it was decided that the cost saved from choosing the smaller column outweighs the cost of the higher pressure drop. Therefore, Design A was chosen as the most feasible design due to its realistic, economically favorable parameters. Design A was used to design distillation column. Design A is also selected because of the NRTL-RK thermodynamic package, which successfully forecasts the system's behaviour. The decision made is that, despite Design A having a larger pressure drop than E, the cost savings from selecting the smaller column offsets the expense of the higher pressure drop. As a result, Design A is selected as the most practical design due to its reasonable and advantageous economic criteria, while distillation column design was done using Design A.

4.3.4 Overview of Reactor Design

In order to properly manage dynamic disturbances without reactor runaways, it is essential to provide enough heat transfer surface area or another type of heat removal mechanism for chemical reactors. The operation of reactors is particularly substantially influenced by temperature. Hence, reactors with endothermic or significantly reversible reactions hardly ever have issues with temperature regulation. Heat must be given to endothermic processes in order to obtain products. As a result, they do not experience the risk of 'runaway' phenomena since they are self-regulating, meaning that, as the temperature rises, the response rate speeds up, drawing out more heat and tending to bring the temperature down.

In order to design a reactor, other than temperature needing to be taken into account, the type of reactor needs to be selected. The types of reactors are examined according to their respective advantages, disadvantages, limitations and applications. After the ideal reactor is selected, the general reactor design procedure is taken into account by mapping out the steps needed to design the reactor and finally, the detailed reactor design equations are identified and used to calculate the parameters of the reactor.

4.3.4.1 Types of Reactors and Operation

The type of reactors discussed are laid out in Table 4.3 where the advantages, disadvantages, limitations and applications are listed.

In Table 4.3, three different types of reactors are looked at namely, the batch reactor, continuous Stirred Tank Reactor (CSTR) and the Plug Flow Reactor (PFR). The batch reactor is better for small-scale operations where a high conversion is favorable. Due to the variability between the batches, it would not be a favorable reactor to use. The CSTR has the downfall

Table 4.3 Comparing different reactor types.

Reactor types	Advantages	Disadvantages	Limitations	Applications
Batch	- Perfect mixing - High conversion	- Variability of products per batch - High labour costs	- Mixing never perfect in agitated vessel	- Manufacture of colorants and margarine
CSTR	- Perfect mixing - Handle higher reactant concentrations	- Concentration of reactant in the vessel is the same as the concentration of reactant in the product	- Mixing never perfect in agitated vessel	- Used in pharmaceutical industry
PFR	- No axial mixing and radial gradients	- Poor temperature control - Shutdown and cleaning is expensive	- Flow patterns in real reactor does cause axial mixing	- Used in gasoline production and synthesis of ammonia

that if the mixture does not need to be agitated, it is not the best reactor to use, and the PFR has poor temperature control eliminating the use as the design relies on temperature control.

4.3.4.2 General Reactor Design Procedure

The general reactor design procedure can be seen as follows (Towler and Sinnot 2008);

1. Design specifications: The problem that has to be solved is stated more specifically in the design basis. It often includes information about the main product's production rate, purity requirements and limitations that will affect the design.
2. General design concepts: The vast majority of process designs are built on earlier concepts. Thus, earlier designs from preliminary reports are used as a basis for this work. Each report is evaluated to find the best design.
3. Evaluation and optimisation: Design optimisation can also be done using the economic analysis of the product or process. Each design will be capable of a number of different iterations that, in certain circumstances, are economically viable. Designs that increase waste heat recovery for process reuse by utilising a large amount of heat exchange surface will be appealing in areas with high energy expenditures. Burning more fuel and lowering the plant's capital expense might be more cost-effective in areas with low energy costs.
4. Detailed design and equipment selection: The project then moves on to detailed design once the process or product concept has been decided upon. Here, the precise specifications of tools like instruments, pumps, heat exchangers and vessels are established. The design may yet undergo some changes during the detailed design phase, and optimisation will undoubtedly continue as a deeper understanding of the project's cost structure emerges. However, rather than alterations to the flowsheet, the detailed design considerations frequently centre more on the choice of equipment.

4.3.4.3 Detailed Reactor Design Equations

The temperature and the concentrations of the reactants and products play a major role in determining how quickly a chemical reaction proceeds in homogeneous systems. Other factors, such as pH, the concentration of the catalyst, the concentration of the initiator and the concentration of the inhibitor can alter the speed of reaction. Chemical reactions occur in heterogeneous systems where rates may not be completely controlled by chemical processes, thus making them more complicated as they might additionally be based on heat and mass transfer.

The conversion, X, and a basic reaction such as

$$aA + bB \rightarrow cC + dD \tag{4.3}$$

where the *moles A reacted* $= N_{A0*}X$, and N_{A0} is the moles of initial A. Thus, the conversion for a batch reactor is

$$N_A = N_{A0}(1 - X) \tag{4.4}$$

And that of any flow reactor such as the CSTR and PFR (Fogler 2006) is;

$$F_A = F_{A0}*X \tag{4.5}$$

Table 4.4 Reactor mole balance over a single reactor in terms of conversion.

Reactor	Differential	Algebraic	Integral
Batch	$N_{A0}\dfrac{dX}{dt} = -r_A V$		$t = N_{A0}\displaystyle\int_0^X \dfrac{dX}{-r_A V}$
CSTR		$V = \dfrac{F_{A0}X}{-r_A}$	
PFR	$F_{A0}\dfrac{dX}{dV} = -r_A$		$V = F_{A0}\displaystyle\int_0^X \dfrac{dX}{-r_A}$

Source: Adapted from Levenspiel.

These equations may be summarised and used for the design in terms of differential, algebraic and integral equations as presented in Table 4.3. In addition, two other factors, namely space time and space velocity need to be accounted for. While space time is the time taken to process one reactor volume based on entrance conditions;

$$\tau = \frac{V}{v_0} \tag{4.6}$$

where, V is the reactor volume and v_0 is the volumetric flowrate entering the reactor. The space velocity is used to determine the Liquid Hour Space Velocity (LHSV) and the Gas Hour Space Velocity (GHSV) with the equations (Levenspiel 1999) (Table 4.4);

$$\text{LHSV} = \frac{v_0|_{\text{liquid}}@23.889^0C}{V} = \frac{1}{\tau} \tag{4.7}$$

$$\text{GHSV} = \frac{v_0|_{STP}}{V} = \frac{1}{\tau} \tag{4.8}$$

4.3.5 Detailed Design Description and Solutions

All the units discussed in this section are presented diagrammatically in the PFD of Figure 4.5.

4.3.5.1 Gasification

The gasification of coal is the first step in the plant. A Sasol Lurgi gasification unit represented by R-101 is fed coal, methane and oxygen from streams 1, 2 and 4, respectively. There is also a small amount of additional oxygen being provided from the electrolysis unit R-102. The solids are fed from the top and the gases from the bottom, allowing it to operate in a

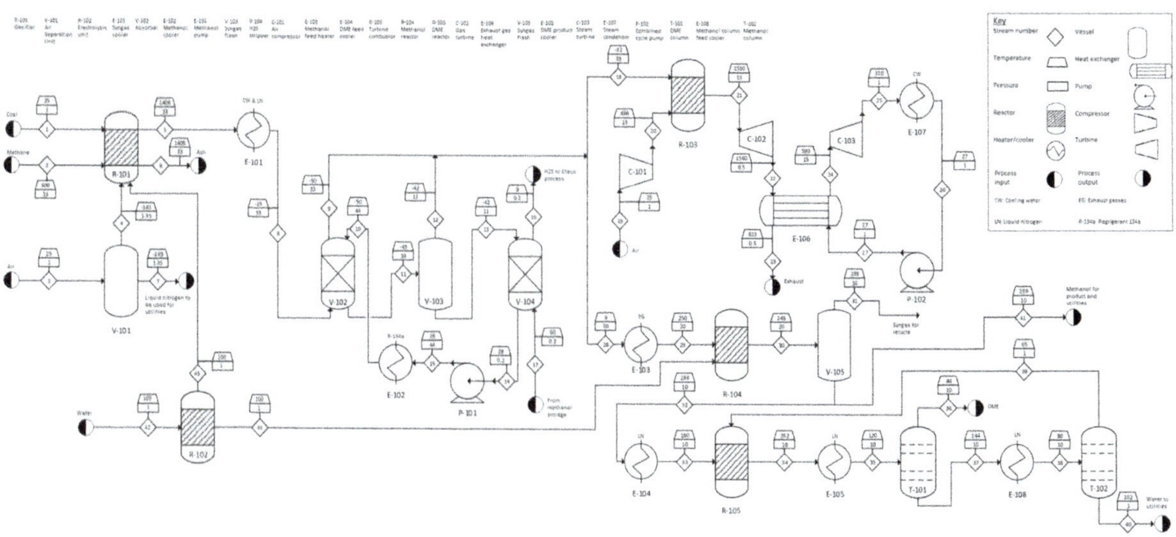

Figure 4.5 Process flow diagram (Detail components are presented in Tables 4.5–4.10).

Table 4.5 Equipment summary.

Reactors		Vessels		Columns		Heaters		Pressure changers	
R-101	**Gasifier**	**V-103**	**Flash**	**V-101**	**ASU**	**E-101**	**Cooler**	**C-101**	**Compressor**
R-102	Electrolysis unit	V-105	Flash	V-102	Absorber	E-102	Cooler	C-102	Gas turbine
R-103	Combustion chamber			V-104	Stripper	E-103	Heater	C-103	Steam turbine
R-104	Methanol reactor			T-101	DME column	E-104	Cooler	P-101	Methanol pump
R-105	DME reactor			T-102	Methanol column	E-105	Cooler	P-102	Water pump
						E-106	Exchanger		
						E-107	Cooler		
						E-108	Cooler		

Table 4.6 Stream summary 1–12.

Stream number	Units	1	2	3	4	5	6	7	8	9	10	11	12
Temperature	[°C]	25	600	25	−181	1408	1408	−193	−25	−50	−50	−43	−42
Pressure	[bar]	1	33	1	1.35	33	33	1.35	33	33	44	33	11
Vapor fraction		Solid	0	1	0	1	Solid	0	1	1	0	0	1
Total mass flow	[kg/h]	58 600	30 300	90 000	21 400	134 000	5050	68 600	134 000	130 000	461 000	402 000	965
Total mole flow	[kmol/h]		1890	8590	1850	9070	0	6740	9070	8910	14 400	12 600	56.8
Component flows [kmol/h]													
N_2		0	0	6790	103	149	0	6690	149	148	0	1.46	9.50E-1
O_2		0	0	1800	1750	0	0	52.9	0	0	0	0	0
H_2		0	0	0	0	4610	0	0	4610	4570	0	32.9	23.9
CO		0	0	0	0	4200	0	0	4200	4150	0	51.8	31.2
H_2O		0	0	0	0	36.3	0	0	36.3	0	36.3	36.3	5.04E−6
CO_2		0	0	0	0	9.28	0	0	9.28	6.67E−2	3.10E−6	9.21	1.06E−1
H_2S		0	0	0	0	19.6	0	0	19.6	1.83E−5	1.03E−3	19.6	5.48E−2
COS		0	0	0	0	7.93E−1	0	0	7.93E−1	4.90E−6	2.29E−5	7.92E−1	2.66E−3
CH_4		0	1890	0	0	41.2	0	0	41.2	38.4	0	2.72	5.77E−1
Stream number	Units	1	2	3	4	5	6	7	8	9	10	11	12
CH_3OH		0	0	0	0	0	0	0	0	4.00E−1	14 400	12 400	1.02E−2
Coal	[kg/h]	58 600	0	0	0	0	0	0	0	0	0	0	0
Ash	[kg/h]	0	0	0	0	0	5050	0	0	0	0	0	0

Stream summary 13–24

Stream number	Units	13	14	15	16	17	18	19	20	21	22	23	24
Temperature	[°C]	−42	28	28	9	50	−32	25	456	2510	1580	613	593
Pressure	[bar]	11	0.2	44	0.2	0.2	33	1	13	13	0.5	0.5	15
Vapor fraction		0	0	1	1	1	1	1	1	1	1	1	1
Total mass flow	[kg/h]	401 000	461 000	461 000	2740	62 700	121 000	549 000	549 000	670 000	670 000	670 000	240 000
Total mole flow	[kmol/h]	12 500	14 400	14 400	92.2	1960	7710	19 000	19 000	23 100	23 100	23 100	13 300

Table 4.6 (Continued)

Stream summary 13–24													
Stream number	**Units**	**13**	**14**	**15**	**16**	**17**	**18**	**19**	**20**	**21**	**22**	**23**	**24**
Component flows [kmol/h]													
N_2		51.5E-1	0	0	5.15E−1	0	126	15 000	15 000	15 000	15 000	15 000	0
O_2		0	0	0	0	0	0	4000	4000	76.7	76.7	76.7	0
H_2		9.00	0	0	9.00	0	4030	0	0	80	80	80	0
CO		20.6	0	0	20.6	0	3520	0	0	80	80	80	0
H_2O		36.3	36.3	36.3	2.19E−2	0	0	0	0	3920	3920	3920	13 300
CO_2		9.11	3.10E−6	3.10E−6	9.11	0	5.66E−1	0	0	3920	3920	3920	0
H_2S		19.5	1.03E−3	1.03E−3	19.5	0	0	0	0	0	0	0	0
COS		7.90E−1	2.29E−5	2.29E−5	7.90E−1	0	0	0	0	0	0	0	0
CH_4		2.15	0	0	2.15	0	32.6	0	0	0	0	0	0
CH_3OH		12 400	14 400	14 400	30.5	1960	3.39E−1	0	0	0	0	0	0

Stream summary 25–36													
Stream number	**Units**	**25**	**26**	**27**	**28**	**29**	**30**	**31**	**32**	**33**	**34**	**35**	**36**
Temperature	[°C]	310	27	27	9	250	245	239	239	180	352	120	44
Pressure	[bar]	1	1	44	20	20	20	10	10	10	10	10	10
Vapor fraction		1	0	0	1	1	1	1	1	1	1	0.704	0
Stream number	Units	25	26	27	28	29	30	31	32	33	34	35	36
Total mass flow	[kg/h]	240 000	240 000	240 000	3190	3190	3500	850	1590	1590	1890	1890	1140
Total mole flow	[kmol/h]	13 300	13 300	13 300	221	221	205	122	49.7	49.7	59.0	59.0	24.9
Component flows [kmol/h]													
N_2		0	0	0	0	0	0	0	0	0	0	0	0
O_2		0	0	0	0	0	0	0	0	0	0	0	0
H_2		0	0	0	116	116	98.8	98.8	0	0	0	0	0
CO		0	0	0	105	105	22.6	22.6	0	0	0	0	0
H_2O		13 300	13 300	13 300	0	0	3.07E−1	3.07E−1	0	0	24.8	24.8	2.29E−2
CO_2		0	0	0	1.91E−5	1.91E−5	3.92E−1	3.92E−1	0	0	0	0	0
H_2S		0	0	0	0	0	0	0	0	0	0	0	0
COS		0	0	0	0	0	0	0	0	0	0	0	0
CH_4		0	0	0	0	0	0	0	0	0	0	0	0
CH_3OH		0	0	0	0	0	82.9	0	49.7	49.7	9.39	9.39	9.85E−2
CH_3OCH_3		0	0	0	0	0	0	0	0	0	24.8	24.8	24.8

Table 4.6 (Continued)

Stream summary 37–44									
Stream number	**Units**	**37**	**38**	**39**	**40**	**41**	**42**	**43**	**44**
Temperature	[°C]	144	80	65	102	239	100	100	100
Pressure	[bar]	10	10	1	1	10	1	1	1
Vapor fraction		0	0.439	1	1	1	1	1	1
Total mass flow	[kg/h]	742	742	298	444	1060	2790	2480	310
Total mole flow	[kmol/h]	34.0	34.0	9.34	24.7	33.2	155	78	155
Component flows [kmol/h]									
N_2		0	0	0	0	0	0	0	0
O_2		0	0	0	0	0	0	78	0
H_2		0	0	0	0	0	0	0	155
CO		0	0	0	0	0	0	0	0
H_2O		24.7	24.7	7.79E−2	24.7	0	0	0	0
CO_2		0	0	0	0	0	0	0	0
Stream number	Units	37	38	39	40	41	42	43	44
H_2S		0	0	0	0	0	0	0	0
COS		0	0	0	0	0	0	0	0
CH_4		0	0	0	0	0	0	0	0
CH_3OH		9.26	9.26	9.26	6.17E−4	33.2	0	0	0
CH_3OCH_3		6.32E−5	6.32E−5	6.33E−5	0	0	0	0	0

Key raw materials and product flows

Raw materials	Stream	Flow [kg/h]	Product	Stream	Flow [kg/h]
Coal	1	58 600	Methanol	41	1060
Methane	2	30 300	DME	36	1140

counter-current fashion. The gasification takes place at 33 bar, so that the pressure does not need to be changed in the future for the Rectisol process. The ash is removed from the bottom of the unit and the syngas is removed from the top at around 1400 °C.

The oxygen is provided by the air separation unit V-101. This separates air at room temperature and pressure into oxygen and nitrogen-rich streams. The oxygen stream has around 94% oxygen with the remainder being made up of nitrogen. The nitrogen in stream 7 is very cold and can be used later in other processes for cooling.

4.3.5.2 Syngas Clean up

The next step is the syngas clean-up stage. This process takes place at high pressure and low temperature. The syngas from the gasifier unit is still very hot and is cooled down to −25 °C in unit E-101. The initial unit is the absorber V-102. This works by contacting the syngas with the chilled methanol to absorb carbon dioxide, hydrogen sulphide and carbonyl sulphide. This operates at 33 bar with the syngas and methanol entering counter currently at −25 and −50 °C respectively. The syngas that leaves the top can be considered clean as the majority of impurities have been removed. The methanol stream leaving the bottom of the unit, represented by stream 11 contains the dirty methanol contaminated with carbon dioxide, hydrogen sulphide and carbonyl sulphide. This normally goes to two flashes, one to remove any syngas that has been trapped in the methanol and the next to remove carbon dioxide. However, due to the operating conditions of the gasifier, very little carbon dioxide is produced, and the carbon dioxide can be left in for later use in the methanol generation step. The flash that is left over is V-103, which operates at 11 bar and separates the syngas and some carbon dioxide to be re-joined with the clean syngas. The remaining methanol only has hydrogen and carbonyl sulphide left, which needs to be cleaned before the methanol can be recycled back into the process. This separation is performed in the stripper V-104, which contacts vaporised methanol at 0.2 bar with the methanol stream, removing the impurities so that they can be sold or processed in the Claus process. The cleaned methanol is then fed back to a pump P-101 and a heater E-102 where it is pressurised and cooled down so that it can be fed into the process again. The Rectisol process is seen to be effective at removing 99.5% of the hydrogen sulphide and 99.6% of the carbonyl sulphide from the initial syngas stream (Figure 4.5).

4.3.5.3 Power Generation

The power generation system consists of a gas turbine and a steam turbine cycle. The gas turbine section takes in air from the atmosphere and uses compressor C-101 to increase its pressure to 13 bar. This high-pressure air is then combusted with the majority of the syngas coming from the Rectisol process. This high pressure, high temperature is then sent to the turbine C-102 which reduces its temperature to 0.5 bar. This remaining exhaust gas is then run through heat exchanger E-106 to transfer some of its remaining heat energy to the water in the steam cycle. This temperature change is enough to completely vaporise the water. This high-pressure steam is then run through steam turbine C-103 which produces more electricity. This steam is then cooled back to liquid form in E-107 and pumped back up to 15 bar using pump P-102.

4.3.5.4 Methanol and DME Synthesis

The methanol synthesis takes the remainder of the cleaned syngas from the Rectisol process, which is heated in E-103 before being fed to the methanol reactor R-103. The vast majority of the syngas is converted to methanol. The resulting stream still has some syngas in it, which is removed in flash V-105 at 10 bar. This syngas is then recycled back into the process. The bottoms of the flash is methanol and needs to be turned to DME. This is first cooled in heater E-104 and is then fed to reactor R-105. The resulting product is a mix of DME, methanol and water. The product stream is separated and cooled in E-105 before being fed to distillation column T-101 which operates at 10 bar. The distillate is 99.99% pure DME which is taken to storage for selling. The bottoms comprise methanol and water and are cooled in E-108 before being fed to distillation column T-102 which operates at 1 bar. The distillate is high-purity methanol and is recycled back to reactor R-105. The bottoms is high-purity water and is saved for replenishment of the water in the rest of the process.

4.3.6 Reactor Design

The preferred method for DME synthesis is the conventional indirect method due to thermodynamics and economic advantages over the direct DME synthesis. Methanol dehydration reactor that allows the production of 10 000 tons/annum of DME and excessive methanol of 5000 tons/annum is the heart of the entire plant for conceptual design. The chemical reaction that takes place in the reactor is represented by the equation:

$$2CH_3OH \longrightarrow CH_3OCH_3 + H_2O \tag{4.9}$$

The conversion of 80% is the leading industrial approach for DME synthesis with various parameters such as reaction conditions, type of catalyst, reactor length, space velocity and pressure drop effects, all of which govern the conversion. Higher conversions are favorable to allow less unreacted material recycling that minimises the entire process's total annual cost. In this reactor design, the same approach in search of suitable design parameters that meet the desired production of DME was chosen.

4.3.6.1 Reactor Type for DME Synthesis

Methanol dehydration is a gas phase reaction that can take place in a fixed bed reactor, fluidised bed reactor, slurry reactor or the recent innovative ways such as catalytic distillation column and membrane reactors.

Slurry phase reactors are the first type of DME reactors to be deployed. These are thought to have good potential to produce DME commercially. These reactors have advantages including improved heat transmission and reduced investment costs. However, because these reactors have a recycle stream, main reactor and liquid–gas separation unit, their equipment design is extremely sophisticated. Another problem that restricts the reactor's use for DME synthesis is the loss of catalyst particles from the reactor due to collisions (Rauch et al. 2013).

A fluidised bed reactor is yet another sort of reactor setup to produce DME. The solid particles in the reactor are held in suspension by the upward flow of the reacting fluid. This promotes rapid mass and heat transfer rates as well as effective mixing (Towler and Sinnot 2008). Due to the catalyst particles natural tendency to wander around in the bed, these reactors are better at removing heat. Due to the intense mixing of the catalyst particles, the mass transfer of gases into solids within the reactor is reduced, improving temperature control. However, fluidised beds can be used with tiny catalyst particles, and the design and scaling up of the design still require empirical design methodologies and this can be cumbersome (Towler and Sinnot 2008). Catalyst is lost because of the reaction between catalyst particles and the reactor walls (Bhaskar et al. 2011).

The fixed-beds reactors are the third choice of reactors for DME synthesis. As a result of their simplicity and minimal initial investment, their applications are either at the laboratory or in pilot. Packed-bed reactors are used for gas and gas–liquid reactions. It is worth mentioning that, where high heat transfer rates are required, fluidised beds should be considered. However, larger diameter for these reactors has poor heat transfer (Towler and Sinnot 2008 and should be considered especially under isothermal conditions. These reactors can be operated isothermally or adiabatically, depending on the conditions of a catalytic process. For catalytic processes with low or intermediate heats of reaction, adiabatic fixed-bed reactors are suitable. Gas–solid contractors are used in these reactors to get over the diffusional constraints. Additionally, they are appropriate for applications that require an ideal longitudinal temperature profile from the reactor inlet to the outlet, resulting in high rates of reaction at the reactor's intake and high conversions at the outflow due to the reactor's decreasing temperature (Nelabhotla et al. 2021). In addition, they have the most consistent residence time compared to other reactors.

At the preliminary design of methanol dehydration reactor, the reactor chosen is the fixed bed reactor due to low capital cost and simplicity for use compared to slurry phase reactors. Although the increased heat transfer rate is the main benefit of a fluidised bed over a fixed bed, fluidised beds are also helpful in operations where it is required to move substantial amounts of solids, such as when catalysts are moved to another vessel for regeneration. However, their design methodologies are mostly empirical, and the process of designing and scaling up big diameter reactors is still unknown (Towler and Sinnot 2008), for these reasons, the fluidised bed reactor type is disregarded.

4.3.6.2 Catalyst Selection

Methanol dehydration reaction is a heterogeneously catalysed reaction with solid acid particles at temperature ranges between 473 K and 573 K (Masudi et al. 2020). The most efficient sites for DME production via the methanol dehydration process, according to theoretical studies, are the Bronsted acid or Lewis's acid sites. The level of acidity at the active sites affects the catalyst's ability to function. However, these catalyst's drawback is that secondary reactions, such as coking reactions, might result in the production of by-products. A decrease in catalytic activity and selectivity for DME production is the result of the observed behaviour (Sang et al. 2013). ZHMS-5 and γ-Al_2O_3 are the most widely utilised catalysts in the industry, and these catalysts will be explored below due to higher activity and stability thus an extensive study can be done to choose the better catalyst suitable for the preliminary design.

Alumina is widely used in industries as a catalyst and catalyst support due to chemical and economic advantages. These applications are feasible since alumina exists in two forms namely the corundum (α-Al_2O_3) and transition (γ-Al_2O_3) forms. The γ-Al_2O_3 is mostly preferred due to its superiority in chemical and physical properties over the γ-Al_2O_3 form (Tokay et al. 2012). However, it is hydrophobic and any contact with water causes a decrease in activity due to competitive adsorption of methanol on the catalyst surface, thus when using this catalyst for DME synthesis, pure methanol should be used as a reagent (Tokay et al. 2012).

Zeolite catalyst variations have a higher affinity for methanol than water molecules because of the weaker proton and zeolite bond strength caused by the partial electron transfer. They consequently deactivate less during the methanol dehydration reaction and when the feedstock contains water. Even though the zeolites in water react to generate hydrocarbon coke, which reduces its porosity and the number of active sites available, the catalyst becomes inactive as a result. The addition of Na or Ti to H-ZSM-5 molecular sieve can improve the selectivity of DME.

It is decided to employ the γ-Al_2O_3 catalyst because, as opposed to H-ZSM-5 as it is cheaper to use than H-ZSM-5. It is also reported to demonstrate a higher activity at lower temperatures that are more advantageous in terms of selectivity for the DME formation than the zeolite. In addition, it produces fewer by-products, however, it is limited to pure methanol only as a regent as water would deactivate the catalyst (Bercic and Levec 1993). Finally, whilst γ-Al_2O_3 deactivates at temperatures higher than 673 K providing a wider operational temperature range in comparison to HZSM-5 that begins to do so at temperatures above 573 K, which further motivates the use of alumina as DME synthesis is best operated whereby the top boundary is already 573 K. Therefore, if operated at 573 K, the zeolite catalyst would undergo deactivation and thus be inefficient.

4.3.6.3 Isothermal Versus Adiabatic Condition

One can model the heterogeneous fixed bed reactor isothermally or adiabatically at temperatures between 473 K and 573 K. The conversion is low in isothermal conditions; however, in adiabatic conditions, the temperature of the catalyst bed rises from the temperature of the feed input to a maximum value. When using cooling utilities, operating expenses for isothermal conditions are higher than for adiabatic situations when using insulators (Towler and Sinnot 2008). Due to slower convective heat transfer from the catalyst bed to the gas bulk flow, adiabatic systems require longer to reach steady state. Prior to reaching thermal equilibrium, the catalyst bed's temperature rises even further (Ghavipour and Behbahani, 2013). Since heat is not lost through the wall in adiabatic reactors, the temperature increases constantly down the length of the reactor, increasing conversion (Ghavipour and Behbahani, 2013). It is disadvantageous, therefore, to operate under adiabatic conditions when utilising γ-Al_2O_3 as at higher temperatures, catalyst deactivation would occur. In that sense, instead of conversion increasing, it would decrease and therefore be unfavorable. Although isothermal conditions may increase the running cost of the equipment, the cost of maintaining, replacing and heat exchanger later on when operating adiabatically outweigh that of isothermal conditions. Finally, due to the reaction's moderate exothermic conditions, it is not difficult to maintain isothermal conditions (Azizi et al., 2014).

4.3.6.4 Modelling of Methanol Dehydration Reactor

The PBR reactor model is chosen to model the methanol dehydration reactor with following assumptions:

- The operation of the reactor is at steady state conditions.
- Radial dispersion of mass and energy is negligible.
- The axial diffusion of mass and heat are negligible due to high gaseous velocity.
- Ideal gas law.
- It is assumed that only methanol enters into the reactor, as once the conversion of syngas to methanol has occurred, the product goes through distillation to obtain a methanol-only stream used in the DME synthesis.
- Catalyst deactivation is ignored due to significant lifetime of the catalyst

The design equation used to determine the relationship between the catalyst weight (W) and conversion is given as:

$$r_i' = \frac{\mathrm{d}F_i}{\mathrm{d}W} = -\frac{F_{io}\mathrm{d}X}{\mathrm{d}W} \tag{4.10}$$

The conversion (X) is defined as:

$$X = \frac{\text{moles of methanol reacted}}{\text{moles of methanol fed}} \tag{4.11}$$

The relationship between the molar flow of each component on the reactor system and the catalyst weight can be deduced from the two equations above as:

$$\frac{\mathrm{d}F_{\text{methanol}}}{dW} = -r' \tag{4.12}$$

$$\frac{\mathrm{d}F_{\text{water}}}{dW} = \frac{1}{2}r' \tag{4.13}$$

$$\frac{\mathrm{d}F_{DME}}{\mathrm{d}W} = \frac{1}{2}r' \tag{4.14}$$

The equations above provide a general overview of the basic reactor modelling, the detailed reactor design section below contains a more in-depth study into the reactor equations and modelling.

4.3.7 Detailed Reactor Design

All detailed calculations, data used and MATLAB codes that are used in determining the results for the sections have been well presented in the sections on appendices in the main report (Bouttell et al. 2022).

4.3.7.1 Reactor Profiles

Previous studies into the reaction kinetics of the catalytic dehydration of methanol to produce DME over γ – alumina provided kinetic equations not based on conditions found in industrial reactors making them inaccurate and invalid. Bercic and Levec (1993) developed an intrinsic rate equation given as:

$$r_{\text{DME}} = kK_{\text{MeOH}}^2 \left[\frac{C_{\text{MeOH}}^2 - ((C_{\text{H2O}}C_{DME})/K_{\text{eqm}})}{(1 + 2\sqrt{K_{\text{MeOH}}C_{\text{MeOH}}} + K_{\text{H2O}}C_{\text{H2O}})^4} \right] \tag{4.15}$$

which is a more realistic representation of the dehydration reaction over the γ – alumina catalyst (Bercic and Levec 1993). Given that the above rate equation is equilibrium limited, the equilibrium constant, which is temperature dependent is given by:

$$\ln(K_{eqm}) = 0.86 \log T + \frac{3138}{T} + 1.33 \times 10^{-3}T - 1.23 \times 10^{-5}T^2 + 3.5 \times 10^{-10}T \tag{4.16}$$

At equilibrium, the rate of reaction is equal to 0, thereby rearranging the rate equation at equilibrium yields the following equation.

$$K_{eqm} = \frac{C_{H2O}C_{DME}}{C^2_{MeOH}} \tag{4.17}$$

Following the reaction stoichiometry in order to solve for the component concentrations in the exit stream, and knowing that at equilibrium, the reaction rate would be equal to zero, both equations can be equated to give the final equation used in obtaining data for the XT-equilibrium curve as follows:

$$e^{0.86\log T + \frac{3138}{T} + 1.33\times10^{-3}T - 1.23\times10^{-5}T^2 + 3.5\times10^{-10}T} = \frac{\left(\frac{1}{2}C_{MeOH0}(X)\right)^2}{(C_{MeOH0}(1-X))^2} \tag{4.18}$$

Using excel solver, the conversions at the various temperatures are found and the data plotted as in Figure 4.6.

The plot depicts the conversion changes as the reaction occurs at various temperatures at equilibrium. It can be seen that as the temperature increases, the conversion decreases. However, the methanol dehydration reaction is optimal when operating between the temperature ranges of 473–573 K at approximately 10 bar (Masudi et al. 2020). A more in-depth analysis into the conversion-temperature profile at equilibrium, within the optimum operating range, is presented in Figure 4.7.

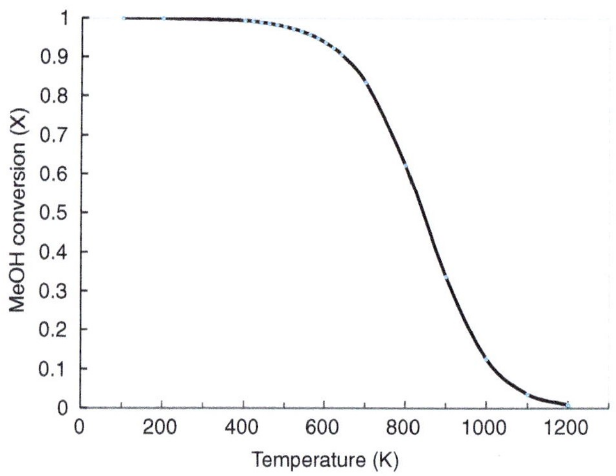

Figure 4.6 Overall conversion-temperature profile at equilibrium.

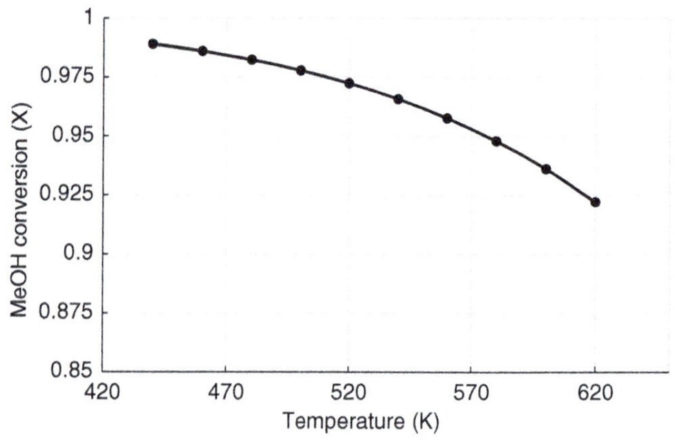

Figure 4.7 Conversion-temperature profile at equilibrium within optimal operating range.

Following the trend, the conversion decreases from 0.98 to 0.95 which is very minimal. Therefore, at equilibrium, it can be concluded that the achievable conversion within the operating temperature range is approximately 96.5%.

In addition to using the reaction kinetics at equilibrium to obtain the conversion temperature plots, an Aspen plus simulation is created whereby isobar plots of conversion and temperature are done in order to compare to those of the reaction kinetics as shown below as well as observe the effects of pressure on the equilibrium reaction kinetics.

It is observed that the results obtained from Aspen follows the same trend, whereby an increasing temperature results in a decreasing conversion (Figure 4.8). A difference can be seen whereby Aspen produced a more linearly decreasing relationship while the reaction kinetics is more non-linear. Additionally, at the same temperatures on both plots, Aspen produced conversions lower than that of the reaction kinetics (decreasing from 0.905 to 0.86), but still fairly accurate. Furthermore, Aspen confirmed the conclusion that pressure has negligible effect on the equilibrium constant as each isobar produced approximately the same profile. The initial simulation operated the reactor at 10 bar and is sufficient following the above analyses.

A Levenspiel plot can be conducted which will provide insight into the required volume of the reactor. From the plot, the sequence of CSTR's and PFR's can be determined, only if multiple reactors are required to achieve the desired conversion. The optimal outcome would be the ability to use a single reactor as this would require less capital and operating costs. The plot is obtained by plotting $\frac{F_{\text{MeOH0}}}{-r}$ against conversion (X), whereby the area under the curve corresponds to the volume of a PFR, while the entire area of the curve corresponds to the required CSTR. In order to obtain the reaction rates, the kinetic parameters need to be determined. The general intrinsic equations for these rate constants were found as summarised in the Table 4.7 (Bercic and Levec 1993).

Using excel, the data required for the Levenspiel plot is obtained, and plotted as per the Figure 4.9.

From the Levenspiel plot, it can be observed that at any given conversion, the area below the curve will always be less than the rectangular area of the curve. For this reason, it can be concluded that a single PFR will be able to convert the methanol into DME sufficiently, without the requirement of any additional reactors, with the added advantage of a lower reactor volume and therefore lower operating and capital costs. The reaction to produce DME is catalysed, therefore, a packed bed reactor is better suited to account for this requirement for conversion of methanol to occur efficiently. An issue encountered from the plot was the fact that after a conversion of approximately 0.9 (as per Figure 4.9) was achieved,

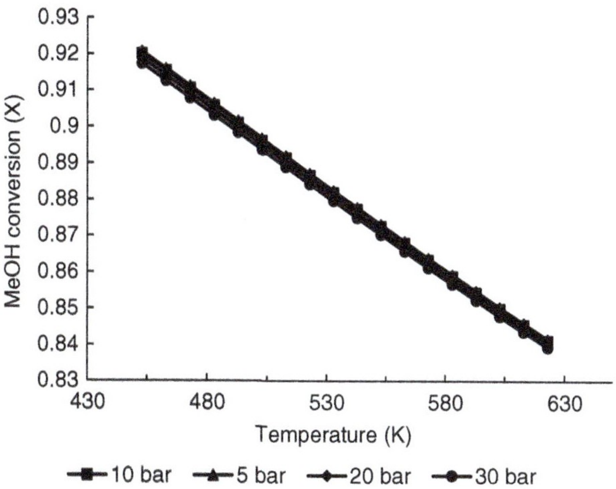

Figure 4.8 Isobar plots for conversion-temperature using Aspen.

Table 4.7 Intrinsic rate constant kinetics.

Parameter	Equation
k	$5.35 \times 10^{13} \exp\left(\frac{-17280}{T}\right)$
K_{MeOH}	$5.39 \times 10^{-4} \exp\left(\frac{8487}{T}\right)$
K_{H2O}	$8.47 \times 10^{-2} \exp\left(\frac{5070}{T}\right)$

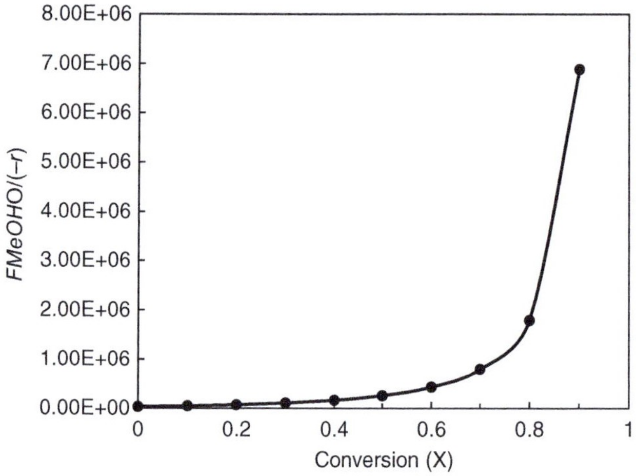

Figure 4.9 Levenspiel plot.

the reaction rates became negative. This is possibly the result of inaccuracies in the reaction rate kinetics at very high conversions. For this reason, it would be logical to aim at achieving conversions between 0.8 and 0.9. These values are out of the range of the previously determined conversions within the optimum temperature ranges; therefore, these values will need to be optimised whereby the effects of reactor length as well as the catalyst amount required would impact the achievable conversion.

4.3.7.2 Sensitivity Analysis

The aim of the reactor is to produce the required 10 000 tons/annum DME. This value is subject to temperature and pressure constraints. A sensitivity analysis is therefore conducted in order to depict the effect of these parameters on the produced DME flow.

A temperature analysis is conducted whereby temperatures within the operating range of 450–600 K are tested and the result is shown in Figure 4.10.

From the sensitivity analysis, a trend is noticed whereby an increasing temperature decreases the production rate of DME. It is clear that once a temperature of 535 K is reached, the amount of DME produced reaches below the required 10 000 tons/annum; therefore it would not be efficient to operate above this temperature. The optimal temperature range would therefore be 473–530 K with the highest amount produced achieved at the lower temperature boundary. Additionally, the catalyst is more favorably operated at the lower temperatures.

The second parameter to vary is the reactor pressure. The sensitivity analysis is conducted, and the results shown in Figure 4.11.

From the sensitivity analysis, the same trend of an increasing pressure and decreasing product produced is observed. However, observing how the DME flow rate changes, it can be seen that up to 100 bar pressure, the rate of DME only

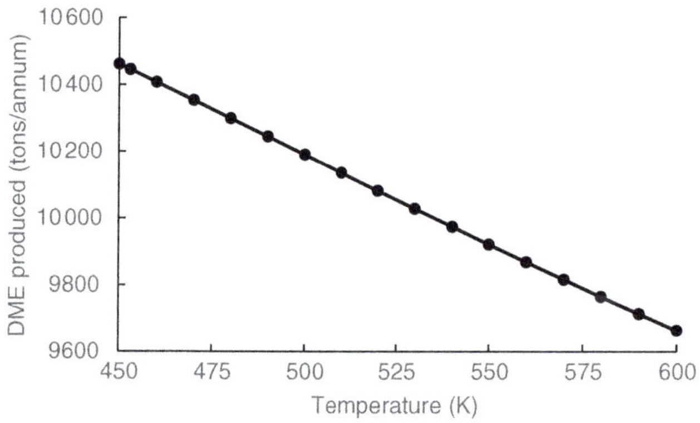

Figure 4.10 Aspen temperature sensitivity analysis.

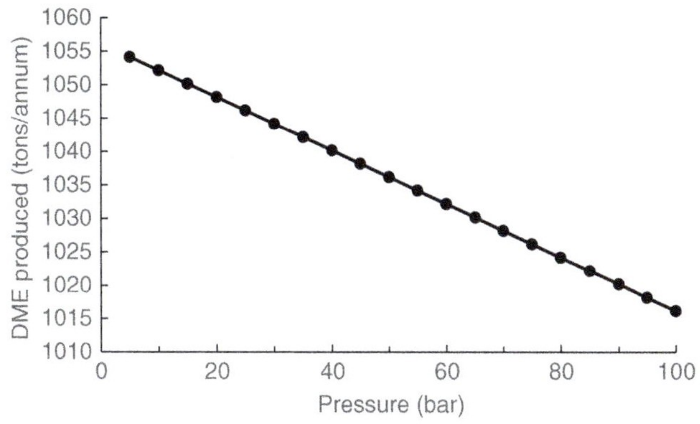

Figure 4.11 Aspen sensitivity analysis for pressure effects.

decreased by approximately 40 tons/annum, which is not much. It can therefore be concluded that pressure does not have a remarkable effect on the flow rate of DME, as will be further confirmed in the section to follow.

Following the sensitivity analysis and preliminary equilibrium profiles, an operating temperature and pressure of 500 K and 10 bar are used in the reactor modelling and optimisation of the reactor.

4.3.7.3 Reactor Modelling and Optimisation

The differential design equations of a packed bed reactor can be used in determining the relationship between various parameters. The relationship between conversion and catalyst weight was the first parameter of concern, governed by the following equation.

$$\frac{dF_{MeOH}}{dW} = -\frac{F_{MeOHo}dX}{dW} = r_{MeOH} \tag{4.19}$$

Which equates to

$$\frac{dX}{dW} = \frac{kK_{MeOH}^2}{F_{MeOHo}}\left[\frac{(C_{MeOH0}(1-X))^2 - \left(\frac{1}{2}C_{MeOH0}(X)\right)^2/K_{eqm})}{\left(1+2\sqrt{K_{MeOH}(C_{MeOH0}(1-X))}+K_{H2O}\left(\frac{1}{2}C_{MeOH0}(X)\right)\right)^4}\right] \tag{4.20}$$

Given the complexity of the above differential equation as well as those to follow, it will be difficult to solve by hand, excel or any other solver. MATLAB has therefore been utilised to code the profiles and obtain the graphical relationship as shown below.

From the plot in Figure 4.12, an exponentially increased relationship between methanol conversion and catalyst weight is observed. A maximum conversion of approximately 96% is attainable as a constant is reached after some time, whereby 75 tons of catalyst would be required. Following the previous analysis that a conversion of 0.9 is the aim, it means that approximately 35 tons of catalyst is required in reactor to achieve the desired conversion.

The governing equation, does however, also provide the relationship between the flow rates and catalyst weight as given by:

$$\frac{dF_{MeOH}}{dW} = kK_{MeOH}^2\left[\frac{(C_{MeOH0}(1-X))^2 - \left(\frac{1}{2}C_{MeOH0}(X)\right)^2/K_{eqm})}{\left(1+2\sqrt{K_{MeOH}(C_{MeOH0}(1-X))}+K_{H2O}\left(\frac{1}{2}C_{MeOH0}(X)\right)\right)^4}\right] \tag{4.21}$$

The profile was coded using MATLAB in order to obtain the following graphical relationship (Figure 4.13).

Previously, the amount of catalyst found was 35 tons to achieve the desired conversion; however, from Figure 4.13, it can be seen at that same catalyst weight, the final methanol flow rate is approximately 35 kmol/hr, making the conversion approximately 30%. The analysis, however, is done on the assumption that only the feed enters the reactors and reacts, when in fact, a recycle stream is also entering the reactor. This would result in the desired conversion being achievable but will need to be further confirmed.

In addition to the methanol flow rate, the concentration profiles of methanol, water and DME against conversion is plotted as obtained from MATLAB shown in Figure 4.14.

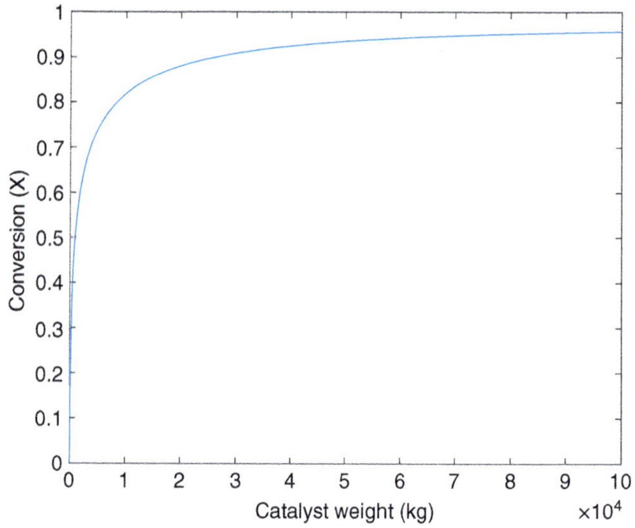

Figure 4.12 Conversion versus catalyst weight. Source: From MATLAB.

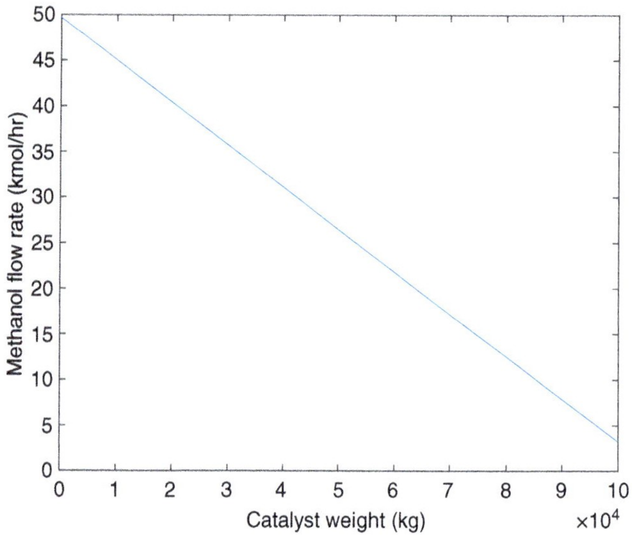

Figure 4.13 Relationship between methanol flow rate versus catalyst weight. Source: From MATLAB.

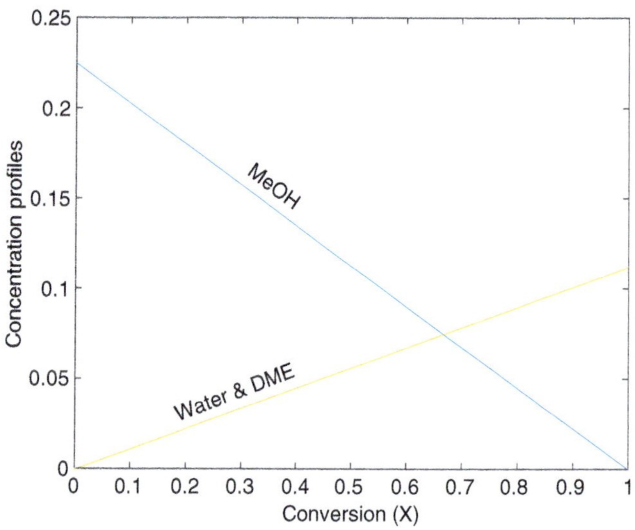

Figure 4.14 Concentration profiles versus conversion obtained. Source: From MATLAB.

As expected, it can be seen that a decreasing methanol concentration with an increasing water and DME while conversion increases. This is due to the fact that as the reaction occurs, methanol gets used up in order to produce DME and water. Therefore, the more reactant that gets reacted, the higher the conversion achieved. Additionally, it is observed that the concentration of DME and water never reaches the highest of methanol, this is due to the reaction kinetics and stoichiometry whereby a single mole of methanol produces half its amount for DME and water each.

The second relationship of interest is that of conversion and reactor length and is governed by the following equation.

$$\frac{dF_{\text{MeOH}}}{dV} = -\frac{F_{\text{MeOHo}}AdX}{dL} = r_{\text{MeOH}} \tag{4.22}$$

Which equates to

$$\frac{dX}{dL} = \frac{kK^2_{\text{MeOH}}}{AF_{\text{MeOHo}}}\left[\frac{(C_{\text{MeOH0}}(1-X))^2 - \left(\frac{1}{2}C_{\text{MeOH0}}(X)\right)^2/K_{\text{eqm}})}{\left(1+2\sqrt{K_{\text{MeOH}}(C_{\text{MeOH0}}(1-X))} + K_{\text{H2O}}\left(\frac{1}{2}C_{\text{MeOH0}}(X)\right)\right)^4}\right] \tag{4.23}$$

The profile was coded using MATLAB in order to obtain the following graphical relationship (Figure 4.15).

From Figure 4.15, an increasing reactor length results in an increasing conversion which becomes a constant after a reactor length of approximately 20 m is achieved. This is the result of the residence time increasing as the longer the reactor, the more time the reactant has to convert. For the desired 90% conversion, it is clear that the reactor should be approximately 13 m in length, with the diameter being assumed as 1 m. An analysis was conducted to assess the effects of the reactor diameter on the reactor length. It is found that an increasing diameter resulted in a longer required reactor length. Therefore, diameters that are too high would increase the capital and operating cost of the reactor rendering the process unfeasible. The choice of diameter is therefore justifiable.

The last relationship of interest is that of pressure drop and catalyst weight, as governed by the Ergun equation.

$$\frac{dP}{dW} = \frac{-\alpha}{2}\left(\frac{P_0^2}{P}\right)\left(\frac{T}{T_o}\right)\left(\frac{F_T}{F_{To}}\right) \tag{4.24}$$

Resulting in the following simplification

$$\frac{dP}{dW} = \frac{-\alpha}{2}\left(\frac{P_0^2}{P}\right) \tag{4.25}$$

where

$$\alpha = \frac{2\beta_0}{A_c\rho_c(1-\Phi)P_0}$$

and

$$\beta_0 = \frac{G(1-\Phi)}{\rho_o d_p \Phi^3}\left(\frac{150(1-\Phi)\mu}{d_p} + 1.75G\right)$$

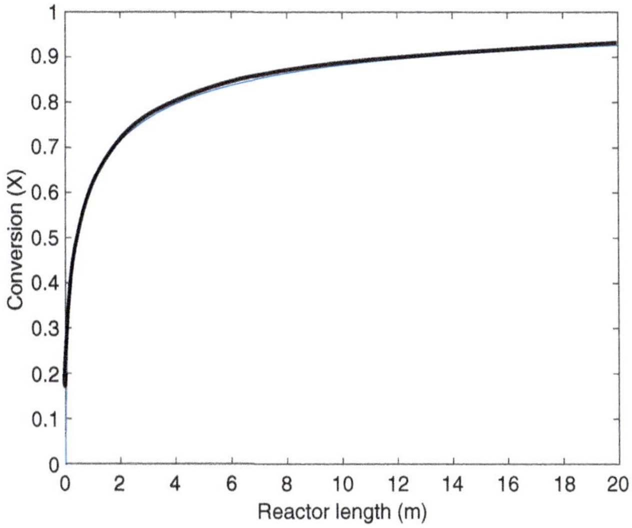

Figure 4.15 Relationship between reactor conversion and length obtained. Source: From MATLAB.

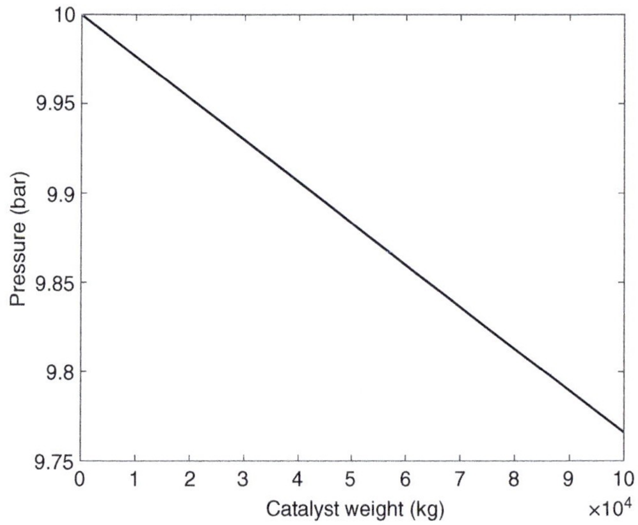

Figure 4.16 Relationship between pressure and catalyst weight. Source: From MATLAB.

The parameters were obtained and calculated, and used in solving a MATLAB code, which produced the following relationship in Figure 4.16.

It can be seen that as the amount of catalyst used increases, the pressure drop incurred increases (pressure decreases). At the required catalyst weight of 35 tons to obtain the desired conversion, the pressure experienced is approximately 0.075 bar which is very minimal and confirms the initial assumption of negligible pressure drop in the reactor.

4.3.7.4 Number of Tubes

Packed bed reactors are fixed with multiple tubes which contains the catalyst and the reactants to flow through. It is therefore empirical that the number of tubes required to achieve the design parameters as found above are calculated. For the design, a single pass tube has been chosen, whereby the reactant flows once through the catalyst, conversion into the product takes place, and the product and unreacted reactant is removed. The size of the tubes usually ranges from 16 to 50 mm in diameter, where smaller diameters are preferred for most duties owing to its compactness and thus feasibility, while larger diameters are generally used for heavily fouling liquids but are advantageous in that they are easier to clean by mechanical means (Towler and Sinnot 2008).

The number of tubes required in the reactor is calculated using the following equation:

$$N_{\text{tubes}} = \frac{A_{\text{reactor}}}{A_{\text{tube}}} \tag{4.26}$$

where N and A are number and area, respectively. A tube diameter of 40 mm has been chosen as they will be feasible, easy to clean and maintain as well as withstanding possible fouling fluids. The determined number of tubes is approximately 600. This value, however, does not account for the spacing pattern between each tube or the clearance between the shell and the tubes. Tubes can either be triangular, rotated square or square pitched in pattern. Triangular and rotated square pitched allow for increased heat transfer rates with the disadvantage of experiencing a higher pressure drop, while square as well as rotated square pitched can be used where mechanical cleaning of the exterior of the pipes is required (Towler and Sinnot 2008). Rotated square pitched arrangement has therefore been chosen as it is a combination of both mentioned advantages. The pitch pattern stipulates that the new tube diameters are now according to $1.25d_0$ with a clearance allowance of 1 cm, bringing the number of tubes required to approximately 380.

4.3.7.5 Heat Exchanger Design

In order to achieve isothermal operations, the reactor is continuously cooled, since the exothermic reaction, which takes place results in an increasing temperature. The choice between a heat exchange jacket and shell and tube type heat exchangers is looked at. A jacketed design is one where the reactor vessel is enclosed by another shell which circulates a cooling medium. Uniform heat exchange occurs between the cooling medium and the walls of the reactor vessel. A disadvantage of this configuration is that the heat transfer area is not large resulting in too low of a heat transfer coefficient (Nitsche and Gbadamosi 2016). The other design of consideration is one in which, like a shell and tube heat exchanger, the reactor contains tubes containing the catalyst in which the reaction takes place, with the cooling medium flowing around the tubes. This configuration provides low maintenance costs and most importantly, provides sufficient heat transfer area as each tube is surrounded by the cooling medium without the concern of specific heat spots not being cooled (Hesselgreaves et al. 2017). The latter of the two configurations has been chosen for this design.

The design of a heat exchanger is governed by the following equation (Towler and Sinnot 2008):

$$Q = UA\Delta T_M \tag{4.27}$$

which determines the heat transferred. However, in order to determine the mean temperature difference, the log mean temperature needs to be determined as per the following equation (Towler and Sinnot 2008)

$$\Delta T_{\text{LM}} = \frac{t2 - t1}{ln\frac{(Tsat-t1)}{(Tsat-t2)}} \tag{4.28}$$

The above equation is not the standard used as the design in consideration is operating isothermally; therefore, the bubble point temperature is used. The design makes use of cooling water at 25 °C with an assumed increase of 15 °C. ΔT_{LM} is determined to be 50.67 °C and the mean temperature is then found using the following equation (Towler and Sinnot 2008):

$$\Delta T_M = \Delta T_{\text{LM}} \times F_t \tag{4.29}$$

where F_t was determined to be approximately 0.95, resulting in a mean temperature of 48.14 °C. The overall heat transfer coefficient is extracted from a table found in Coulson and Richardson (2005) as 1150 W/m^2.°C. Additionally, according to the Aspen simulation, the net heat duty of the reactor is approximately 13 MW. Therefore, instead of acquiring the heat duty as the focus of the calculation, the heat transfer surface area required to achieve these conditions is found to be approximately 245 m^2. The surface area of a single tube is found to be 1.6 m^2, and therefore approximately 150 tubes are required for efficient heat transfer to occur. The design of the number of tubes found that 380 tubes are used inside the reactor. Therefore, it can be concluded that the reactor has sufficient heat transfer area and does not need extra pieces of equipment or an upscale of the reactor which would increase costing.

The amount of cooling water to achieve the heat transfer requirement is calculated from the following equation:

$$\text{Utility mass flow rate} = \frac{\text{Heat duty}}{(Cp)(\Delta Tm)} \tag{4.30}$$

whereby approximately 65 kg/s of cooling water is required to flow through the exchanger/reactor.

Furthermore, the above details can provide an optimisation in that only 150 tubes are actually required in the reactor, this means that savings on tube material is eminent with the possibility of decreasing the reactor size to further decrease costs. Additionally, the tubes could be increased in size to allow for a greater surface contact area for the reactants and catalyst and therefore increasing conversion.

4.3.7.6 Reactor Control

It is imperative that the temperature at which the reactor operates remains constant due to the isothermal decision. If the reactor temperature is exceeded, as previously discussed, it could have a negative effect on the conversion of methanol and thus the amount of DME product produced. The temperature can be controlled by implementing a cascade control for the reactor and cooling water temperature while ensuring that the level in the cooling water surge tank is also maintained. Additionally, some cooling water may not be returned back into the reactor and therefore a cooling water makeup stream will be required. The pressure of the reactor will also need to be maintained.

Therefore, in order to control the reactor temperature, the flow rate of the cooling water makeup stream is manipulated while in order to control the level of water in the surge tank, the removed cooling water flow rate will be manipulated. To control the reactor pressure, the feed flow rate is manipulated. A basic P&ID diagram for the reactor control is given in Figure 4.17.

4.3.7.7 Mechanical Design and Material of Construction

When analysing the multi-tubular reactors make-up, it is clear that the sections through which the reactants flow and the catalyst is contained would be the tubes within the reactor. The tubes are therefore the most important when deciding on the material to use. Grade 304 stainless steel is the optimal choice as it is rust resistant and therefore the cooling medium flowing around it would not impact the tubes in addition to having a high corrosive resistance which is required as methanol is a corrosive component. This stainless steel is also very sturdy and should be able to withstand the weight of the catalyst packed in each tube as well as the ability to withstand the high operating temperature of the reaction.

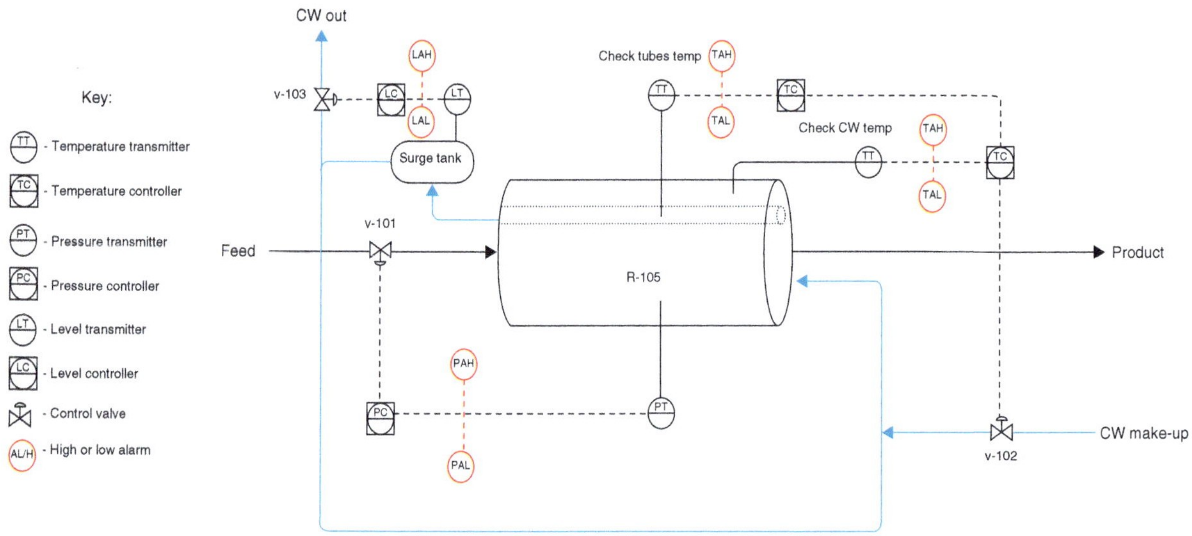

Figure 4.17 P&ID diagram for the control of the reactor.

The shell of the reactor will be under the influence of environmental effects and in contact with the cooling medium being circulated. Carbon steel cannot be used even though corrosion would not be an issue, but when in contact with moisture, carbon steel rusts making it an inviable option. Grade 316 stainless steel has therefore been chosen as it does not easily rust and it is a strong material so would need minimal maintenance and replacement, as with the tubes.

4.3.7.8 Vessel Thickness

The thickness of the reactor shell should be able to withstand the effects of the weight of the reactor and its contents as well as any internal or external stresses and loads. The minimum shell thickness can be calculated as per the following equation (Coulson and Richardson 2005).

$$t = \frac{P_i D_i}{2f - P_i} \tag{4.31}$$

According to Coulson and Richardson (2005), grade 316 stainless steel vessels as for the design stress at the reactor operating temperature is found to be 117.5 N/mm^2. The minimum shell thickness needs to account for possible corrosion, therefore an allowance of 3 mm is chosen as it will account for possible other loads and stresses. The total shell thickness is therefore determined to be 7.27 mm.

4.3.7.9 Reactor Internals

The internal make-up of the reactor consists of multiple tubes which is surrounded by the cooling medium to ensure isothermal operating condition. This arrangement is considered as a tubular packed bed reactor in the form of a shell and tube heat exchanger. The tubes in which the catalyst is contained and therefore where the reaction occurs is supported by structures known as baffles. These are discs of diameter equal to that of the reactor with holes through which the tubes run. These baffles not only support the tubes but increase the heat transfer by partially blocking the cross-sectional area. Additionally, a section of the baffle (baffle window) is cut which allows shell-side flow. The increased heat transfer is the result of the baffle windows being placed on alternating ends. The cooling medium flow geometry and general structure of baffles in the reactor are shown in Figure 4.18.

There are various arrangements and types of baffles which can be utilised. The most common options being orifice, cut segmental or disc and donut baffles. For the reactor analysed, a cut segmental baffle arrangement has been chosen as depicted in Figure 4.18.

4.3.7.10 Catalyst Replacement

Although catalysts do not react during the reaction taking place, catalyst deactivation can occur. When deactivation of the catalyst occurs, it means its catalytic activity and selectivity is decreasing, resulting in decreases in the efficiency of the process. Deactivation can occur from fouling, sintering, poisoning, coking, attrition and crushing to name a few (Bartholomew, 2001). The need for catalyst replacement is therefore warranted. Possible ways of catalyst unloading include

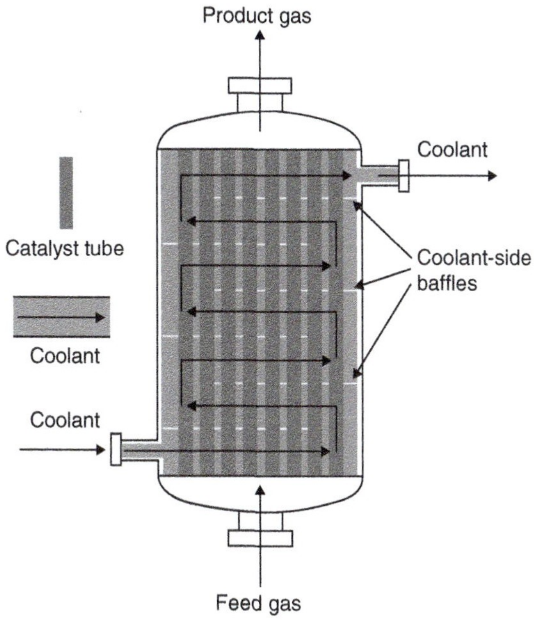

Figure 4.18 Tubular packed bed reactor internals with segmental baffle support.

using a vacuum at the point of entry, dumping the catalyst from a nozzle or flange at the base of the reactor or using a liquid flow through the reactor catalyst which will remove it. It is difficult to determine the exact life span of the catalyst used as the reactor conditions may have a different effect on the catalyst and will therefore need to be experimentally determined, however, to ensure that efficiency is not lost, catalysts should be replaced approximately once a year.

4.3.7.11 Vessel Supports

In order to keep the reactor in place and sturdy, support structures are required. These structures are dependent on the size, shape, weight, internal and external fittings, pressure and temperature as well as the location and arrangement of the reactor. Skirt support is best suited for tall vertical columns, therefore saddle supports have been chosen. These supports are best suited for horizontal vessels and supported at two cross sections. The most optimum positions for these supports are where the least bending moment is experienced and are typically 21% in from each end of the vessel (Towler and Sinnot 2008). A typical saddle support is shown in Figure 4.19.

Saddle support can be constructed from bricks, concrete or steel plate in order to withstand the reactor weight with the saddle dimensions for the required reactor size as summarised in Table 4.8 (Towler and Sinnot 2008).

4.3.7.12 Heads and Closures

The ends of the cylindrical reactor vessel need to be sealed off effectively to ensure no leakages or exposure to atmospheric conditions which will disrupt the operations of the reactor. Flat or domed-shaped heads and closures are available choices. In the category of domed-shaped heads and closures, tori spherical shaped are most commonly used for pressures up to 15 bar. Above 15 bar, hemispherical shaped, being the strongest, prove more economical (Towler and Sinnot 2008). The

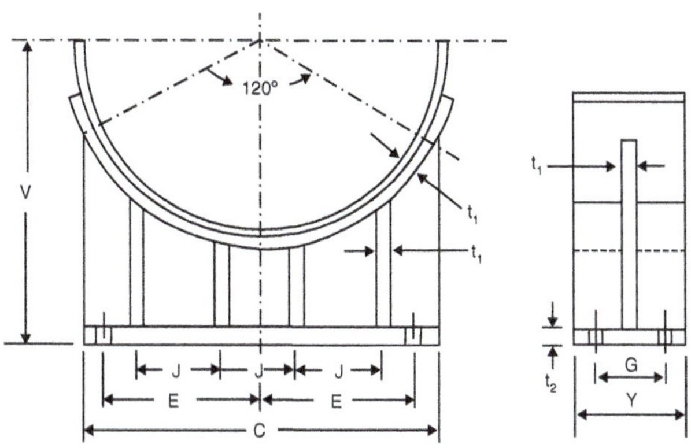

Figure 4.19 Saddle support (Towler and Sinnot 2008).

Table 4.8 Saddle support typical dimensions for a column diameter of 1 m.

Dimensions (m)						mm			
V	Y	C	E	J	G	t_2	t_1	Bolt diameter	Bolt holes
0.68	0.15	0.91	0.39	0.31	0.095	11	8	20	25

Figure 4.20 Torispherical head and closure (Towler and Sinnot 2008).

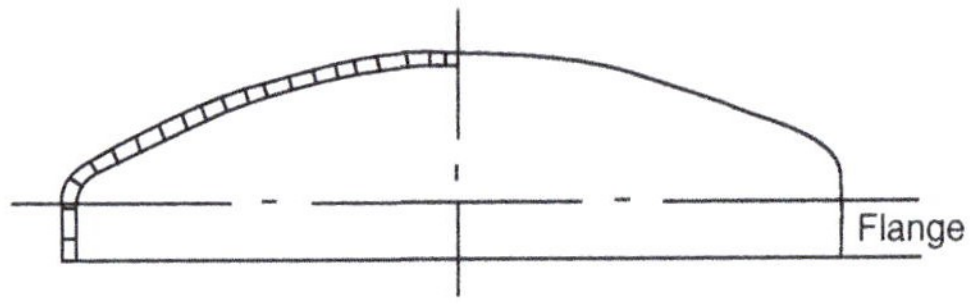

reactor vessel of consideration operates at a pressure of 10 bar and therefore a tori spherical head and closure would prove optimal and feasible. A typical tori spherical head is presented in Figure 4.20.

The required thickness of the heads and closures is calculated as per the following equation ((Towler and Sinnot 2008),

$$t = \frac{0.885 P_i R_c}{f - 0.1 P_i} \tag{4.32}$$

whereby the crown radius is typically equal to the vessel diameter. Accounting for an additional 3 mm corrosion allowance, with the design stress being 117.5 N/mm^2, the thickness of the head and closure required is determined to be 10.54 mm. The head and closure thickness is approximately 3 mm thicker than that required by the shell, the decision has therefore been made to utilise a thickness of 10.54 mm for the whole vessel which would possibly allow for longevity of the vessel and therefore a reduction in costs in the long run.

4.3.7.13 Flanges

The section separating the straight vessel section from the domed head and closure is known as flanges. They are used for connecting pipes and other instruments to the vessel, manhole covers as well as head and closure removal for maintenance and cleaning. Welding neck flanges are the selection choice as they are able to withstand high operating conditions of the vessel as well as vibrational loads if they occur as presented in Figure 4.21 (Towler and Sinnot 2008).

4.3.7.14 Manholes for Maintenance Access

Manholes are used on vessels in order to gain access to the interior of the vessel where cleaning and maintenance is required. Manholes usually range from 0.4 to 0.6 m in diameter for circular covers and 0.3 × 0.3 m to 1.2 × 1.2 m for square covers (Figure 4.21). The reactor vessel of consideration is fitted with a head and closure that can be opened and therefore accessed for maintenance and cleaning. A single square manhole of size 0.6 × 0.6 m is placed in the centre of the vessel for access into the middle section of the vessel for maintenance and cleaning.

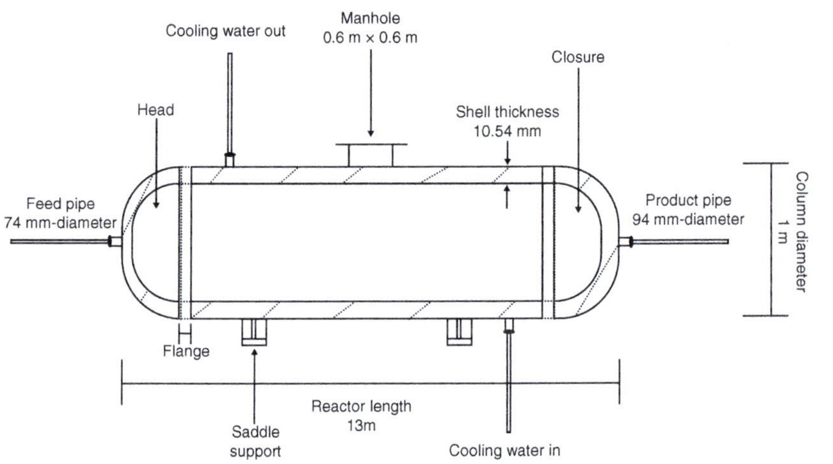

Figure 4.21 Cross-sectional view of reactor.

4.3.7.15 Pipe Sizing and Location

Pipes which are used to transport material into and out of the reactor need to have diameters which can sufficiently accommodate the flows. An effective method of estimating the required pipe diameters for the feed and product streams is through the use of the following equation (Towler and Sinnot 2008).

$$\mathrm{d_{pipe}} = \left(\frac{G}{\rho}\right)^{\frac{1}{2}} \tag{4.33}$$

The required information for $\frac{G}{\rho}$ is determined to be 0.0052 m^3/s and 0.0085 m^3/s for the feed and product, respectively. It is also important that the pipe diameters account for possible effects of corrosion, given the corrosive nature of methanol. A corrosion allowance of 2 mm is used for the pipes. The optimum pipe diameters are therefore determined to be 74 mm and 94 mm for the feed and product, respectively as displayed in Figures 4.22–4.24.

4.3.7.16 Mechanical Drawing Design

Figures 4.18–4.21 above present essential components of the detail reactor design from tubular packed bed to cross-sectional view of the reactor, which include length, width, height, angles, views and other specifications. These are important information to the people who will be assembling and production of the plant, and also to communicate with their clients, investors and other stakeholders. They can serve as guidance in the future construction, manufacturing or maintenance of the plant.

4.3.8 Equipment Information, Sizing, Costing and Power Requirements

A detailed pump needs to be designed. According to (Towler and Sinnot 2008), there are general classifications for pumps, dynamic pumps and positive displacement pumps. The type of pump that is selected depends on the flowrate and the total head of the pump. For this design, pump P-102 is used as a basis. With a flow rate of 250 m^3/h and a head developed of around 150 m (estimated from the Aspen simulation) a single-stage, centrifugal pump running at 3500 rpm will be used.

The power that a pump requires is given by (Towler and Sinnot 2008) in the following equation

$$W = g\Delta z + \frac{\Delta P}{\rho} - \frac{P_f}{\rho} \tag{4.34}$$

where g is the gravitational acceleration, Δz is the difference in height given by z_1–z_2, ΔP is the difference in height given by P_1–P_2, P_f is the friction losses in the pipe and ρ is the density of the fluid (Figure 4.25).

All the information is all readily available, except for the friction term. This is further given by the equation

$$P_f = 8f\frac{L}{\mathrm{d}_i}\frac{\rho u^2}{2} \tag{4.35}$$

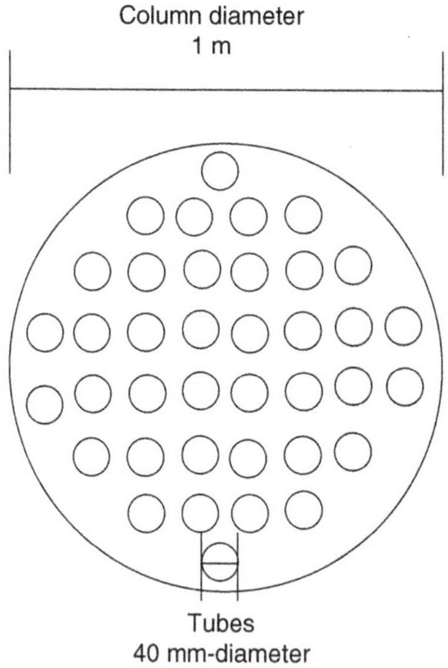

Figure 4.22 Front view of reactor looking at the tubes.

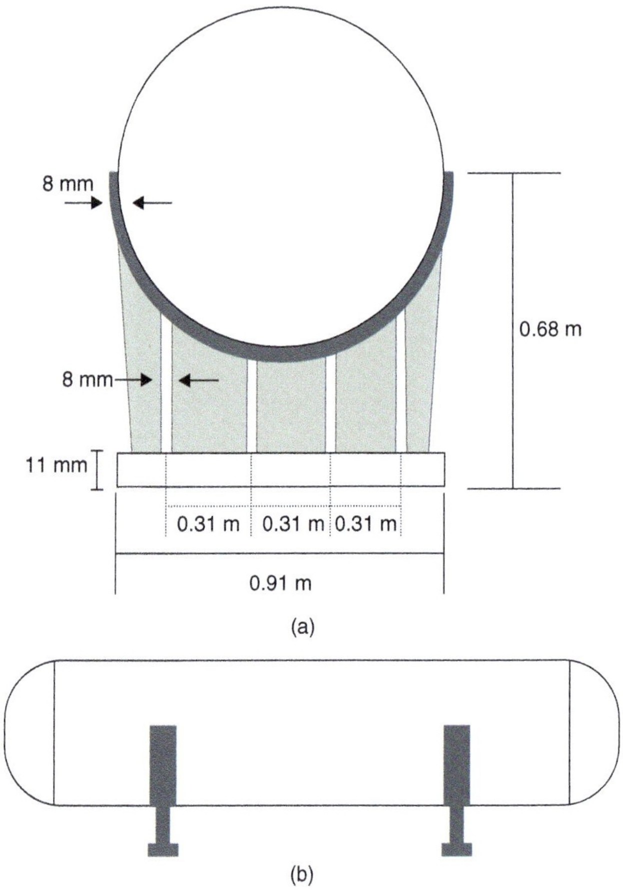

Figure 4.23 Saddle support structures. (a) Cross-sectional view; (b) side view.

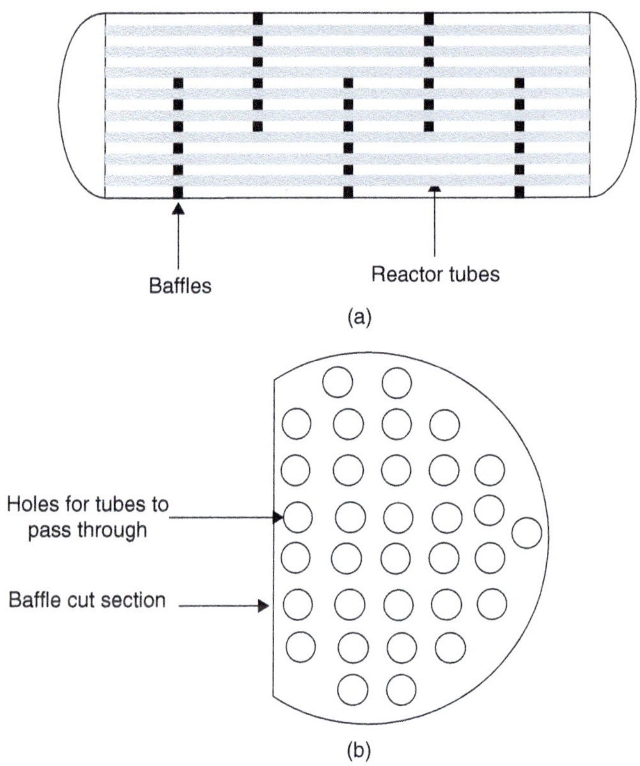

Figure 4.24 Segmental cut Baffle internals. (a) Side view; (b) top view.

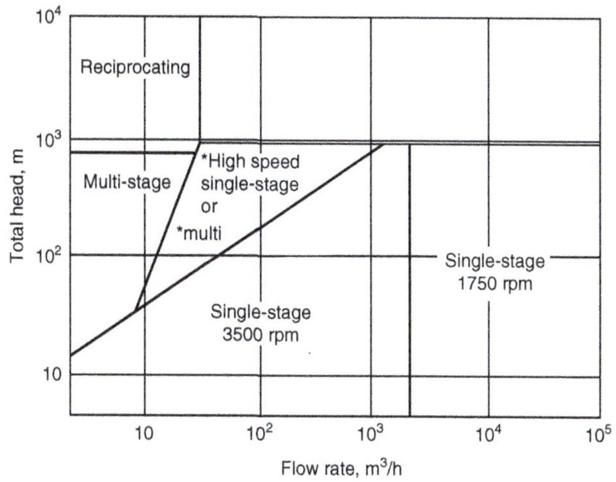

Figure 4.25 Type of pump selection. Source: Retrieved from Towler and Sinnot.

where f is the friction factor, L is the length of the pipe, d_i is the diameter and u is the velocity of the fluid. The pipe diameter will be calculated first. The optimal economic pipe diameter is given in (Towler and Sinnot 2008) as follows

$$d_i = \sqrt{\frac{G}{\rho}} \tag{4.36}$$

The G/ρ term simplifies the volumetric flow of the pipe. The optimal pipe diameter is thus calculated as 0.263 m, while the length of pipe can be assumed to be 10 m. The velocity of the fluid can be calculated by dividing the volumetric flowrate of the fluid by the cross-sectional area of the pipe and is calculated to be 1.27 m/s. The friction factor needs to be calculated next. This is read off of a chart (Figure 4.26) that depends on the Reynolds number of the fluid and relative roughness of the pipe. The relative roughness is calculated by dividing the absolute roughness of the pipe by the internal diameter of the pipe. A cast iron pipe is used since there are no corrosive materials in the fluid, and Towler and Sinnot (2008) recommends an absolute roughness of 0.26 mm for cast iron pipes. Using the diameter calculated earlier a relative roughness of about 0.001 is calculated. The Reynolds number needs to be calculated next and is seen in the following equation

$$\mathrm{Re} = \frac{u \rho d_i}{\mu} \tag{4.37}$$

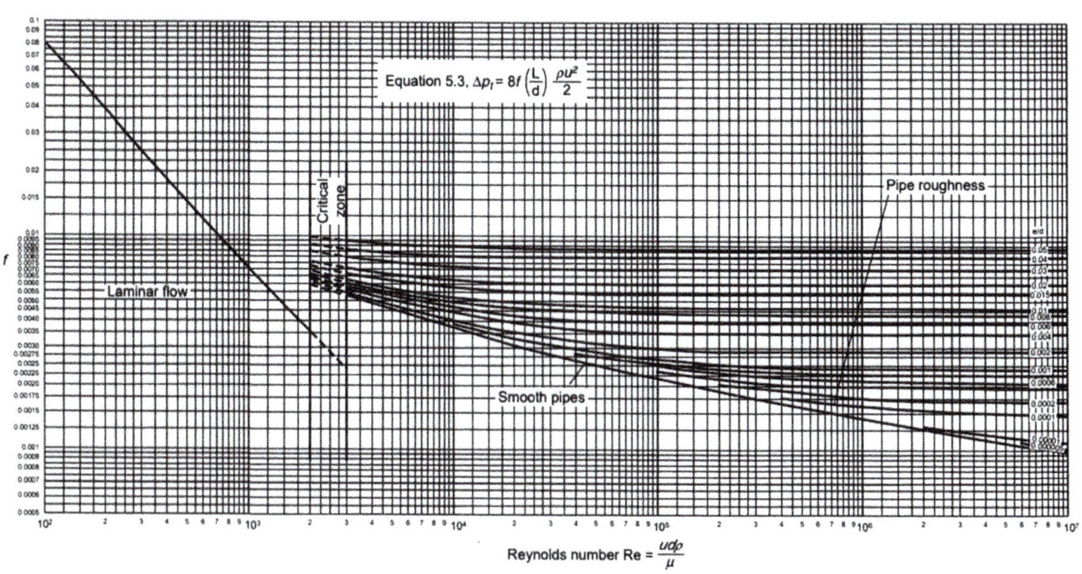

Figure 4.26 Friction factor. Source: Retrieved from Towler and Sinnot.

The viscosity is retrieved from the Aspen simulation, similarly to the rest of the values. A value of 35 000 is calculated. Figure 4.26 is then used and a friction factor of 0.003 is found. With all this, the work of the pump can then be calculated. The work required by the pump is 1479 J/kg. With a mass flow of 66.7 kg/s of water, the total power provided to the liquid is 98.6 kW. The Aspen simulation stated that 125 kW of shaft power is required. This puts the efficiency of the pump at 79.9%.

4.3.8.1 Net Positive Suction Head (NPSH)

To make sure that no cavitation occurs, the net positive suction head of the pump must be above a certain value. This will be determined from the pump manufacturers curve. Towler and Sinnot (2008) states that the available Net Positive Suction Head (NPSH) in a system can be calculated by the following equation

$$\mathrm{NPSH}_{\mathrm{Available}} = \frac{P}{\rho g} + H - \frac{P_f}{\rho g} + \frac{P_v}{\rho g} \tag{4.38}$$

This is all at the entrance to the pump, where P is the pressure above the pump, H is the height of the liquid, P_f is the friction losses and P_V is the vapor pressure of the liquid. Assuming the height of liquid above the pump is 0 m, the friction losses are half the total friction losses for the system calculated earlier, and the vapor pressure of water at 25 °C is 3171 N/m^2 (VR Chemistry 2022), the available NPSH is 10.24 m. As seen on the pumps system curve in Figure 4.27, the required NPSH is 7.9, so no cavitation will occur.

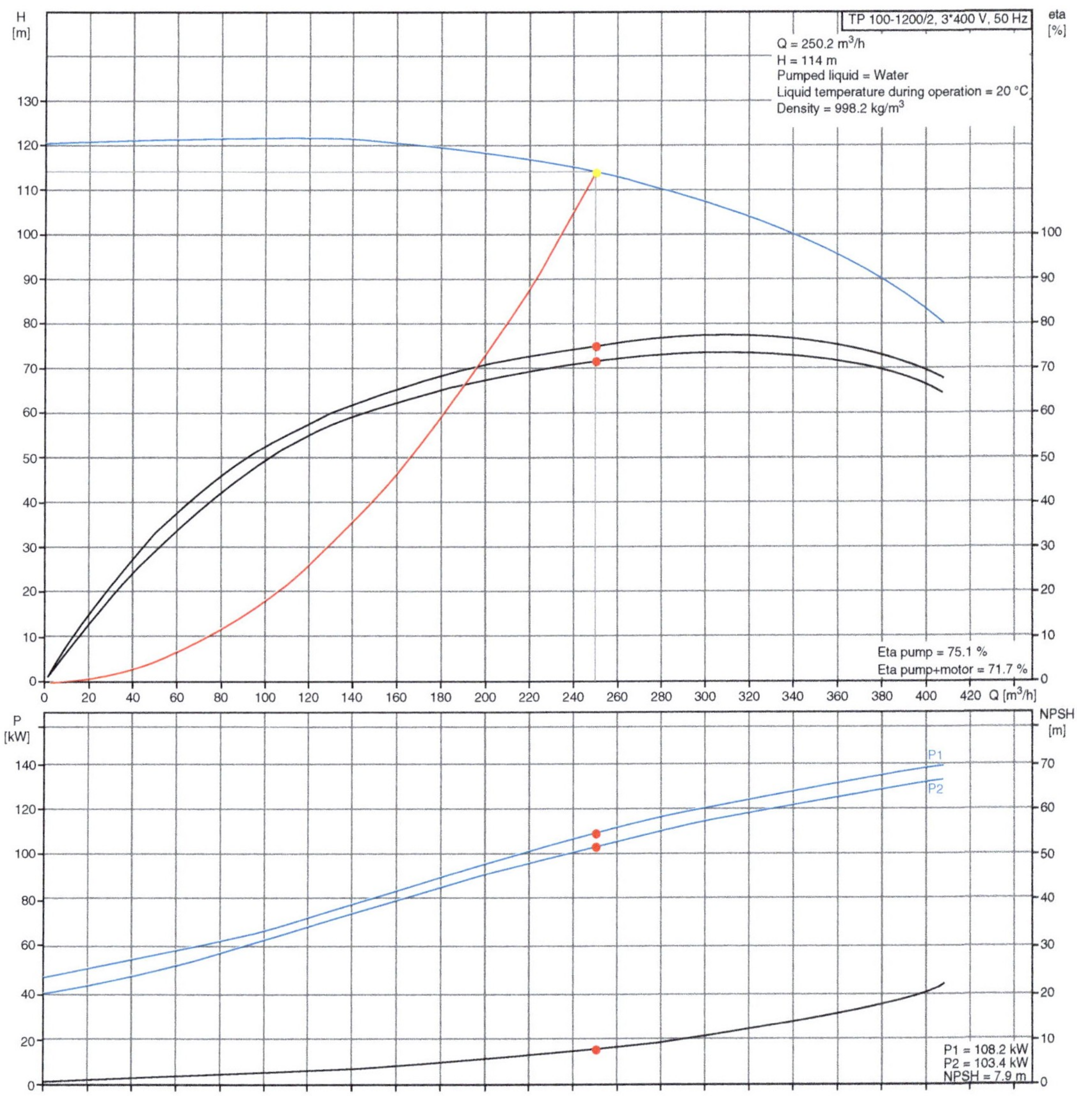

Figure 4.27 Pump curve. Source: Grundfos Academy.

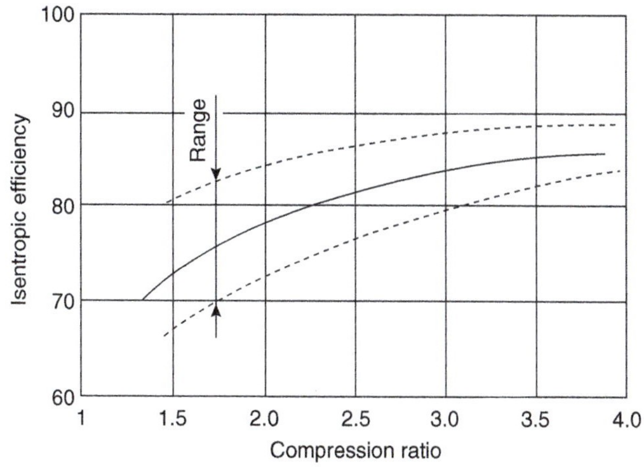

Figure 4.28 Compression ratio versus efficiency. Source: From Towler and Sinnot.

4.3.8.2 Pump System Curve

To find an appropriate system curve for the pump, the pump supplier Grundfos is used. The company supplies a wide variety of pumps and their associated pump curves. To find pumps similar in size to what is required, their catalogue was consulted, and pump model TP 100–1200/2 A3-F-Y-DQQE-2X1 was selected. This has power rating of 132 kW and a flow of 251.8 m^3/h for water at 20 °C, which is almost exactly the requirements of pump P-102. A pump curve and an NSPH curve are provided and also include inputs to specifically input conditions. The flowrate has been previously calculated as 250.1 m^3/h. As seen in Figure 4.27, For a flowrate of 250.1 m^3/h, the head developed is 114 m. The pump also runs at an efficiency of around 75%, compared to the previously calculated 79.9%. The power requirement is between 103 and 108 kW, compared to the assumed power requirement of 125 kW. The pump also runs at a speed of 2984 rpm compared to the 3500 rpm assumed earlier.

The specific speed of the pump can be calculated. This is given by

$$N'_S = \frac{NQ^{0.5}}{(gh)^{0.75}} \times 1.72 \times 10^4 \tag{4.39}$$

This gives a specific speed of 1165 rpm. If this was below 1000 rpm multiple pumps in stages would be recommended by Towler and Sinnot (2008).

4.3.8.3 Detailed Compressor Design

One compressor is used in this plant to increase the pressure of air before being combusted for the gas turbine. This is compressor C-101 and it has to increase the pressure of air at room temperature to 13 bar. One of the most important pieces of information for the compressor is the pressure ratio. This is defined as the ratio between the discharge and inlet pressure of the compressor determined from the equation,

$$P_R = \frac{P_2}{P_1} \tag{4.40}$$

The pressure ratio for this compressor is 13.

As seen in Figure 4.28, higher compression ratios lead to higher efficiencies. However, at higher compression ratios more stages may be needed for the compressor. Towler and Sinnot (2008) state that single stages may only be used for low compression ratios, and INTECH GmbH (2022) states that for a compression ratio between 10 and 15, two stages are normally used. Two stages will thus be used, and the intermediate pressure between the stages needs to be calculated. Towler and Sinnot (2008) recommend that for multistage compressors, the stages be split in a way that gives each stage and equal amount of work. To do this the pressure ratio for each of the stages needs to be the same, and to achieve this the geometric mean pressure will be used to determine the intermediate stage.

$$P_i = \sqrt{P_1 P_2} \tag{4.41}$$

An intermediate pressure of 3.6 bar is calculated. This gives a compression ratio of 3.6 and using Figure 4.28, and isentropic efficiency of 85% is found.

The work for each stage needs to be found. Towler and Sinnot (2008) provide a way to work out the work required to power the compressor.

$$-W = P_1 v_1 \frac{n}{n-1}\left((P_r)^{\frac{n-1}{n}} - 1\right) \tag{4.42}$$

This requires that we calculate n and v_1, which are the polyprotic coefficient and the initial volume, respectively. To work out the polyprotic coefficient, we can use the following equation:

$$n = \frac{1}{1-m} \tag{4.43}$$

where $m = \frac{\gamma-1}{\gamma E_p}$, γ is the ratio of the specific heat. For air, this ratio is 1.4. With the efficiency found earlier, m is equal to 0.34 and n is equal to 1.51. The final missing piece is the initial volume, which can be worked out using the ideal gas law to give 0.86 m^3/kg.

The discharge temperature of the compressor also needs to be calculated. This can be done using the following equation (Towler and Sinnot 2008)

$$T_2 = T_1\left(\frac{P_2}{P_1}\right)^m \tag{4.44}$$

Using the above equation, the temperature of the discharge can be calculated to be around 706 K or 433 °C. The work for each stage is calculated as 137.7 kJ/kg air. For the entire compressor, the work is 275.5 kJ/kg air, and when the isentropic efficiency is accounted for this is 324 kJ/kg air. With an air flow of 549 000 kg/h, or 153 kg/s, the total work required by the compressor is 49.5 MW. This is less than what was originally calculated in Aspen, due to the use of two stages. If the calculations are redone in the same way for only one stage, the work required is very similar to what the Aspen simulation predicted. A more detail calculations have been presented elsewhere (Bouttell et al. 2022).

The auxiliary equipment designed and used in the report are Tables 4.9–4.13, which are Heat exchanger sizing, Column sizing, Flash vessel sizing, Reactor sizes and Sizing for remaining compressors and pumps, respectively.

4.3.8.4 Vessels, Towers and Columns

Columns T-101 and T-102 are distillation columns and can easily be calculated. T-101 will be used as a basis to create a basic column, and then T-102 will be scaled accordingly. The absorber and stripper are also sized in a similar manner.

Table 4.9 Heat exchanger sizing.

Heat exchanger	Units	E-101	E-102	E-103	E-104
Type		Shell and tube	Shell and tube	Shell and tube	Shell and tube
Exchanger function		Cooler	Cooler	Heater	Cooler
Utility		Cooling water and Refrigerant 134-a	Refrigerant 134-a	Exhaust gas	Liquid nitrogen
Area	[m^2]	1210	3090	1.07	2.67E−1
Duty	[cal/s]	−29 000 000	−6 610 000	157 000	11 400
Material of construction		Stainless steel	Stainless steel	Stainless steel	Stainless steel
Input temperature	[°C]	1408	28	9	239
Output temperature	[°C]	−25	−50	250	180
Heat exchanger	Units	E-105	E-106	E-107	E-108
Type		Shell and tube	Shell and tube	Shell and tube	Shell and tube
Exchanger function		Cooler	Exchanger	Cooler	Cooler
Utility		Liquid nitrogen	None	Cooling water	Liquid nitrogen
Area	[m^2]	4.48	350	1320	6.98E−1
Duty	[cal/s]	130 000	59 410 000	−50 200 000	1430
Material of construction		Stainless steel	Stainless steel	Stainless steel	Stainless steel
Input temperature	[°C]	352	–	310	144
Output temperature	[°C]	120	–	27	80

Table 4.10 Column sizing.

Parameter	Units	T-101	T-102	V-102	V-104
Reflux		2.71	1.50	—	—
Number of stages		10	12	7	4
Feed stage		7	9	—	—
Pressure	[bar]	10	1	33	0.8
Column diameter	[m]	0.571	0.817	2.36	3.35
Tray height	[m]	0.3	0.429	1.24	1.76
Column height	[m]	3.3	5.58	9.93	8.81
Column thickness	[mm]	5.05	3	34	22
Vessel weight	[kg]	305	430	26 700	23 900

Table 4.11 Flash vessel sizing.

	Units	V-103	V-105
Pressure	[bar]	11	10
Column diameter	[m]	4.34	0.894
Column height	[m]	6.80	1.4
Column thickness	[mm]	22	8
Vessel weight	[kg]	32 800	508

Table 4.12 Reactor sizes.

Characteristics	Units	R-103	R-104	R-105
Pressure	[bar]	13	20	10
Vessel volume	[m^3]	1890	14.7	10.2
Material of construction		Stainless steel	Stainless steel 304	Stainless steel 304

Table 4.13 Sizing for remaining compressors and pumps.

Characteristics	Units	P-101	P-102	C-101	C-102	C-103
Flow	[kg/h]	461 000	240 000	549 000	670 000	240 000
Fluid density	[kg/m^3]	593	960	1.16	1.62	3.78
Shaft power	[kW]	1120	125	68 700	259 000	38 800
Type		Centrifugal	Centrifugal	Centrifugal	Centrifugal	Centrifugal
Efficiency	[%]	84.4	77.8	72.0	72.0	72.0
Material of construction		Stainless steel 304	Cast iron	Cast iron	Cast iron	Cast iron
Temperature in	[°C]	28.5	26.8	25	2510	593
Pressure in	[bar]	0.2	1	1	13	15
Pressure out	[bar]	44	15	13	0.5	1

The sizes of the flash vessels also need to be calculated. V-105 is used as a base and V-103 is scaled accordingly. The final sizes for the two flash drums can be seen in the following table.

4.3.8.5 Reactor Sizing

The reactors are all sized based on the size of reactor R-105 and scaled based on the residence time.

4.3.9 Process Control, Safety and Environmental Impact

In industrial processes, many of the components such as a distillation column, condenser or reboiler are only efficient if run under specific operating conditions. If deviations in these conditions occur, as with the distillation column, a major error would be the inefficient separation of the component mixture which would decrease the desired product purity and therefore cannot be sold or used in another process. The need for process control is therefore dire. Process control ensures processes are run efficiently and consistently while maintaining the product quality, throughput, yield and energy efficiency (Couper et al. 2012).

In addition, given the nature of industrial process, process control allows for automation thereby providing quick-acting responses to changes in the control variables. When dealing with dangerous materials, these sudden changes can be detrimental to the environment as well as the personnel, process control introduces an element of safety in which the errors can be corrected, and if all else fails, emergency shut down mechanisms are in place to ensure these hazards are neutralised (Couper et al. 2012).

4.3.9.1 Distillation Column Process Control System

A process control system is composed of several control loops which dynamically monitor and change certain operational conditions in response to disturbances and fluctuations. Control loops are made up of four major components: a sensor that measures a specific process variable to be monitored; a transmitter which sends the sensor's data to the control room; a controller that makes sure the measured & set point data correlate; and a control valve that alters a manipulated variable to guarantee the measured & set point data correlate. A distillation column's process control system strives to maximise column efficiency while minimising extra energy expenses (Ponton 1994).

The process of which is of concern is the distillation column for the separation or DME from the methanol–water mixture, as well as it's auxiliary condenser and reboiler components. The system comprises multiple control and manipulated variables as outlined in Table 4.14.

A conflict which is clear is the control of the reflux ratio as well as the top column temperature as both have the manipulated variable being the reflux rate. For this reason, a trade-off will have to be made to control only one. The reflux ratio has a more major impact on the column as it changes the amount of reflux back into the column and therefore changes the composition of the distillate. With that being said, reflux ratio would then also change the temperature of the top of the column, so by controlling the reflux ratio, the temperature should also be controlled and would be the optimal variable to control instead of temperature.

Table 4.14 Control and manipulated variables.

Objective	Control variable	Manipulated variable
Maintain the pressure in the column	Column pressure	• There are two options for the manipulated variable: • Cooling water flow into condenser • Flow into reflux drum from condenser
Maintain the temperature at the top and bottom of the column (distillate composition)	Top and bottoms temperature	• Steam flow rate into reboiler • Reflux rate
Maintain the level of liquid in the column	Liquid level in distillation column	Bottoms flow rate
Maintain the level of liquid in the reflux drum	Liquid level in reflux drum	Distillate flow rate
Maintain reflux ratio	Reflux ratio	Reflux rate

There are various mechanisms/types of control loops which can be implemented to ensure effective control. Open loop, feedback, feedforward, cascade and ratio control are the main schemes of note as presented in Table 4.15.

When observing the type of control scheme to implement for the control variables affecting the distillation column, condenser, reboiler and reflux drum, almost all of the control variables require the output to first change before measurement and then error correction is conducted. The best control scheme is therefore feedback control. Additionally, it was found that cascade control is often employed to control the temperature at the top and bottom of the distillation column, however, possible conflicts in controlling the other variables may arise, and will therefore be further explored. Furthermore, the reflux ratio would be best controlled using ratio control.

Table 4.15 Various control schemes available.

Control scheme	Description	Graphical representation
Open loop	Also known as non-feedback control, the output of the process has no effect on the control variable (Berry and Schott 2017)	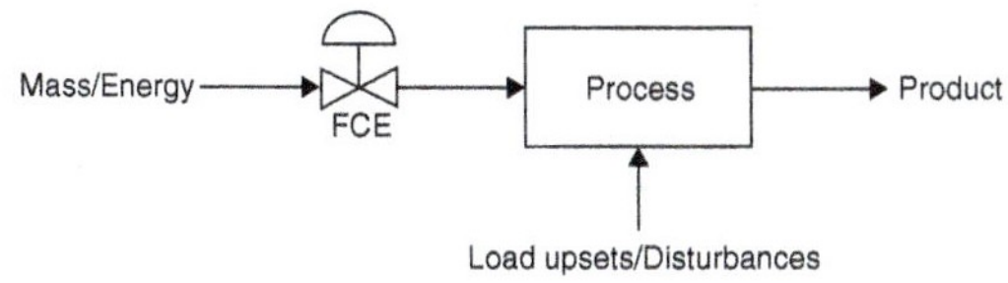
Feedback control	Widely used in automated systems, uses the output of the process to calculate the error from a specified set point, and adjusts the manipulated variable to achieve the set point (Berry and Schott 2017)	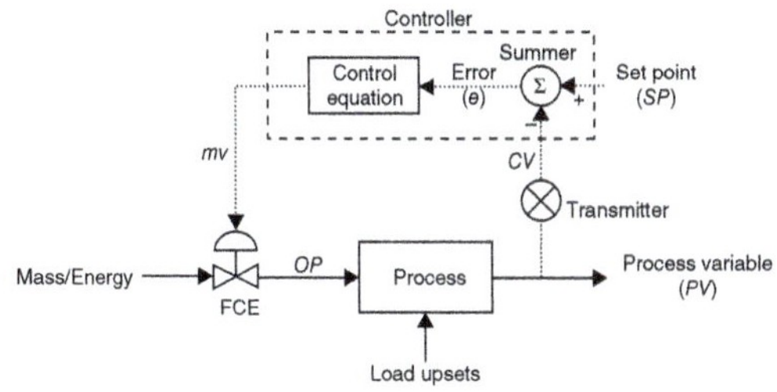
Feedforward control	The process disturbance is typically used instead of the process output to determine the error from the set point, and then adjust the manipulated variable (Poe and Mokhatab 2017)	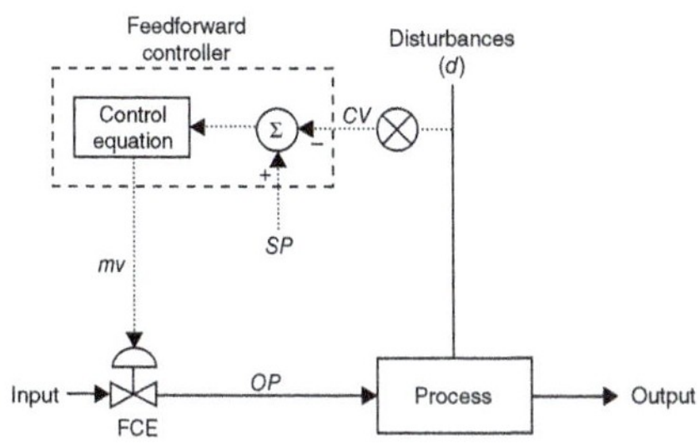
Cascade control	Two or more controllers are used, whereby the output from the first controller is used as the set point for the second controller to then adjust the manipulated variable (Dimian et al. 2014)	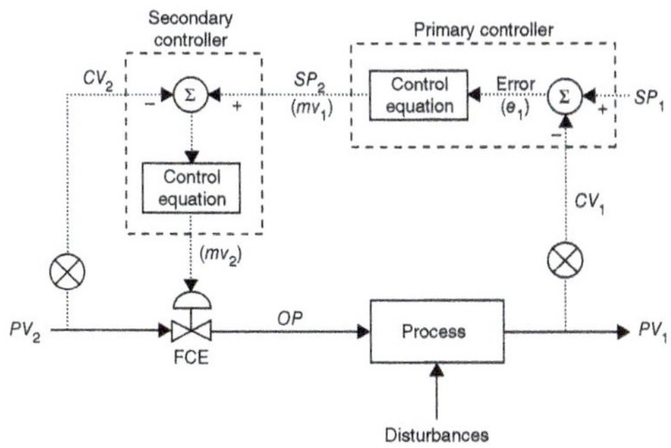
Ratio control	Used to maintain the ratio of two control variables at a specified set point, usually by manipulating one variable to achieve the desired ratio. These control are often best for flow control (Visioli 2005)	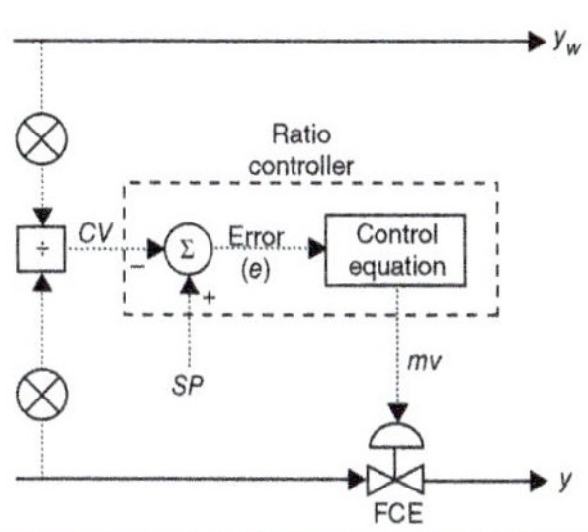

The type of controller used is as important as the control scheme implemented as the controllers are what minimises the offset from the set point and process variable of the system. Proportional, integral and derivative controllers can be used, however, are often most optimal when in combination with each other. For this reason, proportional integral derivative (PID) controllers are the focus. These controllers provide a faster response time as its able to predict the disturbances present, decreased/no offset from the set point, improved stability, easy to implement and although sometimes expensive, provide high accuracies (Toulson and Wilmshurst 2012).

A control system is essential in ensuring that the uniformity of the process output is not jeopardised by process disruptions. Based on HAZOP study, the integrated control loops presented below were meant to limit the impacts of variable process conditions & the safety threats they can generate. The first stage in designing a control system is determining the number of control loops necessary. The distillation column needs 6 control loops, according to J.W. Ponto's approach (Ponton 1994).

4.3.9.2 Flow Control

Since the feed flow rate has such a large impact on column performance, a control loop with a feed line valve is used to keep the feed at a predetermined set point. The flow rate coming from the upstream column serves as a disturbance, necessitating this control loop (Towler and Sinnot 2008).

4.3.9.3 Temperature Control

The primary goals of controlling the temperature in a distillation column are to preserve the column composition profile, regulate disturbance sensitivity and keep the column operating temperature stable (Towler and Sinnot 2008). Temperatures are generally observed at the top & bottom of the column and are controlled by altering the reflux & boil-up ratios. The temperature of the distillate is controlled for this column since the product of the distillate is the main product, DME. The feed temperature, boil-up rate and pressure are all disturbances to the control loop when the reflux ratio is used as the manipulating variable. Other disruptions, such as controller malfunction, are also conceivable.

4.3.9.4 Level Control

Distillation column level controls are critical for preventing flooding as well as drying up, especially the fluid level within the column's reflux drum and sump. The distillate product's control valve is controlled to increase or reduce the fluid level inside the reflux drum. When the distillate product flow is reduced by closing the valve, the fluid levels inside the reflux drum rise and the opposite is true. However, by lowering the amount of overhead product removed, the vapor flow rate along the column is reduced, encouraging a rise inside the bottom fluid level inside the column. As a result, a level controller for the column bottoms is however necessary to prevent the column from drying out or flooding. Both flow controllers will work together to readjust the levels within the column & reflux drum to a unique steady-state set point that accounts for the disruptions generated by the column feed & and reflux flow rates.

4.3.9.5 Interaction Control

In isolation, the execution of a single control measure has no effect on the target control variable. Any modification will influence the entire process unit; hence, it is necessary to observe the interactions among different control loops in order to increase the overall efficacy of the entire control system. Adequate control loop interaction management ensures operation stability and consistent operating conditions. The interactions of control loops are maintained by monitoring the column's overall mass and energy balances (Table 4.16). The mass balance of the column is managed by flow rate adjustment, meanwhile, the energy balance of the column is handled through the operation of the condenser & re-boiler.

Table 4.16 Mass and energy balance control loops and control variables summary.

Mass balance		Energy balance	
Control loop	**Control variable**	**Control loop**	**Control variable**
Bottom level	Column level	External reflux temperature	Columns overhead temperature
Feed flow	Production rate	Overhead condenser temperature	Column cooling
Reflux drum	Distillate	Re-boiler temperature	Columns bottom temperature

4.3.9.6 Pressure Control

Since pressure control directly affects the temperature, compositions, volatilities and overall efficiency of the column, pressure is defined as the primary control variable. Large variations in the column's operating pressure can result in significant problems in the column and column trays (Lieberman 2009). A minor drop in column pressure causes the saturated liquid on the trays to evaporate, resulting in froth formation and entrainment of trays in the overhead (Lieberman 2009). In addition, a rise in pressure reduces the vapor flow rate, resulting in weeping. Direct pressure control methods are costly and are rarely used. Thus, the column pressure is managed by managing the reflux drum & condenser duty (Ponton 1994).

The column is supposed to be free of non-condensable gases, and the control valve used to manage the flow rate coming from the condenser into the column has a negligible pressure drop, so the condenser operating pressure is much like the columns. The cooling flow rate of water going into the condenser is used to manage the column pressure. The cooling flow rate of water, which is described as the dependent variable, influences the condenser duty, which is classified as the control variable. Activating the cooling water valve raises condenser duty, lowering the reflux drum temperature & pressure, and therefore lowering the column temperature to the expected set point.

4.3.9.7 Emergency Shutdown Control

For both safety and economic reasons, it is critical to account for severe failures such as flooding, weeping and column blowout. Flooding often happens due to the re-boiler heat duty set high, increasing the volume and movement of the vapor up the column and facilitating liquid entrainment. This restores separation and significantly reduces product purity. A feedback loop that exploits the steam flow going into the re-boiler is used to prevent this. The control valve shuts down to reduce the rate of steam, so lowering the vapor flow up the column to the appropriate set point. When the vapor flows up the column is inadequate, the liquid leaks through the plate holes, resulting in weeping. The control valve is opened in this case to raise the steam rate by exploiting the steam into the re-boiler, returning it to the correct conditions.

When pressures are high, blowouts are possible. When the pressure exceeds the limitation, the valve of the distillate stream clears to relieve some of the pressure till steady state is restored. For example, in temperature monitoring, the distributed control system checks whether the temperature at a certain site will remain in the green zone. When the temperature begins to drift out of the green zone, it is regulated to keep it there by either rising or reducing its value in response to feedback and/or feedforward control. The distributed control system (DCS) adjusts the temperature to return to the green zone. When the temperature crosses into the yellow zone, the reaction system supervisor may artificially regulate the reaction and the temperature to return to the green zone (Kazuhiko Sano 2020).

4.3.9.8 Process Instrumentation Diagrams

The piping and instrumentation diagram (P&ID) describes the procedures in place to assure product quality and consistency, as well as ensuring that the process's objectives are continuously reached. The basic goals of developing plant instrumentation & control systems are to achieve maximum plant safety, maximum production rate & product quality, maximum cost effectiveness and maximum production stability. The pipe and instrumentation diagram contains a comprehensive description of the plant's instrumentation and control techniques. The P&ID consists of all instrumentation, plant equipment, pipelines, valves and fittings, as well as the layout of all components. The extensive instrumentation and control systems are put in place to guarantee that all production domains function within predetermined safety thresholds, beyond which alarms will be triggered to activate countermeasures to reduce the hazard. Dangerous operating circumstances are promptly detected by plant operators using integrated alarms and automated shutdown systems, resulting in immediate manual or automatic actions (Towler and Sinnot 2008).

4.3.9.9 Choice of Control Mechanism

The control system structure is denoted by the preliminary P&ID. The bypass stream that results in the reflux drum is used to manage the pressure of the column at 10 bar; if the pressure of the column rises above the set point value, the valve in the bypass stream closes, pushing more vapor to the condenser, lowering the pressure directed to the reflux drum and thus lowering the liquid pressure in the reflux drum, lowering the pressure of the column back to the set point. If the condenser effluent temperature rises above the set point, the valve opening will be adjusted to enable more cool utility to flow to the condenser, lowering the temperature again to the set point. If the reboiler effluent temperature rises above the set point, the valve opening will be reduced to enable less hot utility to flow through the reboiler, lowering the temperature back to the set point.

If the fluid level inside the reflux drum rises over the set point, the valve opening is raised, allowing more distillate to depart and returning the fluid level to the set point. The fluid level inside the column, however, will be lower than the specified point. As a result, the valve opening will be reduced, allowing less bottom product to leave and raising the fluid level again to the set position. If the feed flowrate exceeds the specified point, the reflux to feed flowrate ratio will change. As a result, both valves are regulated through the ratio controller to maintain a consistent ratio of reflux of 0.42 to feed flowrate.

4.3.9.10 Process Instrumentation

As mentioned, control of the variables of the distillation column as its associated components are important factors in assessing the efficiency and operability of the process. Variables such as the temperature, pressure, level and flow need to be maintained at specific points. Before these variables can be controlled, they need to be measured. Various devices are available for the measurement of these variables as outlined below.

4.3.9.11 Temperature

Temperature sensors can be categorised into thermometers, electro-mechanical, electronic, resistive, optical and acoustic sensors. Upon evaluation of the broad range of sensors, it was found that electronic and resistive sensors are the most optimal when observing the influence of the temperature range, accuracy and size of the sensors (Affan 2019). The two mentioned categories comprise sensors such as thermocouples, residence temperature detectors (RTDs), thermistors and semiconductor based integrated circuits. Semiconductor based integrated circuits have a limited temperature range of −55 °C to +125 °C making them obsolete for the purpose of the high temperatures experienced within the distillation column and its associated equipment (Affan 2019). The three temperature sensors are compared as presented in Table.

From the Table 4.17, a resistance temperature detector is chosen. This is a thin film or wire wound sensor whose resistance changes with a changing temperature. These sensors provide high stability, accuracy generally of ±0.1 °C and repeatability (Hu et al. 2006). Wire wound RTDs are constructed from a small platinum, nickel or copper coil around a glass or ceramic bobbin with helix-style winding for best industrial use. Thin film RTDs consist of the metal of choice deposited onto a ceramic substrate and coated in glass. Platinum is the most efficient material for construction, owing to its accuracy, reliability, chemical resistance and stability, making it less susceptible to corrosion in comparison to other materials which is required when observing the possible corrosiveness of methanol in the distillation mixture of concern. Additionally, platinum has a higher melting point, allowing for higher temperature measurements (Mokhatab et al. 2019) without the risk of failure or inaccurate readings. Due to the cost of platinum being higher than other common metals, RTDs are relatively more expensive than thermocouples and thermistors, however, its many advantages justify the choice as the sensors longer lifespan would be present cost savings.

4.3.9.12 Pressure

In a distillation column, multiple pressure measurements are required as separation occurs at specific temperatures and pressures, if that pressure is not maintained, more or less energy would be required, resulting in inefficient separation or even unsafe in the presence of certain chemicals. Differential pressure sensors have been chosen which comprise an electronic module and sensory diaphragms. These sensors have accuracies as optimal as ±0.06% while being easy to clean and maintain (Cai and Resetarits 2011). In order to prevent corrosion of the tubes and diaphragms, grade 304 stainless steel has been chosen as they are also able to withstand high operating temperatures.

Table 4.17 Comparison of the temperature sensors.

Function Factor	Sensor		
	Thermistor	**RTD**	**Thermocouple**
Temperature range	−100 °C to +300 °C	−200 °C to +850 °C	−200 °C to +2315 °C
Accuracy	±0.05 °C to ±1.5 °C	±0.05 °C to ±0.1 °C	±0.2 °C to ±0.5 °C
Sensitivity	Poor at higher temperatures	Excellent	Good
Linearity	Poor	Good	Fair
Cost	Low	Moderate	Low
Stability	Fair	Excellent	Poor

4.3.9.13 Level

There is great concern about the possibility of flooding within the distillation column. The level of liquid at the bottom of the column therefore needs to be controlled and measured. Float, hydrostatic, radar, ultrasonic, capacitive and radiometric sensors are a few which could be utilised for liquid level measurement (Morris and Langari 2021). A continuous level sensor is preferred over a point level sensor as the liquid level should not reach further than the minimum or maximum stated level. Additionally, by implementing a continuous level sensor, warning levels before exceeding the minimum or maximum are able to be detected without the need for multiple-point sensors (Moran 2017). In a distillation consisting of trays or packings, challenges would be present as devices requiring 'sticks' and possibly waves as a means of measurement would be obstructed therefore providing inaccurate readings. In addition to the exclusion of float sensors or dipsticks as being an option for the distillation column, many other sensors were ruled out as per the reasonings identified in Table 4.18.

The guided wave radar level sensor is the choice of instrumentation as they have been found to be able to measure the level of fluid in typically difficult conditions such as in closed tanks, turbulent liquids, with the presence of obstructions (such as trays) and steam condensate (Morris and Langari 2021). These sensors provide accuracies of ±0.25% while being noncontact, therefore, the effects of corrosion from the fluid are not of concern (Kaineder et al. 2019). Grade 304 stainless steel is the material of choice as it would provide long-term protection against other environmental effects, which it may withstand while being sturdy. Additionally, this higher grade is preferred to the lower 316 as methanol can be corrosive if the sensor comes into contact with the fluid. These sensors can be costly in comparison to sensors like ultrasonic sensors, but their advantages of being able to provide accurate, continuous and noncontact readings make them a worthy investment.

It is also important that the level of liquid in the reflux drum is sufficient to ensure that adequate reflux is returned to the column. Due to the simplicity of measuring the liquid without requiring exact readings but rather accurate approximations, a simple float level sensor will be utilised. These sensors can have accuracies of ±2–8 mm while being of low cost (Zhang 2008). Float sensors can be constructed from PVC or stainless steel, among others. Due to the high operating temperatures of the fluid, PVC cannot be used as the possibility of the float melting is eminent. Grade 316 stainless steel is therefore optimal as the float would last longer without requiring regular maintenance. DME being non-corrosive also ensures that the float will not corrode or degrade inside the drum.

4.3.9.14 Flow

The flow of material into and out of the column, as well as the reboiler, condenser and reflux need to be measured and controlled to ensure that not too much or too little flow is provided which will result in inefficient separation. Positive displacement, mass or velocity flow sensors can be utilised for the flow through the pipes. For the plant design of concern, positive displacement flow sensors are used for the measurement of the streams entering and exiting from the condenser, reflux drum and reboiler, while mass flow sensors are used when observing the feed, distillate and bottoms product flow rates.

Rotary vane meters are employed as they have excellent accuracies of ±0.1% along with reputable repeatability. These meters have the added advantages of the accuracy being unaffected by pulsating flow, additional power supply not being required as well as lengths of straight pipes not being a necessity. These meters are able to withstand high temperatures and pressures therefore making them effective for the system at hand (Martyr and Rogers 2021).

A coriolis mass flow meter is utilised owing to its high accuracies of ±0.1% and can further be found to be ±0.05%. Even though these are on the costly side in comparison to many meters, they are worthy investments as they require

Table 4.18 Exclusion of certain level sensors.

Sensor	Reason for exclusion
Hydrostatic	Provides good accuracy (±0.5%) but requires accurate knowledge of the fluid density being measured (Morris and Langari 2021). Mixture densities can be difficult to determine and given the nature of distillation, these sensors would not be reliable
Ultrasonic	This noncontact sensor proved best suited for highly viscous liquids as well as solids with a decent accuracy (±1%). These sensors are, however, sensitive to changes in ambient temperature, and therefore would result in inaccurate readings when temperature changes occur (Morris and Langari 2021)
Nucleonic	This noncontact sensor proved reliable for the measurement of hot, molten metals as well as powders or solids. A restriction is placed on these sensors in that strict safety precautions and permits need to be followed and obtained (Morris and Langari 2021). The plant design of consideration does not deal with the mentioned measurable medium and therefore this sensor would not be applicable

less maintenance resulting in operating cost savings in the end without needing replacement or repairs like many other inefficient, cheaper meters. These meters account for a wide flow range, can handle many fluid types and can be recalibrated without removal, while requiring low maintenance (Morris and Langari 2021). The separation process of concern produces the desired DME product which is need to be accurately measured to ensure sufficient production, thereby furthering the motivation for the use of the coriolis flow meter. Given the corrosive nature of methanol, as previously mentioned, both meters used are constructed from grade 304 stainless steel to account for the corrosive effects, as well as providing a strong material to improve the meter lifespan and reduce how often maintenance may be required.

4.3.9.15 Valve Selection

In process control, the type of valves used is as important as the controllers and sensors. Even though high-accuracy sensors and controllers have been utilised, they become redundant if inefficient valves have been chosen, as the valves implement the changes inputted by the controllers. Globe, gate, ball and butterfly valves are some of the common valves available. The control valves need to be able to regulate the flow of material and therefore butterfly and globe valves are of interest. Due to the non-linearity of the flow through butterfly valves, they can be difficult to control when compared to ball valves. Additional mechanisms would be required to improve the control of the butterfly valves. Ball valves are able to provide air-tight shut-off while being reliable in preventing back flow of fluid through the pipe, as well as reliability in flow regulation. V port ball valves in particular are best suited for flow regulation. These valves are cost effective, easily maintained and operated, while being able to withstand corrosive and aggressive fluid mediums (Parisher and Rhea 2022). Ball valves can be constructed from stainless steel, PVC, brass and bronze among others, however, due to the corrosive nature of methanol, grade 304 stainless steel is utilised.

In addition to the control valves, pressure relief valves are required as safety precautions on the distillation column when excessive pressure would result in system failure. A typical spring-loaded safety valve will be used as they are the most versatile and reliable. The spring would compress when pressure increases causing the disc the move, resulting in the release of pressure. These valves also have the advantages of requiring low maintenance, competitive pricing, compatibility with many chemicals and high temperatures (McAllister 2014). Provided that these safety relief valves will be placed at the top of the column where methanol (being corrosive) is of minimal composition, tungsten steel will be used over carbon and stainless steel due to its ability to withstand higher temperatures, its durability and not being susceptible to rust like some carbon steels. Although being slightly more costly, its longevity means less maintenance or replacement is required making it a worthy investment (Bahadori 2014).

Furthermore, in the case of automated control valve failure or where emergency shutdown is required, a manual globe valve is installed on the main pipeline to prevent further flow of material into the process as they are known to provide excellent shutoff.

4.3.9.16 Actuator

Controlling processes manually is time-consuming and often results in errors. The valves used in process control are therefore automated to have timeous and efficient control of the process variables when required. The control valves are automated through the implementation of actuators. Pneumatic, electric and hydraulic actuators are some types which may be utilised. Outline in Table 4.19 is some of the differences between the main actuator systems mentioned.

Pneumatic actuators have therefore been chosen due to their being able to provide excellent linear accuracy and repeatability, being able to withstand high temperatures, meeting safety regulations through the use of air and not hazardous fluids, they are light weight, require minimal maintenance and comprise durable components making them cost effective (Kececi 2019). The recent development shows that these systems can be combined with electrical components which further improve their efficiency and cost but will have to be further explored and researched. Pneumatic actuators can be constructed from steel pressings or aluminium castings. Aluminium has been chosen due to its versatile nature, corrosive resistance, electrical and thermal conductivity (able to withstand high temperatures), flexibility and sturdiness (Hussain et al. 2020).

4.3.9.17 Instrument Siting

Following the decisions of control schemes and instrumentation, the following instrument placements have been decided as presented in Table 4.20.

The analysis of instrumentation siting as well as the process control analysis allows for the construction of the P&ID diagram as shown in Figure 4.29.

Table 4.19 Types of actuators.

Actuator system	Description
Pneumatic	These systems use compressed air as the working fluid. These systems are cost effective while proving good response times. They do, however, require additional pieces of equipment such as compressors which would require energy, but are overall a low-cost option. Additionally, these systems are very simple and easy to operate allowing for a wide range of forces, and they are able to operate under many conditions such as extreme temperatures due to the use of air instead of hazardous fluids (Kececi 2019)
Electric	These systems use electricity which is converted into kinetic energy as the working fluid. These systems offer the highest precision, good response times and easily programed. They do, however, often require large electric motors making the system costly and are often not suitable for many environments like high temperatures (Leitch 2007)
Hydraulic	These systems use incompressible fluids from a pump as the working medium in the form of a piston-cylinder configuration. These systems are able to provide high speeds and constant forces without requiring additional fluid, however, there is possibility of fluid leakages resulting in decreased efficiency and environmental issues. These systems are high maintenance and do require a variety of additional equipment such as pumps and motors making them more costly (Leitch 2007)

Table 4.20 Siting of various control instruments.

Sensor	Siting	Reason
Pressure sensor	Top of the column	• Ease of access as structures would be built at the top of the column to gain access for maintenance and cleaning • The column pressure is mainly measured at the top of the column
Temperature sensor	Top and bottom of the column	• Easily accessed as mentioned above • The temperature is different at the top and bottom and therefore needs to be controlled at both points
Level sensor	Bottom of distillation column and float inside reflux drum	• Easily accessed • Reflux drum will have manholes therefore making access, removal and cleaning of the float easy
Flow sensors	On the pipes which require measurement	Given the fact that pipes will be easy to access for maintenance and cleaning, the sensors too are easily accessed

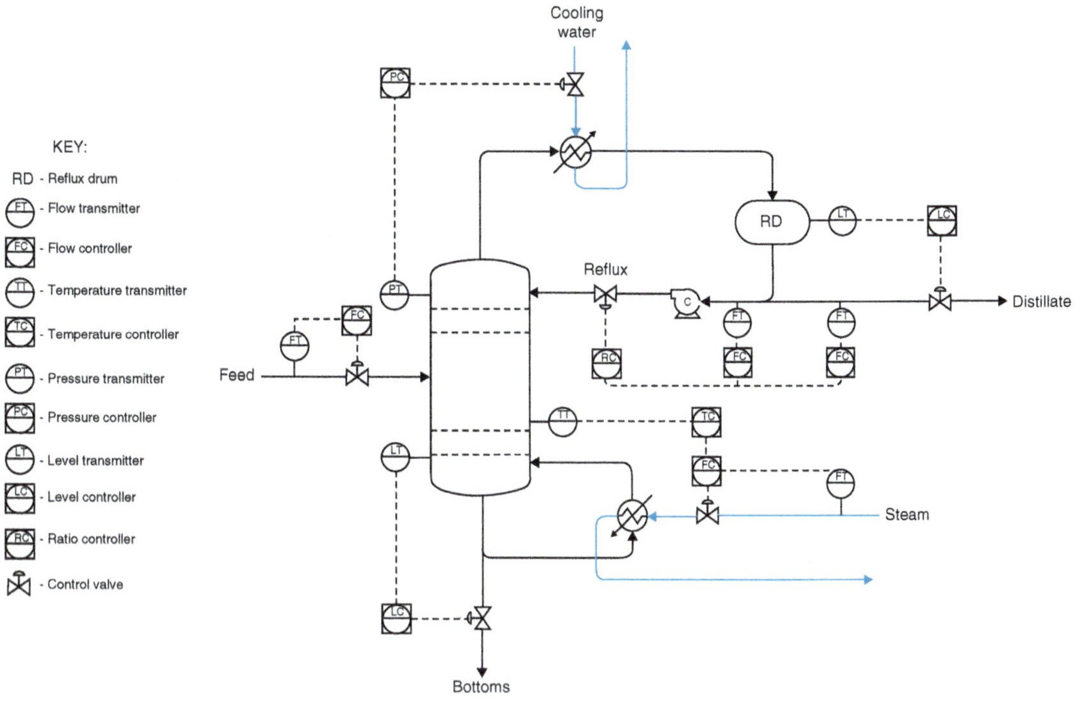

Figure 4.29 Pre-HAZOP P&ID.

4.3.10 HAZOP Analysis

Chemical process facilities are planned and run with profit in mind, but safeguarding the safety of plant staff, the public and the environment are critical (Towler and Sinnot 2008). The running of process plants involves risks owing to the many process variables and the characteristics of the chemical species engaged in the processes. HAZOP analysis is one of the most important tasks engineers may take to guarantee a safe plant operation (Towler and Sinnot 2008). HAZOP is a methodical way of analysing and reviewing process aspects in order to discover any deviations from the original design that may result in risks and operability issues. Following the identification of potential hazards and operability issues, the following phase in a project is to identify ways in which the risks might be reduced (Towler and Sinnot 2008).

The HAZOP analysis is conducted on the multicomponent distillation column in the DME synthesis section of the plant. The control measures integrated into the system are based on the process control measures discussed previously and used to develop a P&ID. The P&ID is then further modified by implementing additional safety control measures to try and mitigate the negative impacts of the column.

4.3.10.1 HAZOP Evaluation

The following guide words are used to describe the HAZOP methodology and are used to develop the P&ID post-HAZOP analysis (Towler and Sinnot 2008):

- Intentions: Clarifies how a certain portion of a process is planned to work.
- Deviations: These are deviations from the designer's original intention, which may be detected by the guiding words shown in Table 4.21.
- Causes: These are the causes for or how deviations may develop in a particular process. Only plausible reasons should be considered relevant in the process.
- Consequences: These are the outcomes or repercussion that occur from a meaningful deviance detected within the system.
- Hazards: These are the effects that may cause damage, injury or death.

As strategic design components in the balance of plant, HAZOP around distillation column, condenser and reboiler are undertaken as presented in Table 4.22.

4.3.10.2 Control Structure Post HAZOP

Control measures are put in place and illustrated in Figure 4.30 to show approximately where transmitters and controllers are situated in order to prevent accidents and damage to the equipment. The diagram includes all of the HAZOP mitigation measures mentioned in the analysis above as well as the process control measures described that would ensure optimal operation of the column. The instrumentation implemented allow for easy control and would allow the process to operate within required parameters.

In Figure 4.30, manual globe valves are installed in the event of an automated control valve failure or when an emergency shutdown is required. This can be used to prevent further flow of materials into the pipeline.

High- and low-level alarms regarding the temperature, pressure and level have been implemented, which are used to help maintain running conditions of the column to avoid damage to the equipment, overflowing, under flowing or temperature and pressure inconsistencies within the pipelines. If the pressure, temperature or level within the pipeline is inadequate, the process control measures put in place will mitigate the problems. This, in turn, controls the automated valves to alleviate the problem. If this is not done or not done in time, the alarms are triggered and relevant parties get the issues tended to. If severe enough, it can cause an emergency shutdown of the plant until the matter is resolved.

An emergency shutdown allows for a pause in the process in cases such as weeping, flooding or drying out of the column. All equipment in the distillation system has the potential to operate in highly flammable atmosphere and thus it is of utmost

Table 4.21 Guide words for HAZOP analysis.

Guide words	Meanings
More	Quantitative increases or declines
Less	Less than required
No or not	Full negation of these intentions

Table 4.22 HAZOP evaluation to separate DME from methanol and water.

Vessel	Distillation column, condenser and reboiler			
Intention	Separate DME from methanol and water			
Parameter	Guide word	Causes	Consequences	Actions required
Temperature	More	• Insufficient cooling water supply into condenser • Heat exchanger tube failure • Steam control valve failure for reboiler • Increased steam flowrate	• Since the heavy key component and undesirable products will rise to the top, separation will be challenging. • Decrease in the purity of the product	• Install a high-temperature alarm. • Regular inspection of valves • Install a temperature indicator
	Less	• Cooling control valve failure • Low inlet steam temperature • Less steam flowrate	• The desired separation will not be feasible • Distillate product and purity will decrease • More bottom product	• Regular inspection of valves • Install a temperature indicator
Pressure	More	• Pressure indicator malfunction • Excessive vapor • High vapor pressure	• Increased fluid retention • Flooding • Separation efficiency will decrease	• Install a high-pressure alarm • Install pressure relieve valve
	Less	• Low vapor flow • Vapor pressure applied insufficiently	• Excessive weeping • Separation efficiency will decrease	• Restart the column • Install a low-pressure alarm
Level	More	• A tray's increased vapor pressure drop • Blockage in settling pipes	• The splitting will be destroyed • Flooding	• Corresponding reduction in feed and return flow
	Less	• Leakage in settling pipes	• Loss of product	• Install low level alarm
Flow	No	• Pump failure • Pipe blockage	• The column will fail to work • The column dries up	• Install low flow alarm • Shut down column
	Less	• Pump failure • Control valve malfunction	• Product will decrease • The column dries up	• Install low flow alarm • Shut down column
	More	• Control valve malfunction • Control valve left fully open	• Operating temperature will decrease • Flooding	• Install high flow alarm

importance to ensure an emergency shutdown is implemented to avoid fires or explosions from occurring. It is important to recognise that the process control measures that were investigated can be implemented in the system after doing the HAZOP analysis to minimise possible incidents from occurring.

The transmitters, controllers and alarms are placed in order to combat the issues addressed in the HAZOP analysis.

4.3.11 Plant Location, Layout and Environmental Assessment

As previously discussed, successful coal mines are selected based on coal quality and geology. It is also important to take note of the accessibility to raw materials, labour, the utilities infrastructure and the transport requirements. South Africa has 19 coalfields in use. They are in the north-eastern part of the country, namely Mpumalanga, Limpopo and extend to KwaZulu-Natal and the Free state (Department of Energy 2016).

Sasol Mining for instance has six mines (the Mooikraal, the Bosjesspruit, Impumelelo, Shondoni, Syferfontein and the Thubelisha collieries) (Sasol 2022) in operation. These coalfields supply feedstock to the Secunda (Sasol Synfuels) and Sasolburg (Sasolburg Operations) complexes in South Africa (Sasol 2022). The coal mined by Sasol Mining is then transported to the Secunda petrochemical.

As mentioned, Impumelelo (Impumelelo Mine location (2023) is said to be Sasol's deepest mine and the farthest away from Sasol operations (Sasol 2022). This makes it a safe option as it is far from communities and therefore there is less

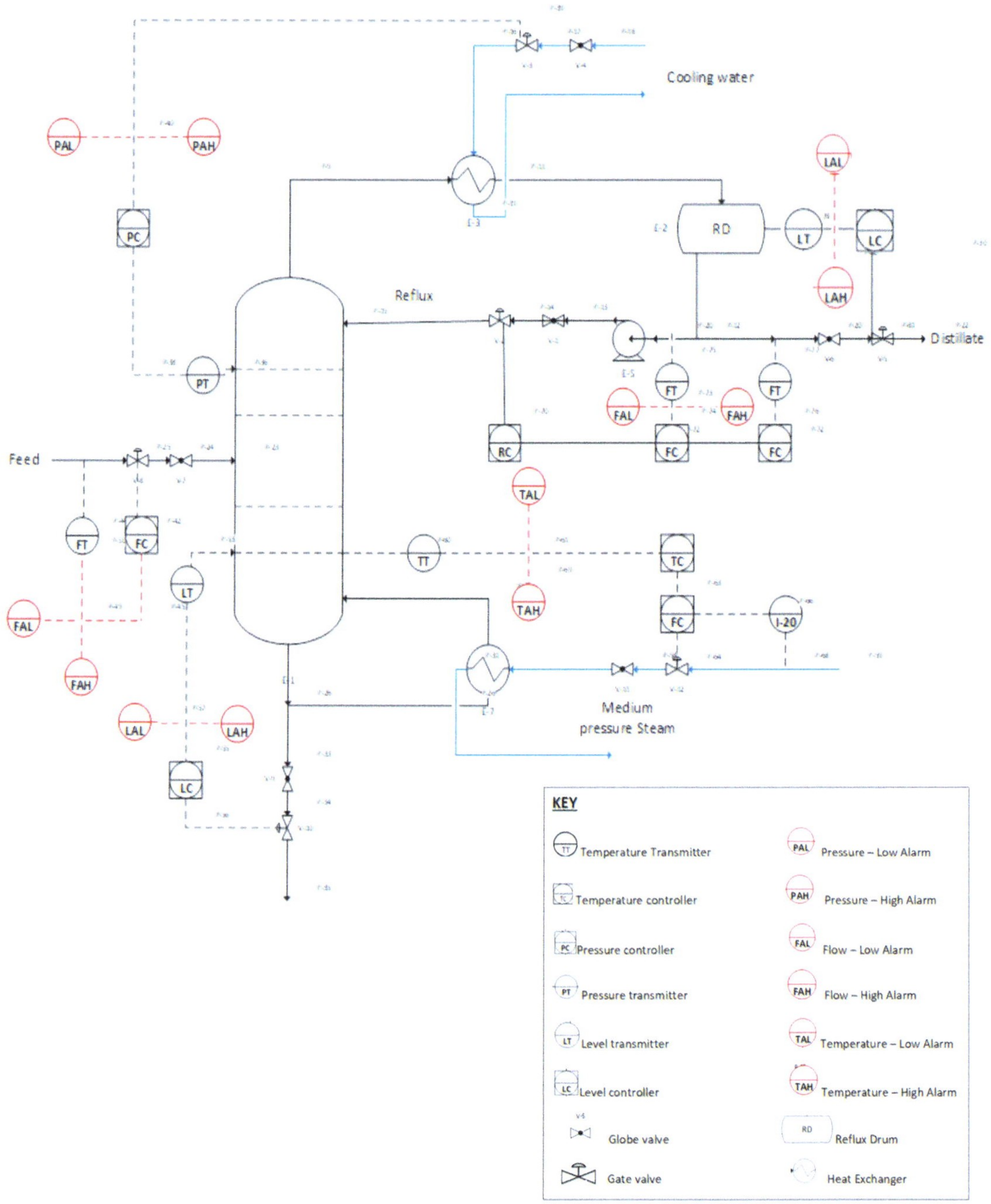

Figure 4.30 Post-HAZOP P&ID.

pollution to nearby communities. It has a 27 km single flight conveyor to Sasol coal supply and has been shown to have the capacity to supply 10.5 million tpa, all of which is supplied to the Sasol Secunda Synfuels Operations (Sasol 2022). This would make the best site to use as the process for this design and requires the need for synthetic fuels to be produced and utilised (Department of Energy 2016), thus it becomes beneficial to set up a plant close to this mine for the design required.

4.3.11.1 Transportation

Sites need to have adequate transportation surrounding them to allow for people working on the mine, companies/people utilising the mine and delivery services to and from the mine to travel to and from the mine. It is always good to have more than one means of transport to utilise when looking at transportation regarding the site at hand such as roadways, railways, airports, waterways, etc. A variety of transportation is always good to have as the prices differ between different means of transport. For example, public transport may be cheaper than using one's own vehicle that would require one to pay for their own fuel which may be more expensive in the long run as compared to the shared cost implemented in public transport (seeing as the price of fuel is continually increasing).

The Impumelelo Mine location (2023) illustrates roadways that lead right to the border of the site. This means that public and private transportation to and from the site is possible. The figure also illustrates an airfield in Secunda which can be utilised. The airfield can be located approximately 41 km from the site.

At the moment, the nearest airport to the site would be O.R. Thambo International Airport which is approximately 130 km from the site.

4.3.11.2 Site Layout

Plants are generally constructed in such a way that would allow for ease of economic flow and personnel around the site (Towler and Sinnot 2008). It is important to strategically structure a plant based on cost, economic feasibility, safety, utilities, storage, transportation and operation. This ensures that the basic requirements for a plant to operate are set up, such as the schematics presented in Figure 4.31.

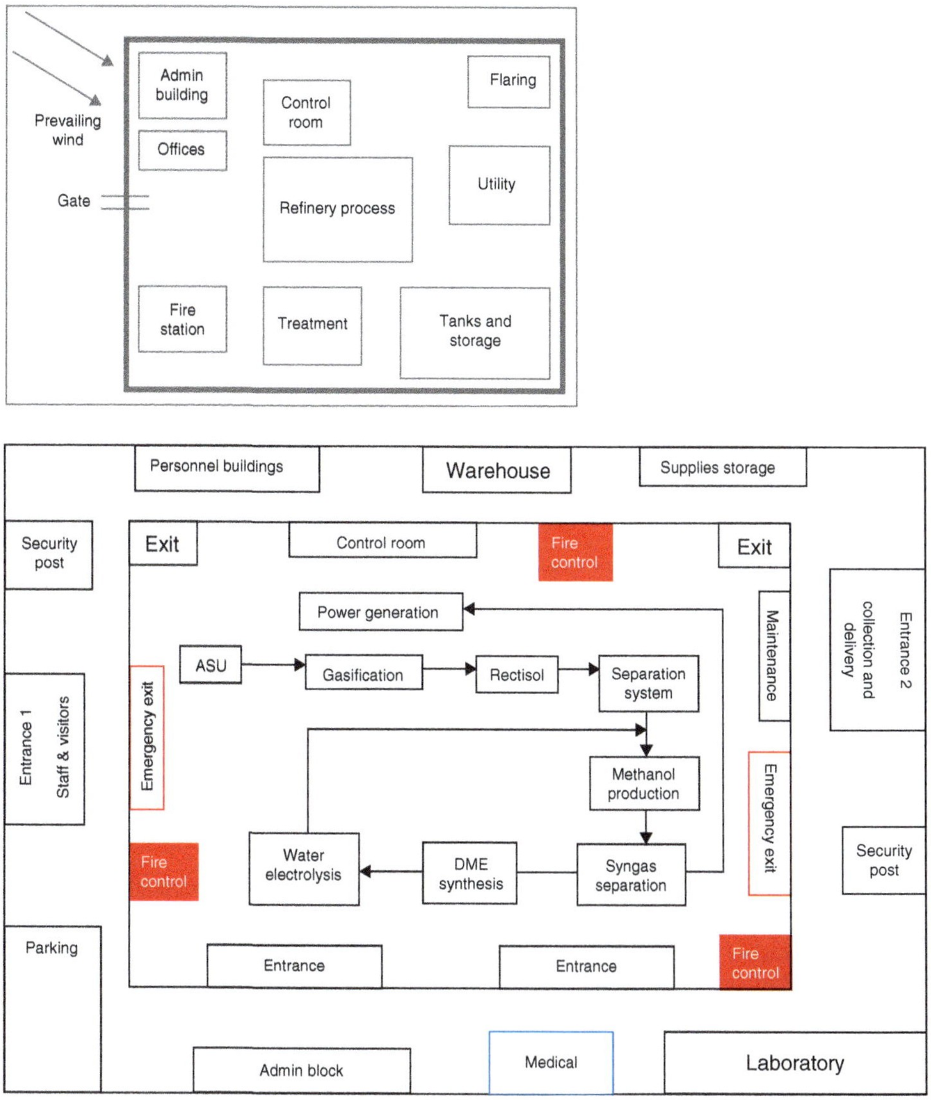

Figure 4.31 Basic plant layout for process adapted from Fahim et al., which include the following features. 1. Security posts, 2.Entrances and exits (for staff and visitors and for collection and delivery); 3. Parking facilities, 4. Administration block (offices), 5. Medical Facility, 6. Laboratory (quality control), 7. Supplies storage, 8. Warehouse (house utilities), 9. Personnel buildings (canteen, offices, change rooms etc.), 10. Emergency exits, 11. Fire control stations, 12. ASU (air separation unit), 13. Process units required.

The figure depicts a very basic outline of how the plant may look, which in turn illustrates that there are separate entrances for the staff and visitors and for collection and delivery. This would allow for better control of who enters and exits the plant and for what reason. There are fire stations located around the plant for good access in case of a fire. Aside from ordinary entrances and exits, there are also emergency exits around the main process of the plant to ensure workers can escape easily in case of an emergency. The control room is also located close to the personnel buildings to allow for remote control. The air separation unit (ASU) and gasification process are also close to an emergency exit in case of the escape of gas effluents. The entrances and exits located around the main process of the plant also allow for easy access as well as efficient ventilation around the plant. Storage is also situated close to the entrance 2 to allow for easy access to products to be transported. There is also easy access to a medical facility in case of injury or illness on the premises.

4.3.11.3 Utilities

Most often, the most needed utility would be water. Although this process seeks to decrease water usage, water may still be needed for the heat exchangers for cooling and heating purposes. The water required may be sourced from a nearby municipality.

Electricity for the process is provided by the plant itself as it has a power generation section as well. There are three well-known power stations in the area of Mpumalanga namely the Kusile Power Station, the Komati Power Station and the Eskom Hendria Power station that can be used if additional power is required.

The fuel required for power generation will be DME which is extracted from the DME synthesis process of the plant. As there is no need to retrieve fuel from a supplier, which will not contribute to the overall annual utility cost.

4.3.11.4 Environmental Impact Assessment

The upside to constructing a chemical plant is that it helps create job opportunities and uplifts the community. It allows for surrounding areas to develop with new buildings and housing. With that being said, the plant still needs to undergo an environmental impact assessment (EIA) to identify effects on the environment, the atmosphere and the community around it. Thus, it is important to develop measures to reduce negative impacts that could possibly be caused by the plant. Such examples of these impacts would include noise and waste impacts.

4.3.11.5 Waste Management

Plant processes usually result in a variety of waste being generated that need proper disposal. Waste generated can come in the form of gas waste, liquid waste, solid waste and in many cases is presented in the form of wastewater and chlorofluorocarbons (CFC) emissions. In plants, such as this, ash is also considered to be a waste product.

Gaseous emissions are said to be one of the hardest waste products to dispose of. The only way to mitigate this would be to implement the right strategies and cleaning processes to control the amount of gas that would be released or stored that are kept within the standards regulated.

Liquid waste which may mostly come in the form of wastewater may go through a purifying/cleaning process or may be distilled and reused in the process. It would be beneficial to recycle the water back into the process to avoid wastage or the dumping of polluted water. Releasing polluted water into the environment may lead to the pollution of surrounding water bodies and land.

Solid waste may be disposed of in legal landfill sites. At these sites, the waste is further broken down or compressed and stored in the ground. Solid waste produced by the plant should be kept at a minimum as much as possible.

4.3.11.6 Noise

Large equipment such as turbines, fans and compressors may be used on the plant and produce loud noise to nearby communities. Such equipment could be placed away from the plant boundaries and/or fit with silencers or fitted with coverings. Noise may also affect the employees on the plant that may result in damage to their hearing. The correct personal protective equipment (PPE) would be required to be worn by the employees to avoid permanent damage.

4.3.11.7 Environmental Impacts

The following impacts should be taken into account as presented in Table 4.23 as they may have a direct effect on human life and the environment. These are important to identify as they allow for the correcting preventative and protective measures.

Table 4.23 Environmental impacts analysis.

Effluent	Description	Impact
Carbon dioxide	Gaseous emission	– Greenhouse emission – Air pollution that contributes to global warming – Human impact: eye irritation, dizziness and headaches
Carbon monoxide	Gaseous emission	– Toxic – Greenhouse gas that contributes to air pollution and global warming – Human impact: nausea, dizziness, headaches, brain damage, breathing difficulties, heart irregularities, death (in severe cases)
Sulphur products	Gaseous emission	– Air pollution – Human impact: respiratory issues, convulsions, dizziness, death
Methane	Gaseous emission	– Greenhouse gas that contributes to air pollution and global warming – Human impact: memory loss, nausea, visual problems, headaches
Nitrogen	Gaseous emission	– High levels cause the production of greenhouse gases – Small amounts released does not have severe environmental issues – Human impact: respiratory issues, limits flora growth
Wastewater	Liquid effluent	– Pollutant – Releasing to environment may lead to polluting of bodies of water and land that would disturb the ecology – Human impact: illness if ingested
Methanol	Liquid Effluent	– Pollutant – Flammable – Human impact: nausea, vertigo, headaches
DME	Liquid Effluent	– Flammable – Human impact: irritation to skin and respiratory system. Hugh exposure may lead to headaches and loss of consciousness

4.3.11.8 Risk Analysis

The purpose of a risk assessment matrix is to manage and prioritise risk (Usmani 2022). This may include the probability and severity of the risk. Risk assessment matrices are tailored for the specific project and constructed in such a way that makes it easy for others to understand.

The risk assessment takes the probability of the event and multiplies it by its impact (Usmani 2022). This method has been used for the design described and ranked in Table 4.24.

Table 4.24 Severity and probability of event.

Probability (Frequency)		
Verbal	**Numeric**	**Description**
Frequent	4	Most Likely to Occur
Probable	3	Likely to Occur
Remote	2	Less Chance to Occur
Improbable	1	Least Chance to Occur

Severity		
Verbal	**Numeric**	**Description**
Frequent	4	Potential for Severe Injury
Probable	3	Potential for Moderate to Severe Injury
Remote	2	Potential for Minor Injury
Improbable	1	Negligible Injury

Source: Adapted from Usmani.

Table 4.25 Risk assessment matrix.

		Risk assessment matrix			
		Severity			
		Catastrophic – 4	**Critical – 3**	**Marginal – 2**	**Negligible – 1**
Probability	Frequent – 4	High (16)	High (12)	Serious (6)	Medium (4)
	Probable – 3	High (12)	Serious (9)	Serious (6)	Medium (3)
	Remote – 2	Serious (8)	Serious (6)	Medium (4)	Low (2)
	Improbable – 1	Medium (4)	Medium (3)	Low (2)	Low (1)

Source: Adapted from Usmani.

A standard risk assessment matrix is presented in Tables 4.25 and 4.26, where the risks can be categorised in three groups (Usmani 2022):

- High (Red) – Unacceptable
- Moderate (Yellow) – May or may not be acceptable
- Low (Green) – Considered acceptable

Table 4.26 The risk assessment and impacts described.

Plant process	Severity	Probability	Product of severity and probability rating	Degree of impact
Impact: Air pollution – Gaseous emissions				
Carbon dioxide	2	2	4	Medium
Carbon monoxide	2	2	4	Medium
Sulphur products	3	3	9	Serious
Methane	2	2	4	Medium
Nitrogen	2	1	2	Low
Impact: Water pollution – Liquid effluents				
Wastewater	1	2	2	Low
Methanol	2	1	2	Low
DME	2	1	2	Low
Impact: Land pollution				
From general plant operation	2	2	4	Medium
Impact: Noise pollution				
From general plant operation – Machinery, equipment, reactors, pumps, etc.	2	2	4	Medium
Impact: General safety for humans				
Gaseous emissions	3	3	9	Serious
Liquid effluents	2	3	6	Serious
Land pollution	1	3	3	Medium
Noise	1	2	2	Low

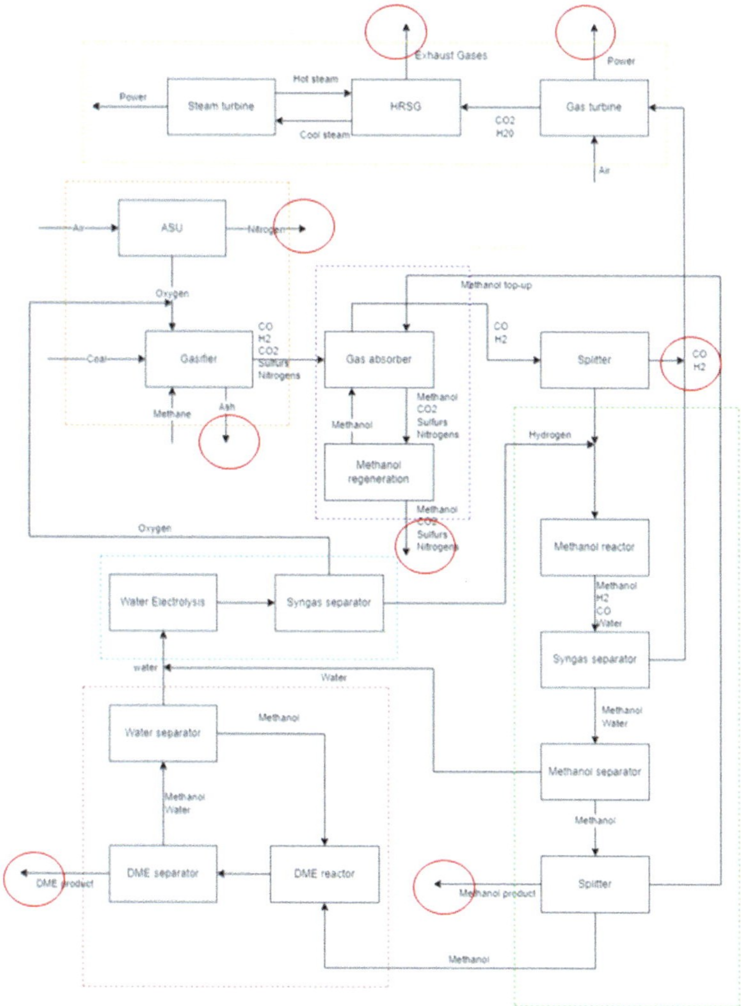

Figure 4.32 Schematic flow diagram of plant, which illustrates approximately where in the process gaseous emissions and fluid effluents described might leave the process. The figure helps to see where safety measures need to be put in place and where staff need to be diligent in wearing the correct PPE.

Determining whether or not the risk is acceptable comes down to how the event affects the plant and safety of the environment and people around the plant.

The correct safety measures must be put in places and surroundings circled in Figure 4.32.

4.3.12 Applicable Acts

It is important for those running the plant to adhere to the laws and legislations pertaining to the operation of the plant. This is done to ensure all operations are legal and law abiding. Compliance with these laws and legislations allow for smooth operation of the plant that causes the least amount of harm to human life and the environment. Failure to comply will result in legal action being taken against the plant (ranging from fines to court cases to the shutting down of the entire plant). The following Table 4.27 illustrates the main laws/acts the plant should abide by.

4.3.13 Costing and Techno-Economic Analysis

There are several pieces of equipment that need to be costed. Towler and Sinnot (2008) have a basic formula for costing the pieces of equipment. This follows the simple equation,

$$C_E = a + bS^n \tag{4.45}$$

Table 4.27 Relevant laws and acts.

Act	Description	Compliance
Occupational Health and Safety Act (1993)	Aims to ensure that employees work in a safe working environment	Place safety measures in place to ensure the safety of employees Provide training for the plant
National Water Act (1998)	Aims to protect water resources. To prevent and control pollution of water bodies. Also aims to ensure the disadvantaged have access to water	Treat water on site before disposal
The Clean Air Act (CAA, 1970)	Aims to control the levels of lead, carbon monoxide, ozone, sulphur dioxide, nitrogen dioxide, PM2.5 and PM10 in the atmosphere	Ensure that the least amount of these gases is produced and released into the atmosphere. Recycle gases back into process where possible
Carbon Tax Act (2019)	Put in place with the aim to reduce the amount greenhouse gases being released into the atmosphere. A tax is paid for these greenhouse gas emissions	Reduce the amount of greenhouse gases produced Recycle gases back into the system Clean gases before releasing into atmosphere
NEMA Air Quality Act (2004)	Place measures in place for air pollution. Aims to prevent and control the pollution in the air and improving the quality of air	Implement safety regulations for the safe emission of gases Try to reduce the amount of greenhouse gases being released into the air Decrease the amount of smoke being released into atmosphere due to burning of coal
NEMA Waste Act (2008)	Aims to prevent and control as much land pollution as much as possible. Aims to minimise consumption of natural resources, reduce waste generation, prevent ecological degradation, recycling and reusing, safe disposal of waste and rehabilitating affected land	Correct disposal of waste in legal landfills if need be Correct storage and pretreatment of waste Placing bins around the plant Promote recycling on plant

This equation can be used to cost almost all the pieces of equipment in the plant. Figure 4.5 show the parameters for the different pieces of equipment. Some pieces of equipment on the plant are too big or too small for the specified values and need to be adjusted using the six-tenths rule found in Towler and Sinnot (2008) to adjust these values up to the actual values.

$$C_2 = C_1 \left(\frac{S_2}{S_1} \right)^{0.6} \tag{4.46}$$

The way this rule is used is either the maximum or minimum size (S) is used to get C_1, then the real price is retrieved using the rule to get the final cost, C_2.

Three different units could not be modelled using the method outlined above. These are the gasifier R-101, the electrolysis unit R-102 and the air separation unit V-101. A method for estimating the costs for the gasifier and ASU was outlined, which used a similar method to the one outlined above.

$$C_{GR} = C_0 \left(\frac{F}{F_0} \right)^{n} \tag{4.47}$$

The gasifier is based off the dry feed into the vessel, and the ASU is dependent on the amount of oxygen produced per hour. This method is very similar to the six tenths rule outlined above, and adjusts the value based off a base cost and capacity, C_0 and F_0.

The cost of the electrolysis unit R-103 also needs to be assumed. To do this it is first determined how much energy is required per kg feed or product. Office of Energy Efficiency and Renewable Energy (2011) describes a 2020 target efficiency of 44 kWh/kgH_2. With a target production of 310 kg/h of hydrogen, this can easily be used to give the energy requirements for the required photovoltaic panels. The costs for these panels are found in Freedom Solar Panels (2021) which describes the installed costs for these panels as $2/W.

Table 4.28 Capital costs for equipment.

Unit	Cost (US$)	Cost in Rand (R)
Heaters		
E-101	100 684	1 877 757
E-102	176 711	3 295 666
E-103	2608	48 641
E-104	1134	21 148
E-105	6158	114 852
E-106	37 387	697 272
E-107	106 080	1 978 393
E-108	111 299	2 075 725
Compressors and turbines		
C-101	2 280 684	42 534 762
C-102	9 558 744	178 270 579
C-103	3 059 990	57 068 815
Vessels		
V-101	9 009 138	168 020 415
V-102	239 792	4 472 115
V-103	272 505	5 082 224
V-104	223 781	4 173 522
V-105	13 943	260 035
T-101	7850	146 402
T-102	11 744	219 019
Pumps		
P-101	103 269	1 925 969
P-102	36 201	675 141
Reactors		
R-101	9 984 152	186 204 436
R-102	28 406 971	529 790 000
R-103	2 142 403	39 955 808
R-104	105 449	1 966 628
R-105	84 592	1 577 644

The base costs for all the pieces of equipment can be seen in Table 4.28. The exchange rate is taken as R18.65 per USD as of 07 August 2023.

4.3.13.1 Adjusting for Inflation

All the values above obtained from Towler and Sinnot (2008) method are determined on a basis from costs in 2006, and the prices from the method were determined using 2017 as their year. These values need to now be brought up to the current year and this can be done by using cost indices.

$$C_{Y2} = C_{Y1}\frac{I_{Y2}}{I_{Y1}} \tag{4.48}$$

The Chemical Engineering Plant Cost Index (CEPCI) is an indicator specific to chemical plants and can be used to bring the price up to today's money. The CEPCI price for April 2022 is found to be 815.7 (Maxwell 2022).

4.3.13.2 Fixed Capital Cost

Now that the equipment costs have been calculated, the fixed installation costs can also be estimated. Towler and Sinnot (2008) give a method that assigns factors for different costs involved, such as piping or installations for every piece of equipment,

$$C = C_{CS}[(1 + f_p)f_m + (f_{er} + f_{el} + f_i + f_c + f_s + f_l)] \tag{4.49}$$

where C_{CS} is the cost of the component in carbon steel, f_p is the piping installation factor, f_m is the material factor, f_{er} is equipment erection, f_{el} is electrical work, f_i is for instrumentation and control, f_c is installation for civil work, f_s is the factor for structures and f_l is the factor for lagging, insulation and paint. These can be done on every piece of equipment to get its installation cost. All pieces of equipment except for the gasifier are fluids, and everything in contact with methanol is made of stainless steel 304.

The last adjustment that needs to be made is to account for offsite (OS), design and engineering (DE) and contingency costs (X). These give the final fixed capital cost of the pieces of equipment as,

$$C_{FC} = C(1 + \text{OS})(1 + \text{DE} + X) \tag{4.50}$$

The calculations of these new factors give the total Fixed Capital Investment for the plant as US$465951742.

4.3.13.3 Manufacturing Costs

Manufacturing costs can be broken up into direct manufacturing costs (DMC), fixed manufacturing costs (FMC) and general expenses (GE). These costs are normally associated with everyday costs and can be seen in Table 4.29.

DMC, FMC and GE are all included in the cost of manufacturing (COM). Direct manufacturing costs are operating costs that change with production rate, while fixed manufacturing costs are costs that are necessary for the operation of the company regardless of production rate (Turton et al. 2018). The equation to calculate the COM can thus be given as,

$$\text{COM} = \text{DMC} + \text{FMC} + \text{GE} \tag{4.51}$$

To calculate the cost of manufacturing, it is necessary to estimate the fixed capital investment (FCI), cost of operating labour (C_{OL}), cost of utilities (C_{UT}), cost of waste treatment (C_{WT}) and cost of raw materials (C_{RM}).

The utilities cost can be considered a once-off payment and not a payment per year as the utilities keep getting recycled while methanol is produced by the plant. Other costs seen through the refrigeration cycles cost of electricity is accounted by the extra electricity produced by the plant.

The cost of operating labour was calculated using the number of operators each shift (N_{OL}). The number of operators each shift can be calculated using the following equation:

$$N_{OL} = (6.29 + 31.7P^2 + 0.23N_{np})^{0.5} \tag{4.52}$$

where N_{np} is the total number of nonparticulate processing stages, including compression, heating and cooling, mixing and reaction, and P is the total number of processing steps including the handling of particulate materials.

4.3.13.4 Economic Evaluation

Using the calculated values, the discounted cash flow plot is presented in Figure 4.33. With Figure 4.33, while assuming that the plant would be operating for 330 days in a year, the rest of the days are used for plant maintenance. In Figure 4.33,

Table 4.29 Manufacturing costs calculated.

Item	Cost (US$)
Raw materials	130 841 163
Wastewater treatment	519 861
Utilities	
Operating labour	88 621
Fixed capital investment	160 857 909

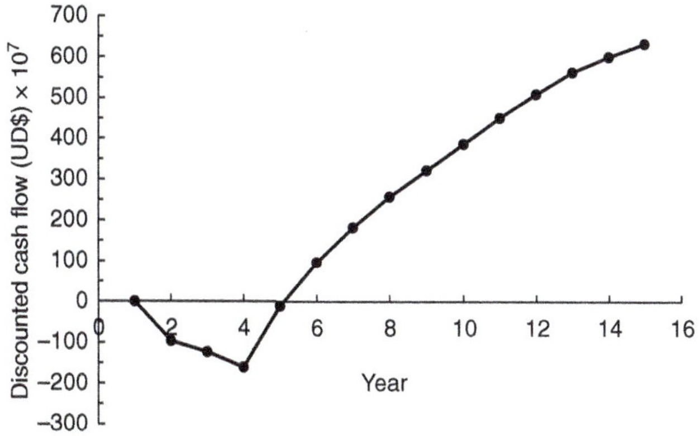

Figure 4.33 Discounted cash flow diagram.

the initial decrease is due to the cost of land and then the fixed capital investments without land sitting at US$1.27 billion while the vertical decline indicates the working capital. Initially, there was little income, but after five years of operation, the initiative began to turn a profit. The net present value is US$661 million at the conclusion of the 12-year period, with a net present value ratio of 5.18. This high NPV ratio indicates that the investment will yield great returns. Additionally, the discount repayment duration is 1.89 years, or roughly two years, which is a pretty short length of time.

4.4 Conclusion

This report describes the design of a chemical plant that is set to retrofit existing Sasol-lurgi gasifiers with the objective of utilising the gasifiers to produce syngas that can be utilised for a multiproduct facility. The design coverts coal and methane to syngas for power generation, liquid fuels and chemicals such as DME and methanol. The plant is set to produce at least 200 MW of electricity, 10 000 tons/annum of DME and 5000 tons/annum of methanol, which account for energy transition from pure coal to less CO_2 emission energy sources and electricity. The purpose of the designed plant in the case study is to reduce the amount of water being used in similar plants such as to investigate greener practices for generating power seeing as people have become more aware of the environmental impact these processes have. Products such as DME are also a greener alternative to diesel. The plant is also to produce its own electricity to sustain itself to reduce cost.

The coal mine selected for this operation is the Impumelelo mine situated in Mpumalanga, which is the deepest of the Sasol mines and far from communities. The mine has the capacity to supply 10.5 million tpa of bituminous coal that would be sent to the Sasol Secunda Synfuels Operations. A plant set up near the mining site would be beneficial as it would reduce the cost of transportation and the plant would provide job opportunities for nearby communities. Even though the plant produces electricity for self-sustainability, nearby power stations (namely the Kusile Power Station, the Komati Power Station and the Hendria Power Station) can be used as backup power sources in case of an emergency. The requirements of the plant were considered and used to design a basic plant layout of how the plant is to be designed to operate efficiently.

The multicomponent distillation column was chosen and based on the design taken by Group of students at the Wits University in South Africa. The approach that the members used is detailed in the report, while the system of the distillation column is modelled for the separation of three components namely, DME, water and methanol with DME being the light key and methanol being the heavy key component. The column is designed using the NRTL-RK thermodynamic model. It has a minimum reflux ratio of 0.42 with 24 stages and a column efficiency of 45%. The column height is 12 m with a top diameter of 0.95 m and a bottom diameter of 0.85 m. The reactor design is a packed bed reactor that is multi-tubular and utilises 35 t of an Al_2O_3 catalyst that should be replaced annually. The reactor runs isothermally, which is 13 m in length with a diameter of 1 m and a vessel thickness of 7.27 mm. It was found that approximately 380 tubes would be required for the reactor. The operating temperature of the reactor is 500 K and a pressure of 10 bar. For the reactor to run isothermally, it would need to be cooled constantly as it deals with an exothermic reaction and thus a heat exchanger is required. The heat exchanger utilises approximately 65 kg/s of cooling water as the medium. It is found that the net duty of the heat exchanger is about 13 MW and the heat transfer area required to achieve optimum conditions would be approximately 245 m^2.

Other equipment designs such as the pump and the compressor have been detailed in the report. The calculations for the compressor show that when isentropic efficiency conditions ae considered, the work for the compressor is 324 kJ/kg of air with a discharge temperature of approximately 433 °C. The pump is found to be a centrifugal pump that will run at 2984 rpm. The head of the pump is developed to be 114 m and a flowrate of liquid flowing through it at a rate of 250 m^3/h. After further calculation on the pump, it is found to have an efficiency of 75%. The pump also requires 125 kW of shaft power.

The process control for the column design is also investigated in the report to show what control measures would be required for the process. Control measures surrounding the pressure, temperature and level of the column and piping need to be kept constant to ensure safe running of the plant and to ensure the composition of the materials flowing are correct. Valves are also decided upon, and they are a key component in controlling these aspects. The HAZOP analysis is conducted to see where control measures are to be placed to further illustrate the importance of control systems and alarm systems that need to be installed in the plant. The environmental and human impacts due to the operation of the distillation column are also assessed along with the risks of these impacts. Certain laws and legislations are also found and should we adhered to by the plant or else legal action will be taken up against the plant.

The costing and economics of the plant showed that while assuming the plant runs for 330 days, the plant begins to produce an income after five years of operation. After 12 years a net present value is shown to be US$661411761, which yields a net present ratio of 5.18 which is said to be high and indicates a good return in investments. The discount repayment is also two years, which is a short period of time. The economic analysis of the plant shows that in a short amount of time, the plant will be profitable.

It can be recommended that a detailed design of each equipment should be done to complete the plant design and process control measures should be done around every piece of equipment, then an accurate HAZOP and cost analyses can be made. Waste could be treated on the plant to reduce waste disposal and control emission to the environment although this may increase the cost of the plant.

Finally, the sustainability of CO_2 mitigation strategies in refineries, petrochemicals and coal power-fired plants should also consider the social dimension alongside economic and environmental trade-offs. Thus, the costs associated with the transition to net-zero should be managed in a socially equitable manner, which should include development of regulatory and policy frameworks that facilitate such a sustainable transition to net-zero for the refining sector. The provision of global comparison of the life-cycle greenhouse gas emissions of combustion engines and electric passenger cars as means to energy sustainability become parts of the sustainability equation. On this basis, the next chapter discusses Carbon footprint of internal combustion engines and mitigations.

References

Affan, F.B. (2019). Temperature sensors development in the chemical industries. *J. Alternat. Renew. Energy Sources* 5 (3): 39–43.

Ateba, B.B., Prinsloo, J.J., and Gawlik, R. (2019). The significance of electricity supply sustainability to industrial growth in South Africa. *Energy Rep.* 5: 1324–1338.

Ayamolowo, O.J., Manditereza, P.T., and Kusakana, K. (2022). South Africa power reforms: the path to a dominant renewable energy-sourced grid. *Energy Rep.* 8 (2022): 1208–1215.

Azizi, Z., Rezaeimanesh, M., Tohidian, T., and Rahimpour, M.R. (2014). Dimethyl ether: a review of technologies and production challenges. In: *Chemical Engineering and Processing: Processing Process Intensification*, 150–172.

Bahadori, A. (2014). Chapter 8: Sizing of valve and control valve. In: *Natural Gas Processing*, 371–439. Gulf Professional Publishing.

Bartholomew, C.H. (2001). Mechanisms of catalyst deactivation. *Appl. Catal. A: Gen.* 212 (1–2): 17–60.

Bercic, G. and Levec, J. (1993). Catalytic dehydration of methanol to dimethyl ether. Kinetic investigation and reactor simulation. *Ind. Eng. Chem. Res.* 2478–2484.

Berry, M. and Schott, N. (2017). 17: Process monitoring and process control: an overview. In: *Applied Plastics Engineering Handbook*, 2e, 377–393. William Andrew Publishing.

Bhaskar, T., Bhavya, B., Singh, R. et al. (2011). Chapter 3: Thermochemical conversion of biomass to biofuels. In: *Biofuels* (ed. A. Pandey, C. Larroche, S.C. Ricke, et al.), 51–77. Academic Press.

Bouttell, D., Ismail, I., Mvula, N. et al. (2022). *Balance of plant design- coal to power design problem*. Students' design project. Johannesburg: University of the Witwatersrand Unpublished.

Brown, D.M., Bhatt, B.L., Hsiung, T.H. et al. (1991). Novel technology for the synthesis of dimethyl ether from syngas. *Catal. Today* 8: 279–304.

Cai, T. and Resetarits, M. (2011). Pressure drop measurements on distillation columns. *Chin. J. Chem. Eng.* 19 (5): 779–783.

Coulson, J.M. and Richardson, J.F. (2005). *Chemical Engineering Design*, 4e, Vol. 6. Elsevier Butterworth-Heinemann.

Couper, J.R., Penney, W.R., Fair, J.R., and Walas, S.M. (2012). Chapter 3: Process control. In: *Chemical Process Equipment*, 3e, 31–51. Butterworth-Heinemann.

Department of Energy, 2016. *South African Coal Sector Report*. [Online]. Available at: www.energy.gov.za/files/media/explained/South-African-Coal-Sector-Report.pdf

Dimian, A.C., Bildea, C.S., and Kiss, A.A. (2014). Chapter 15: Plantwide control. In: *Computer Aided Chemical Engineering*, vol. Vol. 35, 599–647. Elsevier.

Elliott C. 2019. *Planning for a "Just Transition": Leaving No Worker Behind in Shifting to a Low Carbon Future*. March 25. https://www.wri.org/insights/planning-just-transition-leaving-no-worker-behind-shifting-low-carbon-future [Accessed 25 February 2023].

Fahim, M.A., Alsahhaf, T.A., and Elkilani, A. (2010). Chapter 14: Safety in petroleum refineries. In: *Fundamentals of Petroleum Refining*, 17 March, 357–376.

Fogler, H.S. (2006). *Elements of Chemical Reaction Engineering*, 4e. Prentice Hall International.

Freedom Solar Panels. 2021. *Commercial Solar Panel Cost and Pricing* Retrieved from https://freedomsolarpower.com/cost-of-commercial-solar-panels-and-installation. Accessed 15 July 2022.

Ghavipour, M. and Behbahani, R.M. (2013). Fixed-bed reactor modeling for methanol to dimethyl ether (DME) reaction over γ-Alumina using a new practical reaction rate model. *J. Ind. Eng. Chem.* 20: 1942–1951.

Grundfos Academy. 2023. *Basic Principles and Pump Types About Pump Curves*. https://www.grundfos.com/ae/learn/ecademy/all-courses/basic-principles-and-pump-types/about-pump-curves [Accessed 30 May 2013].

Haggerty, J., Haggerty, M., Roemer, K., Rose, J. 2018. *Planning for the Local Impacts of Coal Facility Closure: Emerging Strategies in the U.S. West*. August 2018. http://www.sciencedirect.com/science/article/abs/pii/S0301420717306645

Hesselgreaves, J.E., Law, R., and Reay, D.A. (2017). Chapter 1: Introduction. In: *Compact Heat Exchangers*, 2e, 1–33. Butterworth-Heinemann.

Hu, J., Li, Y., and Yeung, K. (2006). 17: In vivo physiological measurements. In: *Clothing Biosensory Engineering* (ed. Y. Li and A. Wong), 285–300. Woodhead Publishing.

Hussain, M.I., Nawaz, S., Sajid, M.M. et al. (2020). Chapter 5: Corrosion resistance of nanostructured metals and alloys. In: *Corrosion Protection at the Nanoscale*, 63–87. Elsevier.

Impumelelo Mine location (2023). *Impumelelo Coal Mine, Formerly Known as the Brandspruit mine, Embalenhle, Mpumalanga, South Africa*. https://www.gem.wiki/Impumelelo_coal_mine. [Accessed 28 June 2023].

INTECH GmbH. 2022. *Basic Steps to Compressor Unit Calculation and Selection* Retrieved from https://intech-gmbh.com/compr_calculation_and_selection/#compressors_selection.

Kabir, K.B. and Bhattacharya, S. (2011). *Dimethyl Ether Production from Gasification of Victorian Brown Coal - Process Model and Related Preliminary Experiments*. Research Gate https://www.researchgate.net/publication/233959719.

Kaineder, A., Michenthaler, C., and Stelzer, D.H. (2019). Guided wave tank level sensor. In: *16th European Radar Conference (EuRAD)*, 936–939.

Kazuhiko Sano, Y.K. (2020). Emergency shutdown of fluidized bed reaction systems. *J. Loss Prev. Process Ind.* 68: 104277.

Kececi, E.F. (2019). Chapter 11: Actuators. In: *Mechatronic Components*, 145–154. Butterworth-Heinemann.

Leitch, I. (2007). 12: Actuators. In: *Handbook of Valves and Actuators* (ed. B. Nesbitt), 279–310. Butterworth-Heinemann.

Levenspiel, O. (1999). *Chemical Reaction Engineering*, 3e. John Wiley & Sons.

Lieberman, N.P. (2009). Chapter 5: Distillation tower pressure control. In: *Troubleshooting Process Plant Control*, 37–49. John Wiley & Sons. Inc.

Luyben, W.L. (2012). Effect of tray pressure drop on the trade-off between trays and energy. *Ind. Eng. Chem. Res.* 9186–9190.

Majozi, T. (2009). Minimization of energy use in multipurpose batch plants using heat storage: an aspect of cleaner production. *J. Clean. Prod.* 17: 945–950.

Mäki, H. (2013). *Municipal Engineers in South Africa Before 1910*. Tampere University Press ePublications.

Martyr, A.J. and Rogers, D.R. (2021). Chapter 15: Measurement of liquid fuel, oil, and combustion air consumption. In: *Engine Testing*, 5e, 511–535. Butterworth-Heinemann.

Masudi, A., Jusoh, N.W., and Muraza, O. (2020). Recent progress on low rank coal conversion to dimethyl ether as clean fuel: a critical review. *J. Clean. Prod.* 277.

Maxwell, C. 2022. *Cost Indices*, Retrieved from https://www.toweringskills.com/financial-analysis/cost-indices/#chemical-engineering-plant-cost-index-cepci.

McAllister, E. (2014). 7: Control valves. In: *Pipeline Rules of Thumb Handbook*, 8e, 187–225. Gulf Professional Publishing.

Mokhatab, S., Poe, W.A., and Mak, J.Y. (2019). Chapter 20: Gas processing plant automation. In: *Handbook of Natural Gas Transmission and Processing*, 4e, 615–642. Gulf Professional Publishing.

Moran, S. (2017). Chapter 20: Tanks and drums. In: *Process Plant Layout*, 2e, 303–313. Butterworth-Heinemann.

Morris, A.S. and Langari, R. (2021). Chapter 17: Level measurement. In: *Measurement and Instrumentation*, 3e, 535–549. Butterworth-Heinemann.

Neal, A. (2020). *Issue Brief | How Coal Country Can Adapt to the Energy Transition*. Environmental and Energy Study Institute https://www.eesi.org/papers/view/issue-brief-how-coal-country-can-adapt-to-the-energy-transition.

Nelabhotla, A.B., Pant, D., and Dinamarca, C. (2021). Chapter 8: Power-to-gas for methanation. In: *Emerging Technologies and Biological Systems for Biogas Upgrading* (ed. N. Aryal, L.D. Ottosen, M.V. Kofoed, and D. Pant), 187–221. Academic Press.

NET Platform. 2020. *National Economic Transition Platform*. https://nationaleconomictransition.org/platform [Accessed 27 November 2022].

Nitsche, M. and Gbadamosi, R. (2016). Chapter 1: Heat exchanger design. In: *Heat Exchanger Design Guide*, 1–19. Butterworth-Heinemann.

Office of Energy Efficiency and Renewable Energy. 2011. *DOE Technical Targets for Hydrogen Production from Electrolysis*. Retrieved from https://www.energy.gov/eere/fuelcells/doe-technical-targets-hydrogen-production-electrolysis.

Parisher, R.A. and Rhea, R.A. (2022). Chapter 5: Valves. In: *Pipe Drafting and Design*, 4e, 87–117. Gulf Professional Publishing.

Poe, W.A. and Mokhatab, S. (2017). Chapter 3: Process control. In: *Modeling, Control, and Optimization of Natural Gas Processing Plants*, 97–172. Gulf Professional Publishing.

Ponton, J.W. (1994). Degrees of freedom analysis in process control. In: *Chemical Engineering Science* (ed. D. o. Engineering), 2089–2095. University of Edinburgh: Elsevier Science Ltd.

Rajcoomar, A. and Ramjeawon, T. (2017). Life cycle assessment of municipal solid waste management scenarios on the small island of Mauritius. *Waste Manag. Res.* 35 (3): 313–324.

Rauch, R., Kiennemann, A., and Sauciuc, A. (2013). Chapter 12: Fischer-Tropsch synthesis to biofuels (BtL process). In: *The Role of Catalysis for the Sustainable Production of Bio-Fuels and Bio-Chemicals* (ed. K.S. Triantafyllidis, A.A. Lappas, and M. Stöcker), 397–443. Elsevier.

Sang, Y., Lui, H., Shichao, H. et al. (2013). Catalytic performance of hierarchical H-ZSM-5/MCM-41 for methanol dehydration to dimethyl ether. *J. Energy Chem.* 769–777.

Sasol, 2022. *Operations/Locations*. [Online]. Available at: https://www.sasol.com/about-sasol/operating-business-units/mining/operations-locations

Tokay, K.C., Dogu, T., and Dogu, G. (2012). Dimethyl ether synthesis over alumina based catalysts. *Chem. Eng. J.* 184: 278–285.

Toulson, R. and Wilmshurst, T. (2012). Chapter 13: An introduction to control systems. In: *Fast and Effective Embedded Systems Design*, 273–295. Newnes.

Towler, G. and Sinnot, R. (2008). *Principles, Practice and Economics of Plant and Process Design*. San Diego: Elsevier.

Turton, R., Shaeiwitz, J.A., and Bhattacharyya, D. (2018). *Analysis, Synthesis and Design of Chemical Processes*, 5e. Pearson.

Usmani, F., 2022. *PM Study Circle*. Source: https://pmstudycircle.com/risk-assessment-matrix [Accessed 20 April 2023].

Visioli, A. (2005). Design and tuning of a ratio controller. *Control. Eng. Pract.* 13 (4): 485–497.

Visser, H., Alex, G., and Brent, A. (2018). Life cycle cost profitability of biomass power plants in South Africa within the international context. *Renew. Energy* 2019 (139): 9–21.

VR Chemistry. (2022). *The Drive to Make Things Happen*. Retrieved from https://vrchemistry.chem.ox.ac.uk/potential/text/solutions3.htm#:~:text=The%20vapor%20pressure%20of%20water,mm%20Hg%20%3D%201%20atm .

Zhang, P. (2008). 1: Sensors and actuators for industrial control. In: *Industrial Control Technology*, 174–186. William Andrew Publishing.

5

Carbon Footprint of Internal Combustion Engines and Mitigations

The transportation sector is a major contributor to CO_2 release into the atmosphere, which in turn is a key culprit in increasing climate change. In order to meet the long-term climate change mitigation targets as set by the Paris Agreement and for individual countries, policies towards decarbonising road, air and ocean transportation should be built around more efficient transport, society, and larger shares of renewable fuels and faster introduction of electric vehicles (EVs) and engines. Many countries, such as Canada, France, Japan, Mexico and the United Kingdom, have announced targets or plans for phasing out internal combustion engines (ICEs) (United Nations Environment Programme 2019; Wappelhorst 2020). Such a strategy would remove direct fossil fuel use from the passenger car system, transferring decarbonisation concerns to fuel and electricity supply chains (Hill et al. 2019). As a preliminary measure, they calculated on-road CO_2 emissions under different scenarios from 2014 to 2050 and proposed that with accelerated EV adoption, the volume of EVs and thus their impact on reduction of CO_2 emissions will remain until about 2030.

5.1 Internal Combustion Engine and Emission

Since decarbonisation of the transport sector has not been very successful due to high costs of substituting energy-dense liquid fossil fuels, transition to electromobility and a shift from gasoline and diesel cars to EVs may offer the potential for large-scale reduction in CO_2 emissions (Creutzig 2016).

The belief that the reduction of CO_2 emissions is achieved through the use of EVs would require detailed empirical and conceptual studies of nations or regions. In the United Kingdom, for example, which consists of England, Wales, Northern Ireland and Scotland with an estimated population of about 62 million people over an area of 243 610 km^2 due to its insular location, it has been trying to reduce dependencies on international and transit road users.

In 2008, the Climate Change Act 2008 was introduced in the UK with the stated aim of reducing its greenhouse gas (GHG) emissions to 80% of the 1990 levels by 2050 (eurostat 2017). In 2016, the transport sector, including international aviation, accounted for 1080 $MtCO_2e$, which is the total weight of CO_2 equivalent to global warming potential, and contributed 27% to the total GHG emissions across the then EU-28 countries (eurostat 2017). Arguably, the increasing share of GHG emissions from the transport sector follows a structural change dynamics where the shift of economies from industry to services corresponds to a rise in transport intensity in energy use, which is rooted in the supply/demand of vehicles (Schafer 2005; Nanaki and Koroneos 2016).

European Commission (2017) reported the range of policies developed by the EU in order to mitigate the effects of transport on climate change, with the majority aimed at reducing CO_2 emissions, and hence reducing total GHG emissions:

- CO_2 emissions for new cars and vans: Car manufacturers have to ensure that their new car fleet does not emit more than an average of 130 g CO_2/km by 2015 and 95 g CO_2/km by 2020. Recent regulation proposals (European Commission 2017) have improved this to 67 gCO_2/km by 2030, a reduction of 30% over the 2020 levels.
- CO_2 labelling of cars: EU legislation ensures that consumers are informed about the fuel efficiency and CO_2 emissions of new cars.
- Fuel quality: EU legislation requires the GHG intensity of vehicle fuels to be cut by up to 10% by 2020. Actions and inactions by central, regional and local governments will have profound effects as cities across Europe introduce mitigation

Measuring Climate Change to Inform Energy Transitions: Carbon Footprint Calculations, First Edition. Sunny E. Iyuke.

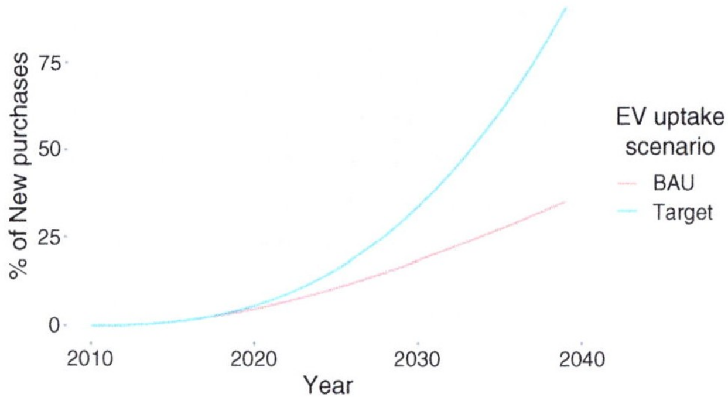

Figure 5.1 Comparison of business as usual and 2040 EV adoption target scenarios, 2010–2040 (Hill et al. 2019).

strategies to tackle the causes of climate change. While the European Parliament has committed its member states to an ambitious climate change strategy reducing GHG emissions and energy consumption by at least 20% by 2020 using a 1990 baseline (DECISION No. 406/2009/EC).

The UK Committee on Climate Change proposed EV characteristics as follows;

- Fifth Carbon Budget (Committee on Climate Change 2015) ascertained that EVs would provide CO_2 abatement scenario from sales reaching 60% of all new car sales by 2030, which would need a growth rate in annual purchases of 33%.
- Introduction of a ban on the sale of new petrol and diesel passenger cars and small vans in 2040 by the UK government in 2017 (DEFRA 2017) has resulted in a series of predictions on the expected EV adoption rate within the region.
- If the results from the two actions above are extrapolated to the 2040 target from existing EV sales rates should lead to an increase of 2074 in 2011 through 5300 in 2017 and subsequently greater EV sales in 2040.
- Figure 5.1 is thus presented to predict EV car sales, which assumes that current trends in EV sales will keep increasing at future rates (Serradilla et al. 2017).

The global transition from internal combustion vehicles to hybrids in 1997 with a small battery and electric motor, as narrated by the Toyota team (Ogiso 2017), has shown an enormous impact on pollution. Nevertheless, it was aimed at by the Europeans to achieve fully electric transportation in this century, which has led to new targets of replacing diesel and gasoline vehicles (King 2007). As of 2009, other electric models with both battery and hydrogen were tested and proved to be practical replacements for ICEs and vehicles, even though they were still expensive at low production rates. As this achievement becomes a turning point in the United Kingdom, as of 2021, the country was able to achieve 1% penetration of battery electric vehicles (BEVs) on the roads, which has continued to double almost every year (Staffell and Kendall 2012). As people find it time consuming to charge larger batteries (e.g. 100 kWh) overnight at residences, and perhaps families lack the necessary infrastructure anyway, Ballard in Vancouver, Canada, introduced the hydrogen fuel cell bus in the 1990s. The hydrogen fuel cell bus has since then been fitted with a large battery to aid in acceleration and braking, otherwise known as the hydrogen fuel cell battery electric vehicle (HFCBEV), which conveniently compares with the pure BEV design. Although the progress made on the HFCBEV concept is many years behind the BEV, the advantages it displays for heavy and highly utilised vehicles have become enormous. This is especially true as the 30 000 HFCBEVs are demonstrated and tested around the world (Carlier 2020).

In 2022, Kendall et al. (2022) highlighted that the key problem in the transition to HFCBEVs is the carbon intensity of hydrogen fuel, as most hydrogen produced worldwide is made from fossil fuels. They emphasised that the dominant global "fossil hydrogen" production, which is mainly sourced from petrochemical plants, rose from 60 Tg/a (where 1 Tg = 10^{12} g) in 2000 to 80 Tg/a by 2022, and that this is subsequently predicted to peak as climate crisis control takes effect. As green hydrogen was sourced from biomass, which was rather costly, businesses and researchers are now using surplus wind and solar energy to power the electrolysis of water to make economical green hydrogen fuel. Although biomass, wastewater and other sources of green hydrogen are still plausible, China plans to produce 100 000–200 000 t of green hydrogen from renewable energy annually and to build 1000 green hydrogen fuelling stations by 2025 (Xu and Patton 2022).

While the search for cheaper sources of hydrogen continues, researchers and students at the University of the Witwatersrand in Johannesburg are beginning to fabricate cheap and biodegradable materials for proton exchange membrane (PEM) and separators for proton exchange membrane fuel cells (PEMFCs), proton exchange membrane electrolysers (PEMEs) and adsorbent glass mat (AGM) batteries, as discussed in three case studies in the next sessions.

5.2 Component Production: PEM Fuel Cell and Electrolyser

(Case Study 1: Reprinted with permission from T. Sibuye & T. Lukhele of the Wits University)

The industrial revolution experienced in 1800's has plunged humanity into overdependence on fossil fuels, resulting in environmental pollution and devastation, global warming, flooding, wildfires, droughts, etc. In efforts to control or mitigate the impacts caused, countries around the world are beginning to set targets in the release and control of GHG emissions, especially carbon dioxide (CO_2), which has been recognised as the cause of global warming (Renau et al. 2021). The world is therefore in search of more sustainable and renewable energy sources to reduce emissions or to stop the resultant devastations. In this vein, fuel cells (FCs) are promising energy conversion and storage technologies with which scientists have conducted experiments to assess and improve their performances for zero-emission in both mobile and stationary electricity power generation.

A FC is a unit that generates energy through an electrochemical reaction, which has a couple of electrodes: an anode and a cathode. They are one of the most environmentally friendly electricity generation technologies because combustion is not necessary and no toxic by-products are released into the atmosphere (EG&G Technical Services Inc. 2004).

Although there are different types of FC s, the most common ones are alkaline fuel cell (AFC), phosphoric acid fuel cell (PAFC), molten carbonate fuel cell (MCFC), solid oxide fuel cell (SOFC) and PEMFC, which is the focus of this study. A typical PEMFC comprises a membrane electrode assembly (MEA) consisting of a proton-conducting polymer membrane functioning as the electrolyte, which separates the anode from the cathode side. Hydrogen is introduced at the anode side, where it comes into contact with a catalyst causing dissociation of the hydrogen into constituent protons and electrons. The protons then pass through the membrane to the cathode, but the electrons cannot pass through the membrane and instead traverse an external circuit, generating electricity, bonding with the protons and forming water, as shown in Eqs. (5.1)–(5.3);

$$H_2 \rightarrow 2H^+ + 2e^- \tag{5.1}$$

$$2H^+ + 2e^- + \frac{1}{2}O_2 \rightarrow H_2O \tag{5.2}$$

Overall reaction:

$$\frac{1}{2}O_2 + H_2 \rightarrow H_2O \tag{5.3}$$

The anode in a typical FC is continuously supplied with fuel, in this instance, hydrogen, while the cathode is supplied with the oxidant, oxygen or air. A FC is comparable to a regular battery, with the exception that when a battery is depleted, it loses its ability to supply energy, but a FC continues to produce power as long as it is provided with fuel (EG&G Technical Services Inc. 2004).

For the PEMFC to function, the PEM should transport protons from anode to cathode without transmitting electrons extracted from hydrogen atoms. The processes can be summarised as follows: the fuel gas is transferred to the reaction sites, electrochemical reaction at those sites, transfer of ions and electrons and their combination at the cathode. Gases must diffuse through the electrode leaving behind any impurities, which may disrupt the reaction, whereas liquid produced at the surface of the electrolyte or added through humidification must either be added to the electrolyte for hydration or drawn away from the reaction sites to prevent blockage or flooding at the reaction sites. Gases move towards the reaction sites based on the concentration gradient between the gas channel (high concentration) and the reaction sites (low concentration). Platinum is used as an electrocatalyst as in most FC s. Since it must be physically supported to allow minimum concentrations (to reduce the cost), the support and the platinum become the electrode (Afolabi et al. 2012), as presented in Figure 5.2.

Nafion is a perfluoro-sulfonic acid co-polymer with a non-polar fluorocarbon foundation and polar sulfonic acid suspended side chains that are commonly used in the PEMFC industry (Sahu et al. 2009). The disadvantages of this exceptional membrane include decreased conductivity at higher temperatures, high costs and a short lifespan. Figure 5.3 depicts the chemical composition of Nafion membrane.

Alternative membranes, on the other hand, have shown the ability to run FCs at greater temperatures with other advantages, such as the ability to use purer hydrogen as fuel for increased efficiency (Zunita et al. 2022). In their review, a polyionic liquids poly([HSO_3-BVIm][TfO]-10 membrane performance was shown to function well at low temperatures, with a proton conductivity of 0.209–0.293 S/cm at 25–40 °C, which is greater than the most often used Nafion membrane.

This work, therefore, is aimed at using hydrogenated and sulfonated polystyrene-butadiene rubber (PSBR). PSBR is a polymer made up of molecules that are arranged in a linear pattern. Its backbone chain is made up of three segments: a long

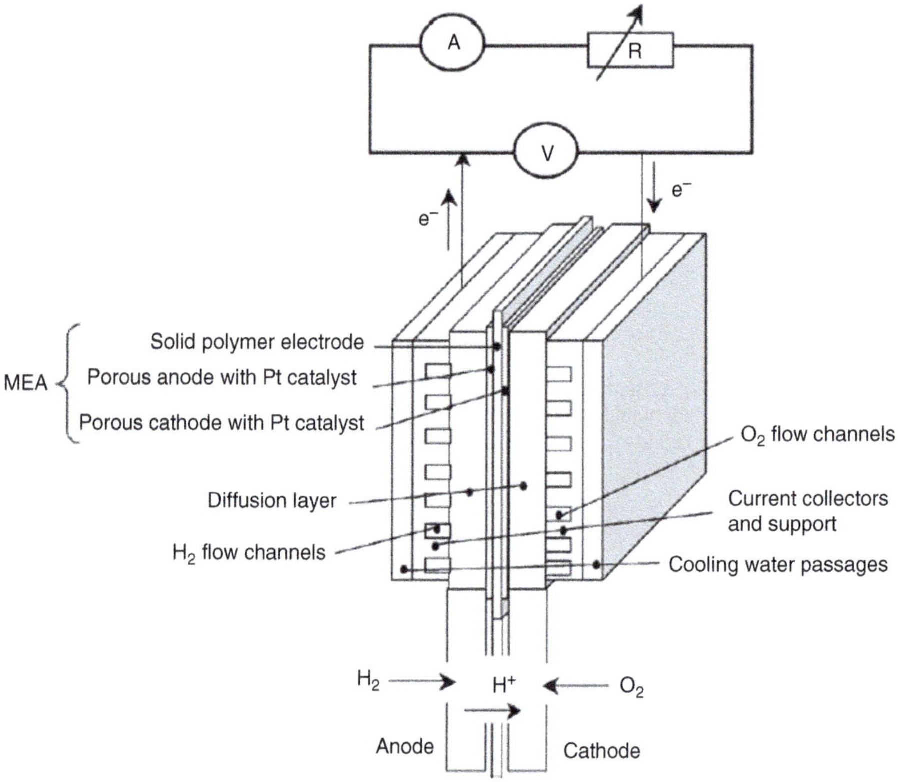

Figure 5.2 MEA of a single-cell PEMFC testing apparatus (Iyuke et al. 2003).

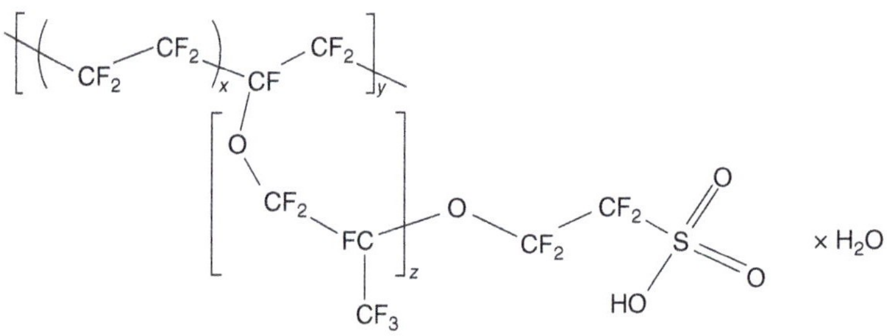

Figure 5.3 Chemical composition of Nafion. Source: Adapted from Sahu et al.

polystyrene chain, a long polybutadiene chain and a long polystyrene chain (PolymerScienceLearning 2022). Because of its excellent abrasion resistance and a customised rate of sulfonation, this polymer is expected to produce a PEM with high ion conductivity suitable for use in FCs (Valentini and Lopez-Manchado 2022). Thus, the research will help to conceptualise generation of electricity and electrolysis of water to produce hydrogen using a PEMFC and PEME that are equipped with synthesised hydrogenated and sulfonated locally available PSBR.

It should therefore be mentioned that using an electrolyte that is produced by hydrogenation and sulfonation of locally available PSBR would lower the cost of FC s and PEME systems. This would be a huge step forward in terms of alternative energy sources and economic viability of PEMFC technology.

5.2.1 Design and Fabrication of a Single Fuel Cell

Figure 5.4 presents the design and assembly of the test rig for a regenerative fuel cell, which comprises the FC and electrolyser single-cell stacks. This was a hybrid project, with an electrolyser stack, a water tank supplying water to the electrolyser and a hydrogen tank collecting hydrogen from the electrolyser. The FC stack is linked to a hydrogen-supplying pipe on the testing rig, which routes hydrogen from the tank through a humidifier before feeding it to the stack's anode. The stack's cathode is linked to a pipe that delivers humidified compressed air before being fed. Flowmeters are used to monitor the flow of the gases, and a light bulb is connected to the FC.

Figure 5.4 Hybrid electrolyser-fuel cell (HEFC) regenerative system, constructed test rig in this study with ±10% precision or experimental error, where (a) is the single-cell electrolyser, (b) humidifiers, (c) water tank, (d) hydrogen tank, (e) water tank to measure water displacement hydrogen gas, (f) is the fuel cell single-cell stack.

The single fuel cell stack was assembled according to the schematic diagram shown in Figure 5.2, while the synthesised membrane was used in the test station consisting of a FC and PEME single-cell stacks.

In this PEMFC operation, the reactants were hydrogen and air (Afrox, South Africa). To quantify the current and voltage generated by the FC, the anode-produced electrons were linked to a compact digital multimeter (Major Tech with DC Current: 200 μA–2 mA-20 mA–200 mA-10A; DC Voltage: 200 mV-2 V-20 V-200 V-500 V; Digital Display: 3½ Digital LCD 1999 max). with a changeable external resistance.

- The anode received humidified hydrogen as a fuel gas, while the cathode received humidified air as an oxidant gas.
- For the catalysed reactions to take place, oxygen and hydrogen are passed through the diffusion layer to the interface between electrode and membrane electrolyte.
- The proton generated is transported to the cathode by the electrolyte membrane, with electrons traversing the external circuit.
- Air and hydrogen from gas cylinders fitted with pressure regulators and flowmeters, while the electrons are tapped as current at the external circuit.
- Humidified hydrogen and air are fed to the FC stack.
- The current flowing in the external circuit and voltage in the stack are measured using a digital multimeter.

The starting material for this experiment was hydrogenated and sulfonated PSBR, as described below.

5.2.2 Hydrogenation of the Synthetic Rubber

The hydrogenation of C=C double bonds and, sometimes, aromatic rings in polymer molecules opens up new possibilities for the synthesis of modified polymers. These hydrogenated polymers have improved mechanical properties such as tensile strength, break elongation, tensile elastic modulus, spinnability from melt and wear resistance (Petrukhina et al. 2019). Figure 5.5 shows the chemical structure and hydrogenation of PSBR used in the work.

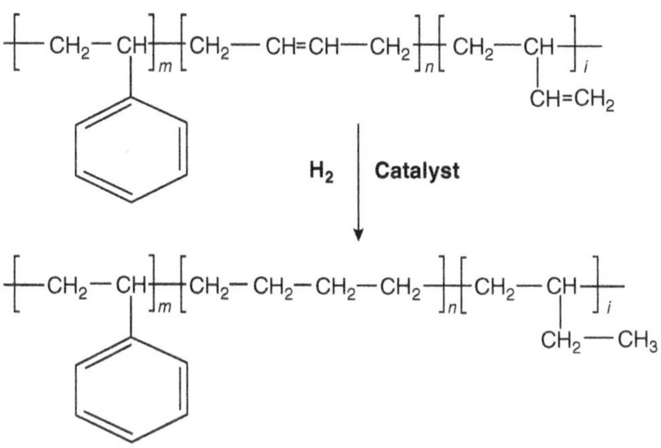

Figure 5.5 Hydrogenation process of PSBR (Guo et al. 2020).

This method consisted of dissolving in a sealed round-bottom flask, 10 g of purified PSBR (25% styrene) in 200 mL of tetrahydrofuran after flushing with nitrogen. About 0.451 g of Pd/C (10%) catalyst was added in a two-necked round-bottom flask with 20 mL of tetrahydrofuran (THF). This solution was stirred for two minutes before mixing it with the rubber solution. The two-necked round bottom flask was then attached to a reflux condenser, and the blend was refluxed in an oil bath with vigorous magnetic stirring at different times of the reaction under selected hydrogen flow rates. The time of the reaction varies between one and seven hours. The flow rates of hydrogen used were 881, 1498, 2159.68 and 2749 mL/min. This process was carried out in a dark fume cupboard to prevent the interaction between catalyst and lights.

5.2.3 Sulfonation of the Synthetic Rubber

Sulfonation of aromatics is a process where a molecule attracts or acquires electrons dependent on the substituents existing on the ring. As a result, aromatic sulfonation is commonly employed in synthesising chemicals and refining of fuel (Idibie et al. 2010). This method is used to create polymers that are proton-conductive and hydrophilic. Its primary application is to increase the proton conductivity of proton-conductive polymers (Mufula 2017). Figure 5.6 presents how sulfonate moiety can be introduced into the chemical structure of PSBR at a very low temperatures of 0 °C with continuous stirring. The sulfonate moiety was introduced into the styrene structure, as seen on Figure 5.6. The styrene structure was able to accept the hydrogen sulphate ion (-HSO_3-) because the ring structure is reactive.

The hydrogenated styrene-butadiene rubber (SBR) of different compositions was sulfonated with chlorosulphonic acid used as the sulfonating agent with concentrations of 0.15, 0.175 and 0.25 M for SBR 23.5 25% styrene content, SBR 40% and SBR 52% styrene content, respectively. The degree of sulfonation was found in the range between 56 and 72% depending on the rubber composition. Trimethylsilyl chlorosulfonate used as the second sulfonating agent was like wise attached to the same polymer backbone, and the degree of sulfonation was between 59 and 74% depending on the rubber's styrene content (Duraczynska et al. 2011).

5.2.4 Results Obtained from Hybrid Electrolyser-Fuel Cell Regenerative System

To characterise the membrane, Fourier-transform infrared (FTIR) spectrometry was used to check the success of hydrogenation and sulfonation. The spectra are presented in Figures 5.6 and 5.7. The FTIR spectrometer was also used to analyse the position and success of hydrogenation. Based on Figure 5.5, the hydrogenation reaction took place on butadiene part of the structure, which leaves room for sulfonation to occur on the styrene structure. Furthermore, hydrogenation on the butadiene structure allows the styrene structure to be more reactive to sulfonation and butadiene structure to be less reactive to sulfonation. The hydrogenation successfully occurred on the butadiene structure instead of styrene structure in the presence of a Pd/C (10%) catalyst and under ultraviolet light (Duraczynska et al. 2011). The conditions favour hydrogenation of unsaturated hydrocarbon butadiene instead of aromatic hydrocarbons. Saturating the butadiene structure enhanced the overall mechanical and chemical features through resisting ageing, thermo-oxidative decay and low swelling (Duraczynska et al. 2011). These conditions favour hydrogenation of unsaturated hydrocarbons and butadiene instead of aromatic hydrocarbons.

From Figure 5.7, the broad signal at about 3500 cm^{-1} correlates to O—H vibrations, while the signals at about 1250 cm^{-1} correspond to C—O bond, which signifies the bond between the hydrogen sulphate ion and the styrene structure. Sulfonate group has a wavelength range of 1000–1400 cm^{-1}, and upon examining Figures 5.7 and 5.8, there are multiple peaks that

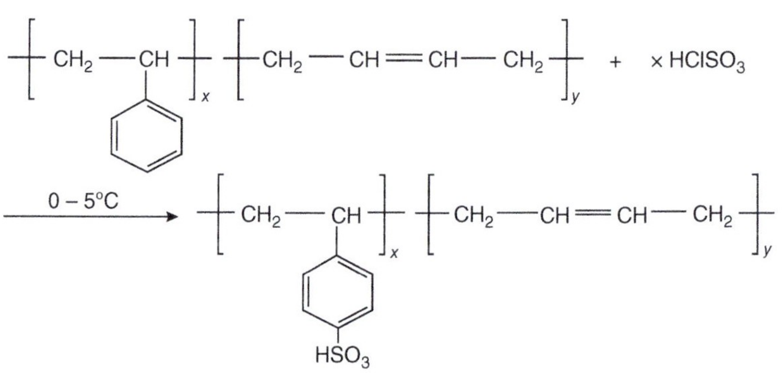

Figure 5.6 Sulfonated PSBR chemical structure. Source: Adapted from Duraczynska et al.

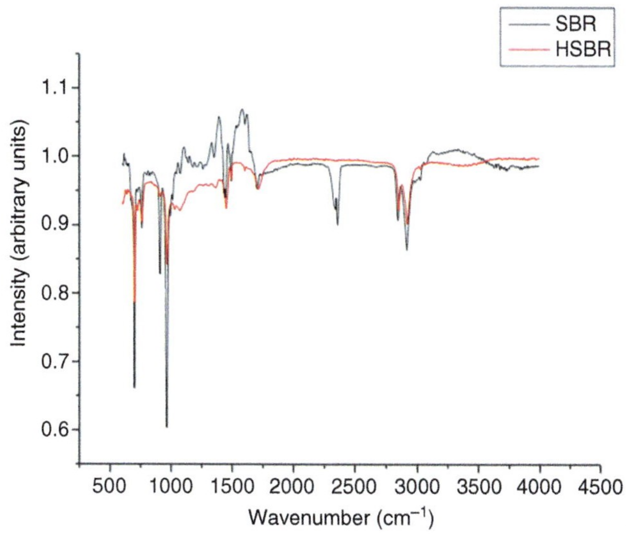

Figure 5.7 FTIR results of the hydrogenated PSBR.

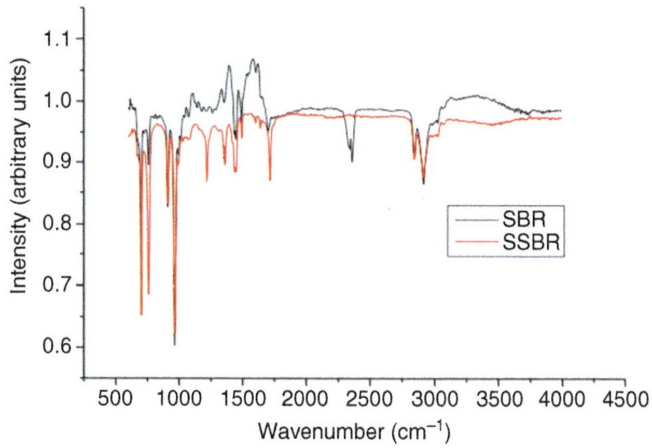

Figure 5.8 FTIR results of the sulfonated PSBR.

prove that sulfonation took place in the MEA. The IR active vibrations responsive to the sulfonation procedure were found at 1040, 1127, 1181 and 1375 cm^{-1} for the smooth side of the membrane (Figure 5.6) and 1039, 1091, 1127 and 1375 cm^{-1} for the rough side of the membrane (Figure 5.8).

5.2.4.1 Proton Conductivity

The proton conductivity of the membrane was evaluated through the alternating current impedance over a frequency range of 1–106 Hz, using H_2SO_4 as an electrolyte. The value at the intersection of high-frequency impedance curve with the real axis was considered as the resistance of the membrane, and the proton conductivity was calculated from Eq. (5.4) (Chai et al. 2010):

$$\sigma = \frac{t}{RS} \tag{5.4}$$

where: σ is the proton conductivity (Ω^{-1} cm^{-1}), S (cm^2) is the surface area of the membrane, t (cm) is the thickness of the membrane and R is the resistance of the membrane.

The proton conductivity of the membrane was therefore evaluated for membrane thickness and the surface area of 156 μm and 25 cm^2, respectively. The average through-plane conductivity of 0.073 Scm^{-1} was found for membranes with 52% styrene SBR, 0.046, 0.038 and 0.036 Scm^{-1} for membranes with 40, 25 and 23.5% styrene SBR. In comparison, Nafion 117 with 0.2 mm and at 30 °C and 100% relative humidity has a conductivity of 0.0133 Scm^{-1}(Sahu et al. 2009), which supports the applicability of the membrane produced in the study.

5.2.4.2 Fabrication and Testing of Membrane Electrode Assembly

The MEA in this work involves casting the two electrodes on both sides of the membranes. The electrodes (4 mg/cm^2 Pt-C on gas distribution layer) were purchased from FuelCellsEtc (USA). The casting of electrodes on the membrane was done by means of hot press at 125 °C for about five minutes at a pressure of 40 kPa. At this stage, the MEA was ready to be tested in single fuel cell stack.

The MEA was tested in a single fuel cell stack at room temperature (approximately 25 °C) and 20 kPa pressure, where humidified hydrogen gas was fed to the anode at a flow rate of 807.8 mL/min, while the humidified oxygen gas was fed to the cathode at the pressure of 15 kPa and flow rate of 538.563 mL/min, as shown in Figure 5.9. It was observed that a 3 V light bulb was easily powered.

The performance test of the membranes synthesised earlier with different styrene contents of 23.5, 25, 40 and 52%, which were sulfonated at the degrees of 74, 71, 60 and 71%, respectively, are presented in the polarisation curves of Figure 5.10. It can be observed from the voltage versus current density curves that the membranes produced initial voltages of 0.63, 0.72 and 0.81 V dropped to 0.23, 0.29 and 0.35 V, respectively, for 25, 40 and 52%. These drops in voltages are attributed to the overpotentials that are normally experienced in PEMFC technology due to polarisation factors emanating from reaction rate, ohmic, gas transport, etc. losses. In comparison with Nafion 112, the voltage also dropped from 0.77 to 0.33 V. It is interesting to observe that 52% of SBR outperformed the Nafion 112. Others are also comparable to the Nafion counterpart. The plots of the power density versus current density for the synthesised membranes also compared very well with that for Nafion 112, where 25, 40 and 52% SBR generated power of 51.41, 64.82 and 78.23 mW/cm^2, respectively, while Nafion 112 generated 74.88 mW/cm^2.

An optimum power density was obtained at the current density of 212.41 mA/cm^2 and further increases in current density resulted in a decrease in power density. This decrease in power density can be attributed to the decrease in the availability of

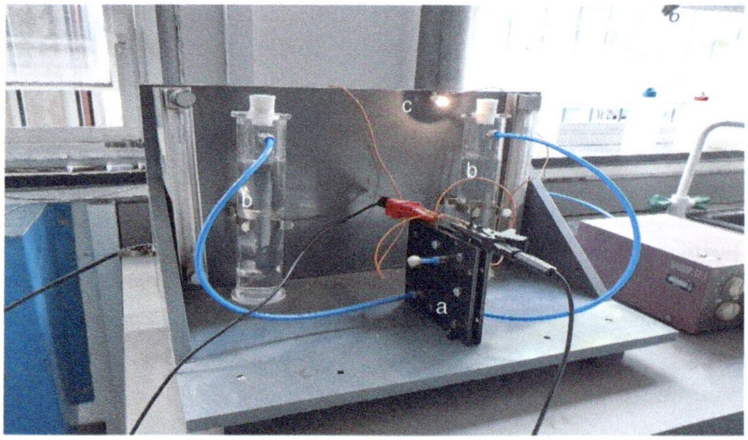

Figure 5.9 A 3 V light bulb was powered by the PEMFC single cell with ±10% precision or experimental error, where (a) is the single-cell stack for the PEM fuel cell, (b) humidifiers, (c) glowing light bulb.

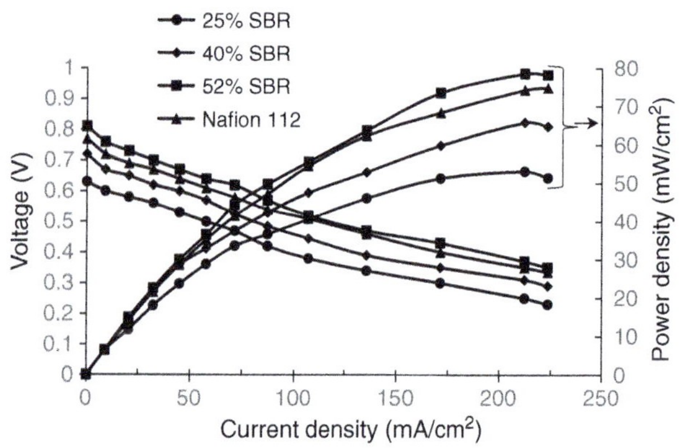

Figure 5.10 The stack's polarisation curves and performance at different styrene content in the SBR, showing reducing voltages due to reaction rate, ohmic, gas transport, etc. losses.

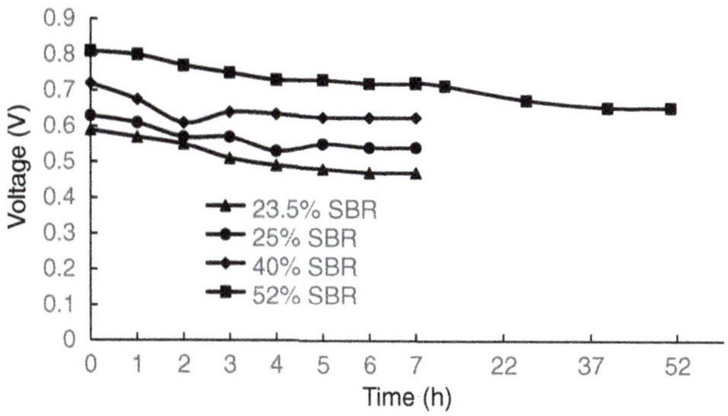

Figure 5.11 The stack performance with time at different at different styrene content in the SBR at the operational conditions of 25 °C and atmospheric pressure.

catalyst active sites and decrease in partial pressures of oxygen and hydrogen gas due to the overpotentials mentioned above (Chaojie et al. 2007). The highest power density of approximately 85 mW/cm^2 was obtained for 52% SBR, while Nafion112 membrane yielded an open cell voltage (OCV) of 0.79 V and a power density of about 77.34 mW/cm^2, which again supports the better performance of the 52% SBR than Nafion 112 membrane that is available in the market.

It is therefore obvious, that increasing degree of sulfonation resulted in higher voltages and power density because it corresponds to the rate of proton transfer through the membrane electrolyte, which in turn results in higher performance of MEA in FC stack.

The four synthesised SBR membranes were tested in the single-cell stack, as presented in Figure 5.11. It was observed that the MEA with 52% SBR gave a voltage of 81 V, which remained almost stable for 7 hours of testing. Although others showed the same stability, but they (23.5, 25 and 40%SBR) gave lower output than the 52% SBR MEA, which was tested for longer time of 52 hours, as seen by the insert in Figure 5.11. However, a slight decrease was seen on the 12, 26 and 40 days, which can be attributed to flooding of the membrane.

5.2.4.3 Testing of the MEA for Electrolyser Applications

In PEM electrolysis, electricity is used from an external power source, such as a clean power source of solar or PEMFC, as presented in Figure 5.4, the water is split into hydrogen and oxygen gases. The gases can subsequently be used as fuel in the FC to produce clean green energy into a cycle of energy and power generation. Thus, the electrolyser and the FC stacks share similar membranes and MEA. In operating the electrolyser, therefore, the external power source reverses the Eqs. (5.1)–(5.3) into the following:

$$H_2O\,(l) \rightarrow 2H^+\,(aq) + ½\,O_2\,(g) + 2e^- \tag{5.5}$$

$$2H^+\,(aq) + 2e^- \rightarrow H_2\,(g) \tag{5.6}$$

With overall reaction as;

$$H_2O\,(l) \rightarrow H_2\,(g) + ½\,O_2\,(g) \tag{5.7}$$

The experimental set-up used in the demonstration of the electrolysis is presented in Figure 5.12, with volume of hydrogen collected in the vessel as marked with an arrow of × height.

In the experimental setup in Figure 5.4, water was introduced to anode, and oxygen gas was released at the same anode side of the membrane, while hydrogen gas was collected at the cathode side. Volume of hydrogen gas was continuously collected as shown and marked ‘x’ in Figure 5.11 and calculated as;

$$V = \pi \times r^2 \times h = 169.89\ \mathrm{cm}^3 \tag{5.8}$$

Figure 5.13 presents collection of hydrogen gas in the cylinder for about the first 8 minutes. The volume was seen to be dependent on the voltage supplied. After 10 minutes, the volume increased by 0.2 cm for 30 minutes repeatedly and remained at 2.5 cm for 30 minutes in a pipe of 9.3 cm in diameter, which came to a volume of 169.82 cm^3. The system was operated at 1.0 V, but when the voltage was increased to 10 V, the MEA was damaged due to overheating. Thus, in comparison to literature (Grigoriev et al., 2011, 2009, Kumar et al. 2018), it was observed that the MEA in this study functioned favourably while producing 169.82 cm^3 H_2 at 1.0 V and in 30 minutes (Table 5.1).

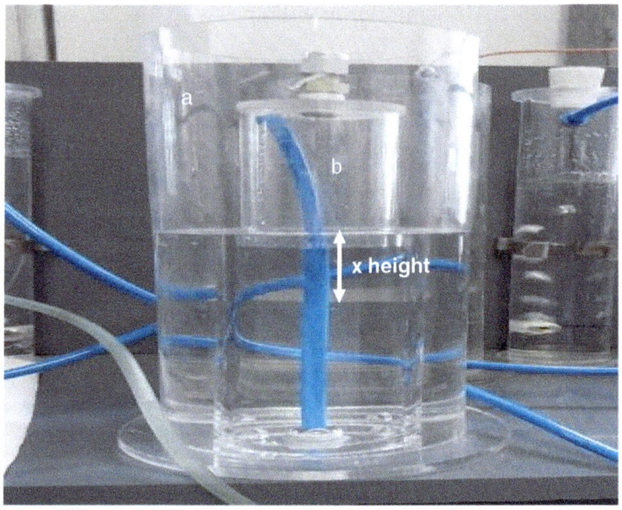

Figure 5.12 Experimental set up for collection of hydrogen during water electrolysis, which is a portion of Figure 5.4, where (a) water tank is to measure water displacement by the hydrogen gas and (b) hydrogen tank indicated as displacement of × height.

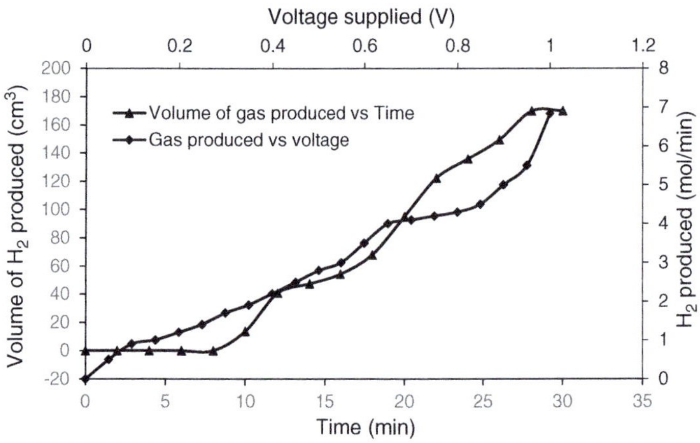

Figure 5.13 Hydrogen produced with time and voltage supplied.

Table 5.1 Comparison of MEA in this study with literature in PEM water electrolysis.

Anode catalyst	Cathode catalyst	Anode loading (mg/cm^2)	Cathode loading (mg/cm^2)	Membrane	Temp (°C)	Voltage at 1 A/cm^2	Reference
RuO_2	30%Pd/P-CNPs	3.0	0.7	Nafion-115	80	2.00	Kumar et al. 2018
Ir-Black	40% Pt/GNF	2.0	0.8	Nafion-115	90	1.67	Grigoriev et al., 2011
Ir-Black	Pt-black	2.0	0.8	Nafion-117	90	1.71	Grigoriev et al., 2009
Pt-C	Pt-C	4.0	4.0	PSBR rubber	25	1.00	This study

5.2.5 Hydrogen and Electricity Production from Water

From the preceding discussion, it can be concluded that electricity was generated from a single-cell stack of hydrogen FCs using hydrogenated and sulfonated locally available polystyrene butadiene rubber. The oxidative properties and proton conductivity of the hydrogenated and sulfonated rubber were tested on an MEA, which was mounted on a typical test rig. The evaluation of the single-cell stacks for both FC and electrolyser with common PEM proved to run very well. The FTIR results proved that the membrane had undergone hydrogenation at 835 cm^{-1} and sulfonation was detected from

1039 to 1375 cm^{-1}, while proton conductivities were found to be 0.036–0.073 Scm^{-1} for membranes with 23–52% styrene SBR rubber.

An O—H and S=O functional groups from the larger moiety of $-HSO_3$ were found on the sulfonated membrane, which further confirmed that sulfonation had occurred. The MEA had enhanced thermal stability due to the hydrogenation and promising proton conductivity for possible industrial fuel and electrolyser applications. The synthesised membranes compared very well with Nafion 112, where 23.5, 25, 40 and 52% SBR generated power of 51.41, 64.82 and 78.23 mW/cm^2, respectively, while Nafion 112 generated 74.88 mW/cm^2. The optimum power density was obtained at the current density of 212.41 mA/cm^2 and further increases in current density resulted in a decrease in power density.

The electrolyser, on the other hand, used similar MEA as its FC counterpart, which conveniently split the water supplied with the current into hydrogen and oxygen gases. The rate of hydrogen production of 7 mol/h does not match up with the theoretical 8.64 mol/h rate that was modelled, perhaps due to the fact that air was used instead of pure oxygen.

5.3 Process Design and Integration of PEM Electrolyser and Fuel Cell

(Case study 2: Reprinted with permission from T. Sibuye of the Wits University)

This section focuses on mathematical modelling of proton exchange membrane electrolyser (PEME) integrated with 500 kW PEMFC to power Richard Ward building of the University of the Witwatersrand. It addresses a continuous one-dimensional mathematical model under steady-state conditions used to simulate the process integration. Optimal performance of this computational model is determined in terms of current density, open-circuit voltage, activation, diffusion and ohmic overpotentials. The optimal design parameters and operating conditions are expressed in terms of operating temperature, electrode thickness, humidity degree and reaction area. Furthermore, an economic evaluation has been performed on the hybrid electrolyser-fuel cell (HEFC) system to determine its economic feasibility. The optimisation was achieved using GAMS software and BARON was used as GAMS solver. The results obtained show that the optimal number of cells in PEME required to support the 500 kW PEMFC is 64.

5.3.1 Hydrogen Energy Source

The global energy demand is rising inordinately and shortage supply of energy remains to be the vast issue for the society at large (Inglesi-lots and Pouris 2016). The energy produced from fossil fuels is swiftly diminishing as a result of exponential growth of human population; hence, the rate at which fossil fuels are formed is disproportional to the rate at which they are consumed. Approximately 74.7% of electricity worldwide is generated from fossil fuels due to their availability and economic viability. However, the drawback of utilising fossil fuels as a main source of energy supply is due to their incapability of fulfilling environmental sustainability goals. Fossil fuels contribute largely to GHG emissions, which are environmentally inimical (Singh et al. 2020).

As a means of tackling this energy crisis, it is mandatory to introduce a renewable alternative energy supply with high energy density and energy conversion efficiency that possesses overall zero GHG emissions. This will not only mitigate climate issues but will also improve air quality to ameliorate public health. Thus, PEMFCs also known as Polymer Electrolyte Membrane (PEM) Fuel cells are suitable replacements for conventional fossil fuels due to their broad diversity of power applications extending from small-scale power systems such as transportation and portable computers, wearable electronic units and military communication devices to large, immovable energy systems such as industries and buildings. The application of PEMFCs was initially adopted by a research programme known as Partnership for a New Generation of Vehicles (PNGVs) in the United States of America in 1993. Until date, the commercialisation and testing phase of PEMFCs has extended globally as a result of their sustainable performance and low mass density. Globally recognised companies such as IBM, BMW and Samsung. have embraced the utilisation of PEMFCs as the new emerging, efficient and future-promising competitive power source (Wee 2006).

5.3.1.1 Electrolysis

The PEMFC requires H_2 fuel and O_2 supplied by PEME in order to provide power. PEM electrolysis is a water electrolysis technology that makes use of an electrochemical method to split water into high-purity H_2 and O_2 (Wang et al. 2022). PEM electrolysis was developed by Grubb and Neidrach at General Electric in 1950s. From then, it was primarily utilised in closed areas such as spaces and submarines for life support purposes, leveraging the production of O_2 instead of concentrating on H_2 formation (Ayers 2019). Electrolysis is a highly recommended technique for the generation of chemical products

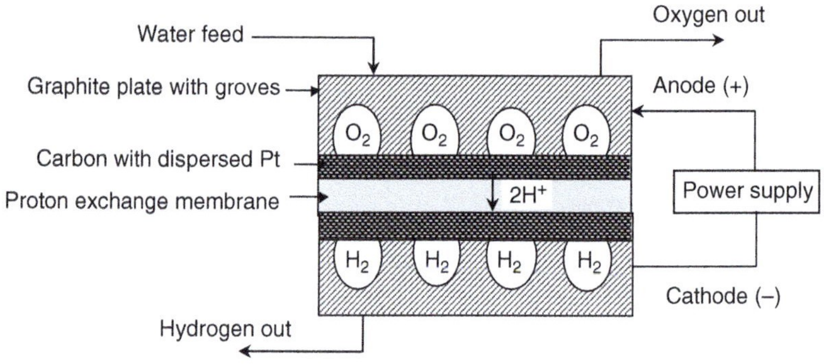

Figure 5.14 PEM electrolysis cell and reactions (Eqs. (5.4)–(5.7)).

from their natural state. The PEMEs not only produce pure H_2 and O_2 but also serve as renewable energy storage as a result of their rapid charging/discharging characteristics, high energy density and compact design. It operates in similar configuration as the PEMFCs since they are constructed and designed with the same materials such as end plates, bipolar plates and catalyst layers, but the electrochemical reactions are in opposite direction (Han et al. 2015). Figure 5.14 depicts the schematic of the PEME cell.

Currently, a broad range of electrochemical processes are implemented in various industries. The accessibility of PEME has warranted a very promising field in energy industry with high capability for future development and for solving different problems encountered by industrially developed countries (Paidar et al. 2016). H_2 fuel has been assessed and declared cost-effective as a prospective technology for shifting the energy industry from carbon-based fossil fuels. Statistically, 96% of H_2 fuel consumed globally is generated from fossil fuels with coal and natural gas being the most utilised primary feedstock towards the production of H_2 through gasification and steam-reforming processes (Singh et al. 2020). As a result of environmental concerns emerging from the application of these processes, which produce much CO_2 in the atmosphere. A shift to a zero-carbon approach for H_2 fuel generation from PEME powered by solar energy offers an energy-efficient alternative route to provide clean power-generation technologies with hydrogen (Smith et al. 2013).

5.3.1.2 Modelling PEM Electrolysers

Modelling of PEME and PEMFC has been conducted individually in the previous years as clean energy conversion systems. The mathematical modelling of PEME is currently performed on the basis of PEMFC models. A comprehensive literature review was conducted by Falcão and Pinto (2020) on different models formulated with the aim of evaluating the PEME performance. Empirical, semi-empirical and analytical models were developed. A one-dimensional PEME model was developed by Han et al. (2015) to assess the influence of design parameters and operating conditions on the cell performance. In that investigation, the sum of open-circuit voltage, diffusion, activation and ohmic overpotentials was examined. The effect of temperature and current density on the degradation of the solid electrolyte membrane was investigated by Chandesris et al. (2015) using a one-dimensional model. It was deduced that anode exhibited less membrane degradation than cathode, and temperature had a greater effect on the performance of the cell. In another study done by Shekhar (2013), a comparison was carried out between one-dimensional and three-dimensional mathematical models. It was established that one-dimensional models are economically viable and more convenient for realistic applications than three-dimensional models.

Various modelling studies based on PEMEs were formulated and optimised using different simulation software. In a simulation study conducted by Prieto-Prado et al. (2018) based on the integration system of photovoltaic (PV) energy, electrolysers and FCs, HOMER simulation software was used to evaluate the technical and economic feasibility of the power system. Abdin et al. (2016) carried out an investigation to assess the effect of humidification, pressure and temperature on the cell performance. MATLAB-Simulink was used for development, and it was deduced that the relative humidity of the reactant gases has an impact on the resistance and membrane hydration. In this simulation study, the optimisation problem was modelled using GAMS software, and BARON was utilised as a solver.

5.3.1.3 Motivation for the STUDY

Modern society requires constant supply of energy on a daily basis because it is the centre of economic development, industrialisation and urbanisation. Various research on water electrolysis has been conducted but was mostly based on experimental work, few mathematical models for optimal design of one-dimensional water electrolysers have been

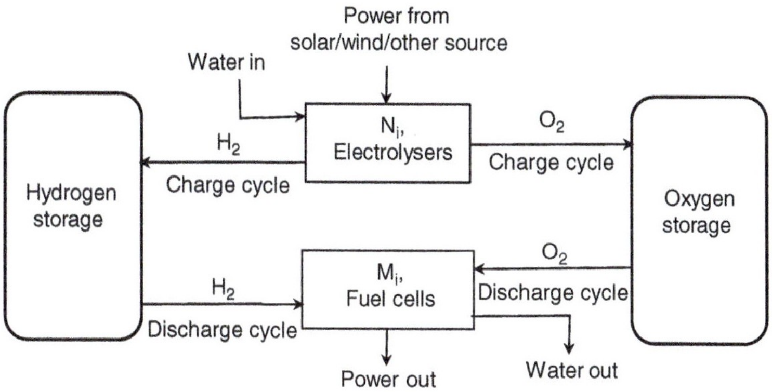

Figure 5.15 HEFC system (Luta and Raji 2019).

formulated. This work focuses on formulation of a one-dimensional non-linear programming (NLP) mathematical model of a PEME to produce H_2 and O_2 to support a PEMFC to produce 500 kW to power Richard Ward building at Wits University. The developed PEME model is then coupled with the PEMFC model and optimised with the aim of extracting optimal performance, design parameters and operating conditions while simultaneously conducting an economic evaluation to assess the feasibility of the proposed HEFC system. A HEFC system is the combination of PEME and PEMFC. An illustrative sketch of the HEFC system is presented in Figure 5.15. In addition, the effect of varying design parameters and operating conditions on the electrolyser cell is assessed and expressed in the form of a polarisation curve.

Advantages of the HEFC system include, but are not limited to, the following:

- No GHG emissions to the environment
- The excess power available from any renewable source of energy is utilised in the background process
- The HEFC system can operate at low atmospheric pressure
- Minimal capital costs
- The system can run off the grid

The objective of the study therefore is to minimise the power consumed by the PEME while extracting its optimal performance, design parameters and operating conditions required to support the PEMFC to produce 500 kW.

5.3.2 Development of Models

The PEME mathematical model formulated in this paper is integrated with PEMFC model formulated in another paper by Singh et al. (2020) with the aim of obtaining optimal performance, design parameters, operating conditions and economic feasibility of the system. It is expedient to mention that this section is a summary of the description in Singh et al. (2020), where more detailed mathematical modelling and richer discussion are presented.

5.3.2.1 Systematic Approach

The procedure outlined in Figure 5.16 is followed in order to formulate the optimisation model for the HEFC system. In all attainable combinations for the NLP optimisation problem, as demonstrated in Figure 5.15, a quantified set of N_i electrolysers can be powered either by hydro, wind, solar or tidal energy to split water into pure H_2 and O_2. The produced H_2 and O_2 are then sent to a FC with a quantified set of M_i cells to produce power.

The HEFC system is composed of two sub-models: the water electrolyser model and the hydrogen fuel cell model. The paper places emphasis on the development of electrolyser model, but the principles are similar to those of H_2 fuel cell model. The water electrolyser model is divided into three parts, namely: voltage, membrane and material balance elements. The voltage losses are included in the voltage component, while the transport and H_2O concentrations are modelled in the membrane element. The material balance component involves mass and molar balances of H_2O, H_2 and O_2 molecules.

The water electrolyser model is solved separately before it is integrated with the hydrogen fuel cell model to evaluate its feasibility by minimising the power consumed. In contrast, the feasibility of the hydrogen FC is determined individually with the objective to maximise the power produced. Once the sub-models are declared feasible, mass balance constraints are incorporated to integrate the sub-models and solve to obtain optimal design parameters that are essential for sizing the inner parts of the HEFC system. Thereafter, techno-economic assessment of the HEFC system is conducted from the

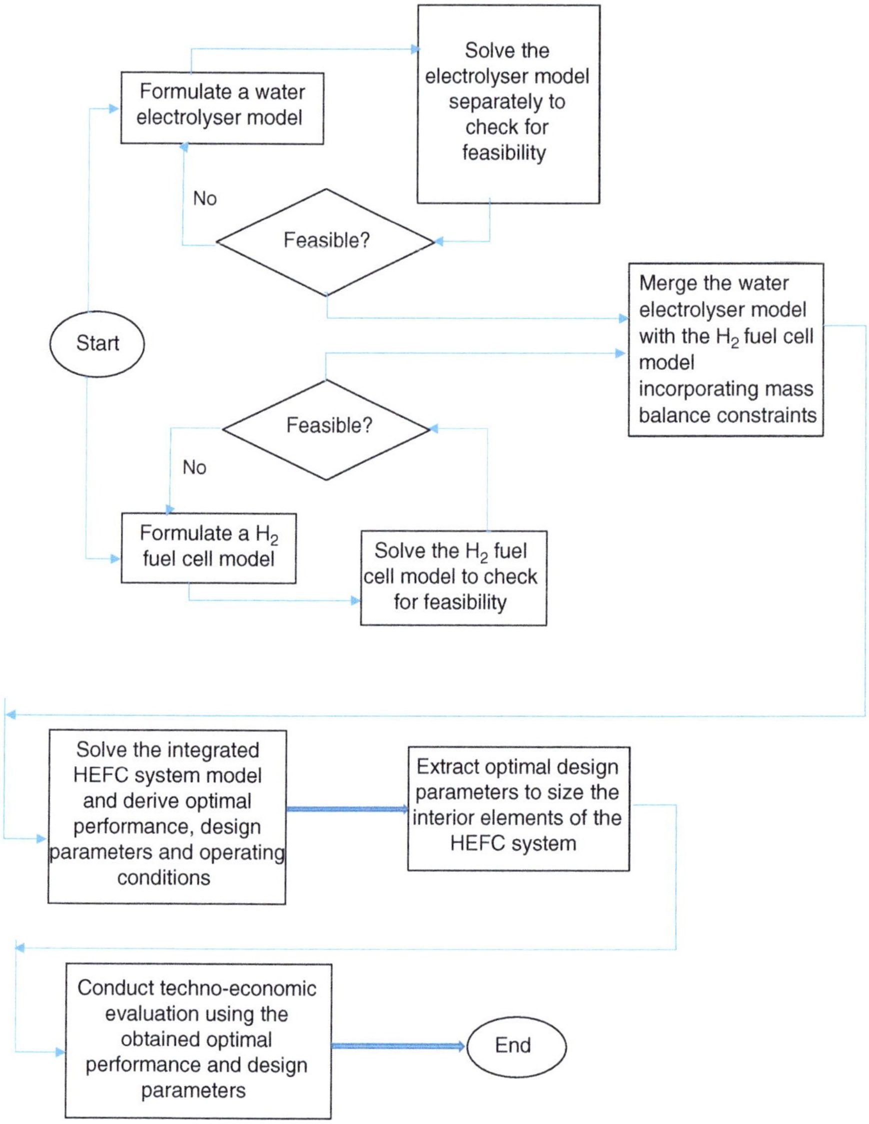

Figure 5.16 Schematic approach for formulating the hybrid electrolyser-fuel cell system.

optimal design variables extracted. The profitability of the system is determined based on internal rate of return, annual cash flow, net present value (NPV) and payback period (PBP).

5.3.2.2 Operation of Water Electrolyser

Water is fed through the inlet channel of the electrolyser cell and flows to the anode for oxidation reaction to occur to form O_2. The oxidation reaction is governed by the following reaction:

$$H_2O_{(l)} \rightarrow \frac{1}{2}O_2 + 2e^- + 2H^+ \tag{5.9}$$

At the anode, the electrons are released and flow via the external circuit to the cathodic site. Concurrently, the protons produced at the anode permeate through the electrolyte membrane to the platinum cathode to form H_2. The solid electrolyte membrane serves as a conductor of protons. The transfer of H^+ ions across the membrane is dependent on various factors such as membrane thickness, number of ions transferred and quantity of water in the membrane. The electrochemical reaction at the cathode occurs as follows:

$$2H^+ + 2e^- \rightarrow H_2 \tag{5.10}$$

The H_2 and O_2 produced then flow to the PEMFC through their respective flow channels to produce power (Singh et al. 2020).

5.3.2.2.1 Voltage Component The mathematical expressions of the electrolyser listed below are obtained from Singh et al. (2020) and are similar to those of H_2 FCs, but the electrochemical reactions are in reverse direction. The cell connections are in series configuration, implying that the current flowing in each cell unit is invariable and the performance of each cell is determined in terms of voltage and current density.

Current The current is expressed in terms of reaction area and current density.

$$I = A^* j \tag{5.11}$$

where I = Current, A = Reaction area and j = Current density.

Open Circuit Voltage (OCV) This voltage is described as the minimal overpotential when all other voltages are negligible. In contrast, a closed-circuit voltage is the maximum voltage of the system when there is no loss in overpotential, and the system enables all its Gibbs free energy to be converted into electrical energy. It is described by the following expression:

$$V_{\mathrm{OCV}} = 1.229 - 0.9 * 10 - 3\,(T - 298) + \frac{R\,T}{zF} \ln\left(\frac{\beta_{\mathrm{H2}}\beta_{\mathrm{O2}}^{0.5}}{\beta_{\mathrm{H2O}}}\right) \tag{5.12}$$

where V_{OCV} is Open circuit voltage, T = Temperature, R = gas constant, F = Faraday's constant, z = Number of electrons transferred, β_{H2} = Partial pressure of H_2, β_{O2} = Partial pressure of O_2, β_{H2O} = Partial pressure of H_2O.

Activation Overpotential This overpotential originates from Butler–Volmer equation. It is described as the voltage loss as a result of electrochemical reactions at the electrode's surface. It is governed by both physical and chemical properties such as reaction site, operating temperature, morphology of the electrode and catalyst property. The expression is shown as follows:

$$V_{\mathrm{act}} = V_{\mathrm{act,an}} + V_{\mathrm{act,cat}} = \frac{R\,T}{\alpha_{\mathrm{an}}F} \ln\left[\left(\frac{j}{2j_{0,\mathrm{an}}}\right) + \sqrt{\left(\frac{j}{2j_{0,\mathrm{an}}}\right)^2 + 1}\right] + \frac{R\,T}{\alpha_{\mathrm{cat}}F} \ln\left[\left(\frac{j}{2j_{0,\mathrm{cat}}}\right) + \sqrt{\left(\frac{j}{2j_{0,\mathrm{cat}}}\right)^2 + 1}\right] \tag{5.13}$$

where $V_{\mathrm{act,an}}$ and $V_{\mathrm{act,cat}}$ = Activation overpotential at the anode and cathode, α_{an} and α_{cat} = Anode and cathode charge transfer coefficient, $j_{0,\mathrm{an}}$ and $j_{0,\mathrm{cat}}$ = Anode and cathode exchange current density.

Diffusion Overpotential Diffusion overpotential is also defined as the concentration overpotential. It is described as the loss of voltage as a result of limitations to mass transport across the membrane–electrode interface. The mass transport of H_2O, H_2 and O_2 across the electrode porosity can be described by Fick's law. This voltage is essential for enhancing reaction rate by preventing O_2 gas bubbles from accumulating on the membrane surface. The equation is described as follows:

$$V_{\mathrm{diff}} = V_{\mathrm{diff,an}} + V_{\mathrm{diff,cat}} = \frac{R\,T}{4F} \ln \frac{C_{\mathrm{O2,me}}}{C_{\mathrm{O2,me,o}}} + \frac{R\,T}{2F} \ln \frac{C_{\mathrm{H2,me}}}{C_{\mathrm{H2,me,o}}} \tag{5.14}$$

where $V_{\mathrm{diff,an}}$ and $V_{\mathrm{diff,cat}}$ = Anode and cathode diffusion overpotential, $C_{\mathrm{O2,me}}$ and $C_{\mathrm{H2,me}}$ = O_2 and H_2 concentration at the electrode–membrane interface and $C_{\mathrm{O2,me,o}}$ and $C_{\mathrm{H2,me,o}}$ = O_2 and H_2 concentration at the electrode–membrane interface at reference conditions

Ohmic Overpotential This overpotential is defined as resistance experienced due to electrons being transferred across electrodes, membranes, bipolar plates and the linkage between the electrolyser and FC. It is made up of two resistances: electrical and ionic. The electrons flowing across the electrically conductive component give rise to electrical resistance, and in contrast, the flow of protons through the membrane induces ionic resistance. According to the investigation conducted by Bernardi and Verbrugge (1991), it was concluded that the ohmic overpotential is governed mostly by the ionic resistance. Hence, due to many unknowns present in the bipolar plates and electrodes, only the ionic resistance in the ohmic overpotential is considered. The equation is shown below.

$$V_{\mathrm{ohm}} = \delta_{\mathrm{mem}} \left[\frac{I}{A * \sigma_{\mathrm{mem}}}\right] \tag{5.15}$$

The membrane proton conductivity is expressed in terms of water content by an empirical expression recommended by Springer et al. (1993).

$$\sigma_{\mathrm{mem}} = (0.5139\lambda_{\mathrm{mem}} - 0.326)\, \exp\left(1268\left(\frac{1}{303} - \frac{1}{T}\right)\right) \tag{5.16}$$

where V_{ohm} = Ohmic overpotential, δ_{mem} = S of the membrane, σ_{mem} = Membrane conductivity, λ_{mem} = Membrane humidity degree.

Total Voltage The input voltage is given as the sum of open-circuit voltage, activation, diffusion and ohmic overpotentials.

$$V = V_{ocv} + V_{act} + V_{diff} + V_{ohm} \tag{5.17}$$

Concentration in the anode–membrane and cathode–membrane interfaces.

The concentration difference on either side of the solid electrolyte membrane is expressed in terms of H_2O content. The water activity at the anode and cathode–membrane interfaces is given as follows:

$$A_{H2O,an} = \frac{y_{H2O,an} * P}{P_{sat}} \tag{5.18}$$

$$A_{H2O,cat} = \frac{y_{H2O,cat} * P}{P_{sat}} \tag{5.19}$$

where $A_{H2O,an}$ and $A_{H2O,cat}$ = Anode and cathode water activity, $y_{H2O,an}$ and $y_{H2O,cat}$ = Molar composition of water, P = Operating pressure, P_{sat} = Saturated vapour pressure.

Furthermore, the saturated vapour pressure of H_2O is dependent on temperature, as shown by the given expression below:

$$P_{sat} = 611 \exp\left[\frac{19.65 * (T - 273.15)}{T}\right] \tag{5.20}$$

The molar composition of water at both electrodes is defined in terms of molar fluxes, as given by the following expressions:

$$y_{H2O,an} = \frac{n_{H2O,an}}{n_{H2O,an} + n_{O2}} \tag{5.21}$$

$$y_{H2O,cat} = \frac{n_{H2O,cat}}{n_{H2O,cat} + n_{H2}} \tag{5.22}$$

where $n_{H2O,an}$ and $n_{H2O,cat}$ = Water molar flux at the anode and cathode, n_{O2} and n_{H2} = Oxygen and hydrogen molar fluxes.

The water molar flux at the electrodes can be described by the expressions as follows:

$$n_{H2O,an} = \frac{\dot{N}_{H2O,cons} + \dot{N}_{H2O,mem}}{A} \tag{5.23}$$

$$n_{H2O,cat} = \frac{\dot{N}_{H2O,mem}}{A} \tag{5.24}$$

Expressions for molar fluxes of H_2 and O_2 in the electrolyser flow channels are:

$$n_{H2} = \frac{\dot{N}_{H2,prod}}{A} \tag{5.25}$$

$$n_{O2} = \frac{\dot{N}_{O2,prod}}{A} \tag{5.26}$$

where $\dot{N}_{H2O,cons}$ and $\dot{N}_{H2O,mem}$ = Molar flow rate of water consumed and across the membrane. $\dot{N}_{H2,prod}$ and $\dot{N}_{O2,prod}$ = Molar flow rate of hydrogen and oxygen produced.

At wide range of water activity conditions, the humidity of water at the electrodes is expressed by the following equations:

$$\lambda_{an} = 14 + 1.4\,[A_{H2O,an} - 1]\; 1 \le A_{H2O,an} \le 3 \tag{5.27}$$

$$\lambda_{cat} = 14 + 1.4\,[A_{H2O,cat} - 1]\; 1 \le A_{H2O,cat} \le 3 \tag{5.28}$$

and/or

$$\lambda_{an} = 0.043 + 17.8^{*} A_{H2O,an} - 39.65^{*}(A_{H2O,an})^2 + 36^{*}(A_{H2O,an})^3\; 1 \le A_{H2O,an} \le 3 \tag{5.29}$$

$$\lambda_{cat} = 0.043 + 17.8^{*} A_{H2O,cat} - 39.65^{*}(A_{H2O,cat})^2 + 36^{*}(A_{H2O,cat})^3\; 1 \le A_{H2O,cat} \le 3 \tag{5.30}$$

The humidity degree of the solid electrolyte membrane is given as an average of the humidity degrees of the anode and cathode, as shown below:

$$\lambda_{mem} = \frac{\lambda_{an} + \lambda_{cat}}{2} \tag{5.31}$$

where λ_{an} and λ_{cat} = Anode and cathode humidity degree.

Concentration of water at the anode and cathode-membrane electrode interfaces is expressed as follows:

$$C_{\mathrm{H2O,mem,an}} = \frac{\rho_{\mathrm{mem,dry}}}{\mathrm{MW}_{\mathrm{mem,dry}}} \lambda_{\mathrm{an}} \tag{5.32}$$

$$C_{\mathrm{H2O,mem,cat}} = \frac{\rho_{\mathrm{mem,dry}}}{\mathrm{MW}_{\mathrm{mem,dry}}} \lambda_{\mathrm{cat}} \tag{5.33}$$

where $C_{\mathrm{H2O,mem,an}}$ and $C_{\mathrm{H2O,mem,cat}}$ = Concentration of water at the electrodes–membrane interface, $\rho_{\mathrm{mem,dry}}$ = Dry density of the membrane, $\mathrm{MW}_{\mathrm{mem,dry}}$ = Dry membrane molecular weight.

The flow channel concentrations of the produced H_2 and O_2 are defined as follows:

$$C_{\mathrm{H2,Ch}} = \frac{P_{\mathrm{cat}} * y_{\mathrm{H2,Ch}}}{RT} \tag{5.34}$$

$$C_{\mathrm{O2,Ch}} = \frac{P_{\mathrm{an}} * y_{\mathrm{O2,Ch}}}{RT} \tag{5.35}$$

where $C_{\mathrm{H2,Ch}}$ and $C_{\mathrm{O2,Ch}}$ = Concentrations of H_2 and O_2 on the channels of electrolyser, P_{cat} and P_{an} = Pressure at the cathode and anode, $y_{\mathrm{H2,Ch}}$ and $y_{\mathrm{O2,Ch}}$ = Molar compositions of hydrogen and oxygen at the channels.

The molar concentrations of O_2 and H_2 at the membrane and electrolyser flow compartments at reference conditions are defined as follows:

$$C_{\mathrm{O2,me,o}} = C_{\mathrm{O2,Ch,o}} + \frac{\delta_{\mathrm{an}} * n_{\mathrm{O2}}}{D_{\mathrm{eff,an}}} \tag{5.36}$$

$$C_{\mathrm{H2,me,o}} = C_{\mathrm{H2,Ch,o}} + \frac{\delta_{\mathrm{cat}} * n_{\mathrm{H2}}}{D_{\mathrm{eff,cat}}} \tag{5.37}$$

Furthermore,

$$C_{\mathrm{O2,Ch,o}} = \frac{P_{\mathrm{ref}} * y_{\mathrm{O2,Ch}}}{RT_{\mathrm{ref}}} \tag{5.38}$$

$$C_{\mathrm{H2,Ch,o}} = \frac{P_{\mathrm{ref}} * y_{\mathrm{H2,Ch}}}{RT_{\mathrm{ref}}} \tag{5.39}$$

where $C_{\mathrm{O2,Ch,o}}$ and $C_{\mathrm{H2,Ch,o}}$ = Channel concentrations of oxygen and hydrogen at reference conditions, δ_{an} and δ_{cat} = Anode and cathode thickness, $D_{\mathrm{eff,an}}$ and $D_{\mathrm{eff,cat}}$ = Anode and cathode effective diffusion coefficient, P_{ref} = Reference pressure, T_{ref} = Reference temperature, $y_{\mathrm{O2,Ch}}$ and $y_{\mathrm{H2,Ch}}$ = Molar composition of oxygen and hydrogen at the channel

The molar composition of H_2 and O_2 in the flow channels is given below:

$$y_{\mathrm{H2,Ch}} = \frac{n_{\mathrm{H2}}}{n_{\mathrm{H2O,cat}} + n_{\mathrm{H2}}} \tag{5.40}$$

$$y_{\mathrm{O2,Ch}} = \frac{n_{\mathrm{O2}}}{n_{\mathrm{H2O,an}} + n_{\mathrm{O2}}} \tag{5.41}$$

Effective diffusion coefficients of the H_2–H_2O and O_2–H_2O mixtures are defined as functions of porosity, percolation and diffusion coefficients of the mixture:

$$D_{\mathrm{eff,H2-H2O}} = D_{\mathrm{H2-H2O}} * \varepsilon * \left(\frac{\varepsilon - \varepsilon_p}{1 - \varepsilon_p}\right)^{\mathrm{r}} \tag{5.42}$$

$$D_{\mathrm{eff,O2-H2O}} = D_{\mathrm{O2-H2O}} * \varepsilon * \left(\frac{\varepsilon - \varepsilon_p}{1 - \varepsilon_p}\right)^{\mathrm{r}} \tag{5.43}$$

Furthermore, the diffusion coefficients are expressed as follows:

$$D_{\mathrm{H2-H2O}} = a\left(\frac{T}{\sqrt{T_{\mathrm{cr,H2}} * T_{\mathrm{cr,H2O}}}}\right)^{b} (P_{\mathrm{cr,H2}} * P_{\mathrm{cr,H2O}})^{1/3} (T_{\mathrm{cr,H2}} * T_{\mathrm{cr,H2O}})^{5/12} \left(\frac{1}{\mathrm{MW}_{\mathrm{H2}}} + \frac{1}{\mathrm{MW}_{\mathrm{H2O}}}\right)^{1/2} \tag{5.44}$$

$$D_{\mathrm{O2-H2O}} = a\left(\frac{T}{\sqrt{T_{\mathrm{cr,O2}} * T_{\mathrm{cr,H2O}}}}\right)^{b} (P_{\mathrm{cr,O2}} * P_{\mathrm{cr,H2O}})^{1/3} (T_{\mathrm{cr,O2}} * T_{\mathrm{cr,H2O}})^{5/12} \left(\frac{1}{\mathrm{MW}_{\mathrm{O2}}} + \frac{1}{\mathrm{MW}_{\mathrm{H2O}}}\right)^{1/2} \tag{5.45}$$

where $D_{\mathrm{eff,\,H2\text{-}H2O}}$ and $D_{\mathrm{eff,\,O2\text{-}H2O}}$ = Effective diffusion coefficients of the H_2-H_2O and O_2-H_2O mixtures, $D_{\mathrm{H2\text{-}H2O}}$ and $D_{\mathrm{O2\text{-}H2O}}$ = H_2-H_2O and O_2-H_2O mixture diffusion coefficients, ε = Electrode porosity, ε_{p} = Percolation threshold, r, a

and b = Empirical coefficients, $T_{cr,H2,}$ $T_{cr,H2O}$ and $T_{cr,O2}$ = Critical temperatures of hydrogen, water and oxygen, $P_{cr,H2,}$ $P_{cr,H2O}$ and $P_{cr,O2}$ = Critical pressures of hydrogen, water and oxygen, $MW_{H2,}$ MW_{H2O} and MW_{O2} = Molecular weights of hydrogen, water and oxygen.

5.3.2.2.2 Membrane Component Mass transport is essential in the solid electrolyte membrane since it plays a crucial role in the production of H_2 and O_2. The driving force is caused by the flow of water across the membrane. This flow of water occurs in three distinct forms:

i. Flow of water due to electro-osmotic drag
ii. Flow of water as a result of pressure difference on either side of the membrane from cathode to the anode
iii. Flow of water as a result of concentration gradient

The flow of water caused by concentration gradient across the PEM is expressed by the equation as follows:

$$\dot{N}_{H2O,grad} = \frac{A\,D_W}{\delta_{mem}}\,[C_{H2O,mem,cat} - C_{H2O,mem,an}] \tag{5.46}$$

Electro-osmotic drag is the transfer of water molecules from anode to cathode electrode due to flowing flux of hydrated protons across the membrane. The equation is defined as follows:

$$\dot{N}_{H2O,eo} = n_d * \frac{I}{F} \tag{5.47}$$

The expression for the molar flow of water caused by pressure difference is given below:

$$\dot{N}_{H2O,pe} = K_{darcy}\,\frac{A * \rho_W}{\mu_W * MW_{H2O} * \delta_{mem}} \tag{5.48}$$

Therefore, the overall flow of water across the membrane is expressed as the sum of molar flow of water due to concentration difference, electro-osmotic drag and effect of pressure.

$$\dot{N}_{H2O,mem} = \dot{N}_{H2O,grad} + \dot{N}_{H2O,eo} - \dot{N}_{H2O,pe} \tag{5.49}$$

where $\dot{N}_{H2O,eo}$ = Molar flow rate of H_2O as a result of electro-osmotic drag, $\dot{N}_{H2O,pe}$ = Molar flow rate of H_2O as a result of pressure effect, $\dot{N}_{H2O,grad}$ = Molar flow rate of H_2O as a result of concentration gradient, n_d = Electro-osmotic drag coefficient, K_{darcy} = Permeability coefficient of water, ρ_W = Density of water, μ_W = Viscosity of water, D_W = Water diffusion coefficient inside the membrane.

5.3.3 Material Balance Constraints

5.3.3.1 Water Balance

Water molecules are consumed in the electrolyser while concurrently keeping the membrane hydrated. The molar flow of water into the electrolyser is equivalent to the molar flow of water entering the anode of the first electrolyser cell.

$$\dot{N}_{H2O,in} = \dot{N}_{H2O,in,an,elect}\,(n)\,\forall n \in N, n = 1 \tag{5.50}$$

where $\dot{N}_{H2O,in}$ = Molar flow rate of water into the electrolyser, $\dot{N}_{H2O,in,an,elect}$ = Molar flow rate of water in the anode. The parenthesis n denotes electrolyser stack.

The equation for the molar flow of water consumed is given as follows:

$$\dot{N}_{H2O,cons} = \frac{I}{2F} \tag{5.51}$$

The water entering the electrolyser is not consumed completely; some leaves the system as excess water. Thus, the expression for the molar flow of water exiting the electrolyser as excess water is expressed as the difference between molar flow rate of water entering the anode, being consumed and that flowing through the membrane.

$$\dot{N}_{H2O,out,an,elect}\,(n) = \dot{N}_{H2O,in,elect}\,(n) - \dot{N}_{H2O,cons} - \dot{N}_{H2O,mem,elect}\,\forall n \in N \tag{5.52}$$

The molar flow rate of water exiting at the cathodic site of the electrolyser is equal to the molar flow rate of water across the membrane.

$$\dot{N}_{H2O,out,cat,elect}\,(n) = \dot{N}_{H2O,mem,elect}\,\forall n \in N \tag{5.53}$$

The total molar flow rate of H_2O exiting a cell in the electrolyser is:

$$\dot{N}_{H2O,out,elect}(n) = \dot{N}_{H2O,out,an,elect}(n) + \dot{N}_{H2O,out,cat,elect}(n) \forall n \in N \tag{5.54}$$

Since the cells are stacked, the following expression describes the molar flow rate of water exiting the previous cell and entering the next anode cell.

$$\dot{N}_{H2O,in,an,elect}(n) = \dot{N}_{H2O,out,elect}(n-1) \forall n \in N, n > 1 \tag{5.55}$$

Hence, the total molar flow rate of water exiting the electrolyser system is:

$$\dot{N}_{H2O,out,elect\ sys} = \sum \dot{N}_{H2O,out,elect}(n) \forall n \in N \tag{5.56}$$

The overall mass balance of the PEME system is given as:

$$M_{H2O,in} = M_{H2O,out,elect\ sys} + M_{H2,out,elect\ sys} + M_{O2,out,elect\ sys} \tag{5.57}$$

where $\dot{N}_{H2O,out,an,elect}$ = Molar flow rate of water out of the anode of the electrolyser, $\dot{N}_{H2O,out,cat,elect}$ = Molar flow rate of water out of the cathode of the electrolyser, $M_{H2O,out,elect\ sys,}$ $M_{H2,out,elect\ sys}$ and $M_{O2,out,\ elect\ sys}$ = Mass flow rate of water, hydrogen and oxygen out of the electrolyser.

5.3.3.2 Oxygen

Only H^+ ions can permeate through the electrolyte membrane, and all the OH^- and O^{2-} ions remain on the anodic site of the water electrolyser.

The expression for the formation of molar flow rate of O_2 in the electrolyser is:

$$\dot{N}_{O2,prod} = \frac{I}{4F} \tag{5.58}$$

The molar flow rate of O_2 exiting the cell of the electrolyser stack on the anode is:

$$\dot{N}_{O2,out,an,elect}(n) = \dot{N}_{O2,prod} \forall n \in N \tag{5.59}$$

$$\dot{N}_{O2,out,elect}(n) = \dot{N}_{O2,out,an,elect}(n) \forall n \in N \tag{5.60}$$

$$\dot{N}_{O2,out,elect\ sys} = \sum \dot{N}_{O2,out,elect}(n) \forall n \in N \tag{5.61}$$

where $\dot{N}_{O2,out,an,elect}$ = Molar flow rate of oxygen out of the anode of electrolyser, $\dot{N}_{O2,out,elect}$ = Molar flow rate of oxygen out of the electrolyser.

5.3.3.3 Hydrogen

The solid electrolyte membrane allows H^+ ions to flow through it from the anode to the cathode, and cross permeation is prohibited since the model being developed is one-dimensional.

The formation of molar flow rate of H_2 in the electrolyser is expressed as follows:

$$\dot{N}_{H2,prod} = \frac{I}{2F} \tag{5.62}$$

The molar flow rate of H_2 exiting the cathode is the same as the molar flow rate of H_2 being formed.

$$\dot{N}_{H2,out,cat,elect}(n) = \dot{N}_{H2,prod} \forall n \in N \tag{5.63}$$

The molar flow rate of H_2 exiting the electrolyser is equal to the molar flow rate of H_2 exiting the cathode of the electrolyser.

$$\dot{N}_{H2,out,elect}(n) = \dot{N}_{H2,out,cat,elect}(n) \forall n \in N \tag{5.64}$$

The total molar flow rate of H_2 leaving the electrolyser system is equal to the sum of all the molar flow rates of H_2 exiting the system.

$$\dot{N}_{H2,out,elect\ sys} = \sum \dot{N}_{H2,out,elect}(n) \forall n \in N \tag{5.65}$$

where $\dot{N}_{H2,out,cat,elect}$ = Molar flow rate of hydrogen out of the cathode of the electrolyser, $\dot{N}_{H2,out,elect}$ = Molar flow rate of hydrogen out of the electrolyser.

5.3.3.3.1 Overall System Performance

Total Power The objective function is to minimise the power consumption by the water electrolyser while maximising the power produced by the PEMFC. Hence, the total power consumed in the PEME is given by the following expression:

$$\text{Power}_{\text{cons}} = \frac{V * I * n_{\text{cell,elect}}}{1000} \tag{5.66}$$

where $\text{Power}_{\text{cons}}$ = Power consumed, $n_{\text{cell,elect}}$ = Number of cells in the electrolyser.

5.3.3.3.2 Economic Evaluation of the HEFC System To perform an economic analysis, the following costs are determined: fixed capital investment, maintenance cost, replacement cost and total annualised cost (TAC). Based on the TAC calculations, revenues and production cost, annual cash flow, present value, NPV and PBP can be determined.

Fixed Capital Investment A fixed capital investment also called initial capital investment is the amount of money required to construct and start a chemical plant. It consists of total direct and indirect capital costs. The expression for the total direct cost of the HEFC system is given by following equation:

$$\text{Cost}_{\text{TDC}} = (\text{Cost}_{\text{stack}} + \text{Cost}_{\text{BOP}})(1 + \text{IF}) + \left(\text{Cost}_{\text{cap}}{}^{*}\text{CF}_{\text{land}}\right) \tag{5.67}$$

The cost of stack and that for the balance of plant (BOP) are expressed as follows:

$$\text{Cost}_{\text{stack}} = \text{Cost}_{\text{cap}}{}^{*}\text{CF}_{\text{stack}} \tag{5.68}$$

$$\text{Cost}_{\text{BOP}} = \text{Cost}_{\text{cap}}{}^{*}\text{CF}_{\text{BOP}} \tag{5.69}$$

where Cost_{TDC} = Total direct cost, $\text{Cost}_{\text{stack}}$ = Capital cost of stack, Cost_{BOP} = Capital cost of balance of plant, Cost_{cap} = Total capital cost, CF_{land} = Land cost factor, IF = Installation factor.

The capital costs of the electrolyser and that of FC are extrapolated from a polynomial function constructed by Thomas (2018), and the expressions are shown below:

$$\text{Cost}_{\text{cap,elect}} = \left(4.6^{*}Y_{\text{purchase}}{}^{2} - 18\,658.2^{*}\,Y_{\text{purchase}} + 18\,920\,787\right) * \text{Power}_{\text{cons}} \tag{5.70}$$

$$\text{Cost}_{\text{cap,fc}} = \left(4.6^{*}Y_{\text{purchase}}{}^{2} - 18\,658.2^{*}\,Y_{\text{purchase}} + 18\,920\,787\right) * \text{Power}_{\text{prod}} \tag{5.71}$$

The following equation below expresses the BOP cost factor:

$$\text{CF}_{\text{BOP}} = \text{CF}_{\text{gms,an}} + \text{CF}_{\text{gms,cat}} + \text{CF}_{\text{dms,H2O}} + \text{CF}_{\text{tms}} + \text{CF}_{\text{Pelect}} + \text{CF}_{\text{cs}} + \text{CF}_{\text{MBOP}} + \text{CF}_{\text{other}} + \text{CF}_{\text{AL}} \tag{5.72}$$

where $\text{CF}_{\text{gms,an}}$ = Oxygen gas management system cost factor, $\text{CF}_{\text{gms,cat}}$ = Hydrogen gas management system cost factor, $\text{CF}_{\text{dms,H2O}}$ = Water delivery management system cost factor, CF_{tms} = Thermal management system cost factor, $\text{CF}_{\text{Pelect}}$ = Power electronics cost factor, CF_{cs} = Control and sensors cost factor, CF_{MBOP} = Mechanical balance of plant cost factor, CF_{other} = Other direct cost factor, CF_{AL} = Assembly labour cost factor.

The total indirect cost is defined by the following expression:

$$C_{\text{TIC}} = (\text{CF}_{\text{construction}} + \text{CF}_{\text{PC}} + \text{CF}_{\text{LC}}) * \text{Cost}_{\text{TDC}} + \left(\text{CF}_{\text{ED}}{}^{*}\text{Cost}_{\text{cap}}\right) \tag{5.73}$$

where C_{TIC} = Total indirect cost, $\text{CF}_{\text{construction}}$ = Construction cost factor, CF_{PC} = Project contingency cost factor, CF_{LC} = Legal and contractor cost factor, CF_{ED} = Engineering and design cost factor.

The fixed capital investment for the HEFC system is the sum of total direct and indirect capital costs.

$$\text{FCI} = C_{\text{TDC}} + C_{\text{TIC}} \tag{5.74}$$

Total Annualised Cost The TAC is defined as the annual cost of maintaining, owning and functioning the HEFC system throughout the plant project life. It is composed of the annualised capital cost, replacement cost and maintenance cost. The following expression describes the TAC equation of the HEFC system:

$$\text{TAC} = C_{\text{cap,annual}} + C_{\text{rep,annual}} + C_{\text{main,annual}} \tag{5.75}$$

where $C_{\text{cap,annual}}$ = Annualised capital cost, $C_{\text{rep,annual}}$ = Annualised replacement cost

$C_{\text{main,annual}}$ = Annualised maintenance cost

Furthermore, the annualised capital cost of the system is defined by the following expression:

$$C_{\text{cap,annual}} = \frac{(\text{FCI} * \text{AI})(1+\text{AI})^{Y\text{plant}}}{(1+\text{AI})^{Y\text{plant}} - 1} \tag{5.76}$$

In the same way as the direct capital cost of the electrolyser and the FC, the replacement cost is derived from a polynomial function proposed by Thomas (2018). The expressions for the replacement costs of both the electrolyser and FC, together with the total replacement cost of the HEFC system, are shown below:

$$\text{Cost}_{\text{rep,elect}} = \left(4.6^* Y_{\text{rep}}{}^2 - 18\,658.2^* \, Y_{\text{rep}} + 18\,920\,787\right) * \text{Power}_{\text{cons}} \tag{5.77}$$

$$\text{Cost}_{\text{rep,fc}} = \left(4.6^* Y_{\text{rep}}{}^2 - 18\,658.2^* \, Y_{\text{rep}} + 18\,920\,787\right) * \text{Power}_{\text{prod}} \tag{5.78}$$

$$\text{Cost}_{\text{rep}} = (\text{Cost}_{\text{rep,elect}} + \text{Cost}_{\text{rep,fc}})^*(1+\text{IF}) \tag{5.79}$$

$$\text{Cost}_{\text{rep,annual}} = \frac{\text{Cost}_{\text{rep}}}{Y_{\text{plant}}} \tag{5.80}$$

It is assumed that the annualised maintenance cost is fixed at 2.5% of the initial capital cost throughout the project life. Where AI = Annual interest rate, Y_{plant} = Plant lifetime, Y_{rep} = Year of replacement, $\text{Cost}_{\text{rep,elect}}$ = Replacement cost of the electrolyser, $\text{Cost}_{\text{rep,fc}}$ = Replacement cost of the FC.

Revenue and Production Costs The total revenue generated by the HEFC system is derived from the main product, power and water produced as by-products. The expressions below show the revenues incurred from the system:

$$\text{Rev}_{\text{elect}} = \text{Power}_{\text{prod}}{}^*\text{Cost}_{\text{elect}}{}^*\text{OP}_{\text{hours}} \tag{5.81}$$

$$\text{Rev}_{\text{H2O}} = \left(\frac{M_{\text{H2O,out,fc,sys}}}{\rho_{\text{H2O}}}\right) * \text{Cost}_{\text{H2O}} * \text{OP}_{\text{hours}} \tag{5.82}$$

$$\text{Rev} = \text{Rev}_{\text{elect}} + \text{Rev}_{\text{H2O}} \tag{5.83}$$

However, the annual interest rate grows at the same proportion as the revenues, and the expression can further be defined as follows:

$$\text{Rev}\,(q) = (\text{Rev}_{\text{elect}} + \text{Rev}_{\text{H2O}})^*(1+\text{AI})^q \; q > 1 \tag{5.84}$$

where Cost_{H2O} = Cost of water in US$ per m^3, $\text{Cost}_{\text{elect}}$ = Cost of electricity in US$ per kWh, $M_{\text{H2O,out,fc,sys}}$ = Mass flow rate of H_2O exiting the FC, OP_{hours} = Operating hours in hours per annum, q = yearly increment.

The following equation give rises to the HEFC system's total production costs

$$C_{\text{TP}}(q) = (\text{CF}_{\text{PR}} + \text{CF}_{\text{DS}} + \text{CF}_{\text{RD}}) * \text{Rev}(q) + \text{FCI}^*\text{CF}_{\text{INS}} \tag{5.85}$$

where C_{TP} = Total production costs, CF_{PR} = Patent and royalties cost factor, CF_{DS} = Distribution and selling cost factor, CF_{RD} = Research and development cost factor, CF_{INS} = Insurance cost factor.

Cash Flow The following series of equations expresses the cash flow of the HEFC system over its entire lifespan.

$$\text{Dep}(q) = \frac{\text{FCI} + \text{Cost}_{\text{rep}}}{Y_{\text{plant}}} \tag{5.86}$$

$$\text{PBT}(q) = \text{Rev}(q) - \text{C}_{\text{TP}}(q) - \text{Dep}(q) \tag{5.87}$$

$$\text{TAX}(q) = \text{PBT}(q) * \text{Tax}_{\text{rate}} \tag{5.88}$$

$$\text{PAT}(q) = \text{PBT}(q) - \text{TAX}(q) \tag{5.89}$$

$$\text{PATPD}(q) = \text{PAT}(q) + \text{Dep}(q) \tag{5.90}$$

$$\text{CASH}(q) = \text{PATPD}(q) - \text{TAC} \tag{5.91}$$

where Dep(q) = Depreciation calculated in straight line method, PBT(q) = Profit before tax, PAT(q) = Profit after tax, PATPD(q) = Profit after tax plus depreciation, CASH(q) = Annual cash flow.

Net Present Value (NPV) NPV is described as the cumulative discounted position at the end of plant's operation. The present value of the future cash flow is calculated from the discount factor expressed in terms of annual interest rate. The equations that govern the NPV are given as:

$$\mathrm{DF}(q) = \frac{1}{(1+\mathrm{AI})^{q-1}} \tag{5.92}$$

$$\mathrm{PV}(q) = \mathrm{CASH}(q) * \mathrm{DF}(q) \tag{5.93}$$

$$\mathrm{NPV} = \sum \mathrm{PV}(q) \tag{5.94}$$

where DF = Discount factor, PV = Present value.

Payback Period (PBP) PBP is the duration needed for an investment to start generating profit in reference to its initial outlay. It is expressed by the equation,

$$\mathrm{PBP} = \frac{\mathrm{FCI} * Y_{\mathrm{plant}}}{\sum \mathrm{Rev}(q)} \tag{5.95}$$

Illustrative Example The water electrolyser is modelled according to the data presented in Tables 5.2 and 5.5. The economic parameters presented in Tables 5.3 and 5.4 are used to assess the economic feasibility of the HEFC system.

The following assumptions hold during the formulation of the models of the PEME:

i. The electrolyser functions under steady-state conditions, and it is a one-dimensional model
ii. High-purity H_2 and O_2 are produced by the electrolyser
iii. All water molecules throughout the system are in liquid phase
iv. The PEME may be powered by any renewable source of energy
v. Current in each cell unit is the same since the cells are connected in series configuration
vi. Pressure difference on either side of the membrane is neglected
vii. Water activity through the membrane is invariable
viii. Excess water exiting the electrolyser does not enter the PEMFC
ix. The H_2O activity passing through the electrolyte membrane is uniform
x. The electrolyser operates for 8000 hours per annum. The remaining hours in a year are reserved for maintenance during shutdown

5.3.4 Analysis of the Outputs from the Different Model Solvers

The formulation and modelling of the PEME were conducted in GAMS version 24.8.5 with the objective function of minimising the power consumed by the electrolyser. The NLP optimisation problem was solved using BARON as a solver. The optimisation problem was conducted in a desktop computer comprising the following specifications: Intel Core i7 processor on Windows 7 Professional 64-bit system software with 8 GB of RAM. All the polarisation curves together with the power conversion and capital cost curves were plotted using MATLAB.

Table 5.2 Water electrolyser parameters.

Parameter	Value
Molar flow rate of water (mol/s)	100
Pressure (bar)	1.20
Anode charge transfer coefficient	0.25
Cathode charge transfer coefficient	0.8
Anode exchange current density (A/m^2)	0.004
Cathode exchange current density (A/m^2)	5
Membrane thickness (μm)	178

Table 5.3 Economic parameters.

Direct costs	Cost factor (%)
Stack	38
Hydrogen gas management system	6
Oxygen gas management system	2
Water delivery management system	5
Thermal management	5
Power electronics	26
Controls and sensors	6
Mechanical balance of the plant	5
Other direct costs	2
Assembly labour	5
Land	4
Installation factor	10
Indirect costs	**Cost factor (%)**
Construction costs	2
Engineering and design	8
Project contingency	15
Legal and contractor	15

Source: Adapted from Colella et al.

Table 5.4 Other cost factors and rates.

Other factors	Numerical value	Reference
Working capital	20%	Turton et al. (2018)
Annual interest rate	5.5%	Trading economics (2022)
Inflation rate	5.73%	Statista (2022)
Maintenance factor	2.5%	Colella et al. (2014)
Purchase year	2020	—
Replacement year	2030	—
Project life (years)	25	—
Tax rate	28%	South African Revenue Service (2022)

Table 5.5 Operating condition boundaries.

Variables	Symbols	Electrolyser range	Reference
Temperature (°C)	T	25–90	Atyabi et al. (2019)
Current density (A/m^2)	J	6000–6019 760	–
Humidity degree	λ	1.40–16.69	–
Electro-osmotic drag coefficient	n_d	0.3–0.895	–
Power consumed (kW)	$\text{Power}_{\text{cons}}$	103.47	–
Reaction area (cm^2)	A	400–794	–
Anode and cathode thickness (μm)	δ	50–7000	–

Platinum and nickel electrodes are recommended as good design materials when constructing PEME since they are known to be highly corrosion-resistant. It is essential to overcome corrosion because it has an influence on the cell's performance.

5.3.4.1 Model Validation

To effectively validate the feasibility of the water electrolyser, its energy requirement is determined. According to Züttel et al. (2010), the minimum theoretical energy needed at ambient conditions is 39.70 kWh/kg H_2. However, the energy requirement of the electrolyser is estimated to be 59.55 kWh/kg H_2 which exceeds the minimum theoretical energy required. Thus, it can be declared that the PEME model is feasible.

5.3.4.2 Optimal Performance, Design Parameters and Operating Conditions

The relationship between number of cells in the electrolyser and power consumed is evaluated in Figure. 5.5. The optimal number of cells in the PEME required to support the PEMFC to produce 500 kW is 64 cells. This is achieved when the PEME is operated at a minimum power of 103.47 kW. Hence, the results presented hereafter are based on the PEME model developed with 64 cells.

From Table 5.6, it is evident that the electrolyser operates at high current density, which is favourable for the cell performance due to its less degrading condition because of lower cell voltage (Knights 2012). The activation overpotential contributes largely to the voltage losses due to the electrolyser being operated at high current density, which triggers the electrochemical reactions at the electrodes to occur rapidly. The ohmic overpotential follows the activation overpotential as the second main contributor to potential losses since the flow of H^+ ions in the membrane is resistive (Singh et al. 2020).

As indicated in Table 5.7, the optimal operating temperature of the electrolyser at atmospheric pressure is 25 °C. However, in literature, it is confirmed that elevated temperatures ranging from 40 to 80 °C improve the cell performance (Han et al. 2015). For this model, this low temperature is desirable because its optimal humidity degree is very low, at 1.40. Humidity degree indicates the quantity of H_2O present in the membrane. Thus, if the model was operating at high temperatures, its membrane would dehydrate and lead to lower performance due to less water molecules passing through it, as represented by the low electro-osmotic drag coefficient of 0.3. Alternatively, at high humidity degree, flooding would occur at

Table 5.6 Optimal performance results.

Parameter	Value
Current density (A/m^2)	6000
Open circuit voltage (V)	1.3048
Activation overpotential (V)	1.6891
Diffusion overpotential (V)	0.1410
Ohmic overpotential (V)	0.1664
Total Input voltage (V)	3.1349

Table 5.7 Optimal design parameters and operating conditions.

Parameter	Value
Temperature (°C)	25
Electro-osmotic drag coefficient	0.3
Anode thickness (μm)	6200
Cathode thickness (μm)	7000
Humidity degree	1.40
Reaction area (cm^2)	786

the cathode and degrade the cell's performance due to increase in resistance of the electrode. Hence, it is vital to manage water adequately on the membrane because humidification degree is crucial to the electrolyser's performance.

5.3.4.3 Effect of Varying Design Parameters and Operating Conditions

The exchange current density is influenced by catalyst material, operating temperature and other physical properties that are not easily quantifiable at the electrodes (Figure 5.17). Based on the optimised model, the exchange current densities for the model at the anode and cathode are 0.004 and 5 A/m^2. Figure 5.18 shows the effect of varying the anode exchange current density at three different values 1, 3 and 5 A/m^2. From the polarisation curve, it can be observed that when the anode exchange current density is increased, the input voltage declines. This indicates that the performance of the PEME decreases.

Similarly, at the cathode, the effect of varying cathode exchange current density is shown in Figure 5.19.

Increasing the cathode exchange current density at three different values 0.001, 0.0025 and 0.004 A/m^2 decreases the input voltage by larger extent in comparison to the total voltage decrease in the anode. This is because the exchange current density at the cathode is higher than that at the anode, indicating that the surface of the cathode electrode is more active. There is lower energy barrier that the H^+ ions must subdue in passing through the electrolyte membrane to the catalyst surface; hence, more current is produced at any overpotential (Barbir 2013). From the results of the analysis, it can be deduced that the anode and cathode exchange current densities have a huge influence on the final performance of the electrolyser.

The solid electrolyte membrane is the fundamental component in the water electrolyser. Its performance is affected by many factors such as permeability, membrane conductivity, thickness, humidity degree and operating temperature. Theoretically, the ohmic overpotential is proportional to the thickness of the membrane and inversely proportional to membrane conductivity, as indicated by Eq. (5.15). The effect of varying membrane thickness is illustrated in Figure 5.20.

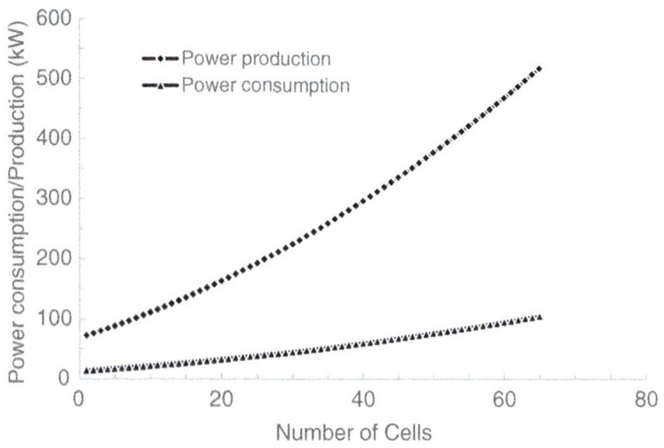

Figure 5.17 Power consumption and power produced by the HEFC system.

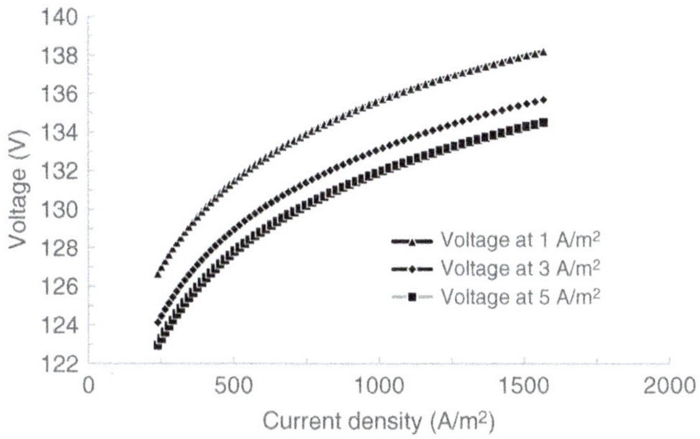

Figure 5.18 The influence of exchange current density at anode on polarisation curve.

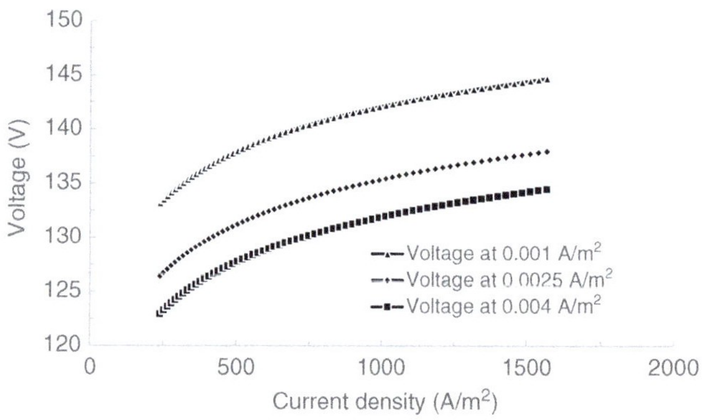

Figure 5.19 The effect of exchange current density at cathode on polarisation curve.

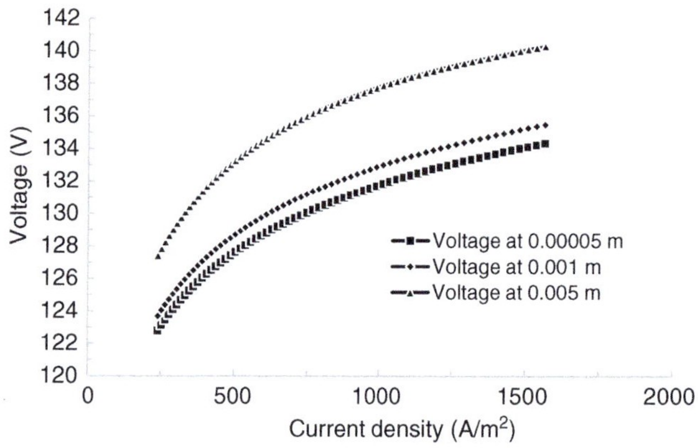

Figure 5.20 Membrane thickness effect on polarisation curve.

Thickening the membrane from 50 to 5000 μm yields a significant increase in total voltage of the electrolyser. This is true because as the membrane gets thicker, a higher ohmic resistance and loss in diffusion overpotential are experienced. This observation is in excellent agreement with the experimental study conducted by Ayers et al. (2012). In addition, the molar flow rate of water due to concentration gradient drops when the membrane thickness is increased. This is best described by Eq. (5.46). Hence, it is essentially to balance the membrane thickness during its fabrication to enhance the electrolyser performance.

Operating temperature affects the open-circuit voltage, activation overpotential, diffusion coefficient and membrane conductivity. Theoretically, the activation overpotential increases when the electrolyser is operated at elevated temperatures (Han et al. 2015). The effect of temperature on the performance of the electrolyser is shown in Figure 5.21. Based on the

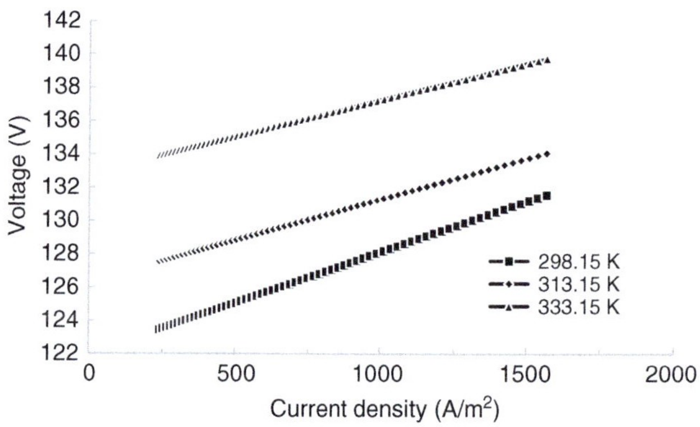

Figure 5.21 Temperature effect on polarisation curve.

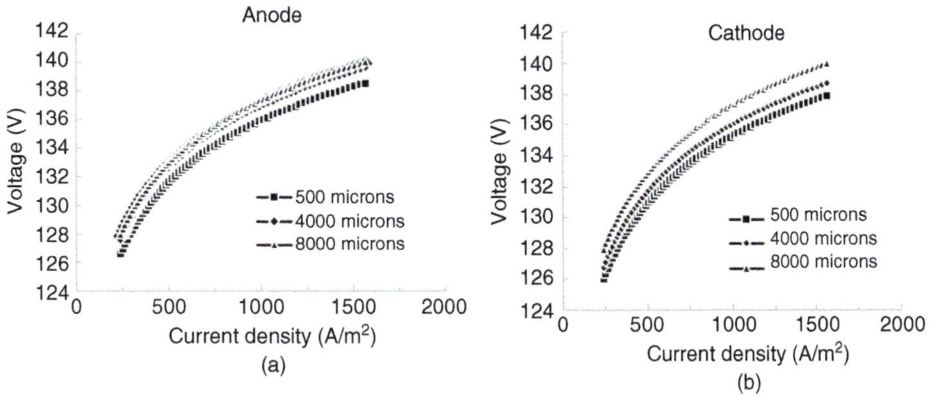

Figure 5.22 Effect of anode (a) and cathode (b) electrode thicknesses.

analysis, increasing the operating temperature from 298.15 K (25 °C) to 333.15 K (60 °C) results in overall voltage increase. However, the voltage difference between the three different temperatures drops at high current densities. This deviates from the results obtained by Han et al. (2015). It could be due to a decrease in electrocatalytic reaction rates at the electrodes, which further leads to increase in charge transfer resistance on both sides of the electrolyte membrane (Li et al. 2021).

In various industrial applications, the electrode thickness in water electrolysers is varied across a wide range of values for optimisation purposes because of their huge influence on mass transport processes and losses in ohmic overpotential. The model is considered to be constructed with platinum electrode and carbon paper material. As illustrated in Figure 5.22, the effect of electrode thickness on the electrolyser for both anode and cathode is examined. There is an improvement in electrolyser performance when the electrode thickness increases. This outcome contradicts with the results from the model developed by Han et al. (2015). According to Han et al. (2015), the performance of the electrolyser drops with increasing electrode thickness because of higher diffusion and ohmic losses. For this model, the electrode porosity used is 0.6, which is twice the electrode porosity used in the model developed by Han et al. (2015). Hence, at high porosities, the electrical resistance of the electrode increases since reactants are delivered at a high diffusion rate to the active area of the electrodes, which enhances the performance of the electrolyser (Larbi et al. 2013).

5.3.4.4 Economic Evaluation

The electrical energy supplied to the electrolyser is assumed to have been derived from solar PV panels. The adoption of renewable energy has resulted in a phenomenal change in the power generation industry. The cost of electricity generated from fossil fuels is US$0.14/kWh, whereas the cost of utility-scale electricity in the solar and PV power generation sector is $0.069/kWh (Strielkowski 2020). According to the international currency exchange rate, 1 USD is equivalent to R18.65 (Xe currency converter, August 2023). This is lower than the cost of electricity derived from fossil fuel. However, in this work, the costs associated with the consumption of electrical power from solar PV are not considered because the model formulated is based on the optimal design. Henceforth, the purpose is to determine the operational feasibility in terms of power consumption. The basic cost factors utilised in the direct and indirect capital costs for different elements in the HEFC system are extracted from Colella et al. (2014). The power production capacity of the HEFC system is colossal; it is modelled to produce 500 kW. Hence, the forecourt costs are recommended to be utilised rather than the centralised production facility (CPF) cost factors due to vast size differences between the HEFC system and the CPF. As illustrated in Table 5.8, the NPV throughout the 25 years of the project life is estimated to be US$16032. This means that the revenues generated by the system can compensate for all business risk. The PBP is 4.54 years. According to Sinnott (2005), PBP of less than 5 years is considered to be cost-effective. In addition, the system operates in an environmentally sustainable way, and no costs are incurred because no pollutants are emitted into the environment.

Measuring capital costs is crucial in any chemical process in order to estimate economic values and quantify uncertainties. Capital cost is the capital required to build up the HEFC system (Turton et al. 2018). The polynomial function plotted in Figure 5.23 is extrapolated from an investigation report conducted by Thomas (2018), and it is described by Eq. (5.79). Both the PEME and the PEMFC are governed by the same capital cost expression because they are constructed with similar structural materials. As a result of vast market uptake, enhanced efficiencies, excessive production capacity and cheaper electrodes and catalysts, the overall capital cost is expected to decline with time.

Table 5.8 Techno-economic results.

Parameter	Value
Total annualised cost (US$)	735
Internal rate of return (%)	5.49
Discount factor (%)	8.52
Payback period (years)	4.54
Net present value (US$)	16 032

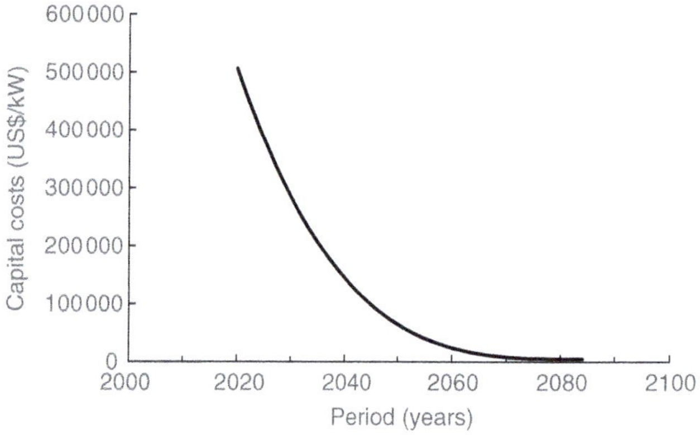

Figure 5.23 Capital cost of PEME and PEMFC from 2020 to 2085.

5.3.5 Integrated PEM Electrolyser and Fuel Cell

This work is based on design and integration of a PEME that can support a 500 kW PEMFC to power Richard Ward building at Wits University. The developed PEME mathematical model is integrated with PEMFC model and optimised in GAMS as an NLP problem with the objective of minimising power consumption by the water electrolyser. Based on the analysis, it is confirmed that the physical and chemical properties of the PEME have a huge influence on its performance and the HEFC system as a whole. This system is eco-friendly, it supplies 100% clean power with zero carbon emissions to the environment, and in addition, it is evaluated to be economically feasible.

Some impediments associated with the PEME model are outlined as follows:

- The PEME model is one-dimensional; hence, cross permeation of H_2 and O_2 is negligible. Two-dimensional and three-dimensional mathematical models can be formulated in future with the aim of evaluating the effect of cross permeation on the performance of the PEME.
- The parameters used in the mathematical formulation of this electrolyser model are obtained from literature, especially from Singh et al.'s (2020) paper. It is indefinite whether these parameters elucidate the water electrolysis operation at the optimal conditions obtained.
- The operating costs of the solar power consumption are considered negligible in this mathematical model.

The mathematical modelling results provide an intuition into the performance of the PEME for optimal design. Furthermore, this research work provides an insight for better comprehension of how water electrolysis and HEFC systems operate.

5.4 Component Production: Platinum Group Metal Recovery from Waste Automotive Catalytic Converter for PEM Fuel Cell Technology (Case Study 3)

Large amounts of waste automotive catalytic converter (WACC) containing platinum group metals (PGM: Pt, Pd and Rh) are generated as the automotive scrap, which not only wastes platinum resources but also causes environmental impacts if the WACC is not adequately managed. In another scenario, a lack of adequate hydrogen sources, widespread refuelling infrastructure and cost of Pt/PGMs are currently the main issues that restrict the ability of hydrogen FCs to be used in many industrial applications including transportation. Despite, the global transition to sustainability and an increasing unified effort of countries aiming to reduce GHG emissions, FCs and PEME are poised to become critical components as energy sources, e.g. electricity and H_2. This proposal is therefore aimed to recover PGMs from WACC with smelting-collection process for PEME/FC industry and general market outlets. In the proposal, which has been adequately researched and reported in the literature, a novel and effective smelting-collection process for recovering platinum from WACC is presented, where carrier materials serve as slag constituents and the addition of extra slagging flux such as SiO_2 and Al_2O_3 is nearly avoided compared to platinum recovery from single WACC carrier. After smelting, platinum is concentrated into Cu–Pt alloy, and the glass slag, which is believed to be safe for the environment, is obtained. The results indicated that platinum content in obtained alloy and glass slag was 0.63 wt% and 9.4 g/t, respectively. In order to recover PGMs at feasibly economic scale, the production capacity is increased from ½ tonne of WACC at Wits laboratory to equal or greater than 1196 tonne per month at Mintek, since the Mintek 1.5 MW furnace can recover in excess of 99.5% PGMs. This business plan is therefore submitted to a Sister Company (SC), which comprises of six milestones involving among others, test running and trial production for 0.2 kg WACC per month, scale-up production levels 1 and 2. These levels involve processing of ½ tonne of WACC per month whereby 3 times 5 kg per batch is processed in a day, while scale-up production level 2, involves processing 1200 tonne of WACC per month and distribution and marketing of PGMs to Wits, SC and other identified users and market. The project takes three years at the cost of US$1.16 million, the 1st processing stage incurs a loss of US$166466, which should be covered by the investment of about US$428954 in the first year. While the processes at the 2nd year of ½ tonne of WACC and 3rd year of 1200 tonne of WACC per month will yield a net profits of US$105797(10%) and US$1638616(62%), respectively, the subsequent financial projections into 4–6 year performances, give profits of US$1662877(62%), US$1739074(63%) and US$1789257 (64%), respectively. In terms of safety, environmental protection and economic feasibility therefore, this project will breakeven in no time and therefore brings huge benefits to the Institution, SC, all stakeholders and the county in general.

5.4.1 Status of PGM Recovery

PEMFC technology comprises water electrolyser (PEME) and a PEMFC. While the PEME is a devise that produces green hydrogen from water electrolysis, PEMFC generates electricity, heat and water, respectively, through electrochemical reactions. FCs are used today in a range of applications, from providing power to homes and businesses, keeping critical facilities like hospitals, grocery stores and data centres up and running, and moving a variety of vehicles including cars, buses, trucks, forklifts and trains. FC systems are a clean, efficient, reliable and quiet source of power. FCs do not need to be periodically recharged like batteries, but instead continue to produce electricity as long as a fuel source is provided. PEME and PEMFC are composed of an anode, cathode and an electrolyte membrane (typically platinum [Pt]-based). A typical FC works by passing hydrogen through the anode of a FC and oxygen through the cathode. At the anode site, the Pt catalyst splits the hydrogen molecules into electrons and protons. The protons pass through the porous electrolyte membrane, while the electrons are forced through a circuit, generating an electric current and excess heat. At the cathode, the protons, electrons and oxygen combine to produce water molecules, while the reverse is the case with PEME, where water is split into H_2 and O_2 as electricity is supplied to the platinum electrodes. Both devices do not possess moving parts and operate silently and with extremely high reliability.

A lack of adequate hydrogen sources, widespread refuelling infrastructure and cost of Pt/PGMs are currently the main issues that restrict the ability of hydrogen FCs to be used in many industrial applications including transportation. Thus, with the global transition to sustainability and an increasing unified effort of countries aiming to reduce GHG emissions, FCs and PEME are poised to become critical components as energy sources. This project is therefore aimed to recover Pt/PGMs from WACC with smelting-collection process for PEME/FC industry and general market outlets.

Liu et al. (2021) have expressed the sentiment that WACC from automotive is generated at a growing rate due to intentional or unintentional reasons such as service life and the car crash. They also reported the situation in the following context and review, that it was estimated that approximately 40 million vehicles were scrapped worldwide in 2010, which accounted for about 4% of total global ownership (Sakai et al. 2014; Yano et al. 2019). PGMs are widely used in automotive catalyst applications due to their excellent heat stability, corrosion resistance, stable electrical properties and catalyst properties. For instance, it is reported that the gross demand worldwide for Pt, Pd and Rh in the automotive industry was 99.1, 266.4 and 26.7 t, respectively. However, in comparison to the gross demand, the recycling of PGMs from WACC amounted to 43.0 (Pt), 80.9 (Pd) and 10.3 t (Rh). The natural resources of PGMs are limited and are mainly found in countries like

Russia, South Africa and Canada. Normally, PGMs are difficult to extract from the mineral ores since their content is low, with average content ranging from 5 to 20 g/t. WACC occupies a significant place among the secondary resources containing PGMs since each tonne of WACC contains about 500–5000 g/t PGMs (Alekseeva et al. 2019). Ordinarily, WACC contains some toxic elements such as Pb, Cd and brominated flame retardants, which can contaminate the local environment without the necessary treatment. As a result, the recovery of PGMs from WACC is necessary, not only in terms of assuring their steady supply but also reducing environmental pollution (Mouza et al. 1995). In the recovery of platinum from WACC Wang and Chen (2013) identified the dominating smelters-refineries as Platinum Metals Resources (Yimen) Co. Ltd. (China), Umicore (Belgium), Heraeus (Germany), BASF/Engelhard (USA), Johnson Matthey (UK) and Nippon/Mitsubishi (Japan). In the recovery process, WACC is first mechanically crushed to evaluate the grade of PGMs and then subjected to extraction of PGMs by hydrometallurgical, pyrometallurgical or other methods, which contain biological and physical separation pre-treatment. In hydrometallurgical methods, PGMs are directly dissolved with a strong oxidising acid, such as hydrochloric acid and aqua regia with chlorine gas. The hydrometallurgical process, featured by low-energy consumption, appears to be effective in extracting PGMs from WACC due to the selectivity of leaching solution on metals. Some inevitable disadvantages, however, including large quantities of solid or liquid waste, high cost of reagent, long processing time and low PGM recovery rate, especially rhodium, limited its further development. In pyrometallurgical process, the pulverised WACC can be melted with the addition of fluxes and appropriate collector metals such as Ni, Cu and Fe at high temperature. PGMs are therefore, concentrated in the metallic phase forming an alloy phase that is separated from the molten slag which contains carrier materials and added flux. The benefits of pyrometallurgical method include higher PGMs recovery rate, shorter processing time, generation of a few waste gases and solutions and industrial feasibility (Chen and Huang 2006). Many researchers have studied the recovery of PGMs from the single carrier WACC by collector smelting collection process under different conditions and achieved a high recovery rate of PGMs (Yoo 1998; Dong et al. 2010). In the lead collection process, the ground WACC with cordierite carrier mixed with litharge for high-temperature smelting at about 1100 °C for two hours in the presence of addition of sodium carbonate, anhydrous borax and potassium bitartrate or other additives. As a result, the 70–80% for Rh and 90–95% for Pt and Pd can be successfully recovered (Compernolle et al. 2011). Much other literature reported that the WACC mixed with cordierite carrier was treated in an electric arc furnace through the addition of flux such as SiO_2 and CaO, metal collector of $CuCO_3$ and reducing agent, resulting in a final overall recovery rate of PGMs up to 99% (Fornalczyk and Saternus 2013; Hoffmann 1988). Iron collection process was also widely used in recovery of PGMs from the single carrier WACC (Benson et al. 2000). The crushed WACC mixed with iron collector, reducing agent and fluxes such as CaO, Al_2O_3 and SiO_2 are smelted in a plasma arc furnace or electric arc furnace at a higher operating temperature, over 1450 °C. The overall recovery efficiency of PGMs by iron collection can approach 99%. However, the methods involved appropriate fluxes such as SiO_2, Al_2O_3, CaO, MgO and Na_2CO_3 to improve the rheological properties of slag, depending on the different components of the catalyst when recovering PGMs from WACC of a single carrier material. Very few papers focus on recovering PGMs from two different WACC carriers, e.g. cordierite and alumina carriers simultaneously via smelting-collection process (Burkhard et al. 1994; Zhao 1998).

5.4.2 Recovery Techniques for PGMs from WACC

The WACC with different carrier material is used in the proposal for optimised results for mass production. The chemical compositions of WACC are measured by X-ray Fluorescence (XRF, ZSX) spectroscopy, particularly, the platinum content is determined by inductively coupled plasma mass spectrometry (ICP-MS), as shown in Table 5.9. Analytical-grade CaO is introduced as an ingredient additive to increase the alkalinity of samples. The carbon, industry grade, is used as reductant to prevent the oxidisation of carrier, e.g. Fe, Cu, Pb, etc. Besides, the analytical-grade of carrier powder is used as a collector metal of platinum. Prior to the smelting process, the different WACC carriers are crushed, milled passed through a 200-mesh sieve and dried at 105 °C for 24 hours to remove its free moisture. Different amounts of the cordierite (if available) carrier material, alumina carrier material, iron or copper powder and CaO are homogenised by a blender mixer and placed in a graphite crucible. The mass fraction of the samples is shown in Table 5.10.

5.4.3 Procedure for Trial Run

The process route for both the trial run and the actual mass (x quantity) PGM recovery is shown in Figure 5.24. The feeding materials are prepared by mixing 200 g (or × qty) of different carriers WACC with an appropriate amount of flux, namely 30–80 g CaO. After homogenisation, the mixture is placed into a high-purity graphite cylindrical crucible, 7–35 g collector

Table 5.9 Chemical compositions of cordierite and alumina carrier materials used for experiments[a] (Liu et al. 2021).

Sample	Content (Wt%)												
	SiO_2	Al_2O_3	MgO	CaO	ZnO	ZrO_2	CrO_3	TiO_2	S	Tcrr	Pt[b]	Other	LOI
Cordierite carrier[a]	62.1	10.4	0.4	0.3	0.33	0.03	0.8	0.7	4.9	0.7	701	2.8	16.9
Alumina carrier[a]	1.2	84	–	0.04	–	0.07	–	–	4	0.5	754	1.49	8.7

Key: Tcrr is total carrier, LOI is loss of mass on ignition.
a) Pre-dried at 120 °C for 12 hours.
b) g/t.

Table 5.10 Designed experimental samples used in the trial run (Liu et al. 2021).

Samples	A:C (wt%)	Chemical compositions of the mixture of C and A (wt%)						
		SiO_2	Al_2O_3	MgO	CaO	S	TCrr	Pt (g/t)
AC50	50:50	31.65	47.2	0.2	0.2	4.5	0.6	727.3
AC60	50:60	37.7	39.8	0.3	0.2	4.6	0.6	722.1
AC70	50:70	43.8	32.5	0.3	0.2	4.6	0.6	716.8
AC80	50:80	50.0	25.12	0.3	0.2	4.7	0.7	711.5
AC90	50:90	56.0	17.76	0.4	0.3	4.8	0.7	706.3

A is alumina carrier WACC, *C is* cordierite carrier WACC, TCrr *is* total carrier.

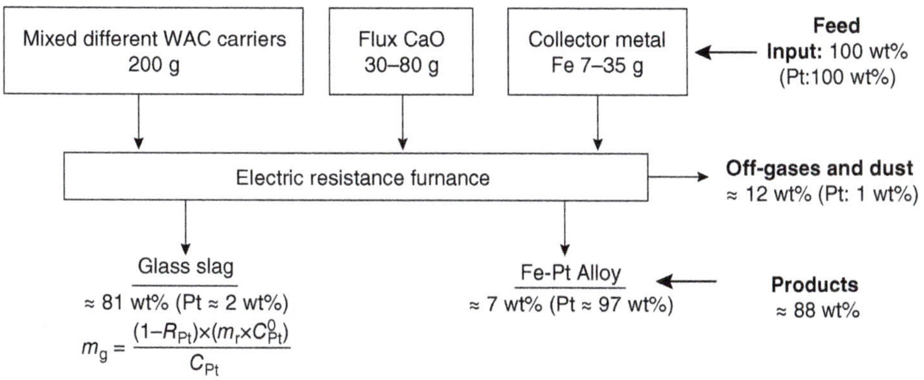

Figure 5.24 Simplified flowsheet of this smelting-collection technology. Source: Liu et al./Springer Nature.

metal of carrier (Cu or Fe) powder is placed on the top of it. Subsequently, this graphite crucible is placed into an electric resistance furnace (Figure 5.25) with eight U-shape $MoSi_2$ heating elements when the internal temperature of the furnace reached the target smelting temperature. The temperature of the furnace is heated to the target temperature of 1250–1600 °C and held for 60–180 minutes. In the trial runs, the smelting temperature are 1450, 1500, 1550 and 1600 °C, at the holding time of 60, 90, 120 and 180 minutes, while the amount of collector of iron or copper powder added is 3.5, 7, 10.5, 14 and 17.5 wt% of the quantity of mixed WACC with the CaO/SiO_2 mass ratio ranging from 0.3 to 0.8. After smelting operation, the upper molten slag is rapidly poured into a graphite crucible (Figure 5.25), and the remaining alloy-bearing melts settle at the bottom of the reactor, which is gradually cooled to room temperature. The metallic Fe or Cu obtained at the bottom of the graphite crucible is taken out and weighed for further assay. The metallic collector is separated from the alloy-bearing slag, cleaned and weighed for further assay. The upper glass slag is milled into powder for a subsequent recovery rate of platinum calculation using Eq. (5.96). The compositions of glass slag are examined by XRF analysis, and the platinum is

Figure 5.25 PGM smelting furnace at the School of Chemical & Metallurgical Engineering, Wits.

confirmed by ICP-MS. The platinum recovery rate is then calculated as;

$$R_{Pt} = \left(1 - \frac{m_g \times C_{Pt}}{m_r \times C^0_{Pt}}\right) \times 100\% \tag{5.96}$$

where m_g is the mass of glass slag, C_{Pt} is the platinum concentration in glass slag, m_r, is the mass of mixture with different carrier material in the samples and $C_{Pt}{}^0$ is the platinum concentration of mixture with the different carrier material.

m_g: the mass of glass slag;
R_p: the mass recovery;
m_r: the mass of mixture with different WACC carriers;
$C_{Pt}{}^0$: the platinum concentration of mixture with different WACC carriers;
C_{Pt}: the platinum concentration in glass slag.

5.4.3.1 Scale-up Production Level 1

The process of PGMs recovery is ongoing at 'trial-running' level with the few WACC sample supplied by SC. This is planned to precede the scale-up production, which involves ½ a tonne of WACC per month as presented in Table 5.11. To reduce waste and save laboratory space SC could de-can and label the samples either diesel or gasoline before delivery to Wits. The initial composition or ratio of PGMs in the supplies should also be indicated if known for ease of analysis, since the objective of this business plan is to recover and obtain PGMs: Pt, Pd and Rh as pure metal sponge using Cu as carrier smelting and chemical refining to obtain metal products. While Table 5.12 presents the calculations for possible feed volume with a value of US$32172 monthly turnover, assuming gasoline WACC cans are supplied, with current capacity of 5 kg per batch.

5.4.3.2 Scale-up Production Level 2

From the 99.5% PGM recovery presented in Table 5.12, scale-up production level 1 may not achieve economic feasibility on the long run. As a result, in order to recover PGMs at feasibly economic scale, the production capacity is increased from

Table 5.11 Processing and the cost estimate based on 2015 purchase model $/Can exclusive of margins.

Can collection	Pre-treatment: Physical materials handling	Pyro-metallurgy smelting	Collector removal	Refining
PGM content %	0.1–0.25	0.4–10	50–70	98–99.5
Feed volumes per month	370–500 kg WACC	312–12.5 kg	2.5–1.8 kg	1–1.25 kg
Assuming can weight of 1.2 kg	80–100	78-3 kg	1–0.5	250 g
Estimated process timing	cans per week	2–3 days	1 day	4–7 days
Costs of purchase ($/can)	13.00	12.35	7.25	7.25

Table 5.12 Feed volume with a value of US$32172 monthly turnover, assuming gasoline WACC and 99.5% PGM recovery.

Metal	0.5 tonne WACC (kg)	Initial % composition of PGM	Spot price PGMs/toz (US$)	PGM price per kg (US$)	Initial qty of PMGs on WACC (kg)	Qty of PGMs recovered (kg)
Pt	500	0.05–1.5	859	26 696	0.25–7.5	0.25–7.46
Pd	500	0.09–2.0	2083	64 736	0.45–10.0	0.45–9.95
Rh	500	0.01–0.2	1400	43 509	0.05–1.0	0.05–9.95

Figure 5.26 Alloy tapping from the Mintek's 1.5 MW furnace. Source: Jones/University of the Witwatersrand, Johannesburg.

½ tonne of WACC to equal or greater than 1196 tonne per month. This can be achieved if the Mintek 1.5 MW furnace presented in Figure 5.26 is used, which can recover in excess of 99.5% PGMs. In the literature, Jones (2015), reported that in March 2009, the Mintek's 3 MW furnace smelted, 1196 ton, 3208 PGM oz. (4e) in alloy, 0.28 g/t PGM (4e) in slag and 99.7% of built-up recovery (i.e. Mass of PGM in alloy + Mass of PGM in slag) in a total of 15 000 ounces of PGMs in alloy form.

5.4.3.3 Analytical Instruments

The compositions of WACC, glass slag and alloy are measured by XRF analysis, and the platinum content is confirmed by ICP-MS, as discussed in Liu et al. (2021). All experiments or trail runs are repeated twice, and the mean values are used to calculate the recovery rate of Pt, Pd & Rh.

The X-ray diffraction (XRD, CuK radiation on PAN analytical B.V./X'Pert Pro diffractometer) is used to characterise the crystalline phases of WACC, the iron or copper alloy and glass slag. Microstructural characterisations of the alloy are measured by a scanning electron microscopy (SEM, Ultra Plus, Zeiss, Heidenheim, Germany) with an accelerating voltage of 15 kV, equipped with energy-dispersive X-ray spectroscopy

5.4.4 Business Ownership

This business is referred to as SC Chair of Energy Transition to Zero Emission (ETZE) planned for the Wits University. It is a planned partnership between SC and the University of the Witwatersrand, Johannesburg (Wits). It is to be domiciled in the School of Chemical & Metallurgical Engineering, but run administratively under the Wits business unit known as Wits Enterprise.

With the consortium proposed above, the project will run successfully to completion. However, the provision of two graduates from the Presidential Youth Employment Intervention (PYEI) Graduate Programme who are currently undergoing training in the Microporous Lab at Wits working on PEME and FC technologies is acknowledged. Although the term of these graduates at Wits has lapsed in December 2022, the graduates will continue to be engage in the project for continuity and because of the experience, they have earned on the project, which makes them handy in a proposed spin-off company till 2025.

5.4.5 Project Plan and Budget

The process flow for the project involves Can collection → De-canning → Separation → Pre-treatment → Smelting → converting → Collector removal → Refining → Packaging, with a US$32172 monthly turnover. The delivery of this project is broken down into six key milestones which broadly entails the manufacturing and testing of the recovered PGMs over 36 months. The author, an expert in the field, provides technical advice and supervision for the project. In total, undertaking this project will cost US$1.16 million including human resources, chemical and material costs and all overheard costs.

5.4.5.1 Project Milestones, Targets and Pricing

These financial projections are proposed for the partnership between SC and Wits in the PGM: Pt, Pd and Rh recovery. As mentioned above, the financial model assumes metal recovery of 99.5% metal sponge, sold off at prevailing spot prices. The metal selling prices for year 1 (FY1) during trial run and scale up level and subsequent years (FY1 – FY6) are listed in Table 5.13, which are obtained by averaging the Johnson Matthey price chart (http://www.platinum.matthey.com/prices/price-charts) London spot prices for the YTD. The price forecasts are extrapolated from daily metal spot price data collected over 5 years with an annual average basis for the forecast year.

The project is divided into the following six milestones/steps, and Table 5.14 provides the details in terms of each milestone, activities, deliverables and estimated cost.

Stage 1: Supplies of waste auto catalytic converter by SC
Stage 2: Establishment of funding mechanism and release of fund. Wits and SC will sign in place the memorandum of understanding (MOU) to include the actual funding timetable.
Stage 3: Test running and trial production for 0.2 kg per month.
Stage 4: Scale-up production level 1. This involves processing of ½ tonne of WACC per month whereby 3 times 5 kg per batch is processed in a day.
Stage 5: Scale-up production level 2. This involves processing 1200 tonne of WACC per month.
Stage 6: Distribution and marketing of PGMs: to Wits, SC and other identified users and markets.

5.4.5.2 Cost of Purchasing Feed Material, Financial Forecast and Projections

Although, the waste auto catalytic converters are supplied by SC, they would attract price as part of the contribution from them at a cost. The price therefore is based on the total mass of precious metals in the converters and an agreed discount to the spot price. It is assumed, for the projections, that WACC will be purchased at a 15% discount to the contained value of platinum and palladium and 20% discount to the contained value of rhodium. This estimate is informed by quotations obtained from existing returns and refunds to customers.

Table 5.15 summarises the 6-year financials for the project as per the detailed model drivers and assumptions presented in the calculations.

5.4.6 Project Feasibility

A PGM recovery business plan from WACC is submitted to SC, which involves combined smelting and collection process with different carrier WACC. Although, the PGM metals are recovered into Cu–Pt alloy, the clean slag, namely glass slag,

Table 5.13 PGM sales price forecasts.

For year	Price/g ($)	Price/g ($)	Price/g ($)	PGM average price (US$/kg)
FY1(2023)	856	2242	13 425	17 361
FY2(2024)	1091	2894	14 096	18 998
FY3(2025)	1099	3335	14 801	20 211
FY4(2026)	1108	3776	15 541	21 461
FY5(2027)	1116	4217	16 318	22 749
FY6(2028)	1124	4639	17 134	24 058

Table 5.14 Milestones and costs including electricity and water.

Milestone	Activities	Deliverables	GO/No GO Flag	Duration	Start date	Expected completion	COST (US$) incl VAT
Milestone 1: Supplies of waste auto catalytic converter	SC supplies de-canned waste auto catalytic converters to Wits and analysis	WACC including ones with Silicon carbide	Go: High quality WACC with PGM coating No Go: Low PGM coating or no PGMs at all	6 months	Mar 2023	Aug 2023	468 352
Milestone 2: Establishment of funding mechanism and release of funds	Signing of MOUs Travels & workshops	MOUs between (1) Wits and SC (2) Wits and Mintek	Go: Adequate funding for consultants, analysis & maintenance of equipment No Go: Full cooperation from MOU partners	4 months	Sep 2023	Dec 2023	10 724
Milestone 3: Test running and trial production	Smelting, refining and analysis of PGMs at Wits lab, materials & chemicals	Pt, 1.194–35.82 g Pd, 2.149–47.76 g Rh, 0.239–4.78 g	Go: Adequate funding No Go: Irregular or no fundi ng	2 months	Jan 2024	Feb 2024	207 143
Milestone 4: Scale-up production level 1	Smelting, refining and analysis of PGMs at Wits l lab, materials & chemicals & packaging design	Pt, 2.985–89.6 kg Pd, 5.373–119.4 kg Rh, 0.597–11.9 kg, and packaging	Go: Consistent supplies of WACC No Go: Lack of supplies and labour	12 months	Mar 2024	Feb 2025	207 143
Milestone 5: Scale-up production level 2	Smelting, refining and analysis of PGMs at Mintek & Wits lab, materials & chemicals & packaging	Quantity ≥ Pt, 0.60–17.85 tonne Pd, 1.07–23.80 ton Rh, 0.12–2.38 tonne and packaging	Go: Consistent supplies of WACC No Go: Lack of functional facilities at Mintek and loadshedding	10 months	Mar 2025	Dec 2025	207 143
Milestone 6: Distribution and marketing of PGMs	• Trial use of Pt on PEM fuel & electrolyser at Wits • Delivery of PGMs to SC & the markets	• Production of H_2 and electricity at Wits • PGM markets	Go: Compliance with government policies No Go: Poor or no appropriate security	2 months	Jan 2026	Feb 2026	57 672
Total required from SC							1 158 177

is generated simultaneously. With 80 wt% of cordierite carrier, flux (CaO/SiO_2) and WACC additions, a high platinum, palladium and rhodium recovery rate over 99.5% is successfully obtained by controlling the copper collector addition, CaO/SiO_2 mass ratio, smelting temperature and holding time at 10.5 wt%, of 0.5, 1550 °C, and 120 min, respectively. Compared with other treatments of single carrier WACC or sole hydrometallurgical method, the key advantage of this method is that it achieves platinum or metal recovery and glass slag synthesis from different WACC carriers by adding only copper powder and CaO. It therefore creates economic value by decreasing the amount of flux addition including SiO_2 and Al_2O_3, and increases the treatment types of WACC by treating two different WACC carriers simultaneously. The method also provides a reference for comprehensive treatment of different WACC carriers and different types of hazardous waste for future works.

The business plan also comprises of six milestones, which involve among others, test running and trial production for 0.2 kg WACC per month, scale-up production levels 1 and 2. These involve processing of ½ tonne of WACC per month whereby 3 times 5 kg per batch is processed in a day, while scale-up production level 2, involves processing 1200 tonne of WACC per month and distribution and marketing of PGMs to Wits, SC and other identified users. The project takes three years at the cost of US$1.16 million, the 1st processing stage incurs a loss of US$166466, which should be covered by the investment of about US$428954 in the first year. While the processes at the 2nd year of ½ tonne of WACC and 3rd year

Table 5.15 PGM recovery project: Six-year income statement projections (US$).

Item	Pax	Year 1	Year 2	Year 3	Total Cost	Year 4	Year 5	Year 6
Professional fees								
Professional consultation on Stack PGM recovery, sales, securities	1	24 129	26 059	28 144	78 331	28 650	29 166	29 691
Program administrator	1	10 724	11 582	12 508	34 814	12 733	12 963	13 196
Research associate	1	12 869	13 898	15 010	41 777	15 280	15 555	15 835
Post-doctoral fellow	1	9651	10 424	11 257	31 333	11 460	11 666	11 876
Sub-total		**1 070 000**	**1 155 600**	**1 248 048**	**3 473 648**	**68 124**	**69 350**	**70 599**
Postgraduate student Support/bursaries								
MSc Student Support (US$4021 per student)	4	16 086	17 373	18 762	52 221	19 100	19 444	19 794
PhD student support (US$4290 per student)	2	8579	9265	10 007	27 851	10 187	10 370	10 557
Sub-total		**460 000**	**496 800**	**536 544**	**1 493 344**	**29 287**	**29 814**	**30 351**
Equipment								
Replacement of crucibles and capacitors		7603	8211	16 423	32 238	16 423	16 423	16 423
Sub-total		**7603**	**8211**	**16 423**	**601 232**	**16 423**	**16 423**	**16 423**
Lab operating costs								
Research lab, facilities, chemicals, gases,		60 054	90 080	270 241	420 375	270 241	270 241	270 241
Trial use of Pt on PEM fuel & electrolyser at Wits		0	0	62 559	62 559	62 559	62 559	62 559
Delivery of PGMs to SC & the general markets		3217	3475	6949	13 641	6949	6949	6949
Utilities: electricity and water for 1.5 MWh/t at US$0.14 per kWh		21 487	1152	3	5 776 987	309 758	16 609	891
Sub-total		**1 201 487**	**3 663 300**	**10 173 333**	**15 038 120**	**545 487**	**545 487**	**545 487**
Other expenses								
Conference and travel		3217	3475	7505	14 197	7505	7505	7505
Office expenses		1609	1737	3752	7098	3752	3752	3752
Books, journals and patenting		4021	4343	4691	13 055	4691	4691	4691
Search and recruitment costs		4290	4633	10 007	18 929	10 007	10 007	10 007
Sub-Total		**13 137**	**14 188**	**25 955**	**53 279**	**25 955**	**25 955**	**25 955**
Total expense without WACC (US$)		**167 200**	**307 423**	**683 553**	**1 158 177**	**685 276**	**687 029**	**688 814**
Waste auto catalytic converter at US$55/kg		131	669 426	328 150		328 150	328 150	328 150

Table 5.15 (Continued)

Item	Pax	Year 1	Year 2	Year 3	Total Cost	Year 4	Year 5	Year 6
Grand total (US$)		167 332	976 850	1 011 703		1 013 426	1 015 179	1 016 964
Average quantity produced (kg)		0	45	108		108	108	108
Gross margin % at average sales of		0	0	2		3	6	8
Operating income at US$24059/kg		**0.14**	**1 082 647**	**2 650 320**		**2 676 303**	**2 754 254**	**2 806 221**
Net profit		**866**	**105 797**	**1 638 616**		**1 662 877**	**1 739 075**	**1 789 257**
Net profit/sales (%)		**166 465**	**0.54**	**3**		**3**	**3**	**3**

of 1200 tonne of WACC per month will yield a net profits of US$105797 (10%) and US$1638616 (62%), respectively, the subsequent financial projections into 4–6 year performances, give US$1662877 (62%), US$1739075 (63%) and US$1789257 (64%), respectively.

From these observations in terms of safety, environmental protection and economic feasibility and quick profitability, this project will breakeven in no time and therefore brings huge benefits to Wits, SC, all stakeholders and the county in general.

Finally, as mentioned in the preceding discussions and the subject on application of carbon footprint to climate change solutions that is discussed in detail in the next chapter, four zero-emissions technologies can be identified, which are bio/synfuel, hydrogen ICEs, hydrogen FC and battery electric have their different advantages and disadvantages. The CO_2 emissions generated in the process of producing electricity from hydrogen or synfuel can vary significantly, while BEVs are carbon neutral if charged solely with renewable power. Their use currently however, leads to high carbon emissions when charged with grid electricity in most regions, because of high carbon intensity of the global grid facility. Green hydrogen on the other hand can be produced from 100% solar and wind power and is delivered to any refuelling station, while the carbon intensity for biofuel and synfuel depends on the sourcing of biomass and carbon, respectively or both.

References

Abdin, Z., Webb, C.J., and Gray, E.M.A. (2016). PEM fuel cell model and simulation in MATLAB-Simulink based on physical parameters. *Energy* 116: 1131–1144.

Afolabi, A.S., Abdulkareem, A.S., Iyuke, S.E., and Pienaar, H.C.V.Z. (2012). Performance and kinetics of Pt–CNT catalyst electrodes in a PEM fuel cell. *J. Mater. Res.* 27 (11): 1497–1505.

Alekseeva, T.Y., Karpov, Y.A., Dal'nova, O.A. et al. (2019). Current state and problems of analytical control of spent automotive catalyst (review). *Int. J. Inorg. Mater* 54 (14): 1421–1429.

Atyabi, S.A., Afshari, E., Wongwises, S. et al. (2019). Effects of assembly pressure on PEM fuel cell performance by taking into accounts electrical and thermal contact resistances. *Energy* 179: 490–501.

Ayers, K. (2019). The potential of proton exchange membrane-based electrolysis technology. *Curr. Opin. Electrochem.* 18: 9–15.

Ayers, K., Capuano, C., and Anderson, E.B. (2012). Recent advances in cell cost and efficiency for PEM-based water electrolysis. *ECS Trans.* 41: 15–22.

Barbir, F. (2013). Chapter three-fuel cell electrochemistry. *PEM Fuel cell.* 2: 33–72.

Benson, M., Bennett, C.R., Harry, J.E. et al. (2000). The recovery mechanism of platinum group metals from catalytic converters in spent automotive exhaust systems. *Resour. Conserv. Recycl.* 31 (1): 1–7.

Bernardi, D.M. and Verbrugge, M.W. (1991). Mathematical model of a gas diffusion electrode bonded to a polymer electrolyte. *AIChE J.* 37: 1151–1163. https://doi.org/10.1002/aic.690370805.

Burkhard, R., Hoffelner, W., and Eschenbach, R.C. (1994). Recycling of metals from waste with thermal plasma. *Resour. Conserv. Recycl* 10 (2): 11–16.

Carlier, M. (2020). Global hydrogen fuel cell road vehicles by type. Energy & Environment. Source: https://www.statista.com/statistics/1291480/hydrogen-fueled-road-vehicles-worldwide/ (accessed 21 March 2023).

Chai, Z., Wang, C., Zhang, H. et al. (2010). Nafion-carbon nanocomposite membranes prepared using hydrothermal carbonization for proton-exchange-membrane fuel cells. *Adv. Funct. Mater.* 20 (24): 4394–4399.

Chandesris, M., Medeau, V., Guillet, N. et al. (2015). Membrane degradation in PEM water electrolyzer: numerical modeling and experimental evidence of the influence of temperature and current density. *Int. J. Hydrog. Energy* 40: 1353–1366. https://doi.org/10.1016/j.ijhydene.2014.11.111.

Chaojie, S., Yanghua, T., Jian, L.Z. et al. (2007). PEM fuel cell reaction kinetics in the temperature range of 23–120°C. *Electrochim. Acta* 52: 2552–2561.

Chen, J. and Huang, K. (2006). A new technique for extraction of platinum group metals by pressure cyanidation. *Hydrometallurgy* 82 (3–4): 164–171.

Colella, W.G., James, B.D., Moron, J.M. et al. (2014). Techno-economic Analysis of PEM Electrolysis for Hydrogen Production. Strategic Analysis, incorporated.

Committee on Climate Change (2015). Sectoral Scenarios for the Fifth Carbon Budget [Internet]. CCC; 2015 Nov. Available: https://www.theccc.org.uk/publication/sectoral-scenarios-for-the-fifth-carbon-budget-technical-report/

Compernolle, S., Wambeke, D., Raedt, D.I. et al. (2011). Direct determination of Pd, Pt and Rh in fire assay lead buttons by laser ablation-ICP-OES: automotive exhaust catalyst as an example. *J. Anal. Spectrom* 26 (8): 1679–1684.

Creutzig, F. (2016). Evolving narratives of low-carbon futures in transportation. *Transp. Rev. Routledge* 36: 341–360.

Decision No 406/2009/EC (2009). Decision no 406/2009/EC of The European Parliament and of The Council of 23 April 2009 on the effort of Member States to reduce their greenhouse gas emissions to meet the Community's greenhouse gas emission reduction commitments up to 2020. *Official J. Eur. Union* L140: 136–148.

DEFRA, DFT (2017). Air Quality Plan for Nitrogen Dioxide (NO2) in UK (2017) [Internet]. Available: https://assets.publishing.service.gov.uk/media/5bb71fdb40f0b664ed47487e/air-quality-no2-plan-directions-2017.pdf [Accessed 16/12/2023]

Dong, H.G., Zhao, J.C., Chen, J.L., and WuL, Y.D.B.J. (2010). Recovery of platinum group metals from spent catalyst: a review. *Int. J. Miner. Process* 145: 108–113.

Duraczynska, D., Bielanska, E., and Litynska, L. (2011). Hydrogenation of acetophenone in the presence of Ru catalysts supported on amine groups functionalized polymer. *Catal. Lett.* 141: 83–94.

EG&G Technical Services Inc. (2004). *Fuel Cell Handbook*, 7ee. Tennesse: U.S. Department of Energy.

European Commission (2017). Regulation Of The European Parliament And Of The Council Setting Emission Performance Standards for New Passenger Cars and for New Light Commercial Vehicles as Part of the Union's Integrated Approach to Reduce CO2 Emissions from Light-Duty Vehicles and Amending Regulation (EC) No 715/2007 [Internet]. Available: http://eur-lex.europa.eu/legal-content/EN/ALL/?uri=CELEX:52017PC0676&qid=1518371681956

eurostat (2017). Energy, transport and environment indicators. eurostat; 2017 Nov. Government HM. Climate Change Act 2008 [Internet]. 2008. Available: https://ec.europa.eu/eurostat/web/products-statistical-books/-/KS-DK-17-001.

Falcão, D.S. and Pinto, A.M.F.R. (2020). A review on PEM electrolyser modelling: guidelines for beginners. *J. Clean. Prod.* 261: 121184.

Fornalczyk, A. and Saternus, M. (2013). Vapour treatment method against other pyro-and hydrometallurgical processes applied to recover platinum from used auto- catalytic converters. *Acta Metall. Sinica* 26 (3): 247–256.

Grigoriev, S.A., Millet, P., Korobtsev, S.V. et al. (2009). Hydrogen safety aspects related to high-pressure polymer electrolyte membrane water electrolysis. *Int. J. Hydrog. Energy* 34: 5986–5991.

Grigoriev, S.A., Mamat, M.S., Dzhus, K.A. et al. (2011). Platinum and palladium nano-particles supported by graphitic nano-fibers as catalysts for PEM water electrolysis. *Int. J. Hydrog. Energy* 36: 4143–4147.

Guo, Y., Yang, J., Zhuang, J. et al. (2020). Selectively catalytic hydrogenation of styrene-butadiene rubber over Pd/g-C_3N_4 catalyst. *Appl. Catal. A: Gen.* 589: 1–9.

Han, B., Steen, S.M., Mo, J., and Zhang, F. (2015). *Electrochemical Performance Modelling of a Proton Exchange Membrane Electrolyser Cell for Hydrogen Energy*, 7007–7015. Elsevier.

Hill, G., Heidrich, O., Creutzig, F., and Blythea, P. (2019). The role of electric vehicles in near-term mitigation pathways and achieving the UK's carbon budget. *Appl. Energy* 251: 113111.

Hoffmann, J.E. (1988). Recovering platinum-group metals from auto catalyst. *JOM* 40 (6): 40–44.

Idibie, C.A., Abdulkareem, S.A., Pienaar, C.H. et al. (2010). Mechanism and kinetics of sulfonation of polystyrene-butadiene rubber with chlorosulfonic acid. *Ind. Eng. Chem. Res* 49 (4): 1600–1604.

Inglesi-lots, R. and Pouris, A. (2016). *On the Causality and Determinants of Energy and Electricity Demand in South Africa*, 626–636. South Africa: University of Pretoria, Department of Economics.

Iyuke, S.E., Mohamad, A.B., Kadhum, A.A.H. et al. (2003). Improved membrane and electrode assemblies for proton exchange membrane fuel cells. *J. Power Sources* 114 (2): 195–202.

Jones, R.T. (2015). Fundamental aspects of alloy smelting in a DC Arc furnace. PhD thesis. University of the Witwatersrand, Johannesburg, pp. 217–219. Available at: https://wiredspace.wits.ac.za/server/api/core/bitstreams/917c3d1f-273e-4711-ae05-c7b89d8dd40c/content

Kendall, K., Ye, S., and Liu, Z. (2022). The hydrogen fuel cell battery: replacing the combustion engine in heavy vehicles. *Engineering* https://doi.org/10.1016/j.eng.2022.11.007.

King, J. (2007). The King Review of low-carbon cars: part I: the potential for CO_2 reduction. 2007 Nov 7. Source: Environmental Science. https://www.semanticscholar.org/paper/The-King-Review-of-low-carbon-cars%3A-part-I%3A-the-for-King/fe41fdd93a92d9298495aecbaa22e3034dc4e7bc (accessed 21 March 2023).

Knights, S. (2012). *Polymer Electrolyte Membrane and Direct Methanol Fuel Cell Technology*, vol. 1, 137–177. Woodhead Publishing.

Kumar, S., Ramakrishna, S.U.B.R., Devi, B., and Himabindu, V. (2018). Phosphorus doped carbon nanoparticles supported palladium electrocatalyst for the hydrogen evolution reaction (HER) in PEM water electrolysis. *Int. J. Ionics* 24: 3113–3121.

Larbi, B., Alimi, W., Chouikh, R., and Guizani, A. (2013). Effect of porosity and pressure on the PEM fuel cell performance. *Int. J. Hydrog. Energy* 38 (20): 8542–8549.

Li, N., Araya, S.S., and Kær, S.K. (2021). Effect of current density, temperature and pressure on proton exchange membrane electrolyser stack. *World Acad. Eng. Technol.* 15 (1): 15–18.

Liu, C., Sun, S., Zhu, X., and Tu, G. (2021). Feasibility of platinum recovery from waste automotive catalyst with different carriers via cooperative smelting-collection process. *J. Mater Cycles Waste Manag.* 23: 581–590.

Luta, D.N. and Raji, A.K. (2019). Optimal sizing of hybrid fuel cell-supercapacitor storage system for off-grid renewable applications. *Energy* 166: 530–540.

Mouza, A.A., Peolides, C.A., and Paras, S.V. (1995). Utilization of used auto-catalytic converters in small countries: the Greek paradigm. *Resour. Conserv. Recycl.* 15 (2): 95–110.

Mufula, A.I. (2017). Production of Ion Exchange Membrane. PhD Thesis. University of the Witwatersrand, Johannesburg.

Nanaki, E.A. and Koroneos, C.J. (2016). Climate change mitigation and deployment of electric vehicles in urban areas. *Renew. Energy* 99: 1153–1160.

Ogiso, S. (2017). The Story Behind the Birth of the Prius, Part 1 [Internet]. Toyota: Toyota Motor Corporation; 2017 Dec 11 [cited 2022 Jul 1]. Available from: https://global.toyota/en/detail/20209700.

Paidar, M., Fateev, V., and Bouzek, K. (2016). Membrane electrolysis—history, current status and perspective. *Electrochim. Acta* 209: 738–753.

Petrukhina, N.N., Golubeva, M.A., and Maksimov, A.L. (2019). Synthesis and use of hydrogenated polymers. *Zhurnal Prikladnoi Khimii* 92 (6): 683–702.

PolymerScienceLearning (2022). SBS Rubber [Online]. Available at: https://pslc.ws/macrog/sbs.htm (accessed 21 October 2022).

Prieto-Prado, I., Del Río-Gamero, B., Gomez-Gotor, A., and Perez-Baez, S.O. (2018). Water and energy self-supply in isolated areas through renewable energies using hydrogen and water as a double storage system. *Desalination* 430: 1–14.

Renau, J., García, V., Domenech, L. et al. (2021). Novel use of green hydrogen fuel cell-based combined heat and power systems to reduce primary energy intake and greenhouse emissions in the building sector. *Sustain. Times* 13 (4): 1–19.

Sahu, A.K., Pitchumani, S., Sridhar, P., and Shukla, A.K. (2009). Nafion and modified-Nafion membranes for polymer electrolyte fuel cells: an overview. *Bull. Mater. Sci.* 32 (3): 285–294.

Sakai, S., Yoshida, H., Hiratsuka, J. et al. (2014). An international comparative study of end-of-life vehicle (ELV) recycling systems. *J. Mater Cycles Waste Manag.* 16 (1): 1–20.

Schafer, A. (2005). Structural change in energy use. *Energy Policy* 33: 429–437.

Serradilla, J., Wardle, J., Blythe, P., and Gibbon, J. (2017). An evidence-based approach for investment in rapid-charging infrastructure. *Energy Policy* 106: 514–524.

Shekhar, K. (2013). An investigation into the minimum dimensionality required for accurate simulation of proton exchange membrane fuel cells by comparison between 1 and 3 dimension models. Master dissertation, University of Cape Town.

Singh, M., Mukoma, P., North, B., and Majozi, T. (2020). A mathematical model for optimum design and synthesis of a hybrid electrolyser-fuel cell system: production of hydrogen and freshwater from seawater. *J. Clean. Prod.* 277: 123899.

Sinnott, R.K. (2005). *Coulson and Richardson's: chemical engineering design. Coulson and Richardon's Chemical Engineering.* Butterworth-Heinemann: Elsevier.

Smith, D.W., Oladoyinbo, F.O., Mortimore, W.A. et al. (2013). A microblock ionomer in proton exchange membrane electrolysis for the production of high purity hydrogen. *Macromolecules* 46 (4): 1504–1511.

South African Revenue Service (2022). Corporate Income Tax. [WWW Document]. https://www.sars.gov.za/types-of-tax/corporate-income-tax/ (accessed 14 Sep 2022).

Springer, T.E., Wilson, M.S., and Gottesfeld, S. (1993). Modeling and experimental diagnostics in polymer electrolyte fuel cells. *J. Electrochem. Soc.* 140 (12): 3513.

Staffell, I. and Kendall, K. (2012). Lower carbon cars by reducing dissipation in hydrogen hybrids. *Int. J. Low Carbon Technol.* 7 (1): 10–15.

Statista (2022). South Africa: Inflation Rate from 1987 to 2027. [WWW Document]. https://www.statista.com/statistics/370515/inflation-rate-in-south-africa/#:~:text=South%20Africa%E2%80%99s%20inflation%20has%20been%20quite%20stable%20for,its%20GDP%20through%20the%20services%20sector%2C%20especially%20tourism (accessed 14 Sep 2022).

Strielkowski, W. (2020). Renewable energy sources, power markets, and smart grids. In: *Social Impacts of Smart Grids*, 97–151. Elsevier Inc.

Thomas, D., 2018. Cost Reduction Potential for Electrolyser Technology. EY P2G Platform, Berlin.

Trading Economics (2022). South Africa Interest Rate. [WWW Document]. https://tradingeconomics.com/south-africa/interest-rate (accessed 14 Sep 2022).

Turton, R., Shaeiwitz, J.A., Bhattacharyya, D., and Whiting, W.B. (2018). *Analysis, Synthesis and Design of Chemical Processes*, 5ee. Pearson.

United Nations Environment Programme (2019). Emissions Gap Report 2019. Nairobi, Kenya.

Valentini, L. and Lopez-Manchado, M.A. (2022). Chapter 1 - Classification of rubbers and components for harsh environmental systems. In: *High-Performance Elastomeric Materials Reinforced by Nano-Carbons*, 1–14. Elsevier.

Wang, L. and Chen, M. (2013). Policies and perspective on end-of-life vehicles in China. *J. Cleaner Prod.* 44: 168–176.

Wang, Y., Pang, Y., Xu, H. et al. (2022). *PEM Fuel Cell and Electrolysis Cell Technologies and Hydrogen Infrastructure Development – A Review*. Energy & Environmental Science.

Wappelhorst, S. (2020). *The End of the Road? An Overview of Combustion-Engine Car Phase-Out Announcements Across Europe*. Berlin, Germany: International Council on Clean Transportation https://theicct.org/publications/combustion-engine-car-phase-out-EU.

Wee, J. (2006). *Applications of Proton Exchange Membrane Fuel Cell Systems*, 1720–1738. Korea University.

Xe Currency Converter (2023). 1 USD to ZAR-convert US Dollars to South African Rand. https://www.xe.com/currencyconverter/convert/?Amount=1&From=USD&To=ZAR (accessed 08 July 2023).

Xu, M. and Patton, D. (2022). China Sets Green Hydrogen Target for 2025, Eyes Widespread Use. Source: https://www.financialexpress.com/express-mobility/china-sets-green-hydrogen-target-for-2025-eyes-widespread-use/2469362/ (accessed 21 March 2023).

Yano, J.Y., Xu, G.C., Liu, H.P. et al. (2019). Resource and toxic characterization in end-of-life vehicles through dismantling survey. *J. Mater. Cycles Waste Manag.* 6 (6): 1488–1504.

Yoo, S. (1998). Metal recovery and rejuvenation of metal-loaded spent catalysts. *Catal. Today* 44: 27–46.

Zhao, H.Z. (1998). Analysis of phases in metallurgical products of plasma treating PGMs secondary resources. *Chin. J. Nonferr. Met* 8 (2): 314–317.

Zunita, M., Raizki, A., Aditya, R., and Wenten, G. (2022). Proton exchange polyionic liquid-based membrane fuel cell applications. *Results Eng.* 16: 100653.

Züttel, A., Remhof, A., Borgschulte, A., and Friedrichs, O. (2010). Hydrogen: the future energy carrier. *Philos. Trans. R. Soc. A Math. Phys. Eng. Sci.* 368: 3329–3342.

6

Application of Carbon Footprint to Climate Change Solutions

As mentioned previously in Chapter 1, forests and oceans play vital roles in regulating our climate, hence increasing the natural ability of forests and oceans to absorb carbon dioxide will help greatly in controlling global warming. As a result, the roles of governments in these efforts to control or stop climate change are to encourage governments and businesses to:

- **Allow fossil fuels to remain in the ground:** Fossil fuels include coal, oil and gas, which exist normally in the ground. Thus, the more they are extracted and burned, the worse climate change will get. Conversely, the more they are left in the ground or effectively returned to the ground, the environment will return to normalcy, as a result, all countries need to move their economies away from fossil fuels as soon as possible.
- **Invest in development and use of renewable energy:** Transiting from fossil fuels to clean and renewable energy sources such as solar, wind, wave, tidal and geothermal power and hydrogen will help with climate change solutions.
- **Adopting sustainable transportation:** Most means of transportation at the moment are petrol and diesel vehicles, planes and ships use fossil fuels, which of course pollute the environment with CO_2, CH_4 etc. Therefore, switching to electric vehicles and adopting hydrogen for plane and ship travel will not only help stop climate change but reduce air pollution.
- **Comfort at home:** Having government policies to help households heat homes with green technologies, e.g. by insulating walls and roofs and switching away from oil or gas boilers to heat pumps. With some changes in lifestyles such as avowing draughty and cold homes, which waste funds and are unnecessary during the winter.
- **Farming and vegan diets:** Agricultural producers, retailers and businesses should improve farming practices and provide more plant-based products to create the ease to help people shift to plant-based meals. Such practices will only help to stop or control climate change by reducing meat and dairy intakes or even going full vegetarianism.
- **Natural carbon sequestration:** Nature is endowed with biogenic and other physical mechanisms for cleaning up emissions from the atmosphere, land and sea, but these processes are overstrained with a higher rate of pollutants release. Therefore, there is need to plant more trees in the right places and encourage reforestation by rewilding schemes to begin with. This is because, as mentioned earlier in chapter 2, photosynthesising plants consume CO_2 as they grow, locking it away in roots and shoots while enriching the atmosphere with oxygen.
- **Advancing forests like the Amazon:** As alluded above, afforestation is a huge solution to climate change, so protecting and advancing similar projects like the Amazon around the world will resolve the problems of climate change to a very large extent. Thus, cutting down forests on an industrial scale for any reason will destroy giant trees that are sucking up carbon and cleaning the environment. The current belief in biofuels from biomass at industrial scale may not only compete with farmlands to grow food but would also result in deforestation depending on the rate of the Scheme 6.1, whereby companies destroy forests to make way for animal farming, soya or palm oil plantations etc.
- **Leveraging on the oceans:** Besides trees, the oceans absorb large amounts of carbon dioxide from the atmosphere, which helps to clean and keep our climate stable, however, many of these waters are overfished, used for oil and gas drilling and production, threatened by deep-sea mining and even rubbish and wastes in waters that lead to the oceans. Thus, protecting oceans will in turn save lives from climate change.
- **Plastic waste and dumping:** Most plastic products are made from fossil oil and gas, and the processes of extracting, refining and turning oil and gas into plastics involve heavy CO_2 release and hence carbon footprint. Plastics burned release greenhouse gasses, because they neither are biodegradable nor break down easily in nature. The demand for

Measuring Climate Change to Inform Energy Transitions: Carbon Footprint Calculations, First Edition. Sunny E. Iyuke.

$$\text{Afforestation} \underset{+CO_2}{\overset{-CO_2}{\rightleftarrows}} \text{Deforestation}$$

Scheme 6.1 Carbon foot printing.

Figure 6.1 Lagos, Nigerian largest plastics dump, where dozens of men gather to sort iron, plastic and nylon atop a 20-ft wall of garbage where hundreds of white herons compete to eat the spoils. Source: AP Photo/Sunday Alamba.

plastics is rising so quickly that creating and disposing of plastics will account for 17% of the global carbon budget by 2050 as required by the Paris Agreement (Tasoff 2019). The demand for plastics has also resulted in overwhelmingly waste dump and human scavengers in poor nations, such as Nigeria, as presented in Figure 6.1 and to feel that climate change problem is too big to solve (Mailonline 2022).

Unfortunately, plastics have surprisingly carbon-intense life cycles, and majority of plastic resins are produced from petroleum, which requires extraction and distillation. These resins are formed into products and transported into markets. These processes are known to emit greenhouse gases, either directly or through the energy required to produce them. Furthermore, the carbon footprint of plastics continues even after they have been disposed of. Most of the disposal methods used currently are dumping, incineration, recycling and composting and they all release carbon dioxide. However, increasing the percentage of bio-based plastics and application of renewable energy in plastics manufacturing could drive down emissions. Thus, while bio-based plastics are produced from plants, which capture atmospheric CO_2 as they grow, they are also renewable sources for bioplastics. It has also been argued that if bioplastics are composted, carbonaceous materials in the materials are released back into the atmosphere as CO_2. This makes the material itself carbon-neutral, although manufacturing still generates a small amount of greenhouse gasses.

6.1 Renewable Energy Consumption and Economic Growth

Global warming and climate change are inseparably the world's most topical issues in the current environment-economy dialogue. Thus, since CO_2 emission is the main greenhouse gas responsible for global warming and climate change, many scholars and policymakers are beginning to tune the relationship between climate change, energy production and sustainable economic growth. As a result, economists can use carbon footprint to measure the economy production possibilities and long-term economic growth (Ricci 2007). These relationships have been accounted for in the mathematical modelling presented earlier in Eq. (1.3). However, many researches in the economics literature investigate the relationship between economic development and environmental pollution, which again will be better quantified in Eq. (1.3). The results can be buttressed by the fact that environmental degradation in an economy initially increases with a nation's prosperity, but at some point, it will eventually decrease with increasing prosperity. It is also interesting to observe that the relationship between a nation's energy consumption and its economic growth can be modelled by the same Eq. (1.3), because the connection between energy consumption, economic growth and pollution emissions can be accounted for with sinusoidal curves of Figure 1.3. The results of these simulations can also be presented as a network depending on the country analysed, which transverse financial development, urbanisation and climate change.

Mongo et al. (2021) used an Auto-Regressive Distributed Lags (ARDL) model to study and analyse the impact of environmental innovations, the consumption of renewable energies, gross domestic product (GDP) per capita and the degree of economic openness on CO_2 emissions for 15 European countries over a 23-year period of time. They observed that in the long run, environmental innovations tend to lower CO_2 emissions, and that 'GDP', 'CO_2 emissions' or 'ARDL model' are the most frequently identified keywords in the network analysis of 1002 keywords (Magazzino et al. 2022), culminating into three clusters of CO_2 emissions, GDP, and energy consumption. Thus, in the comparative study of their impact on climate change, energy consumption is less important than CO_2.

6.2 Pyrolysis to Convert Solid Waste into Useful Energy

Pyrolysis is a process that involves thermal degradation of the waste in the total absence of air to produce recyclable products, including char, oil/wax and combustible gases, including synthesis gases of H_2/CO. Pyrolysis may be seen as an attractive alternative to incineration for solid waste or municipal solid waste (MSW) disposal that allows energy and resource recovery as end products. It has been used to produce charcoal from biomass for many years. If it is used for waste management, MSW can be turned into fuel and safely disposable substances (char, metals, etc.), and the pyrolysis process conditions can be optimised to produce either a solid char, gas or liquid/oil product, namely, a pyrolysis reactor acts as an effective waste-to-energy convertor (Chen et al. 2014). When compared to the conventional incineration plant, which runs in capacity of kiloton per day, the scale of pyrolysis plant is more flexible.

Since treatment, management and disposal of municipal solid MSW is a common problem in every country, MSW pyrolysis can be considered as an innovative alternative for treating MSW. That obtains different chemicals and fuels (Schaefer 1975; Malkow 2004). Very useful energy can be obtained in a cleaner way from pyrolysis than from conventional MSW incineration plants as lower amounts of nitrogen oxides (NO_x) and sulphur oxides (SO_2) are produced as a consequence of the inert atmosphere in the pyrolysis processes and the opportunity to obtain syngas as raw material for electricity generation, methanol and gasoline production. Apart from reduced gas emissions, pyrolysis produces better quality of solid residues in a typical treatment of MSW (Saffarzadeh et al. 2006).

In 2014, Chen et al. (2014) expressed the sentiment that MSW pyrolysis is receiving increasing attention in small cities and towns due to the desire to prevent long-distance transportation; and it is also demanded in big cities as a distributed MSW treatment method due to the increased difficulty in finding new sites for incinerators and landfills. Figure 6.2 therefore presents typical pyrolysis reactors and raw materials. While, Figure 6.2a,b displays the use of pyrolysis reactors for production of charcoal from wood and biomass as earlier highlighted above. They are not meant to support deforestation (Figure 6.2 a and Scheme 6.1) and indiscriminate release of CO_2 and other GHG emissions (Figure 6.2b), and where uncontrolled involvement in charcoal production, as they hardly appreciate the implications of uncontrolled felling of trees and burning of logs for charcoal. Figure 6.2c on the other hand, the Honghoo technology is a 'pure' pyrolysis process with multi-section rotary reactors, which run under temperatures from 300 °C in section I to 500 °C in section III. This demonstration plant is fed with unsorted MSW, while syngas, oil and charcoal are the outputs as products (Chen et al. 2014).

Chen et al. (2014) also draw up the conclusions that adaptability of different reactors to MSW pyrolysis is summarised in Table 6.1, and predict that rotary kiln will still serve as the main reactor type for MSW pyrolysis. However, for small and moderate scale, tubular reactors can be competitive and the operation temperatures are dependent on product choice.

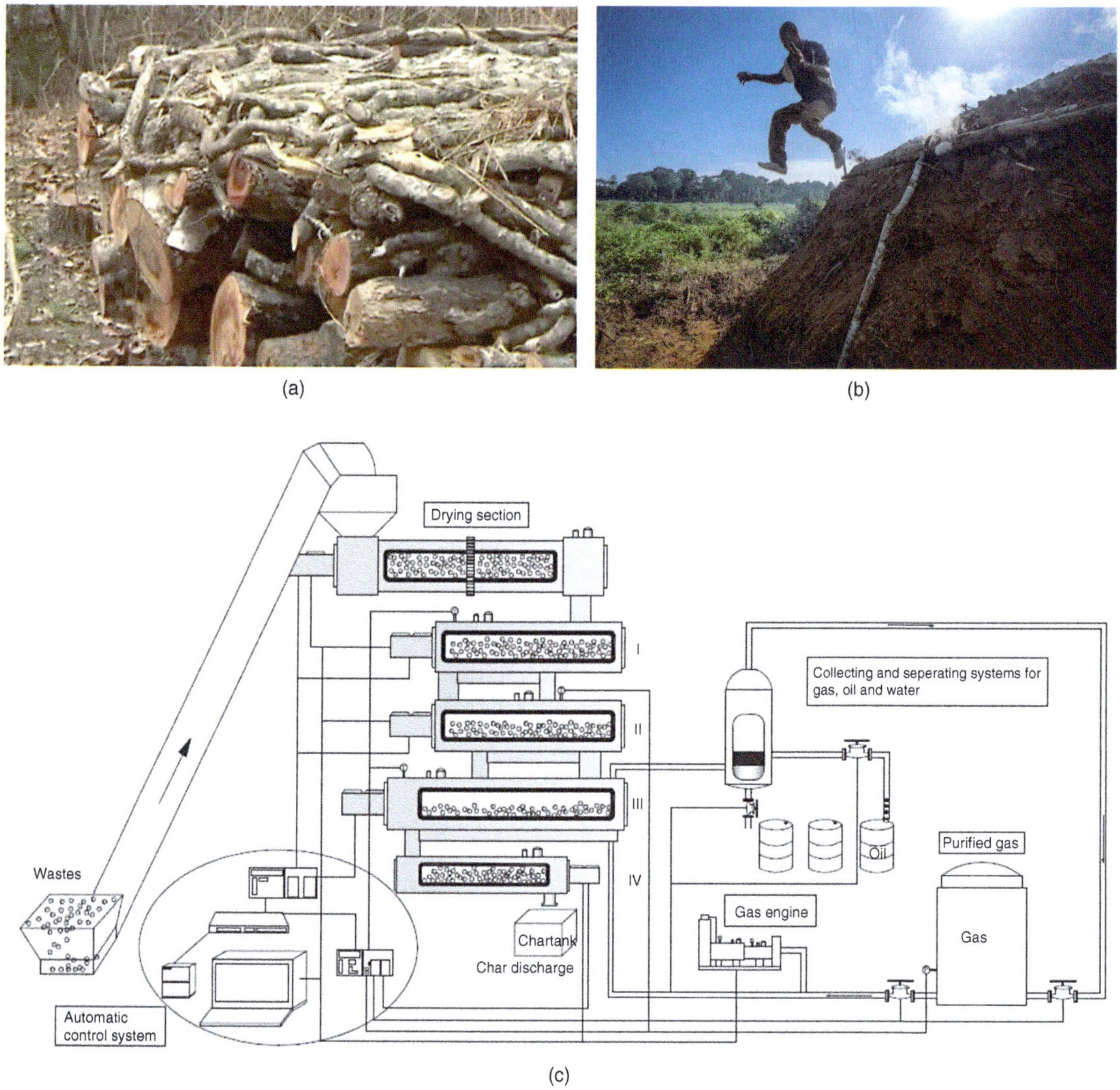

Figure 6.2 Typical pyrolysis reactor and raw materials, (a) Trees cut into woods stacked into earth pit kiln. Source: Adeniji et al./IISTE/CC BY 3.0. in New-Bussa, Nigeria, (b) Solutions 'beyond the kiln', Kiln near Yangambi, Democratic Republic of Congo. Source: Schure et al./Cirad/CC BY 4.0, (c) Honghoo technology, depicted as sections I, II and III are the first, second and third sections of the pyrolysis reactor, respectively, and IV is the char cooling drum. Source: Adapted from Chen et al.

6.3 Factors Affecting Pyrolysis of Used Tyres and Plastics with Respect to Products

MSW is highly heterogeneous as it consists mainly of paper, cloth material, yard waste, (including fallen leaves and branches, etc.), food wastes, plastics and a small amount of leather and rubber, metals, glass, ceramic, earthen materials and miscellaneous, etc. The fractions subjected to pyrolysis would normally include paper, cloth, plastics and yard wastes. Therefore, more heterogeneous in composition and size when compared to waste types such as tyres, plastics and sewage sludge even though these trios may constitute more than 60% of MSW as seen from Figure 6.1. However, for MSW pyrolysis, information on technology development, characterisation of products and correlated pollution might not have been sufficient to support technology application and system design, especially with regards to environmental impacts, which is the fundamental basis for a one-shot MSW pyrolysis application. Even though many studies and reviews have been conducted on pyrolysis of waste tyres and plastics in terms of reactor development and product characterisation,

Table 6.1 Adaptability of different reactors to MSW pyrolysis (Chen et al. 2014).

Reactor-type application recommendation	Running experiences	Requirements on material preparation	Capacity	Maintenance requirement	Flexibility to operation parameters' change	Application recommendation
Fixed-bed reactor	Running in batch, only in laboratory researches	Almost no requirements except for energy saving purpose	Small, may not exceed several tons per day	Low, but batch-by-batch operation demanding manual labour	Excellent flexibility	Not recommended for industrial application
Rotary kiln	Most common	Not rigid	Large, up to 150 000 tpa	Low to moderate	Good flexibility	Recommended but efficiency should be improved; multi-sectional reactor suggested
Fluidised-bed reactor	Only for laboratory researches, no practical experiences for MSW	Very rigid	Large theoretically	Highest	Very limited to size change; good flexibility to temperature, etc.	Heating system should be improved before application to MSW pyrolysis
Tubular reactor	A few	Rigid	Moderate, up to 50 000 tpa	Moderate-to-high	Limited to size and temperature change	Recommended, especially the multi-sectional tubular reactor

conditions for oil production, oil characteristics and upgrading of the heating rate and other governing variables affecting pyrolysis, they have always been isolated from the bulk MSW. Therefore, the pyrolysis products of industrial plastics and tyres (Martínez et al. 2013), the mechanism and kinetics modelling of their pyrolysis are available in the open literature (Al-Salem et al. 2010; Quek and Balasubramanian 2012). About municipal wastes, the pyrolysis of sewage sludge for liquid production has also been addressed for decades (Fonts et al. 2012).

Numerous studies have been performed in laboratories to determine pyrolysis behaviours and product properties of individual MSW fractions, such as paper, plastics, fabric, lignocellulosic materials and putrescibles (Di Blasi 1996; Miskolczi et al. 2013), refuse-derived fuel pellets (Cozzani et al. 1995; Garcia et al. 1995a,b; Lin et al. 1999; Buah et al. 2007; Bosmans et al. 2013) and real MSW (Islam and Beg 2004; Miskolczi et al. 2013). Therefore, there exists interactions between the different individual fractions and the pyrolysis products (Williams and Williams 1997; Ding et al. 2013).

According to Chen et al. (2014), the reactions that take place in a recognised pyrolysis process can be expressed as:

$$C_xH_yO_z + Q \rightarrow \text{Char} + \text{Liquid} + \text{Gas} + H_2O \tag{6.1}$$

where Q is the heat required for the chemical reactions to take place, which includes the following three portions:

6.3.1 Moisture Vaporisation, Q_1

During pyrolysis, the feedstock does not undergo thermal decomposition before its moisture is vaporised, and Q_1 can be calculated as:

$$Q_1 = 1/4\,\mathrm{W} \times 2260;\ \mathrm{kJ/kg} \tag{6.2}$$

where W is the water content of the feedstock to the reactor, %; therefore, to reduce this part of energy, MSW components with high moisture content, such as food wastes, biomass are suggested to be separate before pyrolysis. Furthermore, in order to reduce this portion of energy, a drying step is usually required before the pyrolysis reactor.

6.3.2 Caloric Requirement of Pyrolysis Q_2

According to Raveendran et al. (1996), the caloric requirement of the pyrolysis can be calculated using the following equation:

$$Q_2 = C_{\text{p,M}} \int m_{\text{M}} dT + C_{\text{p,ch}} \int m_{\text{ch}} dT + C_{\text{p,v}} \int m_{\text{v}} dT + Q_{\text{p,}} \text{ kJ}^{-1} \tag{6.3}$$

where $C_{\text{p,M}}$, $C_{\text{p,ch}}$ and $C_{\text{p,v}}$ are the specific heat capacities of the dry materials, char and volatile generated, J/kg/°C; m_{M}, m_{ch} and $m_{\text{p,v}}$ are the mass ratio of the dry materials, char and volatile to the feedstock, %; the former three items at the right side of Eq. (6.3) are the energy required for heating the feedstock and the products to the pyrolysis temperature; and Q_{p} is the reaction heat of the pyrolysis, J/kg. Q_2, can be obtained through differential scanning calorimetry (DSC) or differential thermal analysis (DTA) techniques for different MSW components (He et al. 2006). It has also been found that for reactor design purpose Q_{p} can be neglected and the Q_2 is mainly decided by pyrolysis temperature (Boukis et al. 2007; Wang et al. 2012). The higher pyrolysis temperature does not only result in a higher energy requirement but also cause challenge to the safety of reactors.

6.3.3 The Radiation Loss from the Reactor to the Surrounding Q_3

Since heat is transferred from outside to the inner side of the reactor, Q_3 can be ignored for heat transfer surface design; but on the contrary, when the reactor is heated from the opposite side, the heat transfer surface should bear this portion of energy to maintain the desired temperature of the reactor. This radiation loss, Q_3 can however be reduced with better insulation.

Higher heating rate (HR) corresponds to a higher volatile matter, such as tar and gas yield, which results in less quantity of char (Yi 2007; Velghe et al. 2011). With longer residence time, a higher yield of syngas is obtained due to cracking of tar. While the volatile matter is extracted and cooled, a higher yield of liquid products can be obtained (Font et al. 1995a,b).

Theoretically, the HR can be calculated as:

$$\text{HR} = \frac{\Delta T.\alpha}{m.C_{\text{p}}}, \text{°CS}^{-1} \tag{6.4}$$

where ΔT is the temperature difference between the reactor wall and the feedstock, °C; α is heat transfer coefficient inside the reactor, W/m^2 °C; m is the mass of feedstock heated by per m^2 of heat transfer surface (reactor wall), kg/m^2; C_{p} is the specific heat of the feedstock, J/kg °C.

As seen from Eq. (6.4), HR is decided by ΔT, α and the mass load of the reactor. At laboratory research level, the HR can be set by adjusting ΔT or mass fed to the reactor to realise slow or flash pyrolysis. In a practical and continuously operating reactor, HR is decided by the reactor type, which is characterised by its special temperature difference (ΔT) and heat transfer coefficient (α) features. However, the pyrolysis reactors reported in the literature are mainly fixed-bed reactors, rotary kilns and fluidised-bed reactors (Chen et al. 2014).

In the preceding discussion, it has been reported that a variety of pyrolysis studies and reviews have been conducted on industrial wastes such as tyres and plastics both on the characterisation of the development of pyrolysis technologies, but detailed research is reported on biocrude development and characterisation as a case study (Osayi 2015) in the next section.

6.4 Case Study: Production and Characterisation of Biocrude from Used Tyres and Natural Rubber

(Reprinted with permission from Dr. Osayi Ilawe Julius of the Wits University)

Thermal pyrolysis is one of the viable thermochemical technologies in the management of polyisoprene-based solid waste, which is a major global challenge. This is due to the non-biodegradability of such materials and their continuous usage and increase in demand across all segments of man's daily activities. One of the major advantages of pyrolysis process is the effectiveness in converting such materials under controlled process conditions to optimise the fraction of interest, such as the liquid fraction, also known as biocrude with a near zero environmental pollution impact. Hence, in this study, the production of biocrude from used tyres (UT) and natural rubber (NR) was performed using thermal and catalytic pyrolysis. The operating temperature range from 375 to 750 °C at an interval of 75 °C and feed material particle sizes of 2, 4, 6, 8 and 10 mm at a HR of 15 °C/min was studied. The optimum biocrude yield of 34.40 ± 2.18 and 75.93 ± 2.88 wt.% at a feed particle size of 6 mm was obtained for UT and NR at 600 °C, respectively. The use of Lawani Benin River Nigeria Kaolin, LBK in the synthesis of zeolite NaY using the hydrothermal method at a crystallisation time and temperature of 9 hours and 100 °C was a success. The scanning electron microscopy (SEM) morphology and X-ray diffraction pattern obtained for the synthesised zeolite at an ageing time below one hour was well comparable with the commercial zeolite NaY catalyst used as standard. But the presence of the catalyst in the pyrolysis of both materials gave a yield of 21.80 ± 2.42 and $65.87 \pm 0.73\%$ for UT and NR, which was lower than their thermal pyrolysis yield. The biocrude produced had physical properties comparable with that of conventional fossil fuel, such as diesel and gasoline. The chemical characterisation revealed both biocrude composition to be a complex mixture of aliphatic, aromatics, polycyclic aromatic hydrocarbons (PAHs) and other oxygen, nitrogen, sulphur and chlorinated compounds in small proportions. The non-catalysed and catalysed NR biocrude gave 79.87 and 65.76% of aliphatic, 12.45 and 14.84% aromatic and PAHs concentration of 1.98 and 0.79% while the non-catalysed and catalysed used tyres biocrude had 42.20 and 32.09% of aliphatic, 34.49 and 38.80% aromatic and PAHs concentrations of 17.84 and 22.82%, respectively. The kinetic of the thermal degradation study with the aid of a thermogravimetry and differential thermogravimetry analyser was performed at 15, 20 and 30 °C/min and results showed that an increase in the HR led to an increase in the activation energy and the pre-exponential factor for both UT and NR. Nevertheless, NR displayed higher activation energy with the HRs than UT, an indication that NR was more difficult to thermally decompose than UT. The distillation temperature range of the distillates was within the range of conventional petrol and diesel. The composition of the distillates revealed carbon chain length of C_5—C_{30} with the majority being C_8—C_{10}. A spark ignition generator engine was used to perform the combustion test for the various biocrude distillates and petrol blend in the ratio 0, 5, 10, 15 and 20% successfully without engine modification. For the fuel consumption with respect to generator run time, it was observed that an optimum of 20% natural rubber biocrude distillates-Petrol blend gave comparative fuel consumption behaviour than the others with that of neat petrol. Furthermore, the same composition of distillates gave optimum fuel consumption and power. Hence with further treatment of the biocrude distillates, they could serve effectively as chemical feedstock for industries, or as substitutes for fossil fuel.

6.4.1 Background and Motivation for the Study

The need to meet global energy demand with respect to supply has become a major challenge due to industrialisation and rise in number of road vehicles. This has resulted in price fluctuation for fossil fuels and their products (Kumaravel et al. 2016; Rengga et al. 2015; Martinez et al. 2014). Furthermore, environmental issues and human health hazards arising from the application of fossil fuels have initiated the necessity to explore and develop different sources of renewable energy, which have several advantages.

Frigo et al. 2014; Undri et al. 2014; Hu et al. 2014). Hence, efforts are being made by researchers to harness available technologies in the conversion of different sources for renewable energy (Osayi et al. 2014; Onoji et al. 2016; Quek and Balasubramanian 2013; Liu et al. 2016; Boxiong et al. 2007). Biomass material conversion into hydrocarbon fuel is one of such promising options (Bi et al. 2010; Brammer and Bridgewater 1999; Jahirul et al. 2012). Even though biofuel production from edible vegetable oil through reaction method of esterification and transesterification has been given consideration, high cost of materials and the process leading to a food crisis is a setback. Therefore, research in this area has been re-directed to biofuel production with the use of non-edible vegetable oils (Onoji et al. 2016; Jahirul et al. 2012; Islam et al. 2008; Aguado et al. 2005).

There are various means of converting biomass and non-degradable solid waste to fuels. An alternative option of such conversion is the thermo-chemical process. Several researches on the use of agricultural, industrial and urban solid waste for liquid fuel production have been investigated. Vehicle tyres are an example of such urban solid waste (Brammer and Bridgewater, 1999; Jahirul et al., 2012). After serving their purpose, tyres are disposed of as solid waste. UT make up one of the most problematic hazardous solid wastes in developing countries (Weisz et al. 1979; Bajus and Olahova 2011). In South Africa, more than one hundred and sixty tonnes of waste tyres are generated on a yearly basis (Mastral et al. 2000). This has resulted in the national government formulating a policy of fee charges of about \$223.20 per tonne, including 14% value-added tax on new tyres, introduced by tyre manufacturers and importers into the country (Molewa 2013; Pilusa et al. 2013).

Difficulty in the disposal of this polymer material waste arises from its non-biodegradable nature, accompanied by the challenges of its negative impact on the environment and human health (Osayi et al. 2014). Irrespective of other ways of recycling waste tyres, such as grinding, reclaiming, retreading and so forth being used, these methods do not completely utilise UT as an excellent material for energy recovery (Onoji et al. 2016; Liu et al. 2016). Thus, interest in pyrolysis has been renewed since the optimisation of the process can help in deriving high-value products (Berrueco et al. 2005;

Williams 2013; Harrison and Ross 1996; Bhatia et al. 1993; Fix 1980; Phadke et al. 1983). Basically, pyrolysis encompasses the thermal degradation of organic materials, such as polyisoprene, to low hydrocarbon weight components at high temperature (250–900 °C) in an inert atmosphere (Isayev et al. 1996). Apart from the possibility of performing pyrolysis under atmospheric or reduced pressure, it is also an excellent route for energy recovery. The technique is environmentally friendly and can thermally decompose a wide range of waste, including UT (Schnecko 1998; Midgley and Henne 1929; Beecham 2008; Donatelli et al. 2010; Kar 2011; Chang 1996; De Marco et al. 2001; Yang et al. 1995; Conesa et al. 1998; Senneca et al. 1999; Domínguez et al. 2008; Zhao et al. 2010; Balat et al. 2009). Pyrolysis basic products include liquid fraction (pyrolytic oil), solid residue (char) and gases. Solid residue can find application in the rubber industry as reinforcement agent, activated carbon production, or as solid fuel (with caution due to its sulphur content). The pyrolytic oil or liquid fraction can serve as an additive or substitute for conventional fossil fuel. This is based on its high heating value (range of 41–44 MJ/kg). Similarly, pyrolytic oil can be a source of chemicals that are required for various industrial purposes. The gas fraction can serve as heat energy for the operation of the pyrolysis process (Bridgwater 2007; Bridgwater et al. 2001; Aguado et al. 2002; Cornelissen et al. 2008).

Although pyrolysis of UT has been reported in the literature, there is a need to explore other means of improving the yield and composition of the liquid fraction from tyre pyrolysis. Apart from the pyrolytic oil (liquid fraction) being used as 'dirty' fuel as a result of the presence of tars, sulphur and other inorganic materials, it should be upgraded and refined both as a source of fuel, feedstock for a wider industrial application and for higher chemical value products. This will help increase the market for the pyrolysis products, thereby making the process more economically feasible.

The major component of NR also referred to as *latex rubber* (a term used in order to distinguish it from synthetic rubber produced from fossil fuels) obtained from the species of plants known as *Hevea brasiliensis* (Association of Natural Rubber Producing Countries 2010; Charles 2002) comprises mainly polyisoprene organic compounds, inorganic materials and water. This suggests that NR, which has found application in different areas of man's activities, especially in the plastic and automobile industries (Haydary et al. 2008; Zhang et al. 2008; Pradhan and Singh 2011; Trongkaew et al. 2011), but has not been reported as a source of liquid fuel or pyrolytic oil in literature can be investigated with the pyrolysis method. Hence, the main focus of the research study is to utilise UT, a known source of increasing solid waste currently in Africa with focus on the Sub-Saharan Africa regions as pyrolysis feed material for pyrolytic oil production. The catalyst to be used is zeolite NaY, which will be synthesised using Lawani Benin River Nigeria kaolin (LBK) obtained from Benin City, Edo state Nigeria as precursor for silica and alumina. This will be of great value as this will help increase the market value of such clay due to the many industrial applications of zeolites (Kovo 2010). Characterisation and upgrade of the pyrolytic oil will be performed so as to make it suitable as fuel for internal combustion engines (ICE).

6.4.1.1 Hypothesis

UT and NR are made up of heavy molecular weight hydrocarbon compounds, which can be depolymerised to low molecular weight ones under controlled conditions using a pyrolysis process. It is therefore hypothesised that pyrolysis of NR will produce higher yield of low molecular weight hydrocarbon such as pyrolytic oil than UT.

6.4.1.2 Justification of the Study

In recent years, research on the pyrolysis of UT and other biomass for obtaining different products of interest as highlighted above. Nevertheless, investigation of NR as a source of liquid fuel is yet to be reported in literature; while UT pyrolysis is an area that still needs to be explored and developed in Africa. From available reports, almost 28 million of the UT generated in South Africa on a yearly basis are either burnt or discarded in unauthorised locations (Pilusa et al. 2013). An annual increase of 9.3 million has been projected for this figure (Pilusa et al. 2014).

There is also a rising demand for products made from polymeric materials due to their excellent properties in serving human needs. This has led to an increase in polyisoprene based non-biodegradable solid waste, which has become a negative environmental issue and human health challenge. Hence, knowledge of this project using South African UT for production of pyrolytic oil is of interest in managing such solid waste and turning them into a means of renewable energy while reclaiming and sustaining the environment. This can create a window of opportunity in reducing the unemployment rate as it can lead to job creation when the process is implemented on commercial scale. The success of NR as a means of liquid fuel will give an economic boost and help revamp the farming of NR in countries like Nigeria where such plants are grown. Also, the possibility of LBK as a natural material source of synthesising zeolite NaY will increase the economic value of such material.

6.4.1.3 Research Aims and Objectives

The main aim of this project is to compare the pyrolysis of UT and NR. The accomplishment of aim will be achieved by the following objectives:

- To produce pyrolytic oil from the catalytic pyrolysis of UT and NR using the synthesised zeolite NaY (LBK will be used as natural source for silicon and aluminium silicate in synthesising the zeolite).
- To optimise the pyrolytic oil production process from both UT and NR
- To characterise and upgrade the pyrolytic oil to attain commercial standard as chemicals and substitute fuel energy source.
- To determine the effect of different blends of pyrolytic oil distillates (PD) from UT and NR with conventional gasoline on the fuel consumption and fuel power in an ICT.

The schematic representation of the objectives is shown in Figure 6.3.

6.4.1.4 Report Outline

The report is presented in sections, viz (1) introduction, which contains research background and motivation, hypothesis, justification of the study, research aims and objectives. (2) Methodology, which includes details of the experimental procedures. It also explains the operating conditions of the equipment used in the synthesis of zeolite NaY catalysts, characterisation of the material samples, the analyses and distillation of the produced pyrolytic oils, the properties and performance of the pyrolytic oil distillate/commercial gasoline blend in an ICE. (3) Results and Discussion, where tables, pictures and figures were used to depict the results, which were interpreted and discussion with comparisons to previous research studies. (4) Conclusion and Recommendations. This section provides a summary of the main findings, overall conclusions drawn from the results and recommendations for future work.

6.4.2 Methodology and Experimental Procedure

6.4.2.1 Materials and Procedure Adopted

Zeolite NaY catalyst, car UT and NR (also known as latex rubber) are the raw materials used for the catalytic pyrolysis process. South African made car tyre is collected from a selected, dump site in Johannesburg, South Africa; while the natural rubber lump is sourced from a rubber tree plantation located at Iyanomo village near Benin City, Edo state, Nigeria. The materials are washed to remove debris, stones and sand. The NR and UT samples are cut into different sizes of 2, 4, 6, 8 and

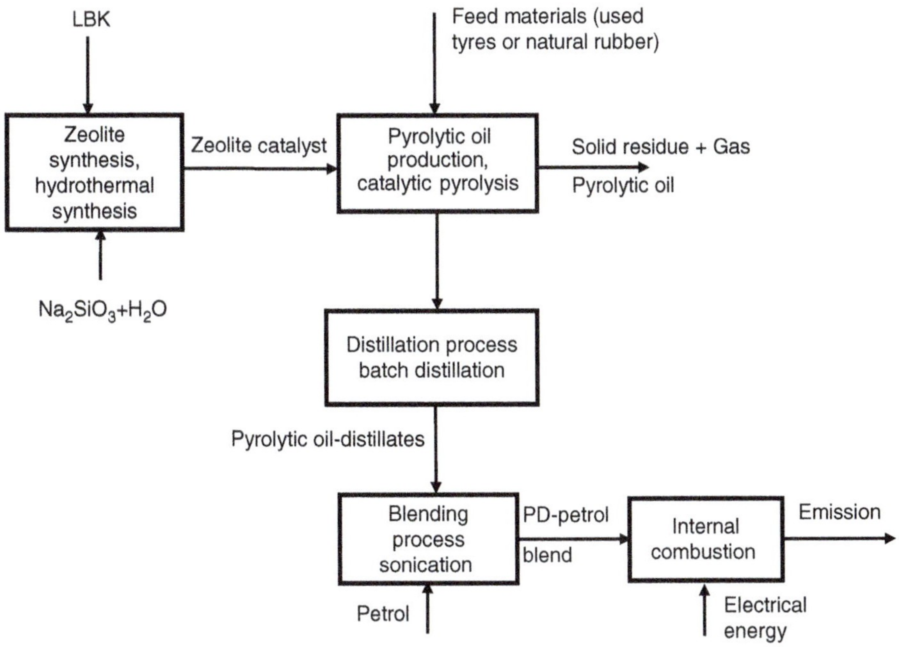

Figure 6.3 Work flowchart of objectives/activities of the study.

10 mm. The chopped pieces represent the UT, which do not contain any wire cord but the fabrics thread. Before pyrolysis, both the NR and UT sample materials are taken for proximate and ultimate analysis.

The zeolite NaY catalyst used for this study is produced in the laboratory of the School of Chemical and Metallurgical Engineering, University of the Witwatersrand by the hydrothermal method. Kaolin (clay) obtained from Lawani River located in Benin City Edo state Nigeria (LBK) serves as precursor for the aluminium and silicon oxides. The produced catalyst is characterised according to an international standard. In order to achieve the aims of this study, the following steps are followed:

- Firstly, the zeolite NaY catalyst is synthesised followed by the production of pyrolytic oil from UT, NR using pyrolysis (with and without catalyst).
- Secondly, the obtained pyrolytic oil from both catalytic and non-catalytic pyrolysis are characterised by physical and fuel properties.
- Thirdly, the pyrolytic oil from the first step is distilled using batch distillation to obtain PD.
- Fourthly, the PD from step three are blended with petrol and then tested in an ICE for fuel consumption and fuel power study.

The grade 95 unleaded petrol used in this study is bought from a Caltex filling station, Braamfontein, Johannesburg – South Africa.

6.4.2.2 Zeolite NaY Synthesis and Characterisation

The raw Lawani River Benin Nigeria Kaolin (LBK) is refined using sedimentation method to remove unwanted and foreign materials such as quartz, which was trapped within the mineral aggregates. The clay is spread on an oven pan and placed in the oven at a set temperature of 50 °C. After 2 hours and 30 min, the oven temperature is raised to 110 °C and maintained at that temperature for approximately 4 hours. The oven pan is then removed and placed in a desiccator and allowed to cool. This is done so as to help reduce the water content in the clay as much as possible before it can be crushed to the desired size and calcined in a furnace. Crushing of the dried clay is done in a ceramic mortal using a ceramic piston. The grounded clay is then sieved to obtain a particle size of 38 μm. A pictorial description of the process is depicted in Figure 6.4.

The ground clay of 38 μm size was placed in a crucible and loaded in a furnace for calcining at 600 °C at a HR of 10 °C/min and kept constant for 2 hours. It was then removed and placed in a desiccator to cool. According to Franco et al. (2003), this would help to remove the hydration water, better disordering of the kaolinite molecule structure and formation of a non-crystalline phase. At this stage, the metakaolin can be said to be an amorphous material comprising of free silica and free alumina (Brindley and Nakahira 1958). The calcined LBK also referred to as LBK metakaolin is characterised with the following equipment: XRF, XRD and SEM.

The synthesis of zeolite NaY from the LBK metakaolin is performed using the aluminosilicate molar ratio $15Na_2O$:Al_2O_3:$15SiO_2$:$450H_2O$ and hydrothermal method (Chandrasekhar and Pramada 2004; Kovo 2010). Sodium hydroxide pellets (99%w/w) and anhydrous sodium metasilicate (Sigma-Aldrich) which served as a source of alkaline medium and additional silica were both supplied by Sigma-Aldrich. LBK Metakaolin served as a combined source of alumina and silica. 28.59 g of deionised water was measured and divided into two equal parts. 0.60 g of sodium hydroxide pellets was dissolved in the first halve of the deionised water, this was followed by dissolving 1 g of LBK metakaolin. 5.45 g of anhydrous sodium metasilicate was measured and added to the second halve of the deionised water. Both sets of precursors were subsequently mixed together to form a homogenous aluminosilicate gel. The gel was then aged with continuous stirring at room temperature. The different ageing times investigated were 0, 0.5, 1, 3, 6, 9, 12 and 24 hours. The aged gel was crystallised for 9 hours at 100 °C in a Teflon-lined stainless steel autoclave. At the end of the hydrothermal treatment, the sample was taken out, filtered and washed with deionised water until the pH was between 7 and 8. The washed sample was subsequently dried overnight at 60 °C. The products obtained were characterised with XRD, SEM and TEM.

6.4.2.3 Physical Characterisation of the Raw Used Tyre and Natural Rubber Samples

Gross calorific values, proximate and ultimate analyses were used to study the physical characterisation of the UT and NR raw material samples. The DryCal modular calorimeter, model 2010, was used to obtain the gross calorific svalues (GCV), the Perkin Elmer precisely STA6000, Shelton USA simultaneous Analyzer was used to obtain the proximate analysis value (with experimental conditions; N_2 gas as purge gas at 40 mL/min, heating rate of 65 °C/min, temperature range of 30 to 800 °C). The Thermo Scientific Flash 2000 elemental analyser was used to obtain the elements present in the samples.

Figure 6.4 Process description of the refinement of raw LBK to particle size reduction before calcination.

6.4.2.4 Thermogravimetric (TGA/DTG) Analysis

Thermogravimetric analysis (TGA) and differential thermogravimetric analysis (DTG) is a thermal analysis method that helps in the study of degradation behaviour of materials during heating. It investigates the mass variation of sample as a function of time and temperature. In this study, TGA was performed in a TG/DTG analyser SDT-Q600 from TA instruments. Sample weights of 10 mg were used. Nitrogen gas at a flow rate of 40 mL/min was used to maintain an inert atmosphere for the thermal degradation. The temperature range for the pyrolysis investigation was 30 to 800 °C using different HRs (15, 20 and 30 °C/min).

6.4.3 Catalytic Pyrolysis Setup

Figure 6.5 shows the schematic diagram of the experimental setup used to produce pyrolytic oil from the catalytic pyrolysis of UT/NR.

Using the experimental setup shown in Figure 6.5, both thermal (without catalyst) pyrolysis and catalytic pyrolysis process were performed on UT or NR separately and in batches so as to be able to compare the quantity of pyrolytic oil obtained from the two sources with and without catalyst. The following parameters were investigated:

- Effect of the catalyst: feedstock (UT and NR) ratio on the pyrolytic oil yield.
- The concentration weight % of 1, 2.5, 5, 6, 7.5 and 10 wt.% of catalyst: feedstock
- Temperature effect on products obtained from the pyrolysis process (ranging from 375 to 750 °C).
- Feedstock particle sizes (2, 4, 6, 8 and 10 mm) effect on pyrolysis products.

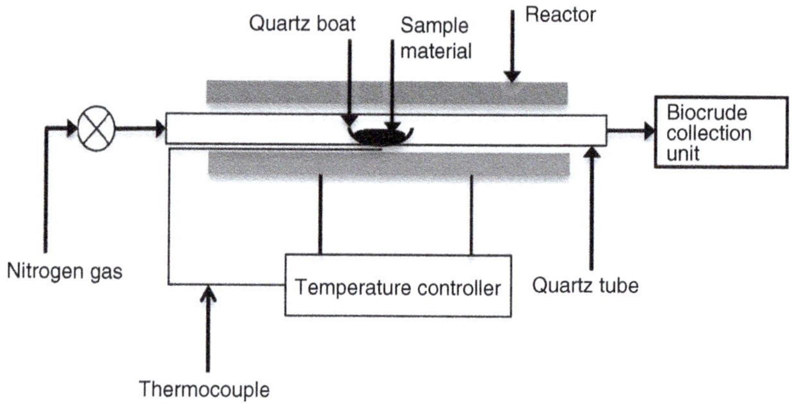

Figure 6.5 A typical schematic diagram of experimental setup for pyrolytic oil production from used tyres and natural rubber in the Nanotechnology laboratory of the University of the Witwatersrand, Johannesburg.

Nitrogen gas baseline supplied by Afrox (Pty) Ltd., South Africa at a flowrate of 150 mL/min was used both as a carrier gas and to keep the process inert. The pyrolytic oil produced was analysed and characterised using suitable equipment and standard laboratory procedures.

6.4.3.1 Pyrolytic Oil Production from Used Tyres and Natural Rubber

The sample material, UT/natural rubber (10 g/ batch) was placed in a glass crucible and introduced into a glass quartz tube positioned in a horizontal fixed-bed reactor chamber. The reactor chamber (which was heated by an electric system) had a K-type thermocouple and a P96 temperature controller connected to it. One end of the glass tube was connected to the N_2 gas cylinder to serve as inlet for the N_2 gas, while the other end served as an outlet for the liquid and volatile fractions (to the collection unit). The N_2 gas helped to ensure an inert environment and act as a carrier gas for the process.

Once the sample was in the reactor, the system was purged with N_2 gas at a flow rate of 150 mL/min to remove air for about 15 min before the experiment commenced. The reactor heater was switched on and the set value or operating temperatures studied were 375, 450, 525, 600, 675 and 750 °C at a HR of 15 °C/min as programmed on the temperature controller. The reactor temperature was kept constant once the process attains the set value temperature for 30 min or till no visible flow of liquid was observed before it begins to cool to 20 °C. This was to allow all possible liquid and vapor products to exit the reactor to the collection unit. Nitrogen gas was continuously supplied in order to keep the reactor inert and also to sweep away the pyrolysed liquid and vapor products to the collection unit. The collection unit was made up of a three-neck round bottom flask, placed in a bowl of ice. Both the pyrolytic oil and the gas fraction entered into the flask from the end connected to the reactor outlet via a flexible hose. The ice helps to retrieve the condensable fraction of the gas fraction while the non-condensable gases was collected for analysis (when needed) or allowed to escape through the outlet of the flask connected to the fume hood.

At the completion of the pyrolysis, N_2 gas flow was stopped and the reactor power was turned off. Upon cooling of the system, the round bottom flask was disconnected from the quartz glass tube reactor and the liquid fraction or pyrolytic oil was collected in a glass bottle and weighed. Also, the char left on the glass crucible in the reactor chamber was collected, weighed and kept in a glass bottle. The amount of gas was determined by the difference between the sum of the pyrolytic oil and char weight from total weight of the feedstock. Afterwards, the system was made ready for the next run by positioning the collection unit.

6.4.4 Characterisation of Used Tyres and Natural Rubber Pyrolysis Products

The chemical characterisation of the pyrolytic oil produced from UT and NR via catalytic and non-catalytic process was performed using the following methods/equipment.

6.4.4.1 Fourier Transform Infrared Spectroscopy (FTIR)

Fourier Transform Infrared spectroscopy (FTIR) was used to analyse the pyrolytic oil produced from UT and NR. The FTIR instrument, Bruker Tensor 27 model with Opus software was used to generate the IR spectra (range from 500 to 4000 cm^{-1}) and identify the functional groups present in the pyrolytic oil.

6.4.4.2 GC–MS Analysis

A GC Agilent© 7890B and MS Pegasus 4D GC × GC from Leco was used for the identification of the compounds present in the pyrolytic oil and the distillates fractions. The GC had a Capillary configuration (Rxi-5silms) of column length 29.77 m (30 m), with internal diameter of 250 μm, with film thickness of 0.25 μm and a maximum temperature of 350 °C. The front inlet type was Split/Splitless with a split mode of 20 : 1 with an inlet temperature of 200 °C. Helium gas of 99.95% purity was used as carrier gas at a constant flow rate of 1.51 mL/min. The initial oven temperature of 70 °C was maintained for 2 min and ramped at 10 °C/min to 300 °C and held at 300 °C for 0.5 min. Transfer line temperature was 260 °C.

The Mass Spectrometer method acquisition delay of 4 min, mass range of between 40 and 450 m/z, with an acquisition rate of 10 spectra/second and electron energy (ionisation energy) of -70 eV was used. The ion source temperature was 250 °C.

Data processing method base line offset of 1 (just above the noise) was used, with a peak width of four seconds and signal-to-noise (S/N) ratio of 500. The library search mode was normal and forward. The number of library hit was 10, while the minimum molecular weight allowed was 40 and maximum molecular weight was 450. The mass threshold was 2% and minimum similarity before name was assigned was 50%. The library used was Replib and mainlb from NIST.

6.4.4.3 NMR Analysis

The ^{1}H NMR and ^{13}C NMR analyses of the pyrolytic oil were performed using the Bruker Advance III 400 equipment operating at 400.132 MHz for ^{1}H and 100.623 MHz for ^{13}C. Samples were prepared in 5 mL NMR tubes using $PDCl_3$ as solvent.

6.4.4.4 Physical Characterisation of Used Tyres and Natural Rubber Pyrolytic Oil

Physical properties such as density, kinematic viscosity, flash point, pour point, calorific value, sulphur content and distillation boiling range of the UT and NR pyrolytic oil were determined according to the American Society for Testing and Materials (ASTM) standard methods as shown Table 6.2.

6.4.4.5 Characterisation of Solid Residue from Used Tyres and Natural Rubber

The proximate analysis was done according to ASTM D3172-73 with the Perkin Elmer precisely STA6000, Shelton USA simultaneous Analyzer and the Thermo Scientific Flash 2000 elemental analyser was used to obtain the elements present in the samples according to ASTM D3176-15. Calorific value of the raw material was found by ASTM E711 using the DryCal modular calorimeter, model 2010.

6.4.4.6 SEM Analysis

UT and NR pyrolytic residue were characterised for surface morphology by Zeiss ultra60 FE-SEM Scanning Electron Microscope located at the School of Chemical and Metallurgical Engineering of the University of the Witwatersrand. The instrument was also used for the LBK and synthesised zeolite NaY catalyst characterisation. The size and shape of

Table 6.2 Standard methods for physical property analysis.

Physical property	Method	Process description
Density	ASTM D1298-12b	Use of hydrometer placed in a constant temperature bath
Kinematic viscosity	ASTM D445-15	Use of calibrated glass viscosity tube submerged in a water bath at 40 °C
Flash point	ASTM D93-15a	Use of Pensky-Martens Closed test cup filled up to its mark with the test sample. The temperature shall be at least 18 °C below the expected flash point for the test sample and the cup
Pour point	ASTM D97-15	Use of test Jar glass which held the sample. Samples are first heated and then cooled at a specific rate at decrements of 3 °C. The temperature 3 °C above the point the oil stops moving is noted as the pour point
Gross calorific value	ASTM-E711	Use of the Parr oxygen bomb calorimeter
Sulphur content	ASTM D7039-15a	Method based on measurement of the specific wavelength from excitation of sulfur, which can accurately measure the concentrations of sulfur from 2 to 500 ppm
Distillation boiling range	ASTM D86-15	120 mL sample is distilled under prescribed conditions and observed temperature and volumes of condensate are reported.

synthesised zeolite crystals were measured using the Zeiss ultra60 FE-SEM Scanning Electron Microscope. A thin layer of gold–palladium was used to coat the zeolite crystals in a BAL-TEC SPD-050 instrument before measurement.

6.4.5 Formulation and Testing of the Oil Blends in Internal Combustion Engine

The pyrolytic oil obtained from the pyrolysis process (with/without catalyst) of UT and NR was distilled according to the setup shown in Figure 6.6a,b.

A volume of 120 mL of pyrolytic oil obtained from the thermal and catalytic pyrolysis of UT and NR at 600 °C at a HR of 15 °C/min was poured into a round bottom flask seated on a heating mantle and distilled (under vacuum) with the aid of a vacuum pump as shown in Figure 6.6a,b. This was done to help reduce pressure build-up in the batch distillation unit. The total volume of distillates obtained at different temperature from the 120 mL of the different pyrolytic oil was noted. The obtained PD were then characterised using GC–MS.

6.4.5.1 Fuel Blend Technique for Pyrolytic Oil Distillates and Petrol

The experimental setup for the fuel blend technique of PD and petrol is depicted in Figure 6.7. It consists of a glass blender, a thermometer and an ultrasonicator. The ultrasonicator horn is 33 cm high with diameter of 0.7 cm, and operates at 24 kHz. The thermometer was positioned at 3 cm from the ultrasonicator horn (which was placed in the middle of the blender).

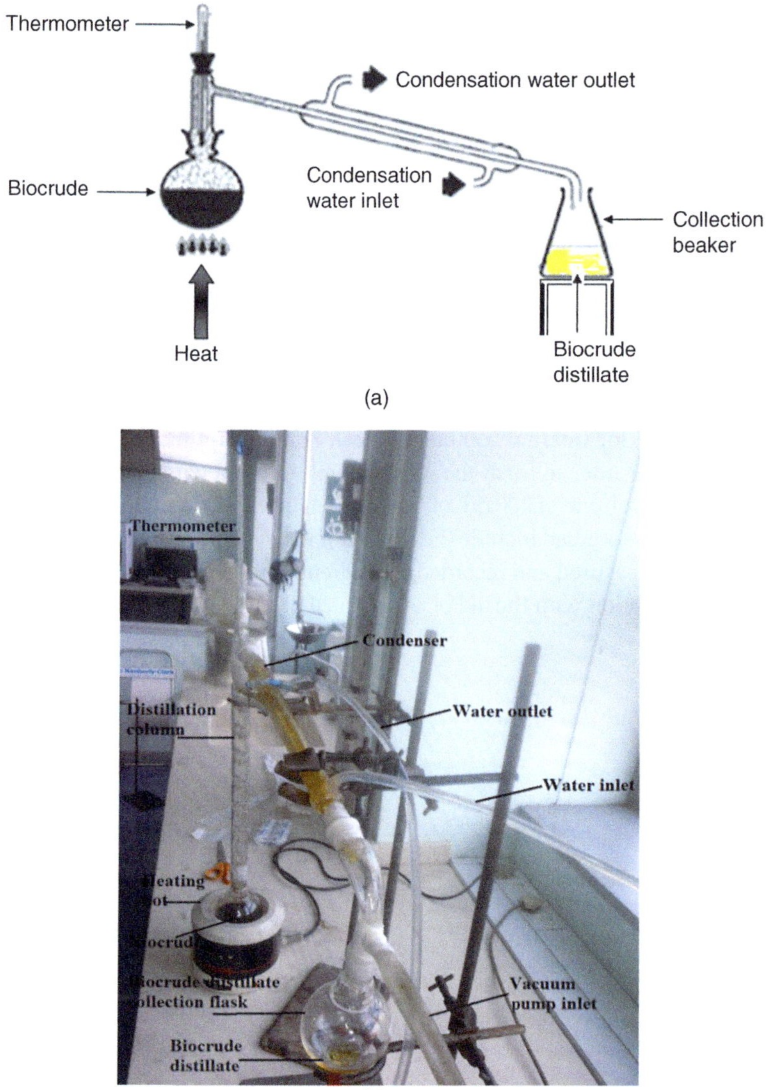

Figure 6.6 Pyrolysis setup, (a) schematic and (b) pictorial set-up of batch distillation of pyrolytic oil from used tyres and natural rubber.

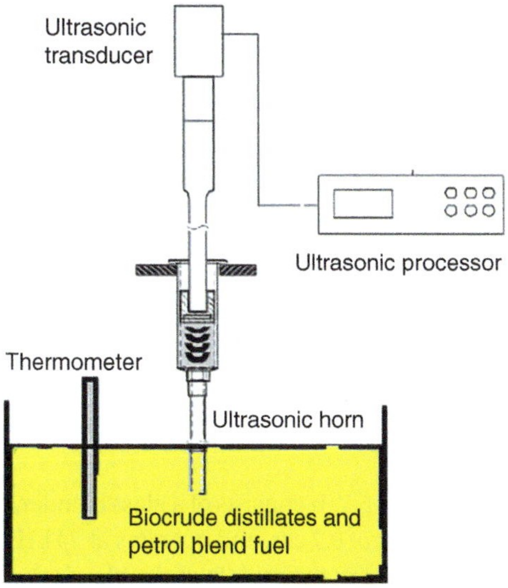

Figure 6.7 Setup for blending of pyrolytic oil distillates and petrol.

PD obtained from UT/NR and petrol were blended in different amounts ranging from 0 to 20% PD. For each blend, the required volume of petrol was first poured into the batch blender. The ultrasonicator horn was then lowered to a depth of 2 cm within the petrol before being switched on. The temperature of the petrol was maintained at 26 °C, and then the appropriate volume of PD was added to the petrol at a constant rate and allowed to blend for a duration of 5 min before switching off the ultrasonicator. The obtained fuel mixture in the batch blender was emptied into a measuring cylinder to be tested in an ICE.

6.4.5.2 Combustion of Pyrolytic Oil Distillates and Petrol Blend in Internal Combustion Engine

A Sinemaster IG2600 BUNDU POWER (AC generator) was used to power a Martlet 375 W heavy-duty 150 mm Bench Grinder, model MD 150BG for the ICE study of the various blended fuels, as shown in Figure 6.8. The generator is a 4-stroke single cylinder, air-cooled spark ignition engine run on 95 unleaded-petrol with a fuel tank capacity of 4.6 L. Its manufacturer's engine specification is presented in Table 6.3.

For the fuel blend test in the ICE, PD and petrol blend with different proportion of distillates (0, 5, 10, 15 and 20%) were prepared using ultrasonication-enhanced blending method for internal combustion test. Before running the blends, the engine was run on neat or un-blended petrol for 60 min in order to flush the fuel line and internal combustion system. The electricity generator improvised tank was filled with a volume of 200 mL of the blended fuel and then the engine was switched on to run for approximately 50 min. The parameters studied include the generator fuel consumption rate and fuel power under load. The volume of the fuel in the tank was measured and recorded at different time interval (0, 5, 10, 15, 20, 25, 30, 35, 40, 45 and 50 min) to determine the fuel consumption with the aid of a calibrated dipstick in the fuel tank.

Figure 6.8 Fuel testing on (a) A Sinemaster IG2600 BUNDU POWER (AC generator). (b) Bench grinder powered during the combustion study.

Table 6.3 Specification and characteristics of the ICE.

Model type	KG166 (single-cylinder, air-cooled 4-stroke, gasoline engine)
Displacement (BorexStroke)	171 cc (66 × 50 mm)
Compression ratio	8.5:1
Rated power (kW/[r/min])	3.3/3600
Ignition system	***Transistor-Controlled Ignition (TCI)***
Spark plug	Bosch - WR7DC
Starting system	Recoil starter
Fuel type	Automotive unleaded gasoline
Fuel consumption (g/kW-h)	500
Lube oil	PD grade or SAE 10W-30, 15W-40
Fuel tank capacity (L)	4.6
Continuous running time	3 h (at rated output)
Overall dimension (L × W × H) mm (inches)	564 × 317 × 453 (22.2 × 12.5 × 17.8)
Dry weight (kg [lbs])	26 (57.2)

6.4.6 Results and Discussion

6.4.6.1 Zeolite NaY Synthesis and Characterisation

Zeolite NaY catalyst used in this study was synthesised using a natural raw material, LBK as precursor for silicon and alumina. After refinement, the calcined LBK at 600 °C with exposure time of 2 hours was sent for analysis. The catalyst synthesised using the calcined LBK was performed at different ageing time (0, 0.5, 1, 3, 6, 9, 12 and 24 hours). A constant crystallisation temperature and time of 100 °C and 9 hours reported in literature (Kovo 2010; Chandrasekhar and Pramada 2004; Zi et al. 1988) as the optimum temperature and time for zeolite NaY synthesis (which is also used in the industry) was adopted for this study (Franco et al. 2003; Sang et al. 2006). The synthesised zeolites at different ageing was characterised using XRF, SEM and XRD.

6.4.6.2 XRF Characterisation of Calcined LBK

XRF analysis was performed on the calcined LBK in order to determine the chemical constituent weight percent present in the LBK. Table 6.4 shows the chemical constituents present in the LBK metakaolin produced at 600 °C revealed by the XRF.

From Table 6.4, the LBK metakaolin main chemical constituent is silicon expressed as silicon oxides, 68.93 wt.% and aluminium expressed as alumina, 24.32 wt.% (with the presence of trace oxides at very low weight percent). The Si/Al ratio of approximately 2.83 obtained in this study was higher than that needed in synthesising commercial zeolite (Si/Al > 1.5) [2–4] and that reported by Kovo (2010) which was approximately 2 when he synthesised zeolite NaY using kaolin as silicon and alumina source. This makes LBK a suitable source for silica and alumina in zeolite synthesis.

Table 6.4 XRF chemical constituent present in LBK metakaolin.

Chemical constituent	Weight %	Chemical constituent	Weight %
SiO_2	68.93	K_2O	0.30
Al_2O_3	24.32	TiO_2	3.09
Na_2O	0.11	P_2O_5	0.10
MgO	0.24	Cr_2O_3	0.05
Fe_2O_3	0.22	NiO	0.0022
MnO	0.02	Total	99.22
CaO	0.08	LOI	2.01

6.4.6.3 SEM and XRD Characterisation of Refined and Calcined LBK

The SEM morphology and XRD analysis pattern of the refined LBK Metakaolin was obtained as shown in Figures 6.9 and 6.10.

From the XRD pattern, the presence of silicon and aluminium was identified. The EDAX result gave a Si/Al ratio of 1.6, which is higher than that required ($\geq$1.5) for the synthesis of commercial zeolite. This made the LBK metakaolin a suitable material for zeolite NaY synthesis. From this characterisation of the LBK, noticeable changes undergone by the LBK Metakaolin (as aluminosilicate gel) in the formation of zeolite NaY is compared with respect to the commercial zeolite.

6.4.6.4 SEM and XRD Characterisation of Synthesised Zeolite NaY from LBK Metakaolin

The SEM images of synthesised zeolite NaY obtained at different ageing time at a constant crystallisation temperature and time of 100 °C and 9 hours, respectively, using LBK metakaolin as a precursor for silicon and alumina is shown in Figure 6.11a–h. Commercial zeolite NaY (supplied by Sigma-Aldrich) SEM image was also obtained to serve as the reference

Figure 6.9 SEM image of LBK metakaolin used in this study (Scale bar = 2 μm).

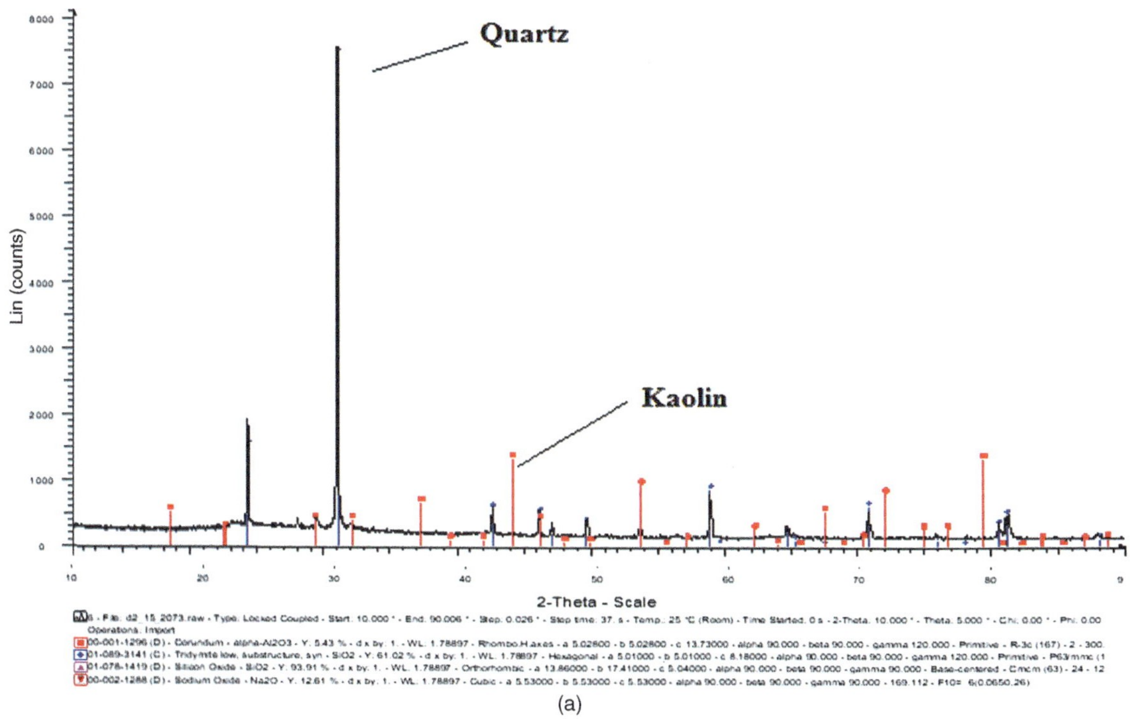

Figure 6.10 (a) XRD pattern of the LBK metakaolin used in this study for zeolite NaY synthesis. (b) XRD pattern of reference commercial zeolite NaY showing its composition.

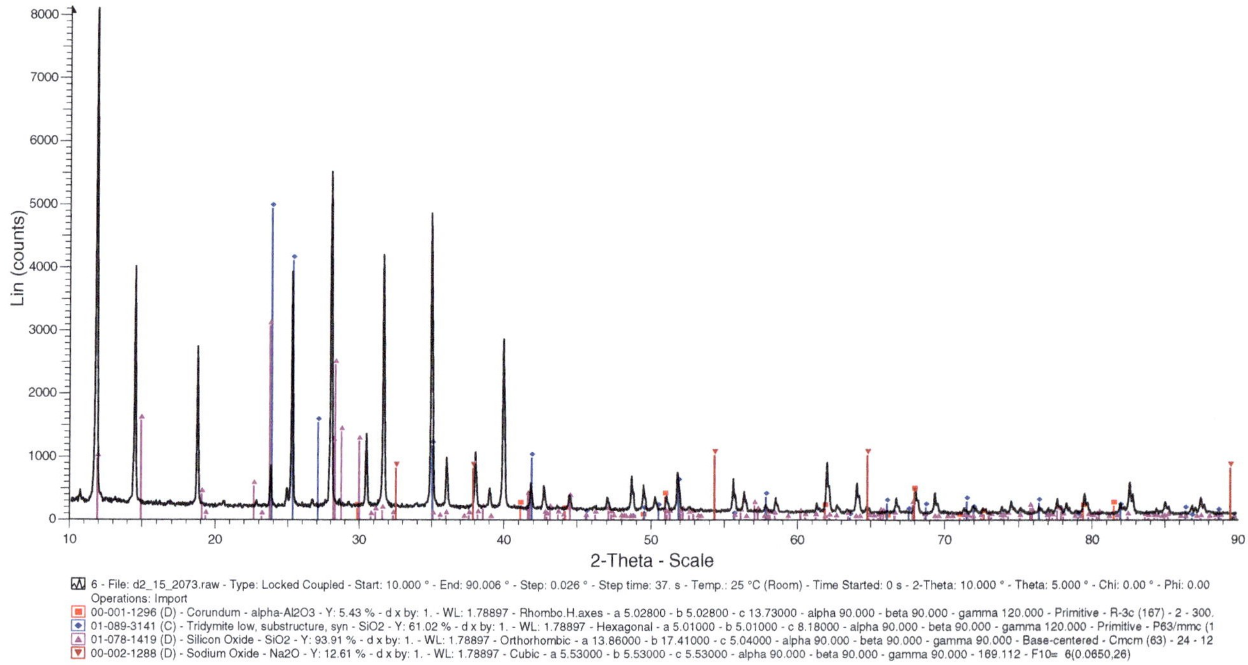

(b)

Figure 6.10 (*Continued*)

Figure 6.11 SEM image of zeolite NaY from LBK metakaolin (effect of ageing time (a) = 0, (b) = 0.5, (c) = 1, (d) = 3, (e) = 6, (f) = 9, (g) = 12, (h) = 24 hours) and (i) is commercial zeolite NaY.

zeolite NaY in this study. The SEM images reveal the presence of crystals even at the beginning of ageing, and with 9 hours ageing and crystallisation at 100 °C without seeding to induce zeolite formation. It was observed that the crystal sizes continued to grow with increasing ageing time. Figure 6.12 presents the role of increased ageing time on the aluminosilicate gel formation in the strong basic environment (NaOH used in this study) in the increase of particle size of the zeolite NaY formed.

From Figure 6.12, the plot reveals that zeolite NaY crystallisation and crystal growth commenced from as low as few minutes ageing time (0 hour) and continued till 6 hours ageing time. Upon further ageing from 6 to 9 hours, it was observed that there was no more crystal growth, which was why the curve became flat. However, from 9 to 24 hours ageing time, the crystals started agglomerating to form globules. This is probably due to hydrogen or van der Waal force of hydrogen and oxygen found in AlO_2 and mH_2O moieties in $Mx/n[(AlO_2)x(SiO_2)y]·mH_2O$, respectively.

Nevertheless, ageing is necessary in zeolite NaY synthesis as it assists in promoting and enhancing the chance of nucleation and crystallisation (Chandrasekhar and Pramada 2004). This was confirmed in this study by synthesising the aluminosilicate gel at 0 hour ageing time at room temperature for 9 hours. The SEM morphology of the synthesised sample displayed the absence of zeolite crystals as presented in Figure 6.13.

The well-defined octahedral crystal shapes obtained for the zeolites NaY obtained at the ageing time lower than 1 hour upward have similarity with that reported in literature by Chandrasekhar and Pramada (2004) and the commercial zeolite NaY used as reference zeolite NaY in this study. It is also worthy to note that the SEM scale bar of 200 nm was used for the capturing of all the synthesised zeolite NaY in this study shown in Figure 6.11a–h while a scale bar of 2 μm was used for

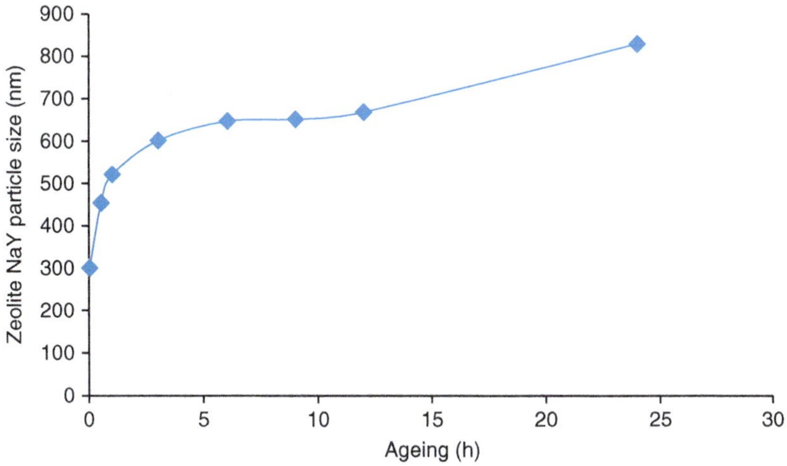

Figure 6.12 Effect of ageing on zeolite NaY particle size.

Figure 6.13 SEM image of 0 hour ageing and 9 hours crystallisation at room temperature of 25 °C.

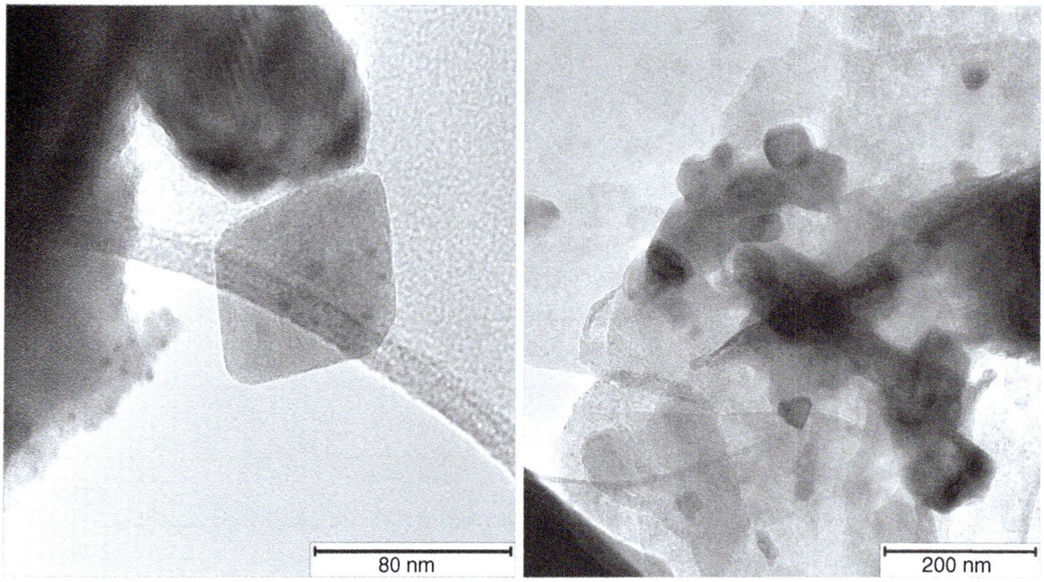

Figure 6.14 TEM image of synthesised zeolite NaY with particles of nanoscale sizes.

the SEM image of the commercial zeolite NaY. This is because the synthesised Zeolite NaY came out as nanoscale particle sizes range as confirmed by the TEM images shown in Figure 6.14.

Also, the X-ray diffraction pattern for the Zeolite NaY produced at different ageing time during this study as presented in Figure 6.15; support the SEM report of the formation of zeolite NaY even at 0 hour ageing time at 100 °C and 9 hours crystallisation time.

All the obtained XRD patterns match well with that of the commercial zeolite NaY supplied by Sigma-Aldrich, which served as a reference standard in this study. From the synthesised zeolite NaY diffraction pattern, the intensity of the peaks for ageing time of below 1 hour and above was a near perfect match with that of the reference commercial zeolite NaY (used as standard). This implies that the calcination of the LBK resulted in the release of free silica and alumina from the kaolin structure, which made the aluminosilicate to be very reactive in the base environment (Zi et al., 1988). The role of ageing can be said not to be responsible for zeolite formation as shown in this study. Other conditions, such as autogenous pressure, crystallisation time, temperature and the reactivity of the aluminosilicate gel formed from the LBK metakaolin and the strong base, NaOH could be the controlling factors (Franco et al. 2003).

6.4.6.5 Summary of Zeolite NaY Synthesised from LBK Metakaolin

Generally, the use of kaolin in the synthesis of faujasite has not yielded much success except for zeolite X (Kovo 2010; Zi et al. 1988; Franco et al. 2003; Sang et al. 2006). This could be attributed to several factors, such as nature of the raw kaolin, quantity, and composition of other materials, such as alkali medium required for zeolite Y synthesis. Nevertheless, in this study, the outcome of the use of LBK metakaolin as a precursor for alumina and silica in the synthesis of zeolite NaY via hydrothermal method was achieved. This was confirmed by the SEM and XRD results. Hence, reaffirming the assertion that the physicochemical property of the kaolin material determines to what extent such kaolin can be converted to zeolite NaY (Chandrasekhar and Pramada 2004). Despite ageing period of more than 9 hours has been reported as necessary for other kaolin in literature (Zi et al. 1988), LBK metakaolin was reactive enough to form zeolite nucleus within a short time of less than 1 hour. Also, the nano sizes of the synthesised zeolite NaY particle were obtained at the first 3 hours of ageing in this study. These synthesised zeolites NaY of different particle sizes would be used in the catalytic pyrolysis of UT and NR.

6.4.6.6 Proximate and Ultimate Analysis and GCV of Used Tyres and Natural Rubber

Both the proximate analysis and heating value of NR and UT rubber samples are presented in Table 6.5. The obtained results for the proximate analysis of NR and UT samples reveal that these raw materials are suitable for pyrolysis, which is in agreement with the data presented by many authors Gonzalez et al. 2001; Chang 1996; Juma et al. 2006; Islam et al. 2008; Kar 2011). Nevertheless, the ash content of NR was much lower than that of the UT in this study as well as those reported in literature (Williams 2013). This is owing to the fact that the NR sample studied had no additive and other inorganic content

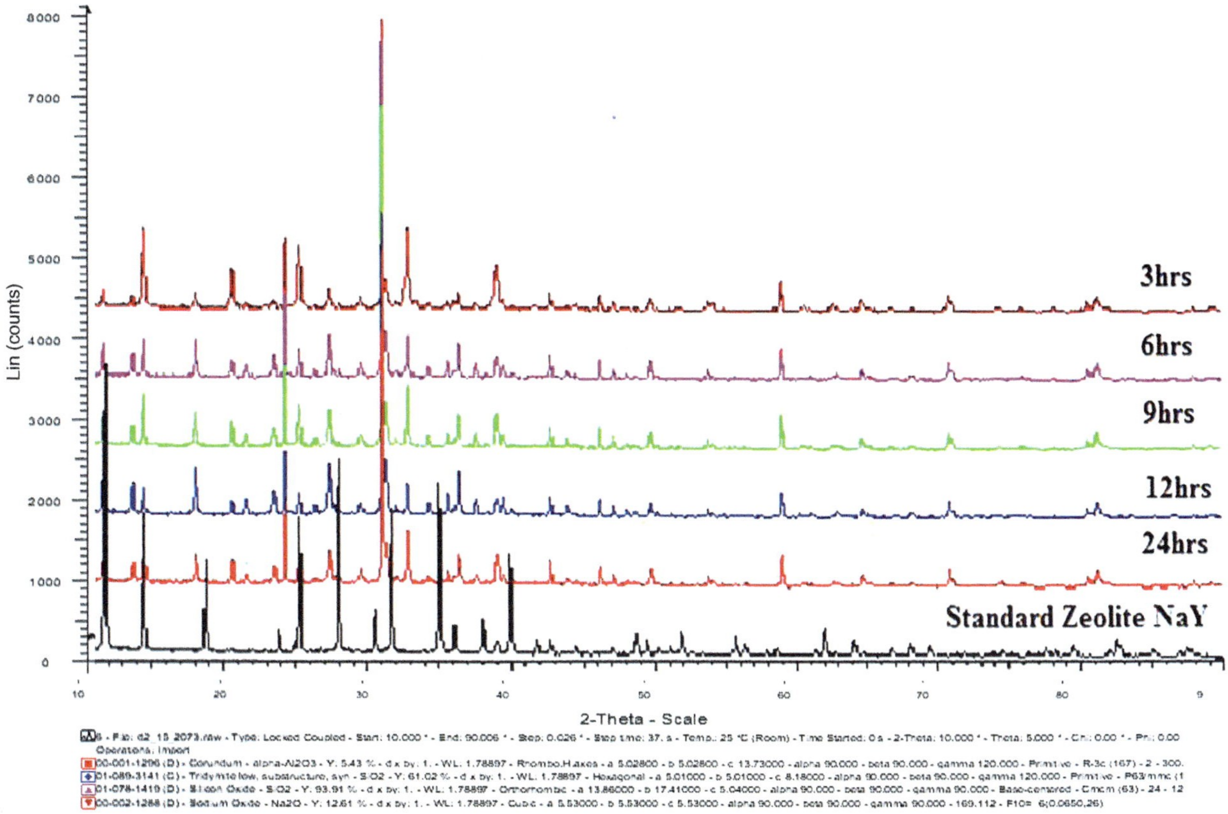

Figure 6.15 XRD pattern for the synthesised Zeolite NaY from LBK produced at different ageing times at constant crystallisation times and temperature.

Table 6.5 Proximate and ultimate analysis of natural rubber and used tyres.

	This study		Gonzalez et al. (2001)
	Natural rubber	Used tyre	Car tyre
Proximate analysis (%)			
Moisture content	1.50	0.50	0.70
Ash	0.70	16.00	8.00
Fixed carbon	89.10	27.10	29.50
Volatile content	8.70	56.40	61.90
Ultimate analysis (%)			
Carbon (C)	87.90	85.00	86.70
Hydrogen (H)	7.90	5.50	8.10
Nitrogen (N)	3.20	0.21	0.40
Sulphur (S)	–	1.20	1.40
Oxygen (O)	1.00	8.09	1.30
GCV (MJ/kg)	39.00	31.22	36.20

(such as zinc, clay and silica) which are usually present in tyre rubber according to manufacturers' formulation (Williams 2013; Quek and Balasubramanian 2013). With the low ash content in NR, a higher mass of product yield should be expected. The fixed carbon content of the NR was almost thrice that available in literature but with a very low volatile content which could be due to the absence of oil-based additives, plasticisers and anti-oxides in the sample (Islam et al. 2008). The GCV of NR (39.00 MJ/kg) and car UT (31.22 MJ/kg) were higher than that of lignite coal (15.00 MJ/kg) and wood (21.70 MJ/kg) making them a better source of solid fuel but lower than that of conventional diesel (44.80 MJ/kg).

6.4.6.7 Degradation of Used Tyres and Natural Rubber

The TGA and DTG analysis curves with respect to reaction temperature, T, at a HR of 15 °C/min for UT and NR samples are presented in Figures 6.16 and 6.17. For the temperature range from 30 to 800 °C, more than one temperature degradation zone was observed for both UT and NR. This is in agreement with reports by different authors (Juma et al. 2006; Karthikeyan et al. 2012; Haydary et al. 2008) that there exists more than one degradation temperature zone during tyre rubber pyrolysis. The TGA curve of both samples shows that the thermal decomposition begins at a temperature of 200 °C for NR, 210 °C for UT and complete at 500°C for NR and 520 °C for UT. The DTG curve (Figures 4.8 and 4.9) also gave a single peak of 386.70 °C for NR and two dominant peaks of 385.64 and 425 °C for UT rubber.

When these results are compared with values of thermal degradation of individual kind of rubbers in literature (Islam et al. 2008; Williams 2013; Haydary et al. 2008) (375 °C for NR, 445 °C for SBR and 465 °C for BR), it is evident that the single peak in NR DTG curve shows the decomposition of only NR in the sample and further confirms the absence of any type of synthetic rubber as the NR investigated was in its raw form. That of the UT, DTG curve reveals that the first peak indicates the degradation of NR and the second dominant peak gives the degradation of SBR and BR, a major type of polymer in car tyre formulation (Islam et al. 2008; Haydary et al. 2008).

The degradation of NR involves two main phases. The first phase can be attributed to the loss of volatile components (moisture and oil), and the second phase as active pyrolytic zone. While, the first stage of degradation of UT includes loss of plasticisers and other volatile additives in the tyre. The second degradation stage involves mainly cracking and rapid degradation of the rubber components present in the UT (Juma et al. 2007). It is at this period that carbon bond scission of weaker chemical bonds and intermolecular associations occur. Heavy-weight molecular hydrocarbons are depolymerised, hydrogenated/dehydrogenated into smaller lower weight hydrocarbon molecules in form of liquid and gas phase. From the dominant peaks revealed by both DTG curves, significant conversion takes place at this stage as well as secondary depolymerisation. Beyond this stage, the sample is mainly left as ash shown by the constant weight displayed in the TGA/DTG curves. This is in accordance with the study done by Williams et al. (1990) on TGA with two distinct areas of weight loss, signifying a lower and higher temperature of degradation.

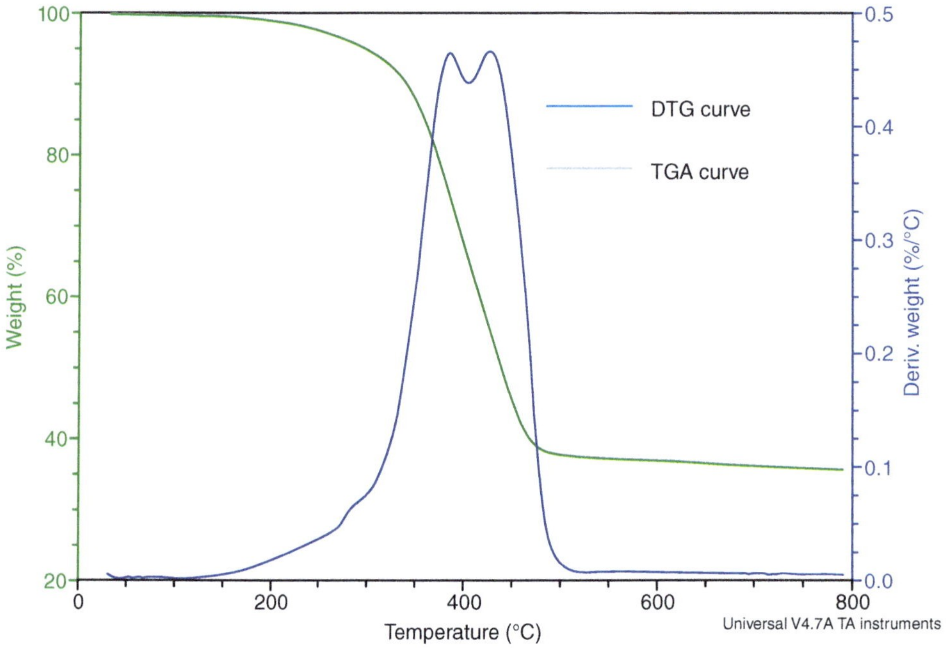

Figure 6.16 TGA/DTG curve of UT at heating rate of 15 °C/min.

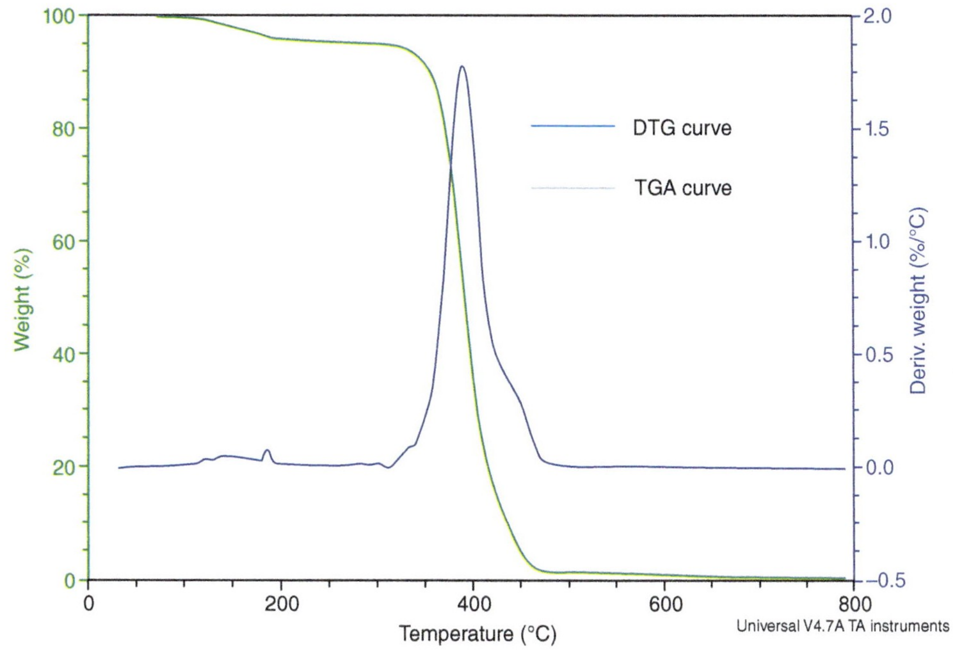

Figure 6.17 TGA/DTG curve of NR at heating rate of 15 °C/min.

6.4.6.8 Thermal Pyrolysis of Used Tyres and Natural Rubber

In this study, optimum HR and mass of the sample used were determined experimentally. This obtained optimum HR and mass of sample were then kept constant in order to determine optimum values for other parameters such as feed material particle size, operating temperature, and catalyst (concentration and particle size) effect on feed material pyrolysis yield and pyrolytic oil composition. Eureqa™ mathematical programming software was then used to propose a fit equation for the pyrolytic oil yield in relationship with the different sets of parameters.

6.4.6.9 Determination of Optimum Particle Sizes on Thermal Pyrolysis of Used Tyres

Figure 6.18a shows the effect of feed size on UT pyrolysis products yield performed at 600 °C at a heating rate of 15 °C/min. The data plotted are the mean value and standard deviation of the pyrolytic oil, char and gas yields for the various feed material particle sizes, which are presented in details elsewhere (Osayi 2015). It was observed that the pyrolytic oil yield increased progressively from 24.5 ± 0.17 to 34.4 ± 2.18 wt.% for 2 to 6 mm feed particle size, respectively. A steady decrease in the pyrolytic oil yield was observed for the feed particle size greater than 6 mm. Nevertheless, there was a steady decrease in gas and increase in solid residue yield from 2 to 10 mm. The increase in solid residue for size 2 to 10 mm may probably be due to less exposed surface area, which reduced the breaking of hydrocarbon molecules and lower thermal conductivity within the UT materials, which prevented even distribution of heat within the entire material for more reaction (Islam et al. 2008; Bridgwater 1999).

Hence, this led to carbonisation or incomplete degradation of the larger UT sizes, resulting in increase in solid char produced and reduction in pyrolytic oil and gas yields. For the smaller sizes, especially 2 mm, the increase in gas yield could be as a result of the wider reaction surface area available thereby enabling rapid spread of heat and speedy degradation of the UT to happen (Pradhan and Singh 2011; Barbooti et al. 2004). Also, the pyrolytic oil and solid residue produced had sufficient time to undergo secondary reaction in the reactor resulting in more gas yield and decrease in solid char and pyrolytic oil produced. The results obtained in this study are in accordance with those reported by Barbooti et al. (2004) where amount of heat on a core of large pieces was restricted from the heat source. Also, Islam et al. (2008) reported that 8 cm^3 produced a higher char yield with corresponding lower gas and liquid yield than those of 4 cm^3. In the same way, Dai et al. (2001) found that there was a higher amount of char produced for a 0.8 mm tyre sample with lower gas and oil yield than that obtained for 0.32 mm sample. Therefore, it can be concluded for this study that 6 mm is the optimum feed size for degradation of UT feedstock to complete with less chances of secondary reaction at the optimum reaction temperature of 600 °C and HR of 15 °C/min. It was this feed size that was applied for the NR pyrolysis in order to compare their product yield.

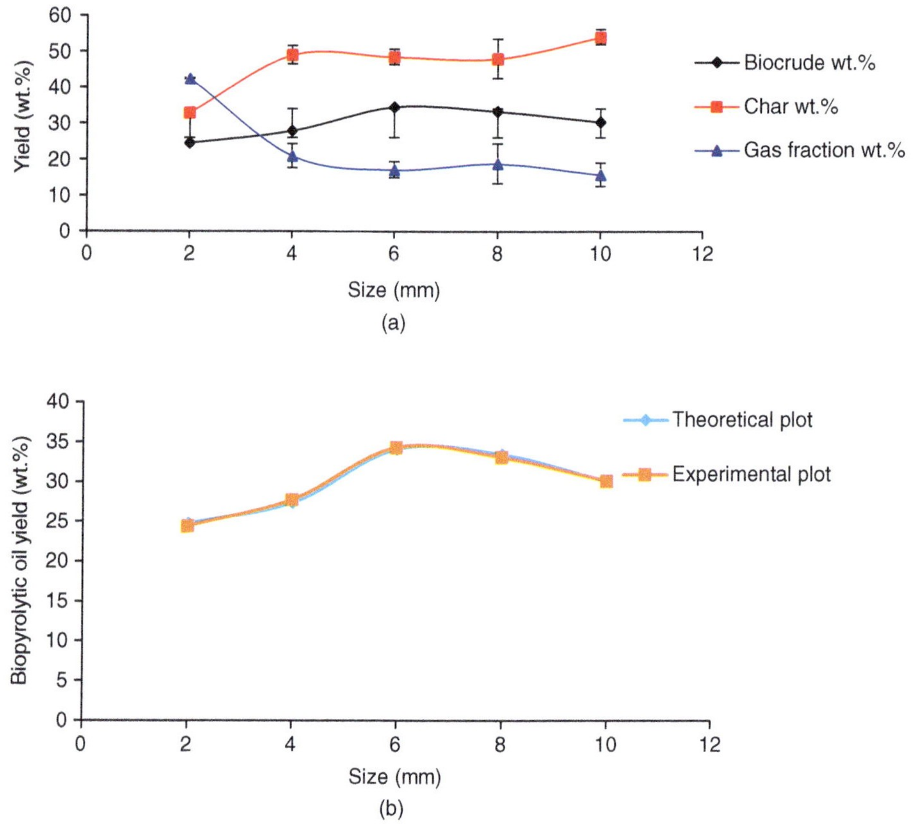

Figure 6.18 Pyrolytic oil production, (a) Particle size effect on UT pyrolysis yield. (b) Experimental and theoretical plot of size effect on UT (STD: Pyrolytic oil = 4.01, Char = 7.92, Gas = 11.05).

To predict the pyrolytic oil yield produced at a given used tyres particle size, Eureqa Newtonian software was used. The proposed Eq. (6.1) satisfied the experimental data as can be seen in Figure 6.18b

$$Y = 21.84 + 1.68\,S + 0.43\,S\,\cos(S) - 4.89e^{-6}\,S^6 \tag{6.5}$$

where, Y = pyrolytic oil yield; S = used tyres particle size.

Figure 6.18b compares the experimental plot and the predicted from Eq. 6.4 with R^2 goodness of fit = 0.99, Correlation Coefficient = 0.99, Maximum Error = 0.41, Mean Squared Error = 0.09, Mean Absolute Error = 0.28 and Complexity = 28.

6.4.6.10 Temperature Effect on Yield of Thermal Pyrolysis of Used Tyres

The product distribution from the thermal pyrolysis of 10 g of UT at a HR of 15 °C/min for a temperature range from 375 to 750 °C at every 75 °C elevation in temperature for feed size of 6 mm was studied. This was performed in order to determine the pyrolysis optimum operating temperature. The process was repeated thrice for each temperature. The process reaction temperature, and weight fractions for the various products yield, pyrolytic oil, char and gas fractions have been plotted together in Figure 6.19a.

From Figure 6.19a it was observed that with an increase in temperature from 375 to 750 °C, the yield of pyrolytic oil increased from 6.49 ± 0.21 to an optimum value of 34.40 ± 2.18 wt.% at 600 °C and then reduces to 14.03 ± 0.13 wt.% at 750 °C. The solid char yield decreased from 59.58 ± 6.98 wt.% to 47.12 ± 0.16 wt.% over the entire temperature range, while the gas yield increased from 33.93 ± 5.46 to 38.85 ± 0.07 wt.%. It is observable that a maximum exists at the temperature where optimum yield of pyrolytic oil was attained possibly as a result of strong cracking of UT at this temperature, 600 °C. Secondary reactions that occurred within the reactor must have led to formation of more gases, which resulted in a decrease in pyrolytic oil yield at higher temperatures of more than 600 °C. This was also in agreement with the optimum temperature obtained by Pradhan and Singh (2011) in their investigation of thermal pyrolysis of UT rubber over a temperature range of 450 to 700 °C.

Hence, it can be proposed that UT rubber is not completely degraded during thermal pyrolysis at a temperature below 600 °C. The low pyrolytic oil yield at, 375 °C is not unexpected as most of the UT rubber components (SBR, BR and NR)

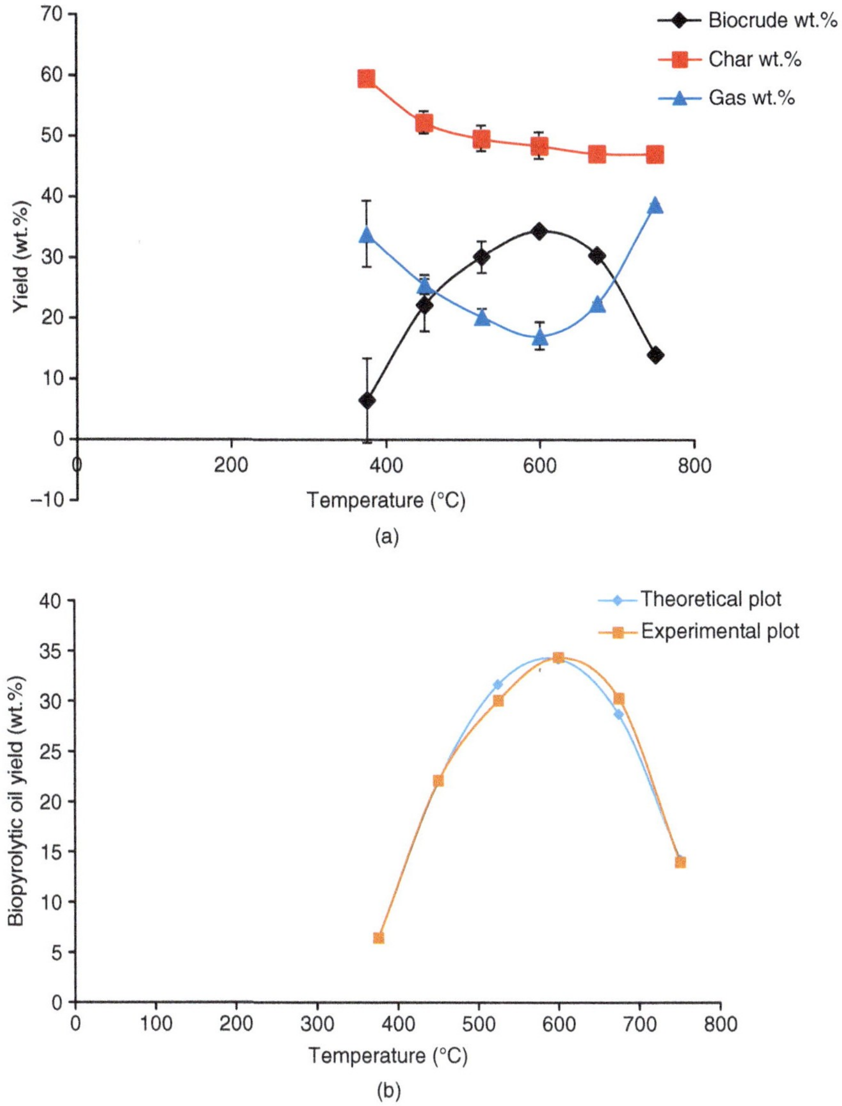

Figure 6.19 (a) Effect of temperature on UT pyrolysis product yield. (b) Experimental and theoretical plot of temperature effect on UT pyrolytic oil production (STD: Pyrolytic oil = 10.84, Char = 4.75, Gas = 8.38).

degradation temperature except for NR is even higher than the operating temperature of 375 °C as revealed in the DTG curve of Figure 6.16. Also, the presence of high amount of gas (33.93 ± 5.46 wt.%) even at low temperature of 375 °C, shows that the UT has high volatile materials such as moisture, plasticisers and other additives which can easily volatilise at low temperature (as obtained from the proximate analysis results in Table 6.5). The solid char yields over the temperature range from 600 to 750 °C are almost equivalent to the values derived from the TGA of UT in Figure 6.15. Since the solid char yield did not reduce over the temperature range of 600 to 750 °C, a conclusion can be reached that UT rubber degradation is completed.

The decline in pyrolytic oil yields and increase in gas above 600 °C could be attributed to the degradation of fractions of the pyrolytic oil into non-condensable gases (Islam et al. 2008), along with secondary reaction involving carbonisation of hydrocarbons in the pyrolytic oil into char (Islam et al. 2008; Cypres and Bettens 1989; Cunliffe and Williams 1998). The further loss in char at higher temperatures could also contribute to high gas yield at higher temperatures making the gas fraction to be more prominent at such temperatures (Brown 1996; Bridgwater 1999). Nevertheless, 600 °C can be said to be the maximum operating temperature for the thermal pyrolysis of UT to obtain optimum pyrolytic oil yield in this study since degradation is completed and gas yield becomes dominating due to the outcome of the effect of temperature on the thermal pyrolysis of UT.

The production of pyrolytic oil from UT pyrolysis at different operating temperature was proposed using Eureqa Newtonian software. The proposed equation fitted the experimental data as shown in Figure 6.18b which described the production yield as a third-order polynomial equation. This can serve as basis for the prediction of pyrolytic oil production using the pyrolysis method. The coefficients of Eq. (6.6) are related to the experimental conditions which need further investigation.

$$Y = 0.41\,T - 126.67 - 3.96e^{-7}\,T^3 \tag{6.6}$$

where Y = pyrolytic oil yield and T = operating temperature.

Figure 6.19b compares the experimental and the predicted plots from Eq. 6.6 with R^2 goodness of fit = 0.99, Correlation Coefficient = 0.99, Maximum Error = 1.58, Mean Squared Error = 0.83, Mean Absolute Error = 0.60 and Complexity = 13.

The third-order polynomial equation was also obtained using the published results of Islam et al. (2008). This was given as;

$$Y = 0.38\,T - 74.30 - 5.48e^{-7}\,T^3 \tag{6.7}$$

with R^2 goodness of Fit = 0.95, Correlation Coefficient = 0.98, Maximum Error = 1.62, Mean Squared Error = 0.54, Mean Absolute Error = 0.38 and Complexity = 13. This is in accordance with that obtained in this study.

6.4.6.11 Effect of Temperature on Yield from Thermal Pyrolysis of Natural Rubber

The thermal pyrolysis of NR (latex rubber) generated three pyrolysis products, which are pyrolytic oil, solid char and gas. The product distribution from the thermal pyrolysis of 10 g of NR at a HR of 15 °C/min for a temperature range from 375 to 750 °C at every 75 °C elevation in temperature for an average feed size of 6 mm were studied. The process was repeated thrice for each temperature. The process reaction temperature, and weight fractions for the various products yield, pyrolytic oil, char and gas fractions are plotted together in Figure 6.20a. The data plotted are the mean value and standard deviation of the pyrolytic oil, char and gas yields for the different temperatures (Osayi 2015).

Considering the product yield results, pyrolytic oil optimum was obtained at 600 °C. This could be as a result of cracking of the NR and dehydrogenation/ hydrogenation of the carbon bond at this temperature due to Diel Alder's mechanism (Osayi 2015). From the results, NR gave pyrolytic oil yield of 75.93 ± 2.88 wt.% which was almost thrice obtained for the thermal pyrolysis of UT as shown in Figure 6.20a). This yield is almost twice the amount obtained from pyrolysis of UT and other polyisoprene-based materials reported in literature. Shen et al. (2006) reported a liquid fraction yield of 52.0 wt.%, char yield of 36.1 wt.% and gas yield of 11.9 wt.% at 500 °C during the thermal pyrolysis of scrap tyres. A related result was also presented by Williams and Brindle (2003) when UT were pyrolysed in a fixed-bed reactor. Roy and Unsworth (1999) got a liquid fraction yield of 56.6 wt.% from the vacuum pyrolysis of UT at 415 °C. Kawakami et al. (1980) reported 53 wt.% of oil yield at a temperature range from 540 to 640 °C from the pyrolysis of UT in a rotary kiln reactor.

The weight percent of the gas obtained in this study increased from 23.63 ± 2.99 to 55.58 ± 1.23 wt.% over the temperature range of 600 to 750 °C, which was also higher than that obtained for UT in this study as reported in Figure 6.19a. This is another potential of NR to provide energy in the form of gas, which can be used as a source of heating/ fuel (Islam et al. 2008). NR gave a low solid residue of 0.43 ± 0.12 wt.% at optimum pyrolytic oil yield temperature of 600 °C, which is within the proximate analysis result. This will make it less challenging in terms of chars volume turn out, which is normally associated with UT pyrolysis process. This further confirms that yields from pyrolysis and their distributions can be greatly influenced by the feedstock composition amongst other factors, therefore making results from different authors vary (Quek and Balasubramanian 2013; Haydary et al. 2008).

In order to predict pyrolytic oil production from NR at different pyrolysis temperature, the Eureqa Newtonian software was used. As shown in Figure 6.20b the theoretical plot obtained from Eq. (6.4) fit the experimental data quite well.

$$Y = 44.45 + 0.03\,T + 22.36 \times \cos(0.23\,T) - 0.06\,T\,\cos(0.23\,T) \tag{6.8}$$

where Y = pyrolytic oil yield and T = operating temperature.

Figure 6.20b compares the experimental and the theoretical plots with R^2 goodness of fit = 0.98, Correlation Coefficient = 0.99, Maximum Error = 2.86, Mean Squared Error = 1.61, Mean Absolute Error = 0.80 and Complexity = 25.

6.4.6.12 Effect of Catalyst Concentration on the Pyrolysis of Used Tyres and Natural Rubber

The catalyst used in this study was zeolite NaY catalyst synthesised as discussed in Section 6.4.2. The aim is to investigate the possibility of using Lawani River Benin Nigerian kaolin (LBK) as a precursor for Si/Al source in synthesising zeolite NaY. The success of LBK as precursor for zeolite can attract industrial usage thereby creating economic value for the Nigerian

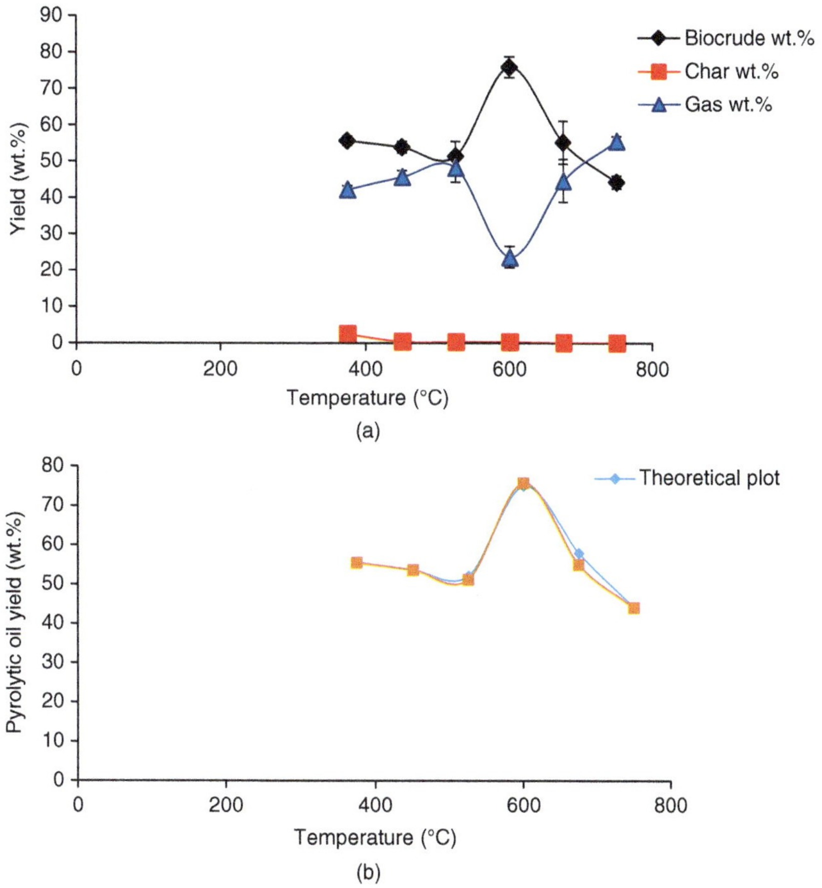

Figure 6.20 Pyrolytic oil production from NR, (a) Effect of temperature on NR pyrolysis products yield. (b) Experimental and theoretical plot of temperature effect (STD: Pyrolytic oil = 10.61, Char = 0.92 and Gas = 10.69).

raw material source. Also, the catalyst was used in order to improve product yield of UT and NRr pyrolysis at optimum temperature. The catalyst to UT and NR ratio range investigated was from 1 to 10 wt.% of the catalyst concentration to the feedstock.

From the study, it was observed that the pyrolytic oil yield for the thermal pyrolysis of UT (as shown in Figure 6.19a and that of NR (as shown in Figure 6.20a) at an optimum temperature of 600 °C was higher than that obtained during catalytic pyrolysis of the two materials at same temperature throughout the entire catalyst for feedstock range studied as reported in Figures 6.21a and 6.22a). The data plotted are the mean value and standard deviation of the pyrolytic oil, char and gas yields for the varied catalyst concentrations (Osayi 2015).

The presence of catalyst only resulted in lower value of pyrolytic oil and gas yield with a corresponding increase in solid char yield for both UT and NR compared to the thermal pyrolysis. It is noteworthy that most of the solid residue obtained during the catalytic pyrolysis of UT and NR contained a deposit of catalyst homogeneously mixed with the feed material before pyrolysis.

From Figure 6.21a, there was a steady increase in gas and solid char yield and a decrease in pyrolytic oil yield till 5 wt.% of the catalyst: UT ratio. A decrease in both gas and pyrolytic oil and an increase in solid char were observed for the catalyst: UT ratio from 5 to 10 wt.%. This could be due to more deposit of catalyst as a result of increased weight percent in the reactor and coke formation from the secondary reaction of the pyrolytic oil. Also for that of catalysed NR pyrolysis, Figure 6.22a shows a steady increase in pyrolytic oil yield from 1 to 7.5 wt.% catalyst: NR ratio with a decrease in the gas yield throughout the 1 to 10 wt.% catalyst: NR ratio. While there was a continuous increase in the solid char yield beyond 7.5 wt.%, a constant yield value of pyrolytic oil (65.87 ± 0.73) was observed for catalyst: NR ratio beyond 7.5 to 10 wt.% indicating no effect of the catalyst on the pyrolytic oil yield.

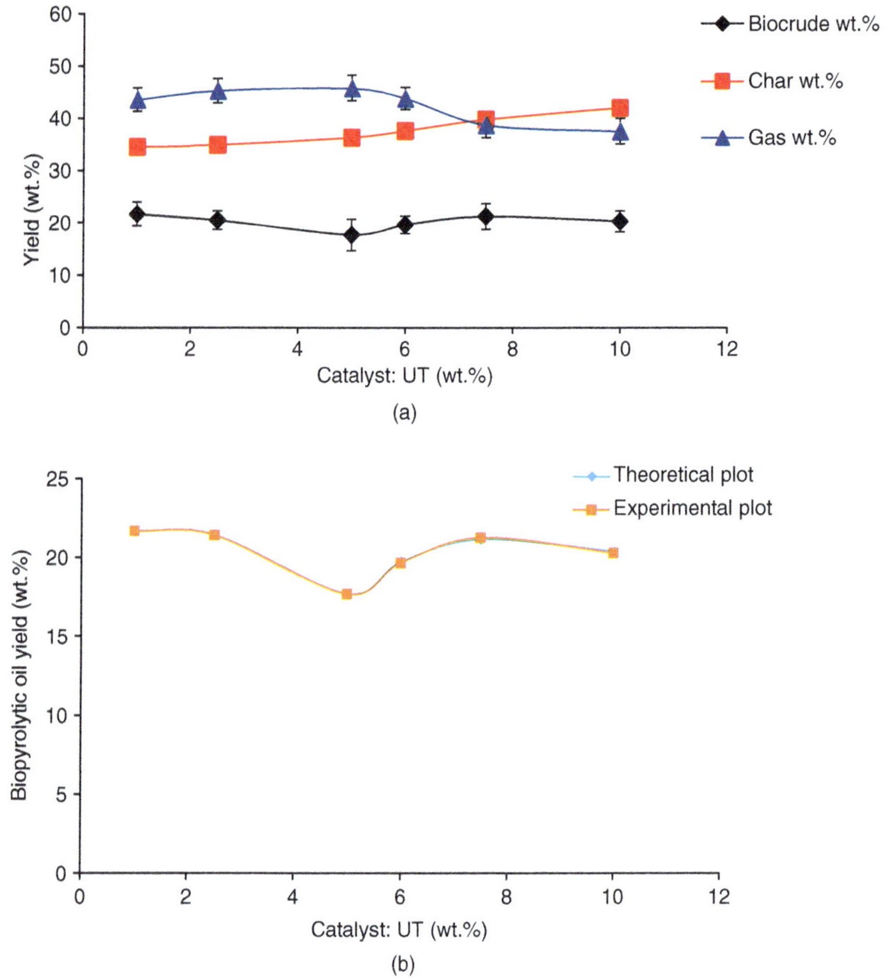

Figure 6.21 Pyrolytic oil production on UT, (a) Catalyst concentration effect on UT pyrolysis at 600 °C. (b) Experimental and theoretical plot of catalyst effect (STD: Pyrolytic oil = 1.41, Char = 2.91 and Gas = 3.46).

This is in agreement with the study done by Pradhan and Singh (2011) on catalytic pyrolysis of UT and tube rubbers at 600 and 700 °C using SiO_2-AlO_2 Kaolin, CaO and MgO catalyst. He reported that none of the catalysts used showed a significant effect in increasing the oil yield compared to the thermal pyrolysis of the studied materials. Similarly, in the two-staged pyrolysis-catalysis of UT by Shen et al. (2006), they observed that the presence of zeolite USY catalyst only caused a reduction in the oil yield and an increase in gas yield with coke formation on the catalyst.

For the effect of catalyst concentration on the production behaviour of pyrolytic oil from UT, the Eureqa Newtonian software was used. A well-fitted theoretical plot with the experimental data was obtained using Eq. (6.6) as shown in Figure 6.21b.

$$Y = 23.16 - 0.49\,A - 3.15\,\sin(4.78 + 1.80\,A) \tag{6.9}$$

where Y = pyrolytic oil yield and A = catalyst concentration.

Figure 6.21b compares the experimental and the theoretical plots with R^2 goodness of fit = 0.99, Correlation Coefficient = 0.99, Maximum Error = 0.12, Mean Squared Error = 0.01, Mean Absolute Error = 0.06 and Complexity = 16.

Similarly, the Eureqa Newtonian software was also used to propose the production behaviour of NR with respect to the catalyst concentration. Both the experimental and theoretical plots as shown in Figure 6.22b fitted well the data.

$$Y = 58.87 + 0.18\,A^2 - 0.001\,A^4 \tag{6.10}$$

where Y = pyrolytic oil yield and A = catalyst concentration.

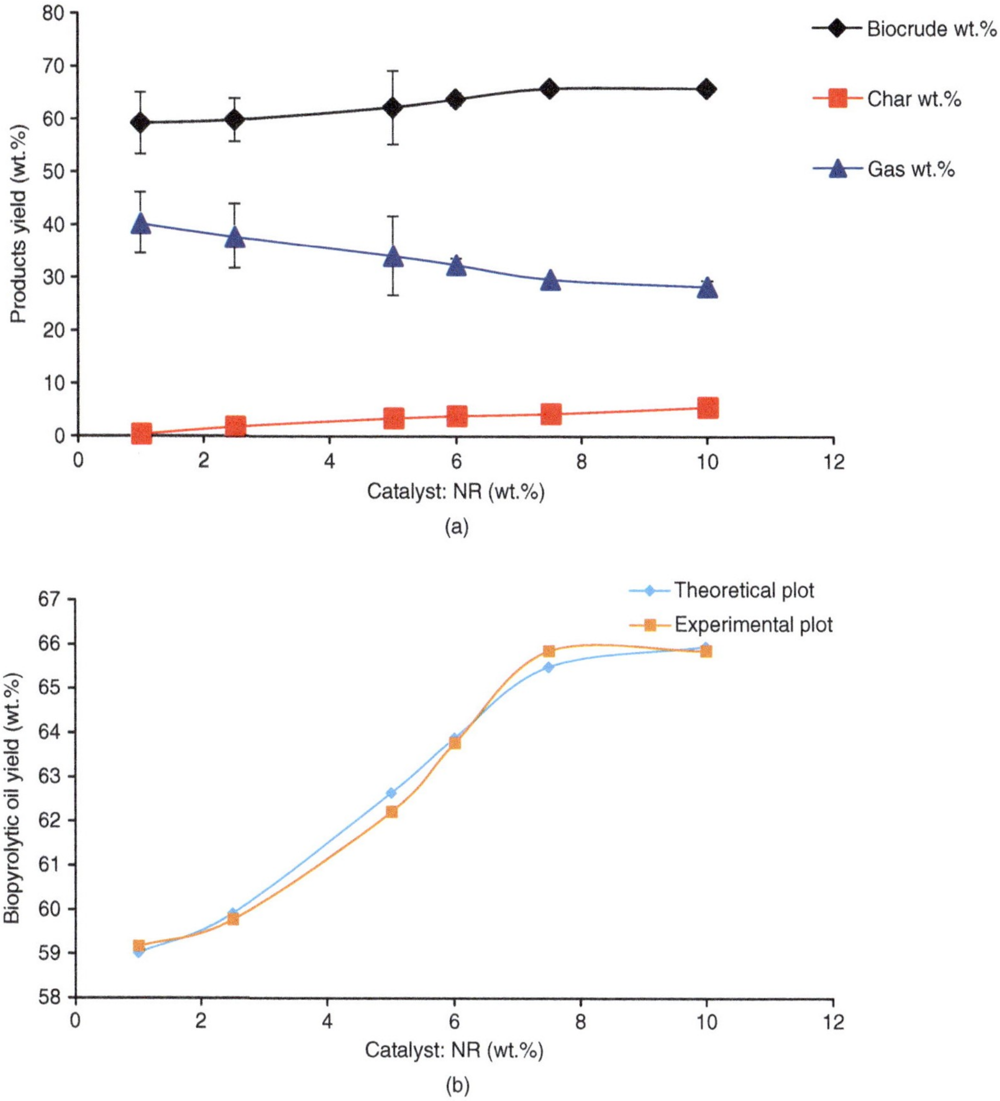

Figure 6.22 Pyrolytic oil production on NR, (a) Catalyst concentration effect on natural rubber pyrolysis at 600 °C. (b) Experimental and theoretical plot of catalyst effect (STD: Pyrolytic oil = 2.90, Char = 1.81 and Gas = 4.62).

Figure 6.22b Compares the experimental and the model plots with R^2 goodness of fit = 0.99, Correlation Coefficient = 0.99, Maximum Error = 0.43, Mean Squared Error = 0.09, Mean Absolute Error = 0.26 and Complexity = 17.

Since in this study, zeolite NaY catalyst had no significant increase in pyrolytic oil yield during the catalytic pyrolysis of NR and UT in comparison with thermal pyrolysis, the effect of the different particle sizes of the synthesised zeolite NaY on the pyrolytic oil composition in terms of gasoline hydrocarbon range was investigated.

6.4.6.13 Effect of Zeolite NaY Particle Size on Bio-Gasoline Yield from Produced Pyrolytic Oil

From the results shown in Figure 6.23, the particle sizes of 0.648 μm and 0.669 μm of zeolite NaY produced the maximum possible yield of 85.73 and 66.17 wt.% bio-gasoline from NR and UT, respectively. The catalytic activity can then be seen from the levels of C_6 to C_{12} hydrocarbon produced from the pyrolysis reaction. Particle sizes greater than the optimum size resulted in lower bio-gasoline yield, this implies that the greater the particle diameter the less its catalytic activity. Similarly, a catalyst with smaller particle size than the optimum size flows easily with the pyrolytic oil produced or are stuck to the wall of the reactor. These catalyst particles which were retained on the reactor wall or flowed along with the pyrolytic oil could reduce the amount of active site of the catalyst available for the reactants resulting in fewer long hydrocarbon chains being broken into short hydrocarbon.

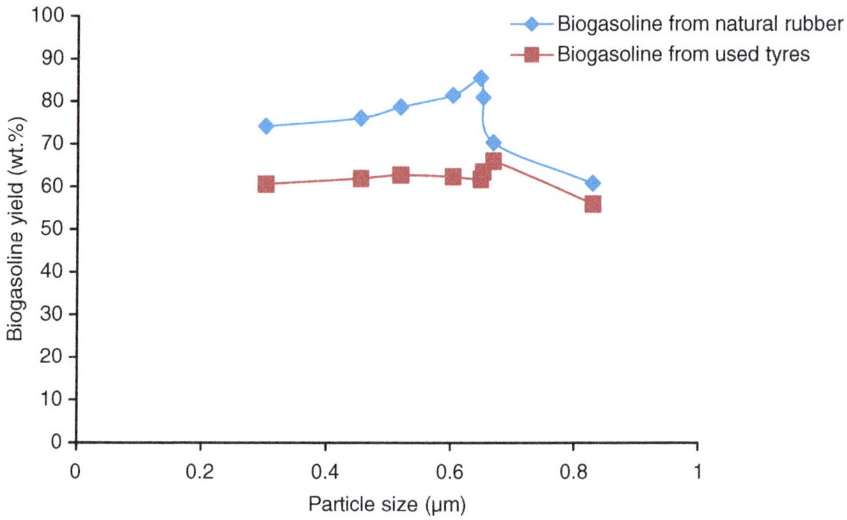

Figure 6.23 Effect of zeolite NaY catalysts particle sizes on pyrolytic oil composition.

The yield of bio-gasoline range obtained at these different sizes of synthesised zeolite NaY was compared with the yield of bio-gasoline range obtained from the use of commercial zeolite NaY (size of 4.113 μm) on NR and UT. From the product formed, 53.17 and 51.77 wt.% of bio-gasoline range was obtained for NR and UT using commercial zeolite NaY. Thus, the synthesised zeolite NaY gave pyrolytic oil with a higher yield of bio-gasoline range than the commercial zeolite. This could be due to less internal mass transfer limitation for the synthesised zeolites associated with their particle sizes which favoured higher catalytic activity as a result of more access to active sites than the commercial zeolite NaY. The sudden drop in bio-gasoline yield observed for both NR and UT from Figure 4.15 can be accounted for by the decrease in available surface area of the catalyst. Also, the catalyst performance in terms of high bio-gasoline yield from NR compared to UT can be attributed to the higher amount of polyisoprene present in NR than in UT.

6.4.6.14 Characterisation of Produced Pyrolytic Oil

The pyrolytic oil produced from both thermal and catalytic pyrolysis of UT and NR as shown in Figure 6.24a,b was an oily dark-brown-colour liquid with acrid smoky smell (Islam et al. 2008; Pradhan and Singh 2011). It was observed that pyrolytic oil from UT had a stronger acrid smell than NR pyrolytic oil probably due to the sulphur content in it (Islam et al. 2008; Pradhan and Singh 2011; Martinez et al. 2014). There was no remarkable phase separation noticed in the NR

(a) (b)

Figure 6.24 Pyrolytic oil from (a) used tyres and (b) natural rubber.

and UT pyrolytic oil. The pyrolytic oil was characterised in terms of both fuel and chemical properties (as discussed in Sections 6.4.6.2 and 6.4.6.3).

6.4.6.15 Physical Properties of Produced Pyrolytic Oil

The physical properties of the pyrolytic oil obtained by thermal and catalytic pyrolysis were analysed for their fuel properties. This is to determine their suitability as liquid fuel and potential for other uses. The fuel properties were determined by using ASTM methods. Table 4.3 shows the fuel properties of the NR pyrolytic oil and UT pyrolytic oil obtained in this study in comparison with that of tyre pyrolytic oil reported by other researchers and commercial automotive (No. 2) diesel. From the table, the following were observed about the pyrolytic oil physical properties.

6.4.6.15.1 Density It is a measure of the mass of a liquid per unit volume at a specific temperature. Density can give an indication of the fuel volume ratio and other fuel properties of a liquid in terms of engine performance. As shown in Table 6.6, the density of NR pyrolytic oil (876 ± 0.01 kg/m^3) was slightly higher than that of the commercial diesel (820–860 kg/m^3) but lower than that obtained from UT pyrolytic oil in this study (890 ± 0.02 kg/m^3) and heavy fuel oil (980 kg/m^3). Since high density is dependent on the molecular weight of the fuel molecules, this implies that NRr pyrolytic oil has heavier molecules than that of conventional diesel and gasoline but lower than that of UT.

6.4.6.15.2 Viscosity The viscosity which is a measure of a fluid frictional property is an important factor in considering the performance of a liquid fuel with respect to resistance to shear, fuel atomisation, and flow in engines and spray burner, must not be too high (Islam et al. 2008). In this study the viscosity of NR and UT pyrolytic oil was 4.65 and 5.18 cSt, respectively, which were both above the maximum range for diesel and gasoline fuel but similar to light fuel oil (Williams 2013). This could be attributed to the presence of heavy compounds in both pyrolytic oil (Pradhan and Singh 2011). As proposed by Islam et al. (2008), this low viscosity is an advantageous characteristic in the handling and transportation of pyrolytic oil.

6.4.6.15.3 Refractive Index It is an important factor to consider as it affects the auto ignition of fuel vapor (Dombrovsky et al. 2001). The refractive index of 1.500 and 1.520 was obtained for NR and UT pyrolytic oil, while 1.508 and 1.514 were obtained for the catalysed NR and UT pyrolytic oil. These values obtained for the refractive index of both the thermal and catalytic NR and UT pyrolytic oil did not deviate much from the acceptable values for that of commercial diesel (1.45–1.475).

6.4.6.15.4 pH Value This is a measure of the acidity or alkalinity level in the fuel. It is important due to the challenges of corrosion such fuel could have on storage systems and the metal components of an engine. The pH value (5.76 and 5.46) of both the thermal and catalytic pyrolytic oil for UT was higher than that for the thermal and catalytic pyrolytic oil of NR (7.68 and 7.64). This revealed the non-acidic nature of NR pyrolytic oil perhaps due to the absence of acidic agents like sulphur.

Table 6.6 Fuel properties of NR pyrolytic oil in comparison to used tyre pyrolytic oil and commercial diesel.

Fuel properties	This study (NR pyrolytic oil)	This study (UT pyrolytic oil)	Pradhan and Singh (2011)	Islam et al. (2008)	Light fuel oil	Commercial No. 2 diesel
Density (kg/m^3)	876 ± 0.01[a)]	890 ± 0.05[a)]	917.9[b)]	970	890	820–860
Viscosity (cSt)	4.65[d)]	5.18[d)]	5.31[d)]	4.90[d)]	21[c)]	2.0–4.5[d)]
Refractive index	1.50, 1.508	1.520, 1.514	—	N/A	—	1.45–1.475
Flash point (°C)	≤36	≤30		≤32	79	>55
Pour point (°C)	−26	−12	−87	−3	—	−40 to −30
pH value	7.68, 7.64	5.76, 5.46	—	4.80	—	N/A
Sulphur content (wt.%)	—	0.035 ± 0.004, 1.35 ± 0.05	1.38	1.36	1.4	1.2
HHV (MJ/kg)	44.13, 45.11	42.72, 42.60	34.61	40.80	44.8	44.0–46.0

a) 20 °C.
b) 15 °C.
c) 30 °C.
d) 40 °C.

6.4.6.15.5 Pour Point The pour point helps to determine the lowest temperature at which the liquid fuel can no longer flow. Reports have shown that pour point is always lower than cloud point. It is the temperature at which fuel gelling occurs resulting in fuel pump failure. Pour point temperature of −26 and −12 °C was obtained for the NR and UT pyrolytic oil, respectively. This implies that the pyrolytic oil would not be problematic for engines in hot and dry regions.

6.4.6.15.6 Sulphur Content The absence of sulphur in the composition of the NR pyrolytic oil makes it a cleaner fuel than UT pyrolytic liquid and fossil heavy fuel oil. The sulphur content obtained for the thermal and catalysed UT pyrolysis pyrolytic oil was 0.035 ± 0.004 and 1.35 ± 0.05 wt.%. The results are within the sulphur content range of 0.11 and 1.54 wt.% for UT pyrolytic oil (Williams 2013) and similar to that of light fuel oil.

6.4.6.15.7 Flash Point The flash point of a liquid is the temperature at which sufficient vapor is evolved from the liquid which can ignite in air. This determines the flammability of the liquid and is a measure of its fire hazard in terms of storage and handling. Considering the flash point of ≤36 °C obtained for NR pyrolytic oil and ≤30 °C for the UT pyrolytic oil, both pyrolytic oil is comparable with that reported for UT pyrolytic oil in literature (Pradhan and Singh 2011; Aguado et al. 2005) but are lower than that of fossil fuels such as commercial diesel, >55 and light oil, which is 79 °C (Islam et al. 2008; Aguado et al. 2005). The low flash points can be attributed to the fact that the pyrolytic oil is composed of a mixture of compounds with a wide distillation temperature range (Islam et al. 2008; Cunliffe and Williams 1998).

6.4.6.15.8 Higher Heating Value (HHV) This is also referred to as GCV or higher calorific value (HCV). It helps to determine the amount of heat that can be released when a specific amount (at 25 °C) of fuel is combusted and its product temperature returns to 25 °C. The HHV of 44.13 ± 1.08 and 45.11 ± 1.22 was obtained for the thermal and catalysed NR pyrolytic oil, which was higher than that of thermal and catalysed (42.72 ± 0.05 and 42.60 ± 0.37) UT pyrolytic oil, and tyre pyrolytic oil reported in literature (Aguado et al. 2005). The NR pyrolytic oil HHV value was within the range for that of commercial diesel (44.0–46.00 MJ/kg) making it a viable alternative source of liquid fuel that can be used directly for electricity generation, industrial furnaces, boilers and static engines.

6.4.6.16 Chemical Properties of the Pyrolytic Oil

In determining the chemical composition of the thermal and catalytic UT and NR pyrolytic oil produced at a temperature of 600 °C at a heating rate of 15 °C/min, recent methods and analytical techniques employed by almost all researchers in identifying and quantifying possible components in pyrolytic liquid derived from different types of organic materials was used. These include FT-IR, NMR and GC–MS analytical techniques as discussed below.

6.4.6.17 FTIR Analysis of UT and NR Pyrolytic Oil

The FT-IR certainly may not be the most suitable analytical technique to check for saturated, aromatic and polar components (Gonzalez et al. 2001; Pradhan and Singh 2011; Islam et al. 2008). However, this analysis operates on the principle of infrared light absorption by most molecules which helps to reveal the chemical properties of the liquid (Pradhan and Singh 2011). Apart from monatomic (helium, neon and argon, etc.) and homopolar diatomic molecules (hydrogen, nitrogen and oxygen, etc.) that do not absorb infrared light, other chemical molecules present in the pyrolytic oil respond to infrared light by bond stretching and contracting in a specific wavelength range independent of the structures of other molecules. Using this principle, the identification of the functional groups present in the pyrolytic oil was achieved.

The FTIR transmittance spectrum for UT and NR pyrolytic oil (thermal and catalysed) are shown in Figures 6.25–6.28, while data results performed on the NR and UT pyrolytic oil compared to those of other authors are presented in Table 6.7. The aliphatic C—H stretched vibrations at 3000–2800 cm^{-1} and C—H deformation vibrations at 1451–1375 cm^{-1} indicate the presence of alkanes. The presence of aromatic compounds is indicated by C—H in-plane bending in 1155–1016 cm^{-1} zones, and the C—H out-of-plane bending peaks in the 965–795 cm^{-1} zone. The peaks between 1650 and 1635 cm^{-1} and 965–795 cm^{-1}, represent C=C stretching vibrations, and are indicative of alkenes. The data reveal therefore that both the thermal and catalysed pyrolytic oil obtained from UT and NR comprise mainly of aliphatic and aromatic compounds with the presence of other hydrocarbon compounds in low proportion. It was also observed from the peaks height in the transmittance spectrum that both pyrolytic oils obtained from NR have more aliphatic hydrocarbon functional groups than that of UT pyrolytic oil. These compounds identified in this study have also been detected by other researchers but in different proportions (Gonzalez et al. 2001).

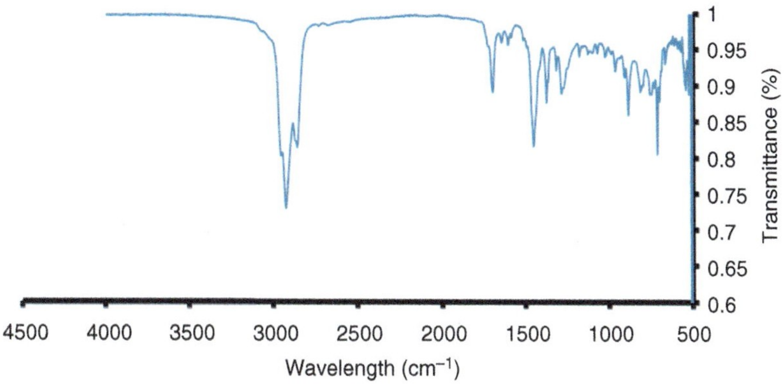

Figure 6.25 FTIR spectrum for crude from used tyres.

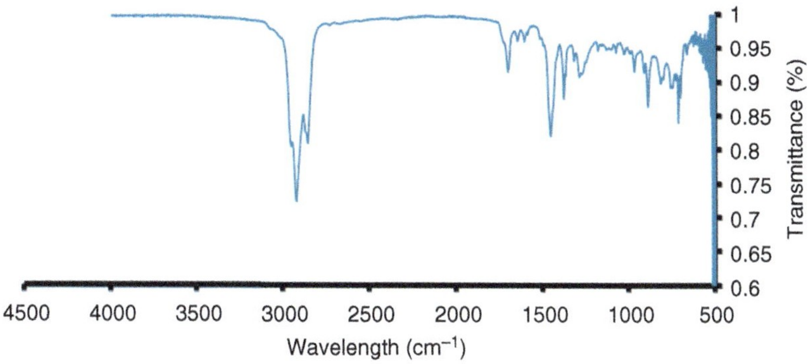

Figure 6.26 FTIR spectrum for biocrude from catalysed used tyres.

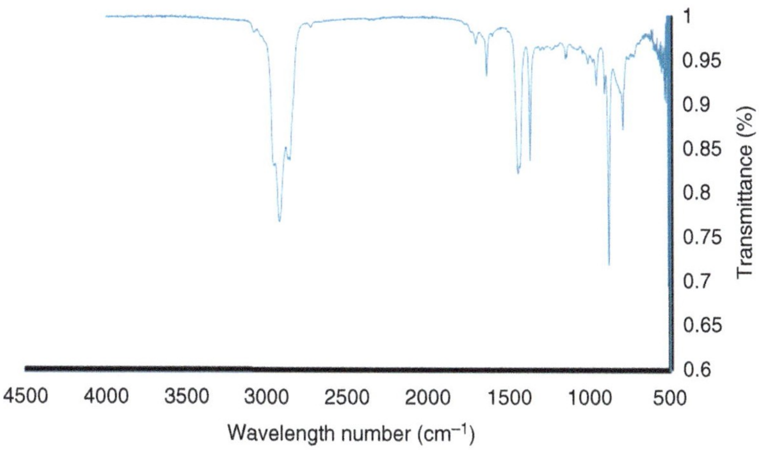

Figure 6.27 FTIR spectrum for biocrude from natural rubber.

6.4.6.18 NMR Analysis of UT and NR Pyrolytic Oil

^{1}H NMR and ^{13}C NMR were performed on the pyrolytic oil produced thermally and catalytically from UT and NR. Figures 6.29–6.32 illustrate the hydrogen and carbon spectrum obtained from the NMR for NR and UT pyrolytic oil with their various protons and carbon content shown in Tables 6.8 and 6.9.

Since quantitative proton NMR analysis was used (inter-pulse delay was greater than $5 \times T_1$ relaxation). Therefore, the area of the NMR signal quantitatively represents the hydrogen in that environment (chemical shift range). The mole percent of hydrogen for each chemical shift range was calculated as follows:

$$\text{mol\%H}_2 = \frac{\text{Area under the peaks of region (chemical shift)}}{\text{Total hydrogen integration}} \times 100 \tag{6.11}$$

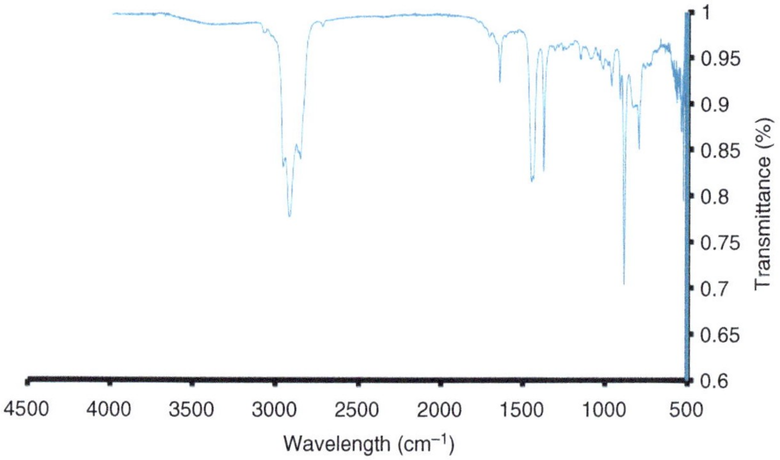

Figure 6.28 FTIR spectrum for biocrude from catalysed natural rubber.

Table 6.7 The FTIR functional groups and compounds found in the pyrolytic oil and other liquids.

Wavelength number (cm^{-1})					
This study	**Islam et al. (2008)**	**Gonzalez et al. (2001)**	**Williams et al. (1990)**	**Functional groups**	**Class of compounds**
—	—	3600–3250	—	O—H stretching	Alcohols, phenols or carboxylic acids
—	—	3250–3100	—	C—H stretching	Aromatic compounds
—	3100–3005	—	—	C=C stretching	Alkenes
3000–2800	3000–2800	3000–2800	3000–2800	C—H stretching	Alkanes
1750–1700	—	—	1750–1650	C=O	Aldehydes or Ketones
1650–1635	1680–1635	1670–1580	1675–1575	C=C stretching	Alkenes
—	1600–1530	1600–1500	1625–1575	C—C stretching	Aromatic compounds
1451–1375	1520–1115	1480–1360	1475–1350	C—H bending	Alkanes
1155–1016	—	1150–1000	—	C—H in plane bending	Aromatic compounds
965–795	1030–835	960–870	950–875	C=C stretching	Alkenes
630–609	830–655	900–675	900–675	C—H out-of-plane bending	Aromatic compounds

Results revealed that the various pyrolytic oil produced is a mixture of aromatics, olefins, paraffins and a small amount of alcohols (species containing Hydroxyl group). The aliphatic protons not bonded to aromatic constitute more than 50% of the total hydrogen (0.4–1.8 ppm chemical shift range), whereas the aliphatic adjacent to aromatic/alkenes group was more than 20% (3.3–1.8 ppm chemical shift range). A higher amount of aromatic hydrogen greater than 7% (9.0–6.5 ppm chemical shift range) was recorded from UT pyrolytic oil than that of NR (which was less than 2.10%) as shown in Tables 6.8 and 6.9. This could be attributed to the polymer nature of NR, which is mainly poly isoprene (most often cis-1, 4-polyisoprene).

From Table 6.8, the NR pyrolytic oil obtained from thermal pyrolysis gave higher aliphatic carbon content, 84.38% and low aromatic carbon content, 15.62% (a source of polycyclic aromatic hydrocarbon, PAHs, which are pollutants) than the pyrolytic oil produced from UT as determined by quantitative ^{13}C NMR analysis (no NOE enhancement for quantitative ^{13}C pulse sequence). The decrease in the aliphatic carbon content from 84.34% to 69.58% revealed by the catalysed natural pyrolytic oil can be related to the role of hydrocracking performed by the zeolite NaY catalyst during the catalytic pyrolysis of the NR (Shen et al. 2006). This confirms that the catalyst helped in cracking the pyrolytic oil hydrocarbon and also favoured the formation of aromatic compounds.

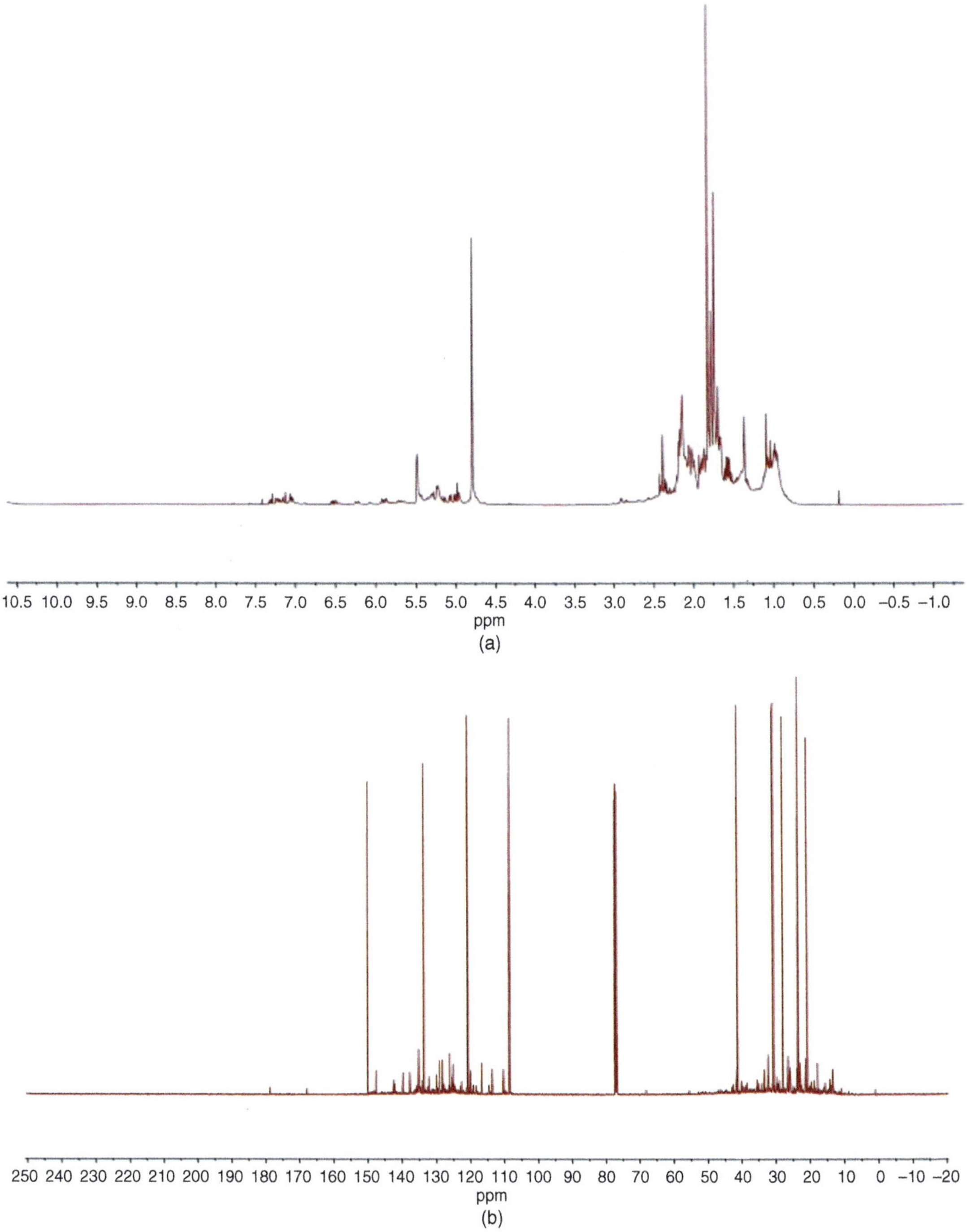

Figure 6.29 NMR spectra of pyrolytic oil for NR, (a) ^{1}H and (b) ^{13}C.

6.4.6.19 GC–MS Analysis of Pyrolytic Oil from UT and NR

The GC–MS analysis was performed on the catalysed and un-catalysed pyrolytic oil obtained from NR and UT in this study with the aim of identifying the nature and type of compounds present in them. The different chromatogram peaks for pyrolytic oil-derived compounds shown in Figures 6.33–6.36 were analysed with NIST search software, with over 400 compounds identified. Considering the percentage area of compounds identified by the GC–MS, compounds with match peaks ≥75% and an area percent ≥0.1 were reported. A few of these compounds are shown in Tables 6.10 and 6.11. The compounds identified in this study are in accordance with that published in the literature (Williams 2013; Shen et al. 2006; Williams and Brindle 2003).

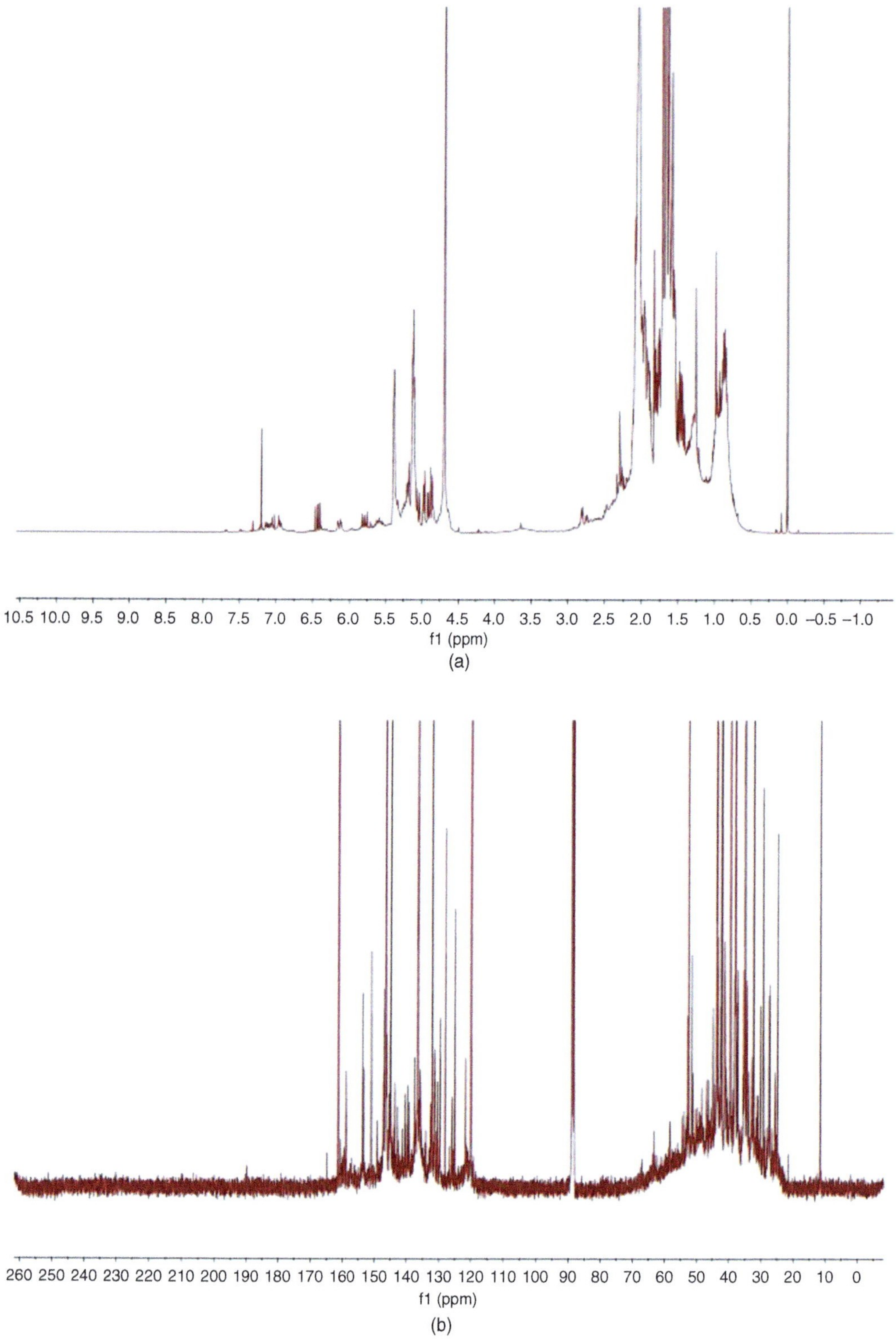

Figure 6.30 NMR spectra of pyrolytic oil for NRZ, (a) ^{1}H and (b) ^{13}C.

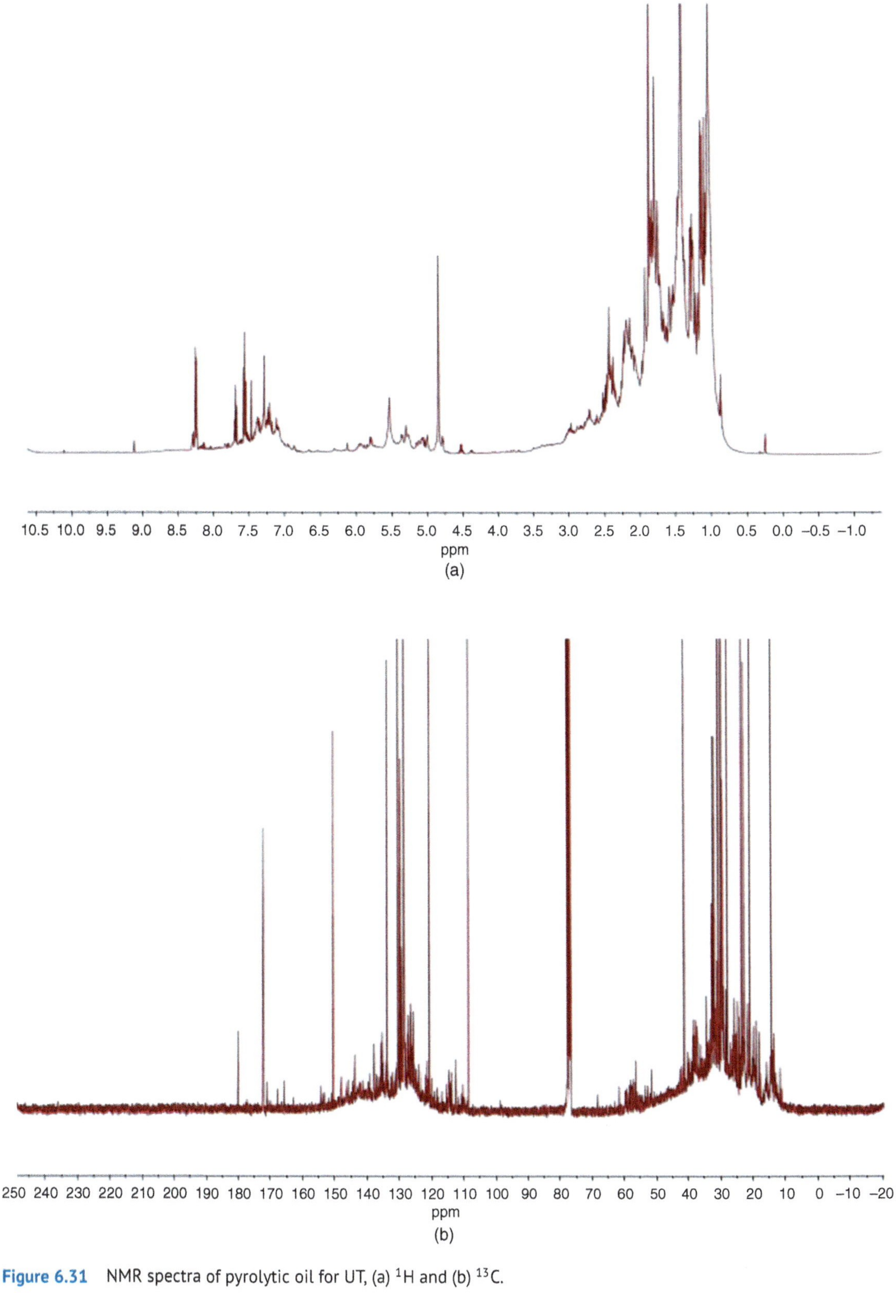

Figure 6.31 NMR spectra of pyrolytic oil for UT, (a) ^{1}H and (b) ^{13}C.

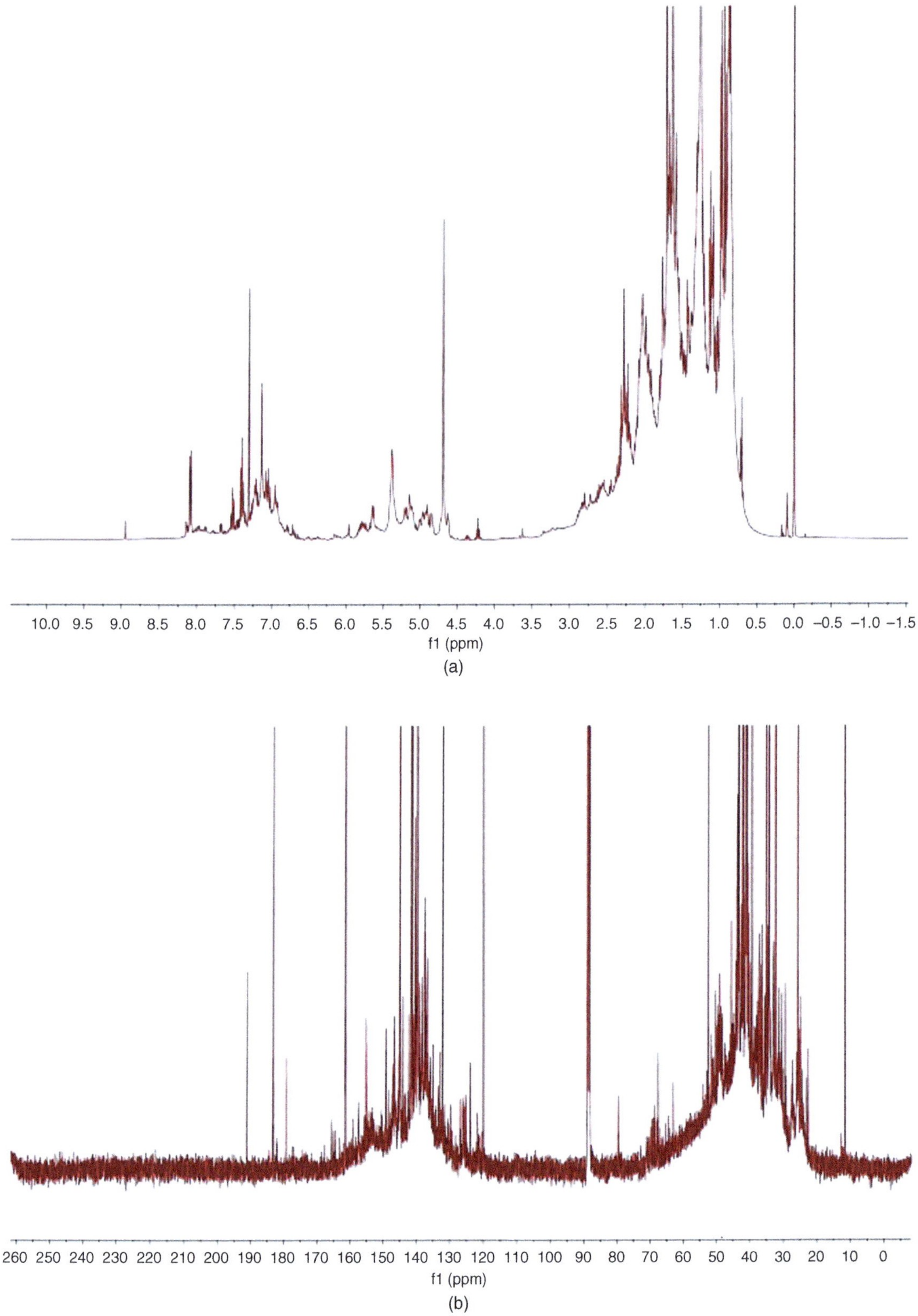

Figure 6.32 NMR spectra of pyrolytic oil for UTZ, (a) ^{1}H and (b) ^{13}C.

Table 6.8 ^{1}H and ^{13}C NMR results for NR pyrolytic oil and catalysed NR pyrolytic oil obtained at 600 °C.

Types of hydrogen	Chemical shift (ppm)	Percent of total hydrogen (mol %)
NR pyrolytic oil		
Aromatic	9.0–6.5	2.08
Phenolic (OH) or olefinic portion	6.5–4.5	12.91
Aliphatic adjacent to oxygen/hydroxyl group	4.5–3.3	0.22
Aliphatic adjacent to aromatic/alkene group	3.3–1.8	25.33
Other aliphatic (bonded to aliphatic only)	1.8–0.4	59.48
Aliphatics	75.0–5.0	84.38
Aromatics	185.0–100	15.62
NRZ pyrolytic oil		
Aromatic	9.0–6.5	0.95
Phenolic (OH) or olefinic portion	6.5–4.5	13.65
Aliphatic adjacent to oxygen/hydroxyl group	4.5–3.3	0.67
Aliphatic adjacent to aromatic/alkene group	3.3–1.8	29.39
Other aliphatic (bonded to aliphatic only)	1.8–0.4	55.34
Aliphatics	75.0–5.0	69.58
Aromatics	185.0–100	30.42

Table 6.9 ^{1}H and ^{13}C NMR results for UT pyrolytic oil and catalysed UT pyrolytic oil obtained at 600 °C.

Types of hydrogen	Chemical shift (ppm)	Percent of total hydrogen (mol %)
UT pyrolytic oil		
Aromatic	9.0–6.5	7.0
Phenolic (OH) or olefinic portion	6.5–4.5	5.60
Aliphatic adjacent to oxygen/hydroxyl group	4.5–3.3	0.66
Aliphatic adjacent to aromatic/alkene group	3.3–1.8	21.75
Other aliphatic (bonded to aliphatic only)	1.8–0.4	64.94
Aliphatics	75.0–5.0	67.59
Aromatics	185.0–100	32.41
UTZ pyrolytic oil		
Aromatic	9.0–6.5	7.12
Phenolic (OH) or olefinic portion	6.5–4.5	6.27
Aliphatic adjacent to oxygen/hydroxyl group	4.5–3.3	0.93
Aliphatic adjacent to aromatic/alkene group	3.3–1.8	21.23
Other aliphatic (bonded to aliphatic only)	1.8–0.4	64.45
Aliphatics	75.0–5.0	68.41
Aromatics	185.0–100	31.59

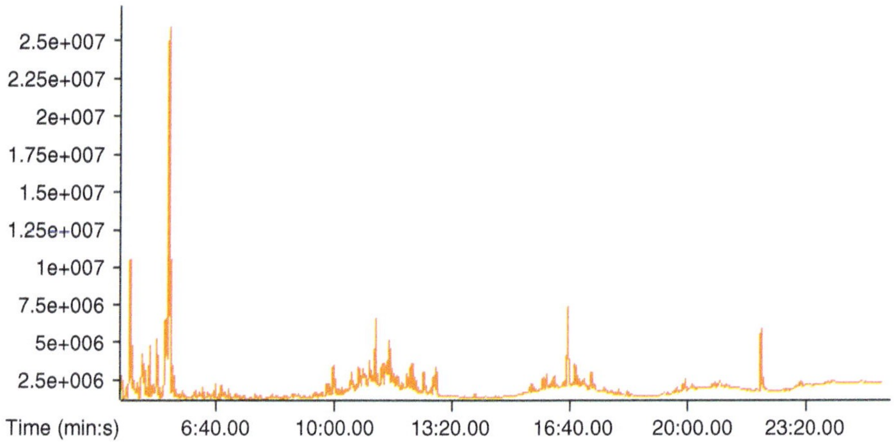

Figure 6.33 GC–MS total compound chromatogram for NR pyrolytic oil obtained at 600 °C.

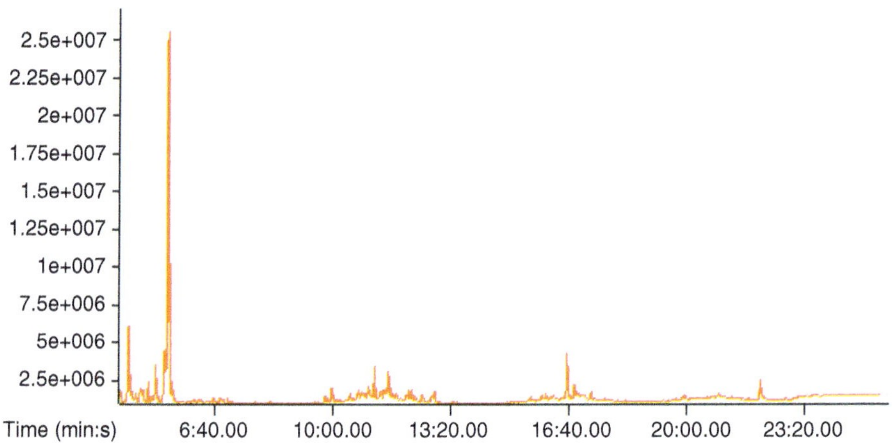

Figure 6.34 GC–MS total compound chromatogram of NRZ pyrolytic oil obtained at 600 °C.

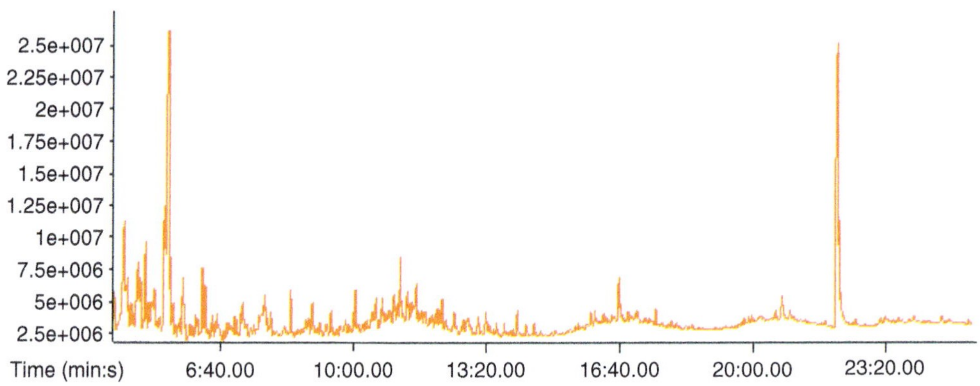

Figure 6.35 GC–MS total compound chromatogram for UT pyrolytic oil obtained at 600 °C.

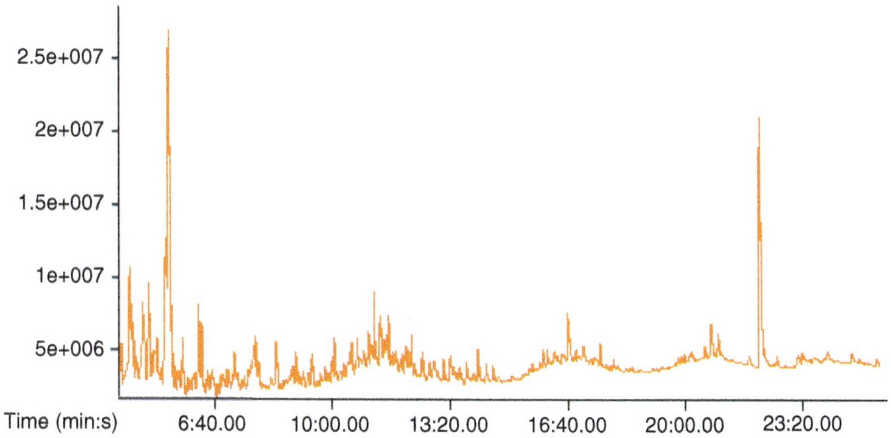

Figure 6.36 GC–MS total compound chromatogram for UTZ pyrolytic oil obtained at 600 °C.

Table 6.10 Some compounds identified by GC–MS characterisation of (a) NR pyrolytic oil and (b) NRZ pyrolytic oil.

Peak #	Name	R.T. (min:s)	Area %	Formula
(a) NR pyrolytic oil				
1	1,3,6-Heptatriene, 2,5,5-trimethyl-	04:01.5	0.4426	$C_{10}H_{16}$
27	2,6-Dimethyl-1,3,6-heptatriene	05:01.5	0.98112	C_9H_{14}
52	1-Penten-3-yne, 2-methyl-	07:01.4	0.20269	C_6H_8
28	1,3-Cyclopentadiene, 5,5-dimethyl-2-propyl-	05:04.9	0.3613	$C_{10}H_{16}$
71	1,7-Octadiene, 2,7-dimethyl-3,6-bis(methylene)-	10:00.2	0.12479	$C_{12}H_{18}$
29	2,6-Dimethyl-1,3,5,7-octatetraene, *E*,*E*-	05:05.5	0.15227	$C_{10}H_{14}$
43	1,3,5-Cycloheptatriene, 3,7,7-trimethyl-	06:06.1	0.30222	$C_{10}H_{14}$
89	Furan, 3-(4,8-dimethyl-3,7-nonadienyl)-, (*E*)-	11:01.1	1.0142	$C_{15}H_{22}O$
2	2-Cyclohexen-1-ol, 4-ethyl-1,4-dimethyl-	04:10.5	0.84296	$C_{10}H_{18}O$
113	cis-à-Bisabolene	12:03.8	0.10683	$C_{15}H_{24}$
90	(3*S*,6*R*)-3-Hydroperoxy-3-methyl-6-(prop-1-en-2-yl) cyclohex-1-ene	11:06.3	0.45876	$C_{10}H_{16}O_2$
91	Naphthalene, 1,7-dimethyl-	11:07.0	0.13136	$C_{12}H_{12}$
131	Squalene	16:02.5	0.87178	$C_{30}H_{50}$
31	Benzene, 1-ethyl-2-methyl-	05:14.6	0.56069	C_9H_{12}
3	Cyclohexene, 1-methyl-4-(1-methylethenyl)-, (*S*)-	04:16.2	8.207	$C_{10}H_{16}$
32	Cyclohexene, 3-methyl-6-(1-methylethyl)-	05:16.0	0.31605	$C_{10}H_{18}$
33	(1*S*)-2,6,6-Trimethylbicyclo[3.1.1]hept-2-ene	05:16.6	1.5327	$C_{10}H_{16}$
92	(3*E*,7*E*)-4,8,12-Trimethyltrideca-1,3,7,11-tetraene	11:10.7	1.5109	$C_{16}H_{26}$
93	ç-Elemene	11:11.0	0.17198	$C_{15}H_{24}$
4	Benzene, 1-ethyl-2-methyl-	04:18.4	0.33184	C_9H_{12}
46	Benzene, 1-methyl-4-(1-methylethenyl)-	06:17.5	0.29391	$C_{10}H_{12}$
7	Benzene, 1,2,3-trimethyl-	04:21.2	0.26541	C_9H_{12}
117	Cyclohexane, 1-ethenyl-1-methyl-2,4-bis(1-methylethenyl)-	12:13.8	0.32085	$C_{15}H_{24}$
118	anti-10-Methyl-endo-tricyclo[5.2.1.0(2.6)]decane	12:14.3	0.76028	$C_{11}H_{18}$

Table 6.10 (Continued)

Peak #	Name	R.T. (min:s)	Area %	Formula
145	Diisooctyl phthalate	22:04.4	6.7325	$C_{24}H_{38}O_4$
8	1,3-Dimethyl-1-cyclohexene	04:23.2	0.243	C_8H_{14}
34	Limonene	05:22.9	42.609	$C_{10}H_{16}$
36	Benzene, 1,2,3-trimethyl-	05:24.9	0.15205	C_9H_{12}
9	Benzene, 1,2,3-trimethyl-	04:26.0	0.18654	C_9H_{12}
(b) NRZ pyrolytic oil				
1	1,3,6-Heptatriene, 2,5,5-trimethyl-	04:01.6	0.30856	$C_{10}H_{16}$
20	2,6-Dimethyl-1,3,6-heptatriene	05:01.5	0.87031	C_9H_{14}
34	2,4,6-Octatriene, 2,6-dimethyl-, (*E*,*Z*)-	07:01.5	0.21984	$C_{10}H_{16}$
37	1,7-Octadiene, 2,7-dimethyl-3,6-bis(methylene)-	10:00.1	0.79883	$C_{12}H_{18}$
21	2,6-Dimethyl-1,3,5,7-octatetraene, *E*,*E*-	05:05.4	0.11845	$C_{10}H_{14}$
44	(3*E*,7*E*)-4,8,12-Trimethyltrideca-1,3,7,11-tetraene	11:00.9	0.85701	$C_{16}H_{26}$
2	2-Cyclohexen-1-ol, 4-ethyl-1,4-dimethyl-	04:10.5	0.21119	$C_{10}H_{18}O$
61	anti-10-Methyl-endo-tricyclo[5.2.1.0(2.6)]decane	12:05.4	0.33091	$C_{11}H_{18}$
73	trans-Farnesol	16:02.5	0.74242	$C_{15}H_{26}O$
22	Mesitylene	05:14.6	0.306	C_9H_{12}
3	Cyclohexene, 1-methyl-4-(1-methylethenyl)-, (*S*)-	04:16.1	6.5379	$C_{10}H_{16}$
23	Cyclohexene, 3-methyl-6-(1-methylethyl)-	05:15.9	0.29224	$C_{10}H_{18}$
63	Bicyclo[3.1.1]heptane, 6,6-dimethyl-3-methylene-	12:08.9	0.32341	$C_{10}H_{16}$
24	(1*S*)-2,6,6-Trimethylbicyclo[3.1.1]hept-2-ene	05:16.5	0.33779	$C_{10}H_{16}$
79	4,8,12-Tetradecatrien-1-ol, 5,9,13-trimethyl-	17:04.6	0.23275	$C_{17}H_{30}O$
46	Cyclohexane, 1-ethenyl-1-methyl-2,4-bis(1-methylethenyl)-	11:10.6	1.3417	$C_{15}H_{24}$
4	Benzene, 1,2,3-trimethyl-	04:18.4	0.14847	C_9H_{12}
30	Benzene, 1-methyl-4-(1-methylethenyl)-	06:17.3	0.12066	$C_{10}H_{12}$
6	Benzene, 1,4-diethyl-	04:21.2	0.22763	$C_{10}H_{14}$
64	Cyclohexene, 1-methyl-4-(1-methylethenyl)-, (*S*)-	12:13.9	0.19425	$C_{10}H_{16}$
65	5-Hepten-1-ol, 2-ethenyl-6-methyl-	12:14.2	0.45075	$C_{10}H_{18}O$
85	Diisooctyl phthalate	22:04.2	3.893	$C_{24}H_{38}O_4$
7	trans,cis-1,7-Dimethylspiro[4.5]decane	04:23.3	0.15007	$C_{12}H_{22}$
25	Limonene	05:22.7	66.274	$C_{10}H_{16}$
47	2,6,9,11-Dodecatetraenal, 2,6,10-trimethyl-, (*E*,*E*,*E*)-	11:19.9	0.13224	$C_{15}H_{22}O$
48	1,5-Heptadiene, 2,6-dimethyl-	11:20.3	0.4911	C_9H_{16}
9	Bicyclo[2.2.1]heptane, 2,2-dimethyl-3-methylene-, (1*R*)-	04:27.4	0.28681	$C_{10}H_{16}$
74	1,3-Bis-(2-cyclopropyl,2-methylcyclopropyl)-but-2-en-1-one	16:16.2	0.28873	$C_{18}H_{26}O$
49	1-Cyclohexyl-1-pentyne	11:21.8	0.11196	$C_{11}H_{18}$
10	3,4-Octadiene	04:29.8	0.1834	C_8H_{14}
27	Cyclohexene, 4-methyl-1-(1-methylethenyl)-	05:29.4	0.12879	$C1_0H_{16}$
11	trans-3-Caren-2-ol	04:35.2	0.26577	$C_{10}H_{16}O$
51	(3*E*,7*E*)-4,8,12-Trimethyltrideca-1,3,7,11-tetraene	11:28.7	0.64462	$C_{16}H_{26}$
82	(1,2,3-Trimethyl-cyclopent-2-enyl)-methanol	17:23.2	0.16199	$C_9H_{16}O$
12	2,6-Octadiene, 2,6-dimethyl-	04:36.5	1.4042	$C_{10}H_{18}$

Table 6.11 Some identified compounds by GC–MS characterisation of (a) UT pyrolytic oil and (b) UTZ pyrolytic oil.

Peak #	Name	R.T. (min:s)	Area %	Formula
(a) UT pyrolytic oil				
1	1,3,6-Heptatriene, 2,5,5-trimethyl-	04:01.9	0.27622	$C_{10}H_{16}$
2	Camphene	04:02.8	0.1468	$C_{10}H_{16}$
46	2-Methyl-1-nonene-3-yne	05:01.9	0.36244	$C_{10}H_{16}$
115	2,4,6-Octatriene, 2,6-dimethyl-, (*E*,*Z*)-	07:01.9	0.25266	$C_{10}H_{16}$
80	p-Cresol	06:03.2	0.3688	C_7H_8O
170	4-Isopropyl-1-phenyl-hexa-1,5-dien-3-ol	09:00.4	0.10126	$C_{15}H_{20}O$
5	1,1,3,3,5-Pentamethylcyclohexane	04:06.2	0.23291	$C_{11}H_{22}$
205	Isomyocorene	10:07.7	0.17025	$C_{10}H_{16}$
118	1,2,3,4,5,8-Hexahydronaphthalene	07:11.0	0.11379	$C_{10}H_{14}$
85	Benzene, 2-ethyl-1,3-dimethyl-	06:12.2	0.32995	$C_{10}H_{14}$
119	Phenol, 2,3-dimethyl-	07:11.4	0.2892	$C_8H_{10}O$
338	Squalene	16:02.8	0.36331	$C_{30}H_{50}$
86	Cyclohexene, 1-methyl-4-(1-methylethylidene)-	06:13.4	0.39596	$C_{10}H_{16}$
51	Benzene, 1,2,3-trimethyl-	05:14.9	1.6238	C_9H_{12}
176	1*H*-Indene, 2,3-dihydro-4,7-dimethyl-	09:11.0	0.26573	$C_{11}H_{14}$
12	Cyclohexene, 1-methyl-4-(1-methylethenyl)-, (*S*)-	04:16.5	2.5857	$C_{10}H_{16}$
88	o-Isopropenyltoluene	06:14.6	0.46476	$C_{10}H_{12}$
324	Sulphurous acid, butyl cyclohexylmethyl ester	14:06.9	0.90262	$C_{11}H_{22}O_3S$
277	1,3,6,10-Dodecatetraene, 3,7,11-trimethyl-, (*Z*,*E*)-	12:09.0	0.21258	$C_{15}H_{24}$
121	Naphthalene, 1,2-dihydro-	07:14.3	0.33148	$C_{10}H_{10}$
13	Cyclohexene, 1-methyl-4-(1-methylethenyl)-, (*S*)-	04:17.3	0.59849	$C_{10}H_{16}$
52	Cyclohexene, 3-methyl-6-(1-methylethyl)-	05:16.4	0.24432	$C_{10}H_{18}$
304	Sulphurous acid, cyclohexylmethyl hexyl ester	13:08.6	0.50938	$C_{13}H_{26}O_3S$
122	5*H*-5-Methyl-6,7-dihydrocyclopentapyrazine	07:15.0	0.12562	$C_8H_{10}N_2$
53	o-Cymene	05:17.3	3.7413	$C_{10}H_{14}$
14	Benzene, 1-ethyl-2-methyl-	04:18.8	0.33273	C_9H_{12}
91	Benzene, 1-methyl-4-(1-methylethenyl)-	06:17.8	0.56752	$C_{10}H_{12}$
54	2-Decene, 5-methyl-, (*Z*)-	05:19.3	0.91225	$C_{11}H_{22}$
55	Limonene	05:19.8	1.3456	$C_{10}H_{16}$
124	Benzene, pentyl-	07:18.4	0.22583	$C_{11}H_{16}$
149	Benzene, 1-methyl-3-(1-methyl-2-propenyl)-	08:17.5	0.13419	$C_{11}H_{14}$
378	Heptacosane	24:01.8	0.14024	$C_{27}H_{56}$
57	Limonene	05:23.5	14.275	$C_{10}H_{16}$
180	Benzene, 4-(2-butenyl)-1,2-dimethyl-, (*E*)-	09:19.9	0.14735	$C_{12}H_{16}$
366	Diisooctyl phthalate	22:07.2	6.6322	$C_{24}H_{38}O_4$
(b) UTZ pyrolytic oil				
1	1,3,6-Heptatriene, 2,5,5-trimethyl-	04:02.1	0.24514	$C_{10}H_{16}$
2	2-Methyl-1-nonene-3-yne	04:03.0	0.11745	$C_{10}H_{16}$
109	2,4,6-Octatriene, 2,6-dimethyl-, (*E*,*Z*)-	07:01.9	0.25173	$C_{10}H_{16}$
79	Benzyl alcohol	06:03.2	0.18381	C_7H_8O
162	4-Isopropyl-1-phenyl-hexa-1,5-dien-3-ol	09:00.4	0.10059	$C_{15}H_{20}O$

Table 6.11 (Continued)

Peak #	Name	R.T. (min:s)	Area %	Formula
163	1*H*-Indene, 2,3-dimethyl-	09:00.9	0.10958	$C_{11}H_{12}$
187	Tricyclo[3.3.0.0(2,8)]octan-3-one, 4-methyl-4-(2-methyl-2-propenyl)-	10:00.4	0.11512	$C_{13}H_{18}O$
110	3,4-Dimethylcumene	07:04.6	0.10434	$C_{11}H_{16}$
188	3-Penten-1-ol, 2,2,4-trimethyl-	10:03.7	1.1644	$C_8H_{16}O$
48	1,3-Cyclohexadiene, 1-methyl-4-(1-methylethyl)-	05:09.5	0.34969	$C_{10}H_{16}$
330	Squalene	16:02.9	0.39697	$C_{30}H_{50}$
84	ç-Terpinene	06:13.4	0.44049	$C_{10}H_{16}$
51	Benzene, 1-ethyl-2-methyl-	05:14.9	1.4766	C_9H_{12}
167	1*H*-Indene, 2,3-dihydro-4,7-dimethyl-	09:11.0	0.27705	$C_{11}H_{14}$
344	Dibutyl phthalate	17:03.2	0.22591	$C_{16}H_{22}O_4$
114	1*H*-Indene, 2,3-dihydro-4-methyl-	07:13.3	0.74024	$C_{10}H_{12}$
10	Cyclohexene, 1-methyl-4-(1-methylethenyl)-, (*S*)-	04:16.6	2.4334	$C_{10}H_{16}$
86	Indan, 1-methyl-	06:14.7	0.45638	$C_{10}H_{12}$
307	Sulphurous acid, butyl cyclohexylmethyl ester	14:07.1	1.0533	$C_{11}H_{22}O_3S$
116	5*H*-5-Methyl-6,7-dihydrocyclopentapyrazine	07:15.1	0.11195	$C_8H_{10}N_2$
225	4,8,12-Tetradecatrien-1-ol, 5,9,13-trimethyl-	11:11.1	0.78057	$C_{17}H_{30}O$
53	o-Cymene	05:17.3	3.6405	$C_{10}H_{14}$
12	Benzene, 1-ethyl-2-methyl-	04:18.9	0.32468	C_9H_{12}
346	Cyclohexane, 1-bromo-4-methyl-	17:06.0	0.13808	$C_7H_{13}Br$
88	Benzene, 1-methyl-4-(1-methylethenyl)-	06:17.8	0.64512	$C_{10}H_{12}$
55	D-Limonene	05:19.9	0.87397	$C_{10}H_{16}$
118	Benzene, pentyl-	07:18.4	0.23006	$C_{11}H_{16}$
142	1*H*-Indene, 2,3-dihydro-1,1-dimethyl-	08:17.5	0.14803	$C_{11}H_{14}$
15	Benzene, 1-ethyl-3-methyl-	04:21.5	0.5109	C_9H_{12}
367	Diisooctyl phthalate	22:06.2	3.888	$C_{24}H_{38}O_4$
56	Limonene	05:23.3	14.059	$C_{10}H_{16}$
359	trans-Geranylgeraniol	20:08.7	0.13648	$C_{20}H_{34}O$

It is observed that pyrolytic oil obtained from the thermal and catalytic pyrolysis of UT and NR are complex mixtures dominated by different aliphatic and aromatic compounds. Examples of such compounds are alkanes, alkenes, alkynes, single-ring benzene and its derivatives, naphthalene, oxygenated compounds, nitrogen-containing compounds and sulphur compounds like thiophene and quinoline. These GC–MS results conform well to the FTIR data of Table 6.7 and ^{1}H NMR of Tables 6.8 and 6.9. The non-catalysed and catalysed NR pyrolytic oil gave 79.87 and 65.76% concentrations of aliphatic, 12.45 and 14.84% aromatic concentrations and PAHs concentration of 1.98 and 0.79% while the non-catalysed and catalysed UT pyrolytic oil had 42.20 and 32.09% concentrations of aliphatic, 34.49 and 38.80% aromatic concentrations and PAHs concentrations of 17.84 and 22.82%, respectively.

Similar to other studies (Islam et al. 2007; De Marco et al. 2001; Kyari et al. 2005), other compound species were also present in the obtained pyrolytic oil but in low proportions. The high aromatic and PAHs recorded in both UT pyrolytic oil could be attributed to the nature of rubber component used in the manufacture of the sample UT that was pyrolysed. A typical example is polystyrene butadiene (SBR) which could be additional benzene ring source and the dehydrogenation reaction that leads to cyclisation of olefin structures during pyrolysis process (De Marco et al. 2001). This is because rubber samples containing a high amount of aromatics will yield liquids with high aromatic content under the same condition (Gonzalez et al. 2001; Cypres and Bettens 1989; Barbooti et al. 2004). Kyari et al. (2005) in their investigation have reported that different types and brands of tyres will yield different concentrations of a benzene ring and PAHs in their products oil.

Islam et al. (2007) also reported UT pyrolysis liquid fraction as a complex mixture composed of 49.54% aliphatic content and 16.65% aromatic content.

In this study, Limonene identified in the GC–MS analysis, is one of the most important compounds (Gonzalez et al. 2001; Williams and Brindle 2003; De Marco et al. 2001; Kyari et al. 2005; Pakdel et al. 2004, 1992, 2001), which could increase the market value of the pyrolytic oil. The amount of Limonene obtained from the pyrolysis of NR is greater than that reported by other researchers for UT pyrolysis (Pradhan and Singh 2011; Islam et al. 2007).

From the FTIR, NMR and GC–MS analyses, the pyrolytic oil produced from the thermal and catalytic pyrolysis of UT and NR at 600 °C at a HR of 15 °C/min can be concluded to be a very complex mixture of organic compounds of aliphatic (C_5—C_{30} majority being C_8—C_{10}), aromatics, PAHs and a small fraction of oxygen, nitrogen and sulphur containing compounds. Pyrolytic oil from NR due to the absence of sulfur, its relatively low amount of aromatic and PAHs could be less source of pollution when combusted directly as liquid fuel than pyrolytic oil from UT.

6.4.6.20 Batch Distillation of UT and NR Pyrolytic Oil

Table 6.12 presents the total percentages of fractional volume of distillates obtained from the thermal and catalytic pyrolysis NR and UT pyrolytic oil at *different ranges* of *temperature* during distillation. As stated above a total volume of 120 mL of the various produced pyrolytic oil was distilled in batches. For the NR pyrolytic oil, a total fractional volume percent of 41.31% was obtained over the distillation temperature range of 60 to 175 °C while the remaining 58.69% formed part of the pyrolytic oil lost to evaporation and the heavy residue in the round bottom flask. Similarly, a batch distillation of the catalysed NR pyrolytic oil gave a fractional yield of 50%, which was 9% more than that obtained from the thermal pyrolysed NR pyrolytic oil but within close temperature range from 50 to 175 °C.

The distillates of the NR pyrolytic oil and NRZ pyrolytic oil can be categorised to be more of light naphtha and middle distillates with no heavy fuel, wax or solid gel produced. For the distillation of the UT and UTZ pyrolytic oil, a temperature range of 30 to 208 °C and 30 to 251 °C was recorded, respectively. This could be attributed to the different rubber components and other compounds used in the tyre formulation. The high distillation temperature range observed for the UTZ pyrolytic oil could be linked to the cracking of the pyrolytic oil by the zeolite NaY to give high boiling point compounds.

The fractional volume percent of distillates made up of liquids and wax gel obtained for the UT and UTZ pyrolytic oil was 62.97 and 71.38%, respectively. The volume % of the light naphtha was less than 34% for the UT pyrolytic oil with 48.9% heavy fuel oil and 17.10% solid gel. That of the UTZ pyrolytic oil had 36.26% of light naphtha and 63.74% of heavy fuel oil. This is also in accordance with the report by Laresgoiti et al. (2004) that car tyre-derived pyrolytic oil had light naphtha, heavy naphtha and middle distillates of 20, 10 and 35%, respectively. Ucar et al. (2005) also found 40, 15 and 45% as the respective values in their truck tyre-derived pyrolytic liquids. The higher amount of heavy fuel oil distillates could be responsible for the wide application of UT pyrolytic oil as diesel fuels and heating oil. Figure 6.37 presents the different distillates, while Figure 6.38 shows the main hydrocarbon group presenting aliphatic, aromatic and PAHs with their different proportions. The GC–MS was also used to determine the compounds present in the various PD. Figures 6.39–6.42. They are similar to the compounds identified in the pyrolytic oil but at different proportions.

Table 6.12 Temperature range and distillate volume % obtained for the batch distillation of the different pyrolytic oil produced.

Temperature range (°C)	Fractional volume (%)	Temperature range (°C)	Fractional volume (%)
NR pyrolytic oil		*UT pyrolytic oil*	
60–87	23.32	30–144	43.12
90–100	15.82	162–198	11.98
105–161	51.12	180–208	5.98
155–175	9.74		
NRZ pyrolytic oil		*UTZ pyrolytic oil*	
50–126	48.48	30–146	42.26
95–148	26.52	158–178	16.90
135–148	7.57	190–251	38.50
140–175	17.43		

Figure 6.37 Different distillates obtained from distillation of NR and UT pyrolytic oil. (a) Clear fraction/light naphtha, (b) middle distillate, (c) heavy naphtha/fuel oil and (d) wax/solid gel.

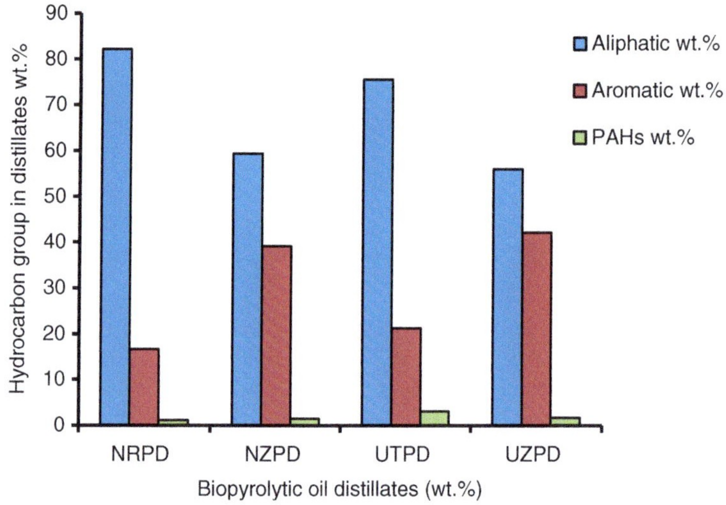

Figure 6.38 Hydrocarbon group present in distillates obtained from the various pyrolytic oil.

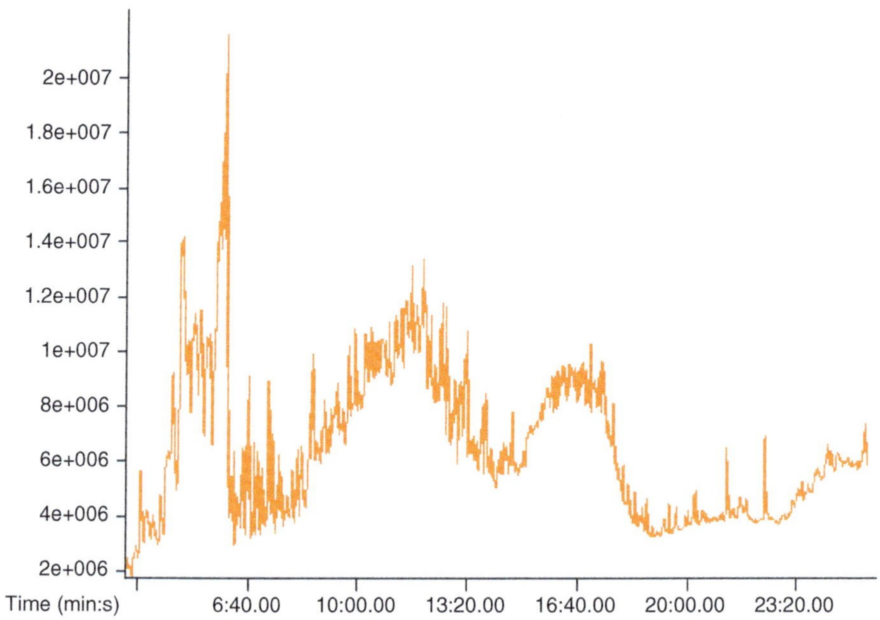

Figure 6.39 GC–MS chromatogram for total compound for NR pyrolytic oil distillates.

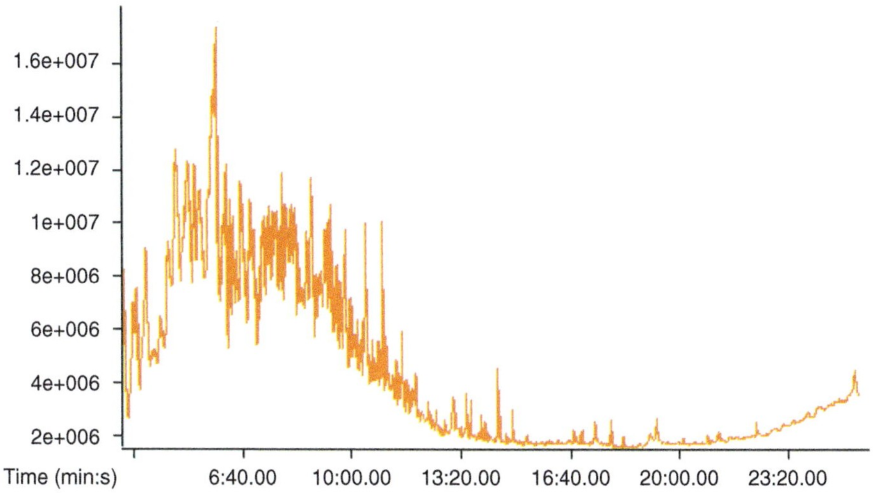

Figure 6.40 GC–MS chromatogram for total compound for NRZ pyrolytic oil distillates.

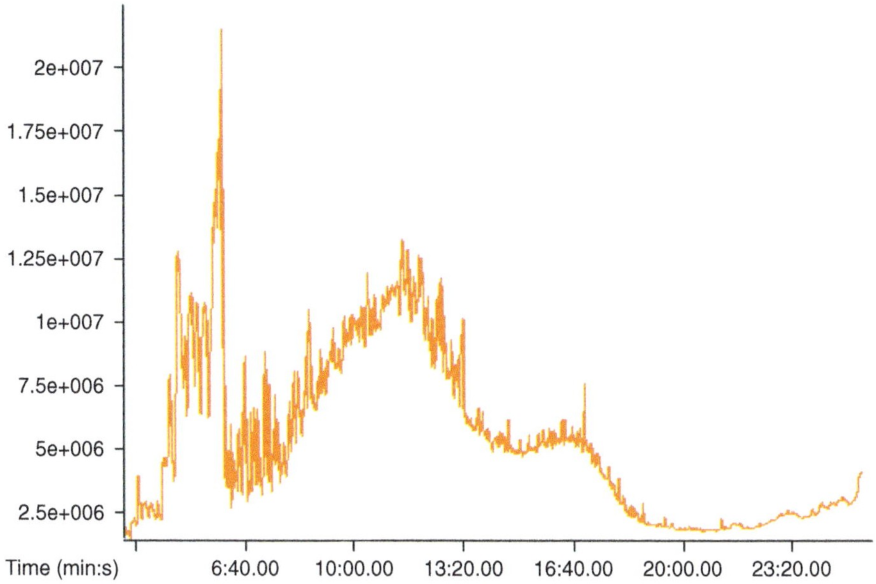

Figure 6.41 GC–MS chromatogram for total compound for UT pyrolytic oil distillates.

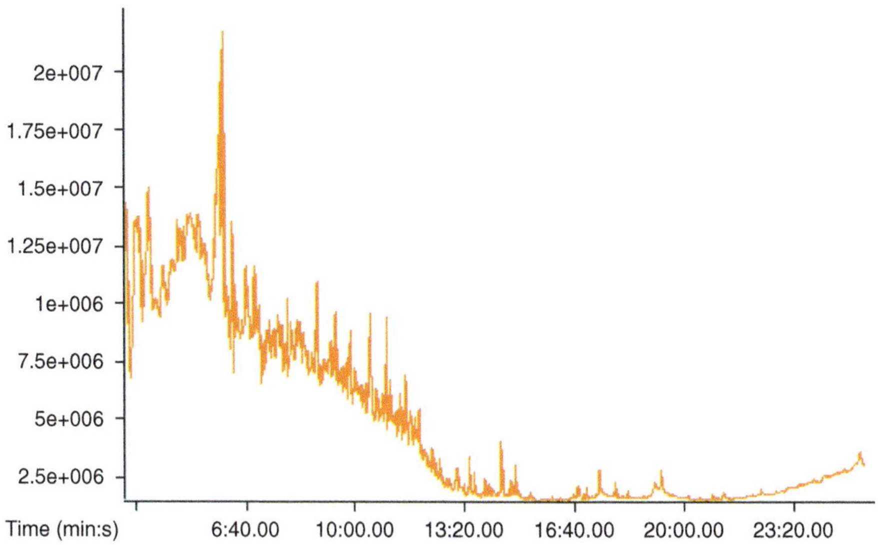

Figure 6.42 GC–MS chromatogram for total compound for UTZ pyrolytic oil distillates.

From Figure 6.38, the percentage in which the different hydrocarbon groups are present in the distillates revealed that NRPD had a higher concentration of aliphatics and lower values for aromatics and PAHs than all the other distillates. This is not unexpected because the NMR analysis results shown in Tables 4.9 and 4.10 also confirm this.

In summary, the NRZ pyrolytic oil and NR pyrolytic oil had more distillate yield within the temperature range for light naphtha compared to that of UTZ pyrolytic oil and UT pyrolytic oil. This is in accordance with the GC–MS analysis results that lower molecular weight hydrocarbon compounds were obtained in the distillates than in the pyrolytic oil. Also, the batch distillation of the UTZ pyrolytic oil gave more heavy oil liquid fraction than the UT pyrolytic oil. This could be attributed to the role of the zeolite NaY catalyst to further crack the heavy hydrocarbon compounds in the pyrolytic oil to lighter hydrocarbon compounds which is in accordance with the use of zeolite NaY as a catalyst in the cracking of pyrolytic oil in the petroleum industry.

6.4.7 Characterisation of UT and NR Solid Residue

The solid residue obtained from the thermal and catalytic pyrolysis of UT and natural rubber at 600 °C and HR of 15 °C/min was characterised using proximate and ultimate analysis. The SEM images were also analysed.

6.4.7.1 Proximate and Ultimate Analysis and Heating Value of UT and NR Solid Residue

The proximate analysis was performed on both chars obtained from UT and NR pyrolysis. This helped in determining the fixed carbon content, volatile material, ash and moisture while the ultimate analysis was used to determine the elemental composition such as carbon, hydrogen, nitrogen, sulphur and oxygen present in the char and the HHV of the char as reported in Table 6.13. The char obtained from both UT and NR pyrolysis had a higher heating value suitable for it to be an alternative source of heating fuel (Pradhan and Singh 2011).

6.4.7.2 Kinetic Study of Thermal Degradation and the SEM Analyses

SEM is an important technique used to determine the microstructure/ surface morphology of solid materials such as carbon. Hence, the solid residue obtained from the thermal and catalytic pyrolysis of UT and NR at 600 °C and HR of 15 °C/min were analysed for their surface morphology at a magnification of 150X as shown in Figure 6.43a–d. The Figure 6.43a,b reveals the presence of pores spread evenly across the residue rough surface with a distribution of particles of different sizes ranging from 132.9 to 632.3 μm for the UT only residue and 149.4 to 267.3 μm for the catalysed UT solid residue. The EDAX results showed that the main element in the UT only residue was zinc with 100 wt.% presence while Na (7.56 wt.%), Al (8.79 wt.%), Si (44.97 wt.%), Fe (7.69 wt.%) and Zn (31 wt.%) was identified for the catalysed UT. This confirms the retaining of the zeolite which was mixed homogenously with the UT sample before pyrolysis. The SEM for the NR solid residue (Figure 6.43c) and catalysed NR solid residue (Figure 6.43d) also revealed the presence of pores distributed evenly across the residue rough surface with a particle size distribution of 72.27 to 188.5 μm.

Table 6.13 HHV, proximate and ultimate analysis of UT and NR.

Property	Natural rubber (this study)	Used tyres (this study)	Li et al. (2004)
Proximate analysis (wt.%)			
Moisture	0.7	0.18	2.35
Volatiles	8.90	17.0	16.14
Ash	4.0	32.3	12.32
Fixed carbon	86.4	50.52	—
Ultimate analysis (wt.%)			
Carbon	89.0	88.28	82.17
Hydrogen	9.4	8.5	2.28
Nitrogen	1.6	2.4	0.61
Sulphur	—	0.82	2.32
HHV (MJ/kg)	39.0	34.8	31.5

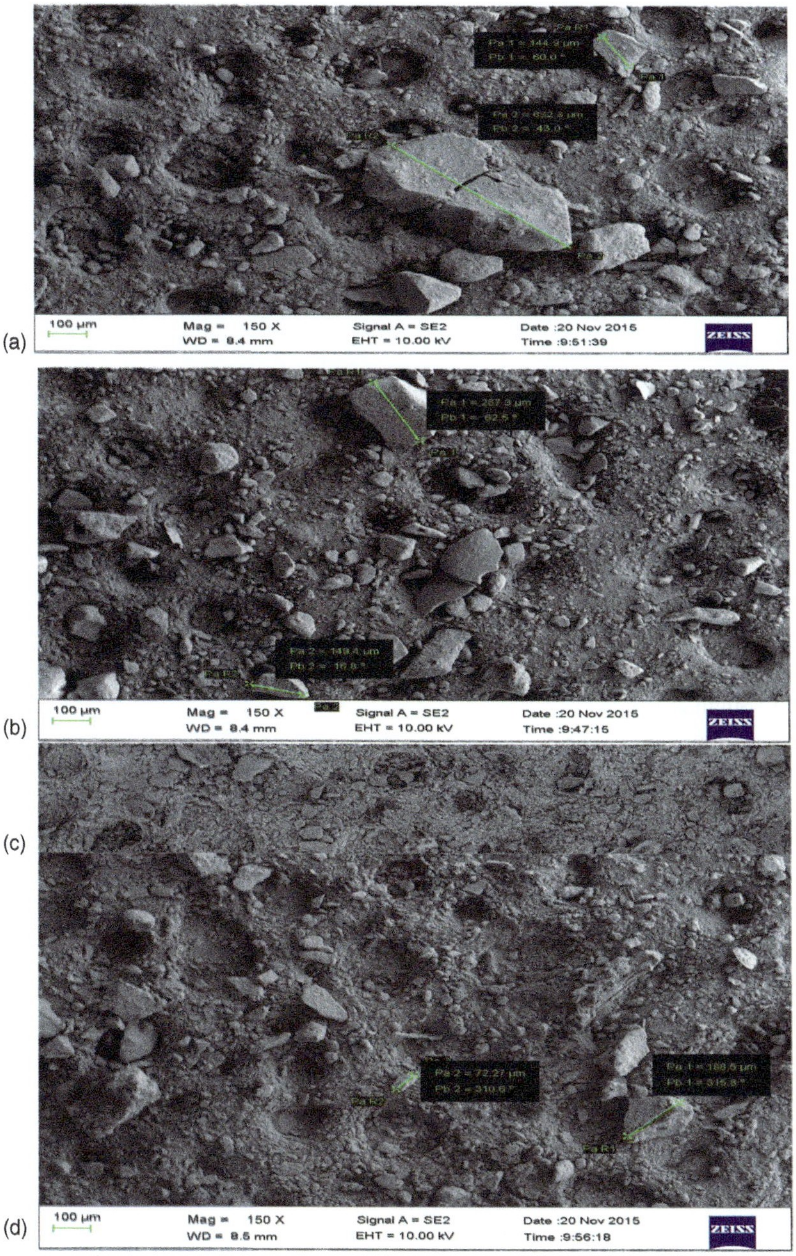

Figure 6.43 SEM image of solid residue at 150X for (a) UT, (b) catalysed UT; (c) natural rubber; (d) catalysed natural rubber.

6.4.7.2.1 Kinetic Study of Thermal Degradation In this study, kinetics of UT and NR thermal degradation study was performed using the TGA/DTG analysis equipment. Different heating rates (15, 20 and 30 °C/min) over a temperature range from 30 to 800 °C were investigated. An understanding of the thermal degradation kinetics of UT and NR can contribute immensely to improve their thermal behaviour. Hence, the study of the UT and NR pyrolysis kinetics will provide significant basis for predicting their behaviour in a pyrolysis reactor. The generally accepted technique for such kinetic study of thermal degradation is the thermogravimetric analysis (TGA/DTG). This is because the main objective of TGA is to get information on the general behaviour of the thermal degradation of UT (Pradhan and Singh 2011).

Also, the activation energy and pre-exponential factors can be determined from the data obtained from the TGA analysis. Since mechanisms of rubber materials pyrolysis involve complex reactions, it is not all reactions that result into the release of volatile component as some only alter the rubber materials mechanical properties (Chen et al. 2001). In TGA, it is only the mass change as a function of temperature and time that is being measured, meaning only those reactions affecting a mass change are considered (Chen et al. 2001; Liompart et al. 2013; Olazar et al. 2008)

6.4.7.3 Mechanism and Determination of Kinetic Parameters

Typically, the thermal degradation of polymeric materials occurs in more than one stage with multiple reactions. With the aid of the TGA, one can find the kinetic parameters and also measure the overall kinetics rather than individual reactions. Hence, Arrhenius equation and obtained thermogravimetric data are used to form the basis of the models. In this study, the kinetic analysis of the data for the UT and NR degradation, a one-step mechanism kinetic model with a first-order reaction and integral method was used to determine their kinetic parameters. The assumption of a first-order reaction in thermal degradation is normally adopted by researchers (Gonzalez et al. 2001; Pradhan and Singh 2011; (Ceylan and Topcu 2014; Gerten et al. 2000; Parikh and Rotliwala 2011; Parekh et al. 2009).

Thus, for the degradation of a material:

$$\text{A (Solid)} \rightarrow \text{B (Solid)} + \text{C (Volatiles)} \tag{6.12}$$

The first-order reaction is given as;

$$\frac{\mathrm{d}X}{\mathrm{d}t} = k(1 - X)n \tag{6.13}$$

where X is defined as the weight loss by mass unit of the initial amount of tyre:

$$X_n = \frac{W_o - W_t}{W_o - W_f} \tag{6.14}$$

where W_o (mg) is the initial mass of the tyre, W_f (mg) is the final mass and W_t (mg) is the mass at a given time, t (min) is the time, n is the reaction order and k is the kinetic rate constant which can also be expressed as;

$$K = k_o \exp\left(\frac{E_a}{RT}\right) \tag{6.15}$$

where k_o is the pre-exponential factor (min^{-1}), R is the universal gas constant (8.314 J/kmol), T is the temperature of reaction (K) and E_a is the activation energy (kJ/mol) with a linear HR A (K/min) given as;

$$A = \frac{\mathrm{d}T}{\mathrm{d}t} \tag{6.16}$$

By Integration, Eq. (6.13) becomes,

$$Ln[-Ln(1 - X)] = Ln\left[\left(\frac{k_o RT}{AE_a}\right) * \left(1 - \frac{2RT}{E_a}\right)\right] - \frac{E_a}{RT} \tag{6.17}$$

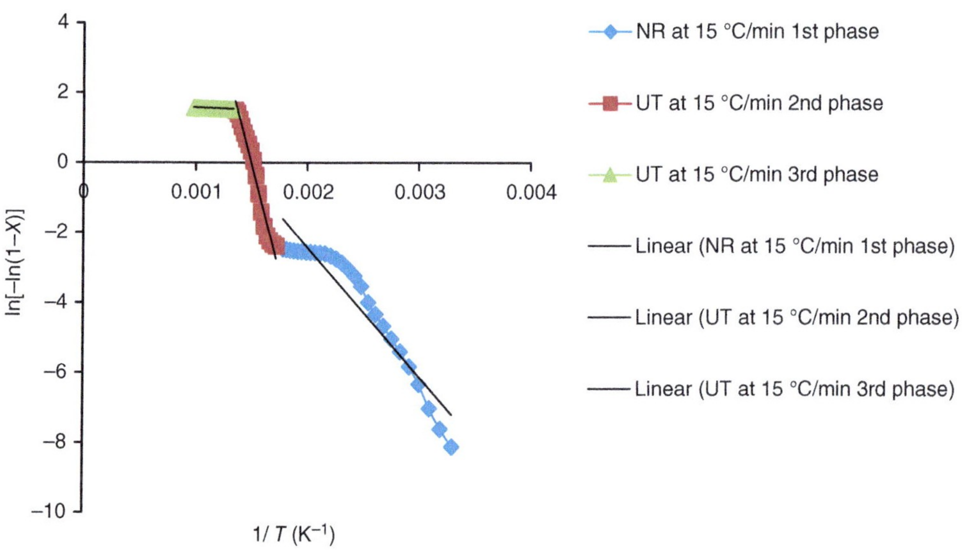

Figure 6.44 Kinetic plots of different phase of NR at heating rate of 15 °C/min.

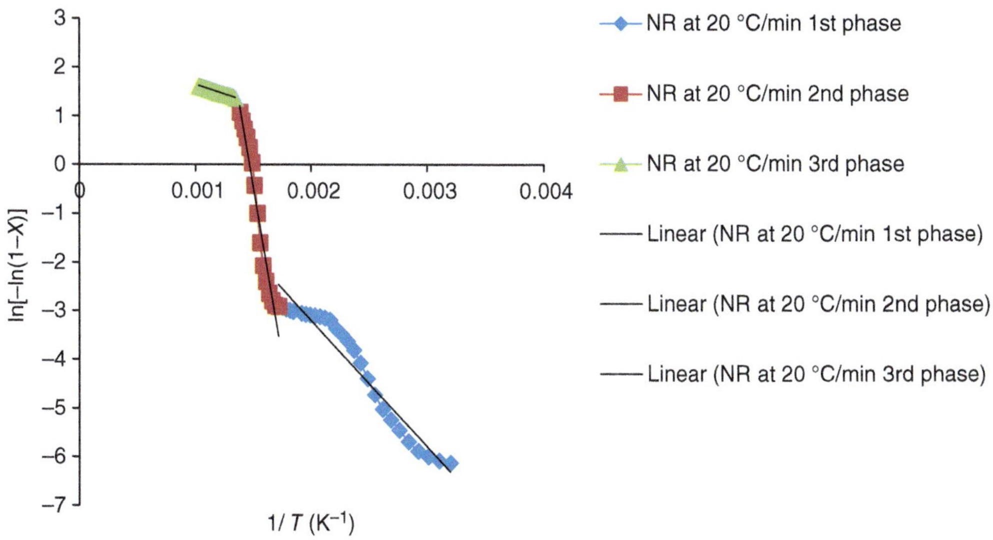

Figure 6.45 Kinetic plots of different phase of NR at heating rate of 20 °C/min.

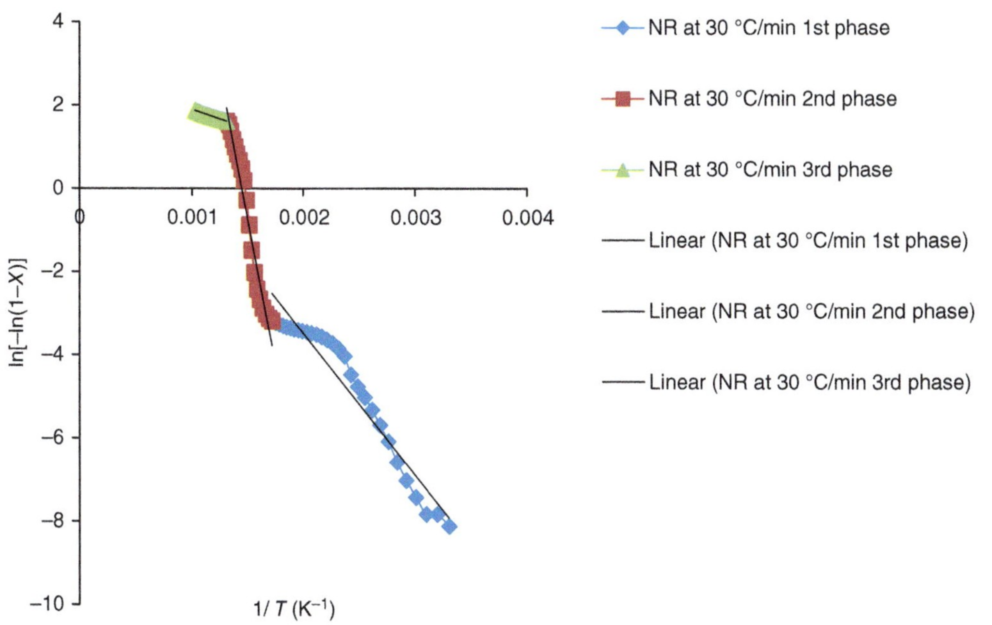

Figure 6.46 Kinetic plot of different phase of NR at heating rate of 30 °C/min.

A plot of the left hand side of Eq. (6.17) against $1/T$ should give a straight line with slope $(-E_a/R)$ which can be used to obtain the activation energy E_a. The pre-exponential factor k_o can be calculated using the exponential energy and the intercept of the plot. Figures 6.44–6.49 represent the kinetic plots of Ln[−Ln(1 − X)] versus $1/T$ calculated from the TGA data for NR and UT at different HRs of 15, 20 and 30 °C/min over a temperature range from 30 to 800 °C in order to determine the activation energy for the different degradation stages. The trend line equations and regression coefficients derived from the kinetic plots for the various stages of UT and NR degradation is shown in Table 6.14.

The determined activation energy for both UT and NR is shown in Tables 6.15 and 6.16, while Table 6.17 represents the average of the three stages of both the activation energy and pre-exponential factor at the different HRs for UT and NR.

From Tables 6.15 and 6.16, it was observed that the value of activation energy for the 2nd stage of UT and NR was higher than that recorded for the 1st and 3rd stages. This could be related to the fact that depolymerisation of the UT and NR in

Table 6.14 Trend line equations and regression coefficients for UT and NR.

Heating rate (°C/min)	1st Stage	2nd Stage	3rd Stage
Used tyres sample			
15	$Y = -3853.3 + 4.17$	$Y = -9092.2X + 13.20$	$Y = -1382X + 3.12$
	$R^2 = 0.98$	$R^2 = 0.96$	$R^2 = 0.99$
20	$Y = -4325.8X + 05.05$	$Y = -9538X + 13.82$	$Y = -1557.7 + 3.19$
	$R^2 = 0.99$	$R^2 = 0.99$	$R^2 = 0.95$
30	$Y = -4368X + 05.05$	$Y = -9621.7X + 13.81$	$Y = -1544.1X + 3.13$
	$R^2 = 0.99$	$R^2 = 0.99$	$R^2 = 0.95$
Natural rubber sample			
15	$Y = -2708.7X + 1.57$	$Y = -13\,414X + 19.83$	$Y = -1208.9X + 3.13$
	$R^2 = 0.99$	$R^2 = 0.96$	$R^2 = 0.99$
20	$Y = -2941X + 2.9045$	$Y = -14\,251X + 20.92$	$Y = -757.97X + 2.4029$
	$R^2 = 0.96$	$R^2 = 0.96$	$R^2 = 0.99$
30	$Y = -3920.4X + 4.73$	$Y = -14\,799X + 21.52$	$Y = -893X + 2.79$
	$R^2 = 0.96$	$R^2 = 0.95$	$R^2 = 0.99$

Table 6.15 Activation energy and pre-exponential factor of used tyres sample.

	Activation energy (kJ/mol)		
Heating rate (°C/min)	1st Stage	2nd Stage	3rd Stage
15	32.04	75.59	11.49
20	35.96	79.30	12.95
30	36.32	79.99	12.84
Pre-exponential factor (min^{-1})			
15	3.20×10^{12}	3.51×10^{16}	2.55×10^{11}
20	2.80×10^{12}	6.66×10^{16}	1.89×10^{11}
30	4.29×10^{12}	9.43×10^{16}	2.65×10^{11}

Table 6.16 Activation energy and pre-exponential factor of natural rubber sample.

Heating rate (°C/min)	Activation energy (kJ/mol)		
	1st Stage	2nd Stage	3rd Stage
15	22.52	111.52	10.05
20	24.45	118.48	6.30
30	32.59	123.04	7.42
Pre-exponential factor (min^{-1})			
15	0.58×10^{12}	0.38×10^{20}	22.7×10^{10}
20	107×10^{12}	1.58×10^{20}	1.50×10^{10}
30	1.09×10^{12}	2.45×10^{20}	25×10^{10}

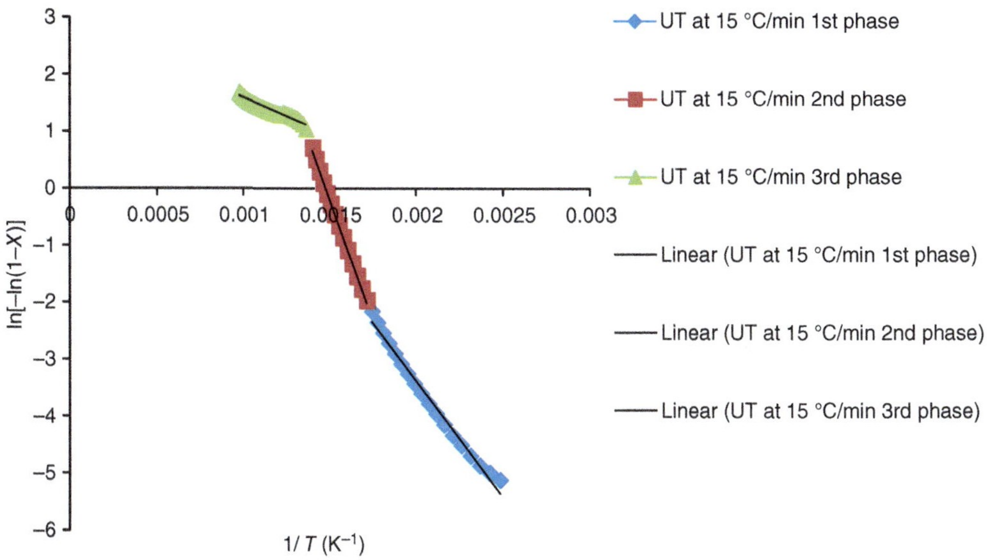

Figure 6.47 Kinetic plots of different phases of UT at heating rate of 15 °C/min.

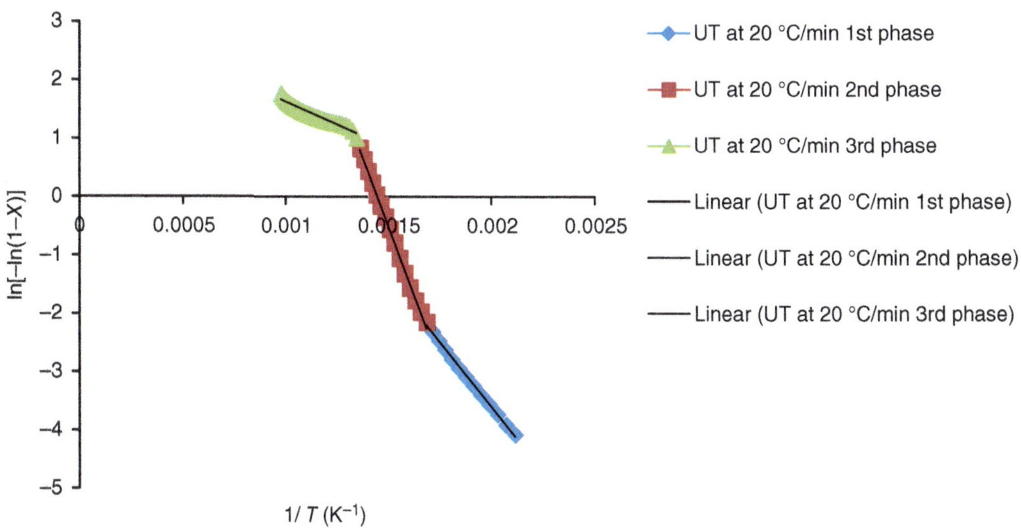

Figure 6.48 Kinetic plots of the different phases of UT at heating rate of 20 °C/min.

form of bond scission and decarbonylation occurred more at this stage requiring more energy for degradation. The result is in agreement with report by Pradhan and Singh 2011) on their use of a TGA/DTG for the investigation of tyre rubber and tube rubber. Also, the average activation energy for NR at the different HRs of 15, 20 and 30 °C/min was higher than that obtained for the UT sample in this study. This implies more energy is needed to depolymerise NR due to its heavy molecular weight of up to a 1 000 000 g/mol than the UT with lower molecular weight (Leung and Wang 1998; Pilusa et al. 2013; Deenanath 2014). Also, from the calculated average results reported in Table 6.17, the activation energy increases as the HR increases. This could explain how the HR affect the peak curve of components (from the DTG graph), which shift further away from their expected peak temperature to higher temperature zone resulting in more energy needed for their degradation. This was also reported by Leung and Wang (1998) from their experimental results that as the heating rate increases, the activation energy and frequency factor also increases due to a shift of the reactions to higher temperature range and the temperature corresponding to peak value of sample weight loss increases making degradation more difficult. Figures 6.50 and 6.51 show the DTG graph variation of the peak temperature for the different degradation/peak temperature of the UT and NR components.

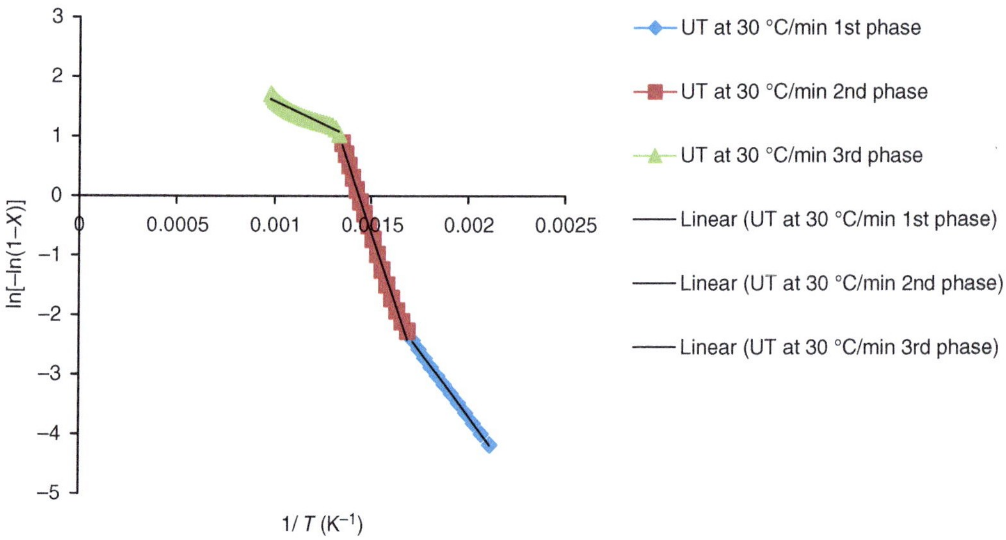

Figure 6.49 Kinetic plots of different phases of UT at heating rate of 30 °C/min.

Table 6.17 Average activation energy and pre-exponential factor for used tyres and natural rubber.

Heating rate (°C/min)	Activation energy (kJ/mol)	Pre-exponential factor (min^{-1})
Used tyres sample		
15	39.71 ± 26.69	1.17×10^{16}
20	42.74 ± 27.51	2.22×10^{16}
30	43.05 ± 27.82	3.14×10^{16}
Natural rubber sample		
15	48.03 ± 45.18	1.25×10^{19}
20	49.74 ± 49.17	5.28×10^{19}
30	54.35 ± 49.65	8.17×10^{19}

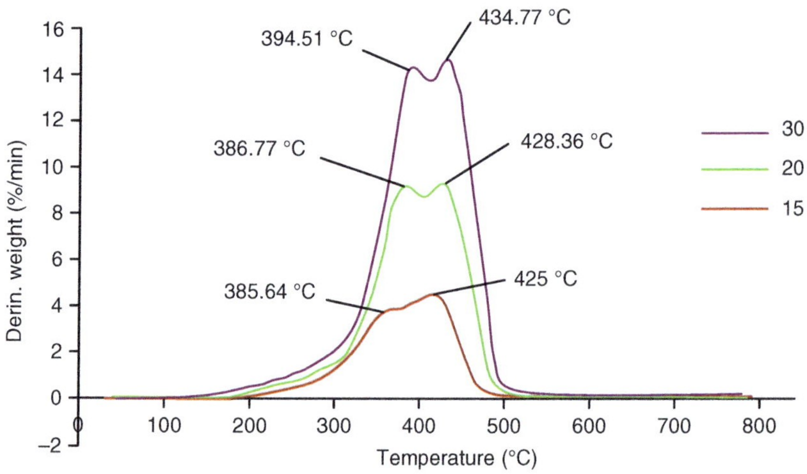

Figure 6.50 Effect of heating rate on temperature degradation of UT.

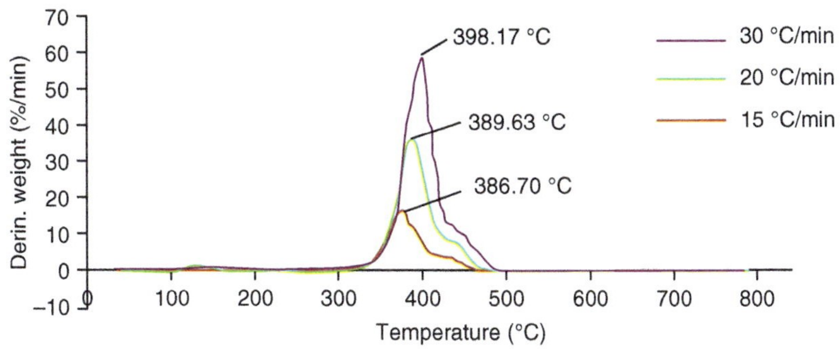

Figure 6.51 Effect of heating rates on temperature degradation of NR.

From the DTG graph for UT, Figure 6.50, as the heating rate increased from 15, 20 to 30 °C/min over the temperature range from 30 to 800 °C, the two prominent temperature peaks shifted from 385.64 and 425 °C for 15 °C/min to 386.77 and 428.36 °C for 20°Cmin to 394.51 and 434.77 °C for the heating rate of 30 °C/min. The same was observed for the NR, Figure 6.51, at different heating rates of 15, 20 and 30 °C/min for the same temperature range of 30 to 800 °C. The single prominent peak in the DTG graph of NR shifted from 386.70 °C at heating rate of 15 °C/min to 389.63 °C at 20 °C/min and 398.17 °C at 30 °C/min. This could be related to the increase in activation energy as the heating rate increases resulting in more energy required to depolymerise the rubber components at higher peak temperature. A similar situation was also reported by Aguado et al. (2005) when they studied the kinetics of scrap tyres under fast heating conditions. It was noted that as they increase the heating rate, the maximum peaks in their DTG curves displaced to higher temperatures (707 K at 2 K/min, 711 K at 5 K/min, 721 K at 10 K/min and 727 K at 20 K/min). Thus, from the activation energy results obtained with the use of the TGA/DTG data in this study, the thermal degradation of UT was found easier than that of NR. Also, it can be concluded from the kinetic study that as the heating rate increases the activation energy and the pre-exponential factor also increases.

6.4.8 Testing of Pyrolytic Oil Distillates and Petrol Blends

The PD obtained from the distillation of the thermal and catalysed pyrolysis of UT and NR pyrolytic oil were blended with neat petrol (in the ratio 0, 5, 10, 15 and 20%) and tested in a generator set. Sections 6.4.8.1 and 6.4.8.2 discuss the investigation performed during the combustion study of the PD/Petrol blend.

6.4.8.1 Fuel Consumption over Time as a Function of Fuel Blend Composition

In order to determine the fuel consumption over time, the generator was made to power a Martlet 375 watt heavy duty 150 mm bench Grinder machine as shown in Figure 6.8 throughout the sets of experiments. First, the neat unleaded petrol was used to run the generator in order to set a reference level of fuel consumed over time against that of the blends. This was then followed by using the blend of 0, 5, 10, 15 and 20% of the NRPD, NZPD, UTPD and UZPD with petrol. This was presented as a change in fuel level with time as shown in Figures 6.52–6.59. During the engine combustion performance test, it was evident that both the distillates from thermal and catalysed pyrolytic oil of UT and NR combusted successfully in the spark ignition engine. This is another success as most reports in the literature are on the use of pyrolytic oil and its distillates from UT in compression diesel engines. Nevertheless, as the ratio of the PD increases in the blend mixture, there were more gaseous emissions associated with that of the non-catalysed UTPD than the others. This could probably be due to the high amount heavier hydrocarbon compounds and aromatics present in it than the others from the NMR and GC–MS analyses results. The NRPD and NRZPD petrol blend of 5 to 20% had similar emission colour with that of neat petrol. From Figure 6.52, an increase in the NRPD from 5 to 20% in blends with neat petrol showed a decrease in the consumption of fuel. The consumption behaviour of the 20% NRDP-petrol blends was comparative with the neat petrol consumption until 25 min of the generator run. Afterwards, the consumption of the neat petrol was higher than that of the 20% NRPD-petrol blend as shown in Figure 6.53. For that of NZPD-petrol blend consumption in Figure 6.54, a similar trend with that of NRPD-petrol blend was observed.

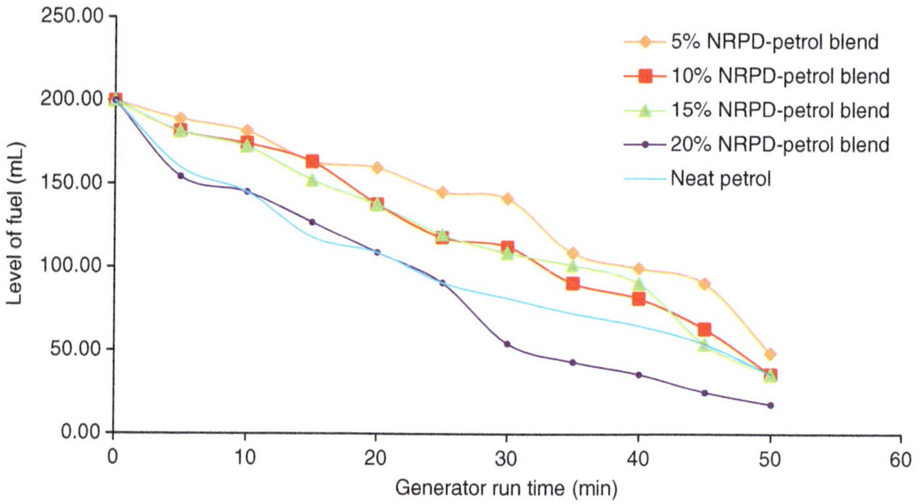

Figure 6.52 Fuel consumption rate as a function of NRPD-petrol fuel composition.

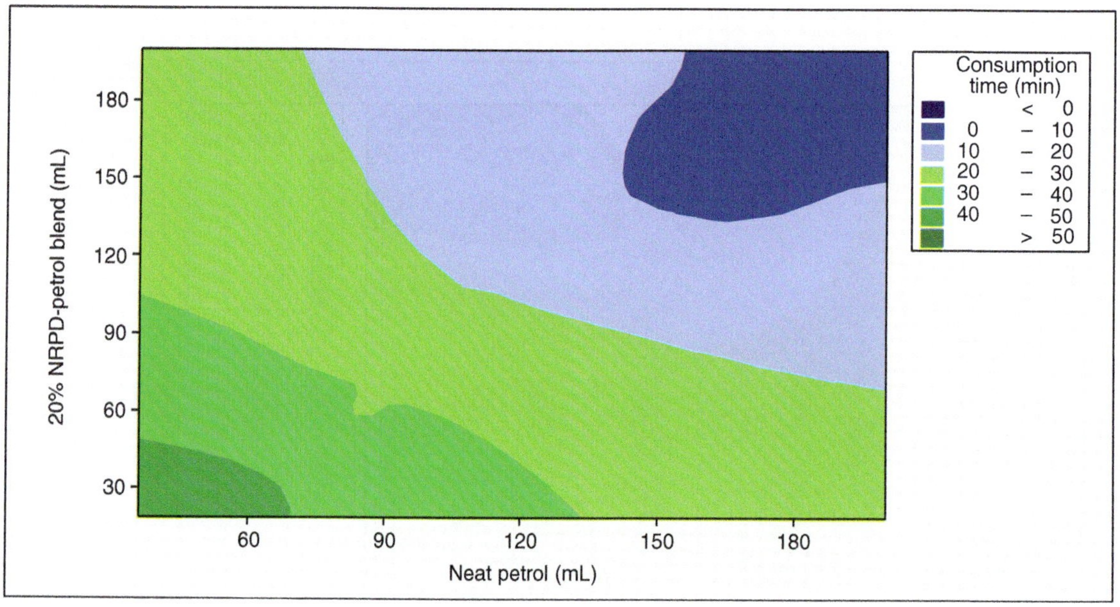

Figure 6.53 Contour plot of consumption time versus 20% NRPD-petrol, neat petrol (mL).

Nevertheless, it was the 10% NZPD-petrol blend that had comparative consumption behaviour with that of the neat petrol till about 30 min of the generator run. Afterwards, the neat petrol displayed faster consumption rate than the 10% NZPD-petrol blend as shown in Figure 6.55.

The consumption rate of the UTPD-petrol blend represented in Figure 6.56 reveals that none of the blends had comparative fuel consumption behaviour with that of the neat petrol until after about 30 min of the generator run. However, a decrease in consumption rate was later observed for the blend than the neat petrol after 30 min run time upward, except for the 5% UTPD-petrol blend. Figure 6.57 displays the comparative consumption behaviour for the 15% UTPD-petrol blend which had a closer behaviour pattern with the neat petrol after 30 min run time.

For the case of the UZPD-petrol blend shown in Figure 6.58, there was not much deviation observed from the UTPD-petrol blend pattern represented in Figure 6.59.

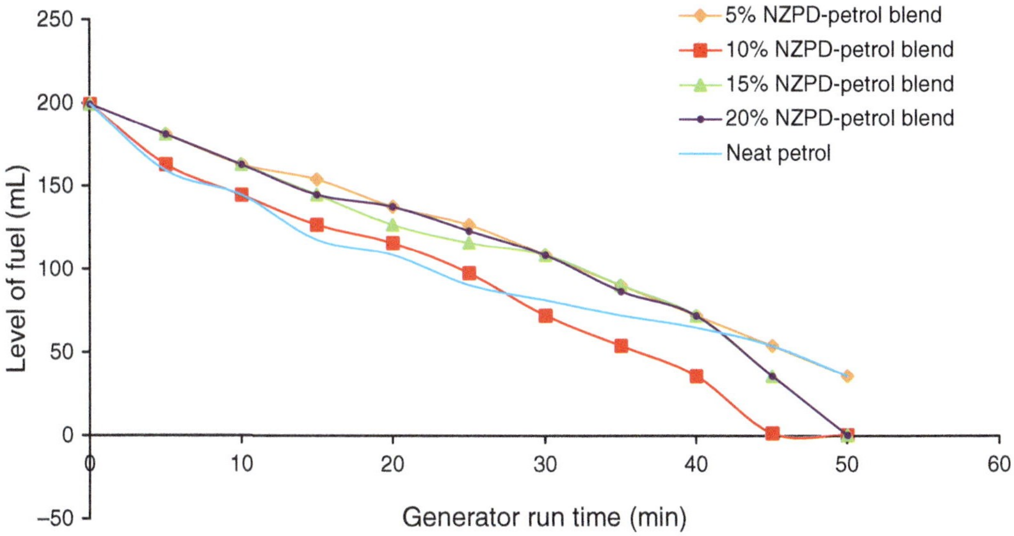

Figure 6.54 Fuel consumption rate as a function of NZPD-petrol fuel composition.

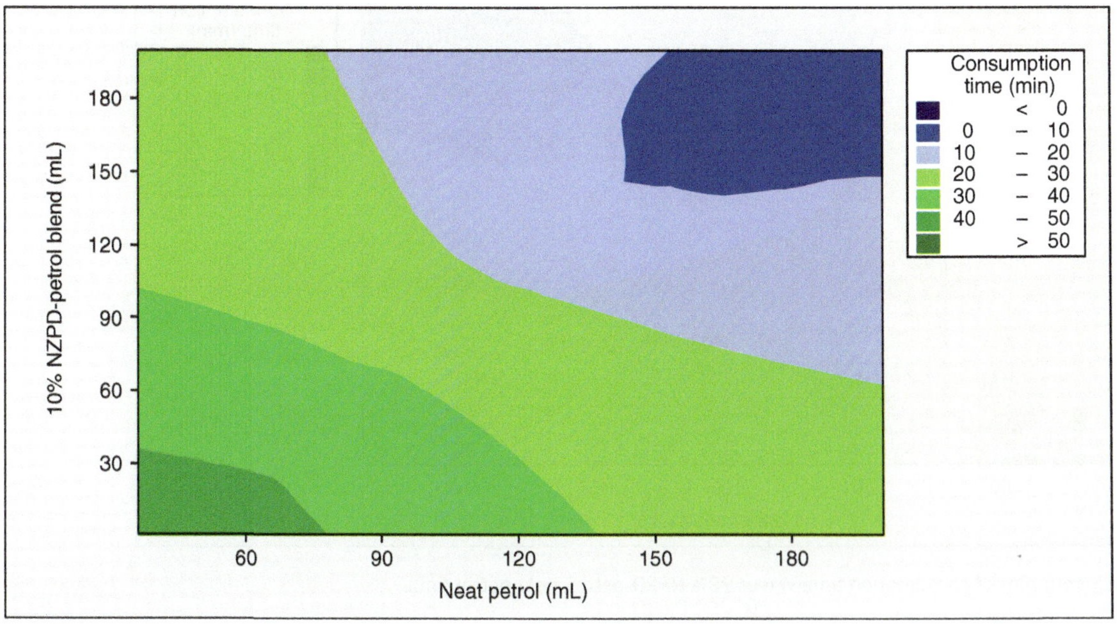

Figure 6.55 Contour plot of consumption time versus 10% NZPD-petrol, neat petrol (mL).

6.4.8.2 Fuel Power as a Function of PD Composition in the Fuel Blend

Figure 6.60 presents the curves of the calculated fuel power as a function of the various concentrations of PD in the PD-Petrol blend fuel. From the curves of Figure 6.60, it was observed that the fuel power increases as the concentration of the different PD increases in the PD-petrol fuel blend. An optimum fuel power performance value of 2.02, 2.20, 2.20 and 2.22 kJ/s was obtained at 20% for the NRPD-petrol blend fuel, 10% for the NZPD-petrol blend fuel, 15% for the UTPD-petrol blend fuel and 15% for the UZPD-petrol blend fuel. When compared with that of the neat petrol, 1.79 kJ/s, the composition of the PD in the fuel mixture resulted in a higher fuel power in the generator.

In summary, the PD obtained from the thermal and catalysed pyrolytic oil from NR and UT combusted successfully in the blend ratio of 0, 5, 10, 15 and 20% with neat petrol in a spark ignition combustion engine without engine modification. This is a different approach from the use of compression ignition engine normally reported by other researchers in the literature (Pradhan and Singh 2011).

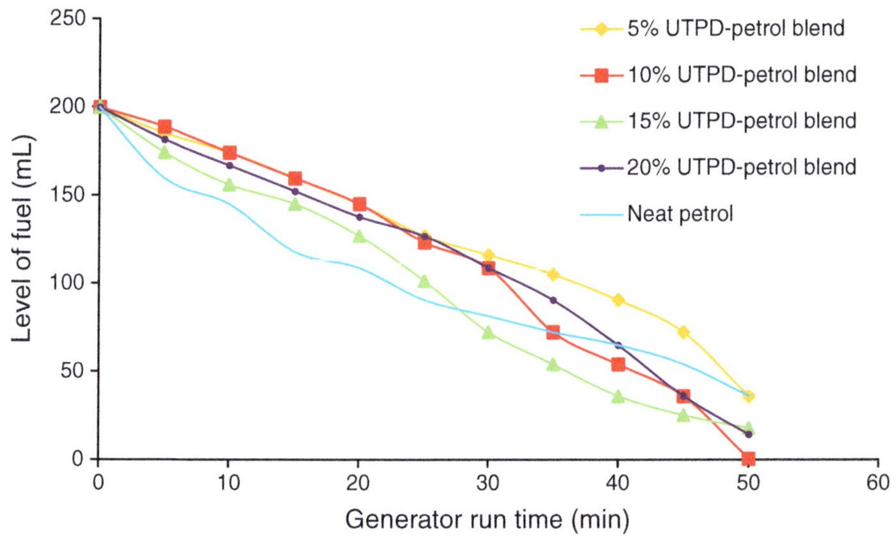

Figure 6.56 Fuel consumption rate as a function of UTPD-petrol fuel composition.

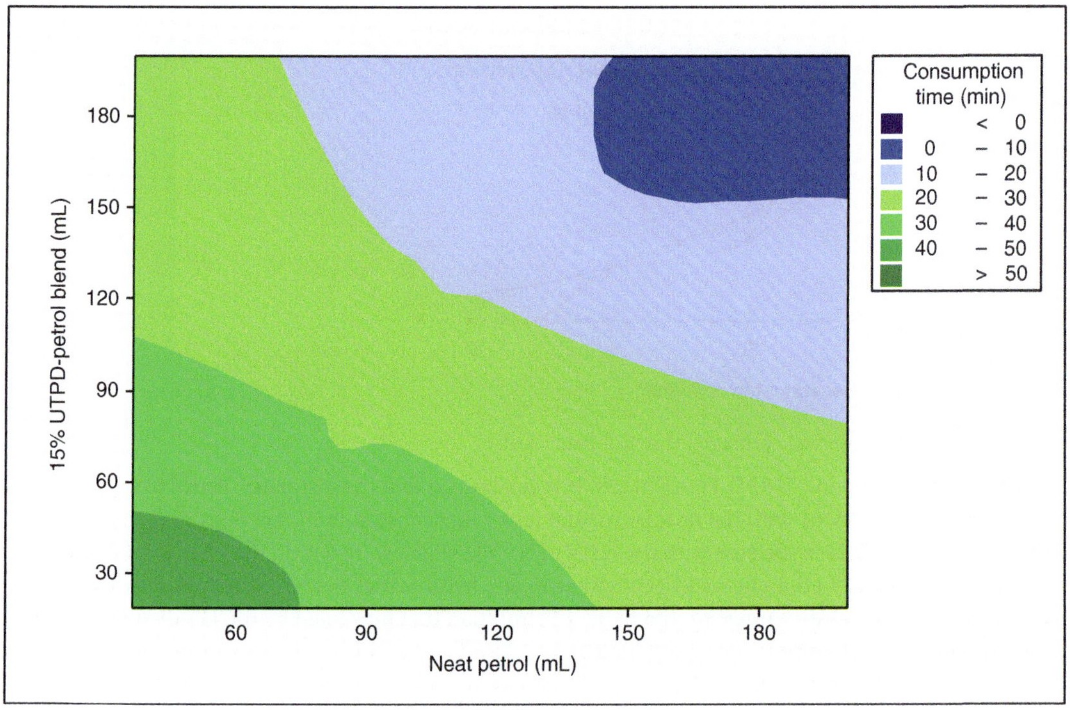

Figure 6.57 Contour plot of consumption time versus 15% UTPD-petrol, neat petrol (mL).

From Figures 6.52–6.58 shown above for the entire PD-petrol blend consumption rate against generator run time, there was an increase in the fuel consumption rate as the percentage concentration of the different PD increases in the fuel mixture. Nevertheless, the difference in the rate of fuel consumption did not deviate much from that of the neat petrol which had a value of 4.03×10^{-5} kg/s while the highest amount of 5.08×10^{-5} kg/s was recorded for UZPD-petrol fuel blend indicating 25.56% increase. This closeness of the different fuel mixture consumption rate with the petrol fuel could be attributed to the closeness in the higher heating value of NRPD, 43.20 MJ/kg, NZPD 43.22 MJ/kg, UTPD, 41.40 MJ/kg, UZPD, 41.41 MJ/kg and 44.39 MJ/kg for petrol fuel. Hence, low amount of neat petrol will be required for blending with PD to obtain similar engine performance.

The fuel consumption rate for the observed optimum PD-petrol blend fuel compared with neat Petrol as represented in Figure 6.59, the order of fuel consumption with respect to generator run time can be concluded as UZPD-Petrol blend

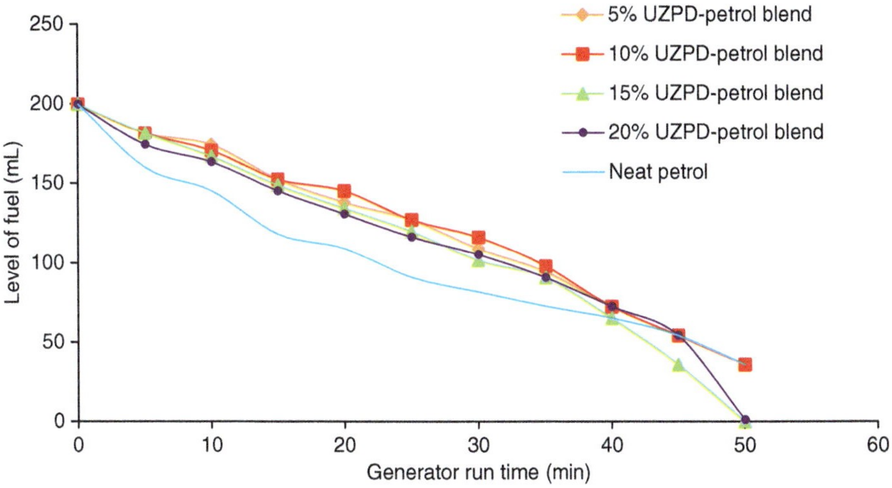

Figure 6.58 Fuel consumption rate as a function of UZPD-petrol fuel composition.

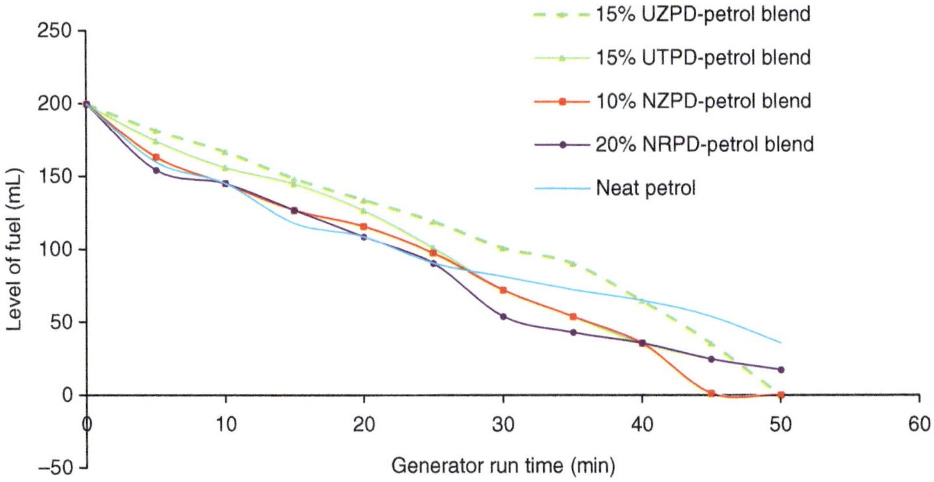

Figure 6.59 Fuel consumption rate for observed optimum PD-petrol blend fuel.

>UTPD-Petrol blend >NZPD-Petrol blend >NRPD-Petrol blend. It is can be said that the distillates-Petrol blends obtained from NR gave better comparative behaviour with the neat petrol and lower fuel consumption rate than that obtained from UT distillates-Petrol blends. This could be associated with the HHV of NR PD which were within the Petrol HHV range and above that of the UT PD. The amount of aliphatic hydrogen and aromatic hydrogen in the NR as revealed by the ^{1}H and ^{13}C NMR analysis are closer in chemical shift range to that of neat Petrol than that presented by the UT distillates. This is in agreement with the ^{1}H and ^{13}C NMR of neat petrol reported by Deavanni and other researchers in literature that neat petrol is composed of a mixture of hydrogen attached to aliphatic and aromatic within the chemical shift of 0.8–1.8 and 2.1–2.8 ppm while the aliphatic and aromatic carbon lie within 10–50 ppm and 100–160 ppm.

Furthermore, the outcome of the fuel power for the different PD composition in the PD-petrol fuel blend in the generator revealed that as the various PD composition increases in the PD-Petrol blend fuel, there is an increase in the generator fuel power. In addition, the optimum fuel power values as a function of the composition of PD in the PD-Petrol fuel mixture was obtained at the same composition that gave optimum fuel consumption rate values for each of the PD in PD-Petrol fuel blend. This satisfies the equation relationship that fuel power is dependent on the product of the mass of fuel consumed and the heating value of the fuel used.

6.4.9 Mathematical Modeling and Simulation of the Pyrolysis of Rubber

TGA is the standard technique used in the study of the thermal degradation behaviour of tyre rubber. It normally relates the weight loss of the material with constant rate (θ) of change in temperature over time. This implies that temperature and time have a linear relationship at a constant HR of θ/time as in Eq. (6.14) and depicted in Figure 6.61.

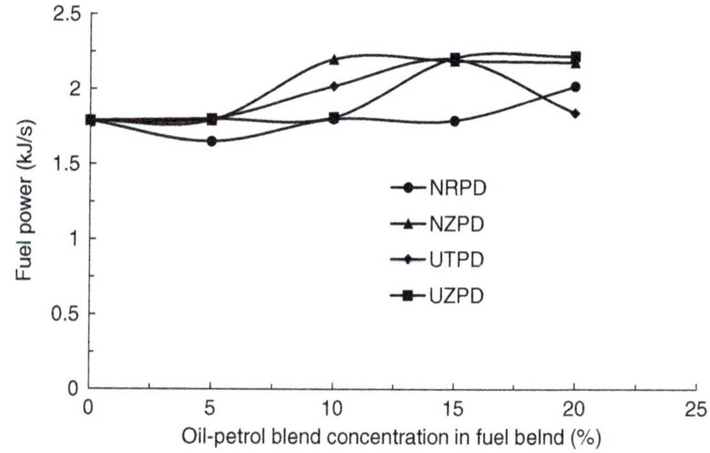

Figure 6.60 Curve showing the fuel power as a function of concentrations of NRPD, NZPD, UTPD and UZPD in the fuel blend.

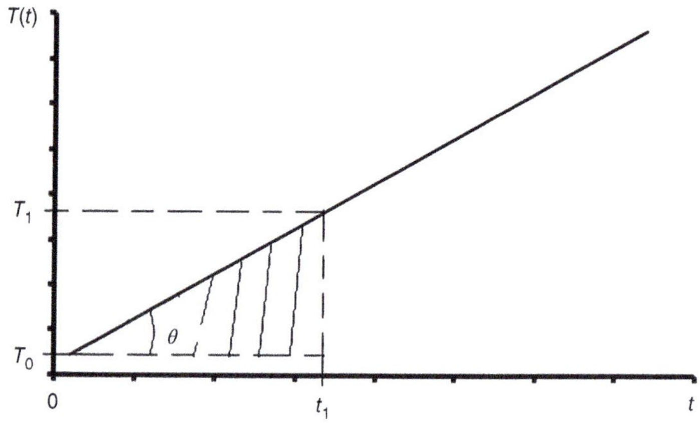

Figure 6.61 Temperature and time relationship in a TGA.

$$T(t) = \theta t + T_0 \tag{6.18}$$

A typical TGA curve for UT, in terms of weight loss against temperature at a HR of 15 °C/min over a temperature range of 30 to 800 °C is shown in Figure 6.62.

As a result, this makes it difficult to study the weight change of a material as it degrades with time at a fixed temperature using the TGA. This section is therefore aimed at developing a mathematical model to adopt TGA as a dynamic apparatus

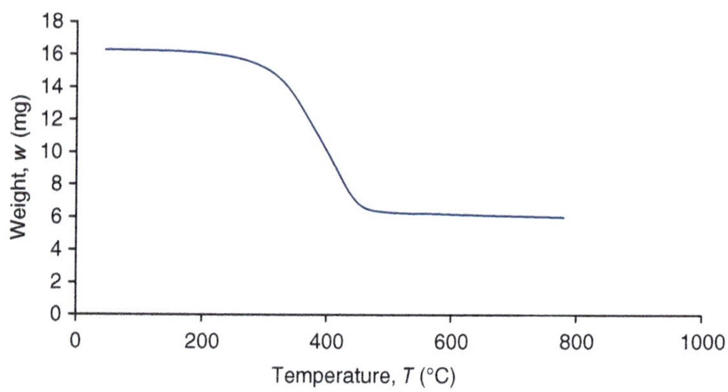

Figure 6.62 TGA curve of UT weight loss versus temperature at 15 °C/min heating rate.

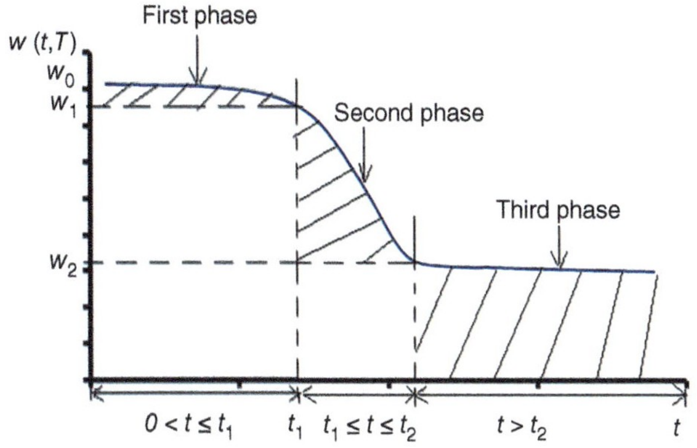

Figure 6.63 Three-phase thermal degradation of weight change with time.

to predict weight change of a material as it degrades with time at a fixed temperature. Thus, experimental data from TGA can be used to simulate the thermal degradation behaviour of UT with changes in time at a particular temperature.

The degradation process as shown in Figure 6.63 is assumed to pass through three phases for the purpose of proper mathematical modelling. The phases are all assumed to be piece-wise continuous mathematically since the end of a phase is the beginning of the succeeding phase. The phases were classified into three time zones $0 \leq t \leq t_1$, $t_1 \leq t \leq t_2$ and $t \leq t_2$ for t_2 being a large time.

6.4.9.1 Mathematical Derivation

First phase degradation model derivation:

$$\frac{\mathrm{d}w(t)}{\mathrm{d}t} = \alpha(t)w(t) \tag{6.19}$$

Equation (6.19) defines the relationship between changes in weight with time where

$$\alpha(t) = -\sigma_1 tT_0 + c_1 \tag{6.20}$$

where $\sigma_1 > 0$, is the rate of change (in degradation) at a fixed temperature, T_0, at this phase.

Imposing the initial condition $\alpha(0) = 0$ gives $c_1 = 0$

$$\therefore\ \alpha(t) = -\sigma_1 tT \tag{6.21}$$

Substituting for c_1 in Eq. (6.20),

$$\therefore\ \alpha(t) = -\sigma_1 tT_0 \tag{6.22}$$

Substituting Eq. (6.22) into Eq. (6.19) gives

$$\frac{\mathrm{d}w(t)}{\mathrm{d}t} = -\sigma_1 tT_0 w(t)$$

$$\int_0^t \frac{\mathrm{d}w(t)}{w(t)} = \int_0^t (-\sigma_1 tT_0)\mathrm{d}t$$

$$\ln w(t) = -\sigma_1 \frac{t^2 T_0}{2} + c_1$$

Taking the exponential of both sides,

$$w(t) = e^{\left(-\sigma_1 \frac{t^2 T_0}{2} + c_1\right)} \qquad \text{where } A = e^{c_1}$$

$$\therefore\ w(t) = Ae^{-\sigma_1 \frac{t^2 T_0}{2}}$$

At $t = 0$, $w(t) = w(0) = w0 = Ae^0$

$$\Rightarrow A = w_0$$

$$\therefore\ w(t) = w_0 e^{-\sigma_1 \frac{t^2 T_0}{2}} \tag{6.23}$$

Equation (6.23) is the general model equation for the first phase of degradation (weight loss of rubber with time) at a fixed temperature, T_0.

At boundary conditions, $w(t_1) = w_1 = w_0\beta_1$ where $0 \le \beta_1 \le 1$ represents the proportion of the overall weight (w_0) remaining at time t_1

Simplifying,

$$-\sigma_1 \frac{t_1{}^2}{2} T_0 = \ln \beta 1$$

$$\sigma_1 = \frac{-2 \ln \beta 1}{t_1{}^2 T_0} \tag{6.24}$$

Recall from Eq. (6.18) that at $t = t_1$,

$$T1 = T(t_1) = \theta t_1 + T_0$$

Then, for a fixed temperature, HR, $\theta = 0$

$$\Rightarrow T1 = T0$$

So, T_1 can be substituted for T_0 in Eq. (6.24).

Second phase degradation model derivation:

At this phase, the weight loss with time is described to have satisfied a logistic model equation of the form;

$$\frac{dw}{dt} \propto T(t)\left(1 - \frac{w}{k}\right) w \tag{6.25}$$

With a minimal weight value, k, where $T(t) = \theta t + T_{0.}$

Introducing constant of proportionality, σ_2, Eq. (6.25) becomes

$$\frac{dw}{dt} = \sigma_2 T(t)\left(1 - \frac{w}{k}\right) w \tag{6.26}$$

Expanding Eq. (6.26) gives

$$\frac{dw}{dt} = \sigma_2 T(t) w - \frac{\sigma_2 T(t) w^2}{k}$$

$$\frac{dw}{dt} - \sigma_2 T(t) w = \frac{-\sigma_2 T(t) w^2}{k} \tag{6.27}$$

The first-order non-homogenous non-linear ordinary differential Eq. (6.27) that is non-linear in $w(t)$, was linearised using the method of change of variable. Let $v = \frac{1}{w} = w^{-1}$

$$\therefore \frac{dv}{dw} = -w^{-2}$$

By chain rule,

$$\frac{dv}{dt} = \frac{dv}{dw}\frac{dw}{dt} = \frac{dw}{dt} = \left(\frac{dv}{dt}\right) / \left(\frac{dv}{dw}\right)$$

$$\frac{dw}{dt} = -w^2 \left(\frac{dv}{dt}\right) = -\left(\frac{1}{v^2}\right)\left(\frac{dv}{dt}\right) \tag{6.28}$$

Substituting Eq. (6.28) into Eq. (6.27),

$$-\frac{1}{v^2}\left(\frac{dv}{dt}\right) - \sigma_2 T(t)\frac{1}{v} = \frac{-\sigma_2 T(t)}{kv^2} \tag{6.29}$$

Multiplying both sides of Eq. (6.29) by $-v^2$,

$$\frac{dv}{dt} + \sigma_2 T(t) v = \frac{\sigma_2 T(t)}{k} \tag{6.30}$$

Using technique of integrating factor for a first-order differential equation of the form,

$$\frac{dv}{dt} + p(t) v = q(t)$$

where integrating factor,

$$\mu(t) = e^{\int p(t)\mathrm{d}t} = e^{\sigma_2 \int T(t)\mathrm{d}t} = e^{\sigma_2 \int (\theta t + T_0)\mathrm{d}t}$$

$$\mu(t) = e^{\sigma_2\left(\frac{\theta t^2}{2} + T_0 t\right)}, \text{ and } q(t) = \frac{\sigma_2 T(t)}{k}$$

By change of variable method, let $z(t) = \frac{\theta t^2}{2} + T_0 t$
Using

$$v(t) = \frac{1}{\mu(t)}\left[\int^{t} \mu(t)q(t)\mathrm{d}t + c3\right] \tag{6.31}$$

Substituting for $\mu(t)$ and $q(t)$ in Eq. (6.31),

$$v(t) = e^{-\sigma_2 z}\left[\int^{t} e^{\sigma_2 z(t)}\frac{\sigma_2 T(t)}{k}\mathrm{d}t + c_3\right]$$

$$v(t) = e^{-\sigma_2 z}\left[\int e^{\sigma_2 z}\frac{\sigma_2 T(t)}{k}\mathrm{d}t + c_3\right] \tag{6.32}$$

From $z(t) = \frac{\theta t^2}{2} + T_0 t$

$$\frac{\mathrm{d}z}{\mathrm{d}t} = \theta t + T_0$$

$$\mathrm{d}t = \frac{\mathrm{d}z}{(\theta t + T_0)}$$

Substituting for $\mathrm{d}(t)$ in Eq. (6.32),

$$v(t) = e^{-\sigma_2 z}\left[\int^{z} e^{\sigma_2 z}\frac{\sigma_2(\theta t + T_0)}{k}\frac{\mathrm{d}z}{(\theta t + T_0)} + c_3\right]$$

$$v(t) = e^{-\sigma_2 z}\left[\frac{1}{k}\int^{z} e^{\sigma_2 z}\sigma_2 \mathrm{d}z + c_3\right]$$

$$v(t) = e^{-\sigma_2 z}\left[\frac{\sigma_2}{k}\int^{z} e^{\sigma_2 z}\mathrm{d}z + c_3\right]$$

$$v(t) = e^{-\sigma_2 z}\left[\frac{e^{\sigma_2 z}}{k} + c_3\right]$$

$$v(t) = \left[\frac{1 + c_4 k e^{-\sigma_2 z}}{k}\right] \tag{6.33}$$

where $c_4 = c_3 e^{-\sigma_2 z}$
Recall, $w(t) = \frac{1}{v(t)}$

$$\therefore\ w(t) = \frac{k}{\left(1 + kc_4 e^{-\sigma_2 z}\right)} \tag{6.34}$$

Imposing initial conditions, $z_1 = z(t_1)$, $w(z_1) = w(z(t_1))$

$$w_1 = \frac{k}{\left(1 + kc_4 e^{-\sigma_2 z}\right)} \tag{6.35}$$

Making c_4 subject of formula, Eq. (6.35) becomes

$$c_4 = \frac{(k - w_1)e^{\sigma_2 z_1}}{kw_1} \tag{6.36}$$

Substituting Eq. (6.36) into Eq. (6.34)

$$w(t) = \frac{k}{1 + \frac{(k-w_1)e^{\sigma_2 z_1} * ke^{-\sigma_2 z}}{kw_1}} \tag{6.37}$$

Multiplying through with w_1, Eq. (6.37) becomes

$$w(t) = \frac{w_1 k}{w_1 + (k - w_1)e^{-\sigma_2(z-z_1)}} \tag{6.38}$$

Recall,

$$z(t) = z = \frac{\theta t^2}{2} + T_0 t$$

$$z(t_1) = z_1 = \frac{\theta {t_1}^2}{2} + T_0 t_1$$

$$\therefore\ z(t) - z(t_1) = z - z_1 = \frac{\theta\left(t^2 - t_1^2\right)}{2} + T_0(t - t_1) \tag{6.39}$$

Substituting Eq. (6.39) into Eq. (6.38) and simplifying,

$$w(t) = \frac{\beta_1 w_0 k}{\beta_1 w_0 + (k - \beta_1 w_0)e^{-\sigma_2\left(\frac{\theta}{2}\left(t^2-t_1^2\right)+T_0(t-t_1)\right)}} \tag{6.40}$$

For a fixed temperature, θ which is the heating rate = 0, Eq. (6.40) becomes,

$$w(t) = \frac{\beta_1 w_0 k}{\beta_1 w_0 + (k - \beta_1 w_0)e^{-\sigma_2 T_0(t-t_1)}} \tag{6.41}$$

Equation (6.41) is the general model equation for the second phase of degradation (weight loss of rubber with time) at a fixed temperature T_0.

In order to determine σ_2, the rate of weight loss with time for the second phase, where $\sigma_2 > 0$, it is assumed that weight loss at $w(t_2) = w_2$,

$$w(t) = w_2 = w_0\beta_2 = \frac{\beta_1 w_0 k}{\beta_1 w_0 + (k - \beta_1 w_0)e^{-\sigma_2 T_0(t_2-t_1)}}$$

Cross multiply,

$$w_0\beta_1 + (k - w_0\beta_1)e^{-\sigma_2 T_0(t_2-t_1)} = \frac{w_0\beta_1 k}{w_0\beta_2}$$

$$(k - w_0\beta_1)e^{-\sigma_2 T_0(t_2-t_1)} = \frac{\beta_1 k}{\beta_2} - w_0\beta_1$$

$$e^{-\sigma_2 T_0(t_2-t_1)} = \frac{\beta_1(k - \beta_2 w_0)}{\beta_2(k - \beta_1 w_0)}$$

$$-\sigma_2 T_0(t_2 - t_1) = \ln\left[\frac{\beta_1(k - \beta_2 w_0)}{\beta_2(k - \beta_1 w_0)}\right]$$

$$\therefore\ \sigma_2 = \frac{-1}{T_0(t_2 - t_1)}\ln\left[\frac{\beta_1(k - \beta_2 w_0)}{\beta_2(k - \beta_1 w_0)}\right] \tag{6.42}$$

Third-phase degradation model derivation

At this phase, it is assumed that the limit of weight loss (for the general Eq. 6.41) is as t tends to ∞

$$\lim_{t\to\infty} w(t) = \frac{\beta_1 w_0 k}{\beta_1 w_0 + (k - \beta_1 w_0)e^{-\sigma_2 T_0(t_2-t_1)}}$$

$$= \frac{\beta_1 w_0 k}{\beta_1 w_0 + 0} = k$$

This implies that at the third phase, k is the value at which change in weight loss with time is negligible

Thus, mathematically, $k = |w_{k+1} - w_k| \to 0$

6.4.9.2 Model Application and Interpretation

Figure 6.64 shows the graph of UT weight change with time for fixed temperature using the first-phase model Eq. (6.23), the second-phase model Eq. (6.41) and given that $w(t) \to k$ as t tends to ∞. Python programming language was used to generate the plotted data for varying temperature (Osayi 2015) from the general equation of the different degradation phases.

At the first phase, w_o which is the initial weight of the UT, t_1 the total time for the first phase of degradation and any fixed temperature, $T_i \in i = 0, 1, 2, 3 \ldots n$ were used where w_0 and t_1 are known values from the TGA. Also, it is assumed that weight changes occur with every other temperature at the same rate of σ_1 calculated from Eq. (6.24) at β_1 (weight fraction remaining at the end of first phase obtained from the TGA data) and T_0. The second phase also has the same rate of σ_2 calculated from Eq. (6.42). The value of t_2, which is the total time for the completion of the second phase, was gotten from the TGA as well. Once these values of σ_1, and σ_2 are known, the time for the completion of the first phase and second phase degradation for the other different fixed temperatures were derived by making time t_1 and t_2 the subject of formula in Eqs. (6.24) and (6.42). Table 6.18 shows the calculated completion time t_1 and t_2 for the different fixed temperature with known values of $w_o = 16.32$ mg, $\sigma_1 = 1.07 \times 10^{-6}$ and $\sigma_2 = 1.534 \times 10^{-3}$.

From Figure 6.64, the degradation of the material occurred faster as the value of the fixed temperature increases. This is in accordance with the reduction in the completion time of degradation observe in Table 6.18 for the first and second phases as the value of the fixed temperature increases.

When the initial weight of UT (16.32 mg) for a temperature of 450 °C in the TGA was compared with the predicted weight at a fixed temperature of 450 °C in the theoretical data generated using the mathematical model, it was 6.89 mg after 42.52 min at 450 °C while the weight of 6.54 mg was recorded after 18 min at a fixed temperature of 450 °C. This is probably due to the material being exposed to a higher temperature (450 °C) from the very onset of the degradation unlike in the case of the TGA where the material was gradually heated at a steady heating rate from a low temperature till it got to 450 °C. Therefore, in terms of percentage with reference to proportion of weight loss over time, it took 42.52 min for the material to lose 57.78% using the TGA while the material had a predicted weight loss of 60% after 18 min, a considerable shorter time at a fixed temperature of 450 °C.

Considering the weight change that occurred in the fixed-bed reactor used for the pyrolysis of UT, experimental value from Figure 6.65 shows that the weight of UT at 450 °C (at a HR of 15 °C/min for 30 min), was 52.3 ± 4.3%. This value was obtained at a lower time using the TGA but higher than that obtained at the predicted fixed temperature of 450 °C. This is

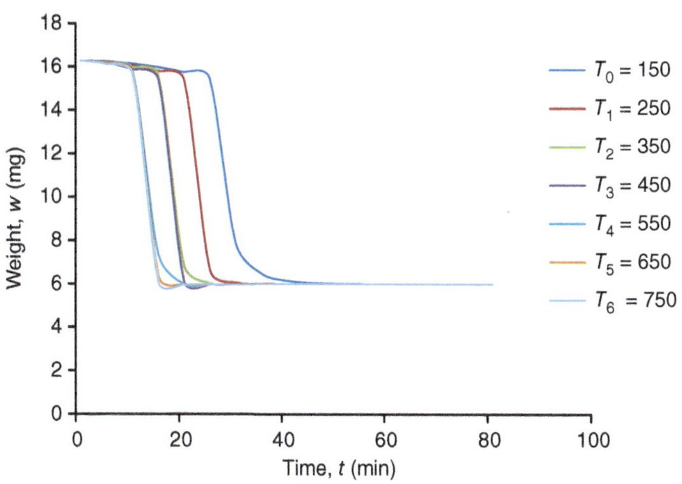

Figure 6.64 First, second and third phase weight loss versus time at different fixed temperature.

Table 6.18 First and second phase total degradation time for different fixed temperature.

Time (min)	T_0 150 °C	T_1 250 °C	T_2 350 °C	T_3 450 °C	T_4 550 °C	T_5 650 °C	T_6 750 °C
t_1	25.97	20.12	17.00	14.99	13.56	12.48	11.61
t_2	75.65	55.78	47.26	42.53	39.52	37.43	35.91

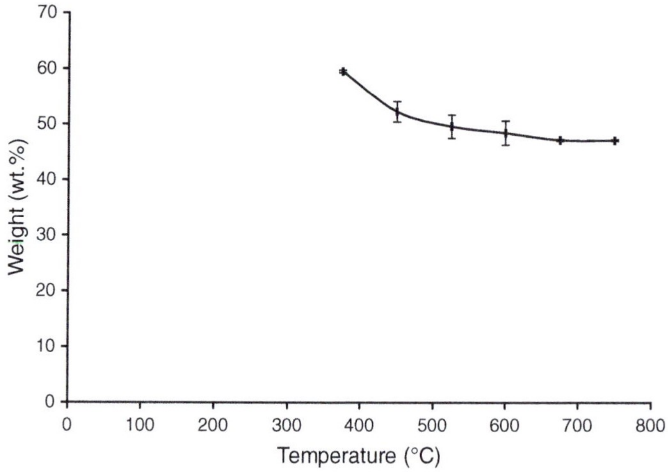

Figure 6.65 UT weight loss over time and temperature in a fixed-bed reactor.

certainly anticipated as the heating of the UT material in the fixed-bed reactor took some time to get to 450 °C as determined by the heating rate (15 °C/min).

Hence, the dynamic weight change for the UT material can be predicted from the chart using a few of the known parameters from the TGA in the mathematical model developed for the different phases of degradation for a varying time at a fixed temperature. This model can be used to obtain different values of UT weight at a particular temperature at any given time *t*.

6.5 Concluding Remarks and Recommendations

Carbon dioxide, being a major greenhouse gas in the atmosphere, emitted from human activities, including burning of conventional fossil fuels and so on, are quickly raising its emission to over 421 ppm (NOAA 2022). While the International Energy Outlook foresees that CO_2 emission will grow from 30.2 BMT (Billion metric tonnes) in 2008 to 43.2 BMT by 2035 (IEA 2017), it is scientifically feasible to reduce carbon dioxide emissions to the atmosphere by using the CO_2 as feed into energy production technologies. Such intervention will certainly mitigate and reduce the anthropogenic influence on the climate change.

As a source of carbon, CO_2, is an important raw material in the synthesis and production of the following:

- Urea, formic acid, and specialty chemicals,
- Fuel, in the production of pyrogas, methanol, bio-oils as alluded to in the preceding sections, etc.,
- Polymer in the production of plastic, rubber, and fibre,
- Food, as an additive and acidity regulator,
- Beverage, in the production of carbonated soft drinks and soda water,
- Construction, as a medium for welding and to produce ready-mixed concrete, etc.

CO_2 is also being used in thermochemical processes like pyrolysis and gasification to valorise wastes like biomass and plastics (Wang et al. 2022; Zhang et al. 2021; Saad and Williams 2017). Since it is one of the main components of the gases formed in the course of pyrolysis and gasification, its reuse as a reactant reduces its release into the atmosphere (Kim et al. 2022). Some other advantages of the use of CO_2 include the promotion of thermal breakdown of volatile components and prevents the synthesis of many dreadful chemical pollutants covering PAHs and benzene derivatives (Xu et al. 2022).

Alherbawi et al. (2021) identified that there are many developments in renewable energy technology in recent decades, which include pyrolysis, gasification, combustion, liquefaction, etc. and most of them have proven to be successful in achieving initially set goals and are suitable for producing high amounts of energy under specified conditions, such as biochar, bio-oil and pyrogas. Their findings also support five key elements of an organic waste source of energy that are sustainable, low price, good quality, sufficient quantity and low emissions. In similar vein, Parthasarathy et al. 2023 gave a detailed review of the operating conditions and outcomes for pyrolysis for many waste materials in previous catalytic pyrolysis studies with CO_2 as presented in Table 6.19. The table highlights the influence of CO_2 on organic waste pyrolysis using catalysts

Table 6.19 Examples of pyrolysis operating conditions and results of previous catalytic CO_2 pyrolysis studies.

Reactor type and its dimensions	Feedstock and its quantity	Operating conditions	Catalyst and its composition	Specific results	Comparison with N_2 pyrolysis	Inference	References
Tubular. Outer diameter: 2.5 cm, length: 120 cm	Swine manure (1.0 g)	Temperature: 20–900 °C Heating rate: 10 °C/min	Steel slag	Catalytic CO_2 pyrolysis improved CO production. The addition of catalyst significantly improved the reaction kinetics associated with the mechanistic role of CO_2 (up to 80%)	Both N_2 and CO_2 pyrolysis produced nearly the same amount of biochar. However, CO_2 pyrolysis yielded less bio-oil	This study experimentally demonstrated that CO_2 could be promising for limiting the formation of toxic pollutants in thermochemical treatment because CO_2 provides an innovative and strategic means of controlling the C/H ratio	Lee et al. (2019)
Tubular. Outer diameter: 2.5 cm, Inner diameter: 2.2 cm, Length: 120 cm	Fishing net waste (1.0 g)	Temperature: 35–600 °C Heating rate: 10 °C/min CO_2 flow rate: 200 mL/min	Ni/SiO_2 (5%).	The formation of pyrogas was accelerated by the application of the catalyst under CO_2 ambience, producing significant amounts of H_2 (770 mmol g^{-1} cat h^{-1}) and CO (1157 $mmol^{-1}$ cat h^{-1})	The mass loss profiles of fishing net waste under N_2 and CO_2 mediums were same, according to TGA results. Due to catalyst deactivation under N_2 influence, the rate of H_2 synthesis considerably decreased. Nonetheless, an additional CO formation was obtained under CO_2 conditions	The CO_2-mediated catalytic pyrolysis produced a higher H_2 yield while maintaining catalyst stability	Jung et al. (2020a)
Tubular	Rice straw (1.5 g)	Temperature: Room: 720 °C Heating rate: 10 °C/min CO_2 flow rate: 200 mL/min	Co/SiO_2 (5%)	Relative to non-catalytic pyrolysis, the use of the catalyst sped up the kinetics of the CO and H_2 reactions by three and six times, respectively. Significant amounts of H_2 and CO were synthesised as a result of the CO_2 and catalyst's synergistic effects	TGA clearly shows that, when compared to the N_2 condition, rice straw underwent more thermolysis at a temperature of about 700 °C. Under N_2 and CO_2 conditions, the final weights of rice straw samples were 32.0 wt.% and 23.0 wt.%, respectively	The combined influence of the CO_2 feedstock and Co/SiO_2 catalyst resulted in an increase in pyrogas production, particularly in the significant formations of H_2 and CO	Jung et al. (2020b)
Tubular. Outer diameter: 2.5 cm	Hen manure (1.0 g)	Temperature: Room: 720 °C Heating rate: 10 °C/min CO_2 flow rate: 100 mL/min	Ni/SiO_2 (5%)	Compared to non-catalytic pyrolysis, CO_2-mediated catalytic pyrolysis substantially increased CO production	In comparison to N_2, CO_2 promoted undue thermal breakdown of volatile species into pyrogas	CO_2-assisted pyrolysis could be viewed as an environmentally benign and commercially feasible method of producing pyrogas from biomass materials	Lee et al. (2020a)

Tubular. Outer diameter: 2.5 cm, Inner diameter: 2.2 cm, Length: 50 cm	Cow manure (1.0 g)	Temperature: Room: 720 °C Heating rate: 10 °C/min CO_2 flow rate: 100 mL/min	Ni/SiO_2 (1.0 g)	CO_2 accelerated the significant production of CO at 600 °C through the homogeneous reaction of VOCs	CO_2-cofeeding catalytic pyrolysis of cow manure produced 25% more pyrogas than catalytic pyrolysis in N_2	Cow manure pyrolysis using CO_2 is a practical method for manure valorisation in the livestock industry without the need for numerous pre-and post-treatment steps	Lee et al. (2020b)
Tubular	Food waste (1.0 g)	Temperature: Room: 900 °C Heating rate: 10 °C/min CO_2 flow rate: 300 mL/min	Pt	The CO_2 reactant and Pt catalyst and had zero effect on the quantity of biochar, but they did have an impact on the generation of bio-oil and pyrogas. At 700 °C, with the catalyst presence, the CO_2 atmosphere, produced roughly 67.3% fewer cyclic compounds, including derivatives of benzene, than under non-catalytic circumstances without CO_2	The TGA curves achieved under N_2 and CO_2 ambiences at 820 °C were identical, there was a noticeable variance at higher temperatures. The quantity of cyclic species in the catalytic reaction diminished from 22% in N_2 to 17% in CO_2	A CO_2-supported catalytic pyrolysis is as ecologically benign disposal approach for food waste	Kim and Lee (2020)
Tubular	Cellulose acetate (1.0 g)	Temperature: 35–710 °C Heating rate: 10 °C/min CO_2 flow rate: 100 mL/min	Ni/SiO_2 (1.0 g)	Owing to the collective effects of CO_2 reactant medium and the Ni-based catalysts, a significant enrichment of CO formation and pyrogas yield was obtained. Additionally, the CO_2 helped to prevent biochar from depositing on the catalyst	The TGA findings showed that CO_2 degraded cellulose acetate more effectively than N_2 did. In the N_2 and CO_2 atmospheres, the residual weights of cellulose acetate were 14.2 and 13.2 wt.%, respectively	In contrast to N_2, the study suggested that CO_2 would be a more advantageous reactive co-feedstock for pyrolysis from a technical and environmental standpoint. As a result, this study demonstrated a straightforward and adaptable technical framework for converting cellulose acetate and CO_2 into pyrogas	Cho et al. (2021)
Tubular. Outer diameter: 2.5 cm, Inner diameter: 2.2 cm, Length: 120 cm	Flood debris (5.0 g)	Temperature: 35–720 °C Heating rate: 10 °C/min CO_2 flow rate: 200 mL/min	Ni/SiO_2 (5%)	In comparison to pyrolysis without catalyts, the production of pyrogas increased by more than one order of magnitude when using a Ni/SiO_2 catalyst. The synergistic influence of CO_2 reactant and Ni catalyst produced almost 50% more pyrogas than catalytic pyrolysis under N_2	Compared to pyrolysis under the N_2 influence, CO production from the pyrolysis of debris waste was improved under the CO_2 reactant medium	The use of CO_2 to aid pyrolysis of flood debris could maximise energy production	Choi et al. (2021a)

(*Continued*)

Table 6.19 (Continued)

Reactor type and its dimensions	Feedstock and its quantity	Operating conditions	Catalyst and its composition	Specific results	Comparison with N_2 pyrolysis	Inference	References
Tubular	Livestock waste (2.0 g)	Temperature: 35–720 °C Heating rate: 10 °C/min CO_2 flow rate: 200 mL/min	Biochar (1.0–2.0 g)	The formations of H_2 and CO were improved by using biochar as a catalyst material in the pyrolysis of livestock waste, and the catalytic ability was directly proportional to the catalyst loading	The concentration profiles of the primary gaseous effluents from pyrolysis under CO_2 and N_2 mediums were almost same at 510 °C. The evolution patterns of the main gaseous products were, however, somewhat different at 510 °C. At 510 °C, a greater amount of CO was produced solely from the CO_2 environment	In comparison to pyrolysis without catalysts, catalytic pyrolysis of livestock waste over biochar produced three times as much pyrogas	Choi et al. (2021b)
Tubular. Outer diameter: 2.5 cm, Inner diameter: 2.2 cm, Length: 100 cm	Pine bark (1.0 g)	Temperature: Room: 720 °C Heating rate: 10 °C/min CO_2 flow rate: 100 mL/min	Ni/SiO_2 (5%)	When the catalyst bed was at 500 °C, each CO_2 and Ni catalyst independently contributed to CO and H_2 formation, respectively, whereas a synergistic influence of both CO_2 and Ni catalyst was demonstrated at 600 °C	The findings imply that there is a greater amount of coke deposition on catalyst used in a N_2 medium than in a CO_2 ambience	The combined effect of CO_2 medium and Ni/SiO_2 catalyst improved pyrogas generation	Lee et al. (2021)
Tubular	Disposable diaper (1.5 g)	Temperature: 40–720 °C Heating rate: 10 °C/min CO_2 flow rate: 200 mL/min	Ni/SiO_2 (5%)	Ni catalyst increased H_2 and CO formations at higher temperature under CO_2 by 4 and 15 times more than pyrolysis without catalyst		CO_2 provided a tactical way to create chemicals with added value and lessen the aromaticity of pyrogenic products	Kwon et al. (2021a)
Tubular	Synthetic textile waste (1.0 g)	Temperature: 40–720 °C Heating rate: 10 °C/min	Co/SiO_2 (4%)	CO_2-assisted catalytic pyrolysis produced three and eight times more H_2 and CO than non-catalytic pyrolysis, respectively. Over 80% of the textile waste was transformed into pyrogas and CH_4 thanks to this process, which also inhibited catalyst deactivation	In contrast to catalytic pyrolysis conducted under the N_2 medium, which produced 7 wt.% of condensable gases, only 1 wt.% of volatiles were left over from the CO_2-assisted method	Waste synthetic textiles could be converted into pyrogas using CO_2-assisted catalytic pyrolysis, which would minimise the production of polycyclic aromatic hydrocarbons and benzene derivatives	Kwon et al. (2021b)
Tubular	Cellulose (20.0 g)	Temperature: Room: 600 °C	Ce-Co-Na (5%)	CO_2 increased ketone yield in bio-oil and CO yield in pyrogas		The study offers an alternative method for pyrolysing biomass to produce pyrogas and ketone platform compounds	Jiang et al. (2022)

Source: Adapted from Parthasarathy et al.

for energy and biofuel recovery. The effect of catalytic CO_2 pyrolysis on the pyrolysis products, yields and characteristics are also discussed elsewhere (Parthasarathy et al. (2023).

However, at the international community, the importance of education and training to address climate change is currently receiving attention, because education is crucial to promote climate actions. It certainly helps people understand and address the impacts of the climate crisis, thus empowering them with the right knowledge, skills, values and attitudes needed to act as agents of change are heavily discussed in the next chapter. As a result, Chapter 6 below presents climate change policy and skills education that is buttressed with a case study from the University of the Witwatersrand in Johannesburg.

References

Adeniji, O.A., Zaccheaus, O.S., Ojo, B.S., and Adedeji, A.S. (2015). Charcoal production and producers' tree species preference in Borgu local government area of Niger state, Nigeria. *J. Energy Technol. Policy* 5: 11. ISSN 2224-3232 (Paper); ISSN 2225-0573 (Online).

Aguado, R., Olazar, M., Gaisán, B. et al. (2002). Kinetic study of polyolefin pyrolysis in a conical spouted bed reactor. *Ind. Eng. Chem. Res.* 41 (18): 4559–4566.

Aguado, R., Olazar, M., V'elez, D. et al. (2005). Kinetics of scrap tyre pyrolysis under fast heating conditions. *J. Anal. Appl. Pyrolysis* 73 (2): 290–298.

Alherbawi, M., Parthasarathy, P., Al-Ansari, T. et al. (2021). Potential of drop-in biofuel production from camel manure by hydrothermal liquefaction and biocrude upgrading: a Qatar case study. *Energy* 232: 121027. https://doi.org/10.1016/J.ENERGY.2021.121027.

Al-Salem, S.M., Lettieri, P., and Baeyens, J. (2010). The valorization of plastic solid waste (PSW) by primary to quaternary routes: from re-use to energy and chemicals. *Prog. Energy Combust. Sci.* 36: 103–129.

Association of Natural Rubber Producing Countries (2010). Rubber rises on demand and supply, currencies and pyrolytic oil oil: gains from China's exchange rate reform. *Nat. Rubber Trends Stat.* 2: 6.

Bajus, M. and Olahova, N. (2011). Thermal conversion of scrap tyres. *Petrol. Coal* 53: 98–105.

Balat, M., Balat, M., Kirtay, E., and Balat, H. (2009). Main routes for the thermo-conversion of biomass into fuels and chemicals. Part 1: pyrolysis systems. *Energy Convers. Manag.* 50 (12): 3147–3157.

Barbooti, M.M., Mohamed, T.J., Hussain, A.A., and Abas, F.O. (2004). Optimization of pyrolysis conditions of scrap tires under inert gas atmosphere. *J. Anal. Appl. Pyrolysis* 72: 165–170.

Beecham, M. (2008). *Global Market Review of Automotive Tyres—Forecasts to 2014.* Pub ID JA1867566. Bromsgrove, UK: Aroq Limited.

Berrueco, C., Esperanza, E., Mastral, F.J. et al. (2005). Pyrolysis of waste tyres in an atmospheric static-bed batch reactor: analysis of the gases obtained. *J. Anal. Appl. Pyrolysis* 74: 245–253.

Bhatia, V.K., Padmaja, K.V., Kamra, S. et al. (1993). Upgrading of biomass constituents to liquid fuels. *Fuel* 72: 101–104.

Bi, J., Guo, X., Liu, M., and Wang, X. (2010). High effective dehydration of bio- ethanol into ethylene using nanoscale HZSM-5 zeolite catalysts. *Catal. Today* 149: 143–147.

Bosmans, A., Vanderreydt, I., Geysen, D., and Helsen, L. (2013). The crucial role of waste-to-energy technologies in enhanced landfill mining: a technology review. *J. Clean. Prod.* 55: 10–23.

Boukis, I.P., Bezergianni, S., Grammelis, P. et al. (2007). CFB air-blown flash pyrolysis. Part II: operation and experimental results. *Fuel* 86: 1387–1395.

Boxiong, S., Chunfei, W., Cai, L. et al. (2007). Pyrolysis of waste tyres: the influence of USY catalyst/tyre ratio on products. *J. Anal. Appl. Pyrolysis* 78: 243–249.

Brammer, J. and Bridgewater, A.V. (1999). A review of biomass drying technologies for thermal conversion. *Renew. Sust. Energ. Rev.* 3 (4): 243–289.

Bridgwater, A.V. (1999). Principles and practice of biomass fast pyrolysis processes for liquids. *J. Anal. Appl. Pyrol. l* 51: 3–22.

Bridgwater, T. (2007). *Pyrolysis of Biomass. IEA Bioenergy Task 34.* Birmingham, UK: Bioenergy Research Group, Aston University.

Bridgwater, A.V., Czernik, S., and Piskorz, J. (2001). An overview of fast pyrolysis. In: *Progress in Thermochemical Biomass Conversion*, vol. 2, 977–997.

Brindley, G.W. and Nakahira, M. (1958). The kaolinite-mullite reaction series: II, metakaolin. *J. Am. Ceram. Soc.* 42 (7): 314–317.

Brown, D. (1996). *Bio-Oil Production and Utilization* (ed. A.V. Bridgwater and E.N. Hogan), 96–100. CPL Press.

Buah, W.K., Cunliffe, A.M., and Williams, P.T. (2007). Characterization of products from the pyrolysis of municipal solid waste. *Process. Saf. Environ. Prot.* 85: 450–457.

Ceylan, S. and Topcu, Y. (2014). Pyrolysis kinetics of hazelnut husk using thermogravimetric analysis. *Bioresour. Technol.* 156: 182–188.

Chandrasekhar, S. and Pramada, P.N. (2004). Kaolin-based zeolite Y, a precursor for cordierite ceramics. *Appl. Clay Sci.* 27 (3–4): 187–198.

Chang, Y.M. (1996). On pyrolysis of waste tyre: degradation rate and product yields. *Resour. Conserv. Recycl.* 17 (2): 125–139.

Charles, S. (2002). *Noble Obsession; Charles Goodyear, Thomas Hancock, and the Race to Unlock the Greatest Industrial Secret of the Nineteenth Century*. Hyperion. ISBN: 9780786867899.

Chen, J.H., Chen, K.S., and Tong, L.Y. (2001). On the pyrolysis kinetics of scrap automotive tyres. *J. Hazard. Mater.* 84 (1): 43–55.

Chen, D., Yin, L., Wang, H., and He, P. (2014). Pyrolysis technologies for municipal solid waste: a review. *Waste Manag.* 34: 2466–2486.

Cho, S.H., Jung, S., Rinklebe, J., and Kwon, E.E. (2021). CO_2-assisted catalytic pyrolysis of cellulose acetate using Ni-based catalysts. *Environ. Pollut.* 275: 116667. https://doi.org/10.1016/J.ENVPOL.2021.116667.

Choi, D., Jung, S., Jung, M.K. et al. (2021a). Strategic disposal of flood debris via CO_2-assisted catalytic pyrolysis. *J. Hazard. Mater.* 412: 125242. https://doi.org/10.1016/J.JHAZMAT.2021.125242.

Choi, D., Jung, S., Lee, D.J. et al. (2021b). A new upgrading platform for livestock lignocellulosic waste into syngas using CO_2-assisted thermo-chemical process. *Energy Convers. Manag.* 236: 114084. https://doi.org/10.1016/J.ENCONMAN.2021.114084.

Conesa, J.A., Font, R., Fullana, A., and Caballero, J.A. (1998). Kinetic model for the combustion of tyre wastes. *Fuel* 77 (13): 1469–1475.

Cornelissen, T., Yperman, J., Reggers, G. et al. (2008). Flash co-pyrolysis of biomass with polylactic acid. Part 1: influence on bio-oil yield and heating value. *Fuel* 87 (7): 1031–1041.

Cozzani, V., Nicolella, C., Petarca, L. et al. (1995). A fundamental study on conventional pyrolysis of a refuse-derived fuel. *Ind. Eng. Chem. Res.* 34: 2006–2020.

Cunliffe, A.M. and Williams, P.T. (1998). Composition of oils derived from the batch pyrolysis of tyres. *J. Anal. Appl. Pyrolysis* 44: 131–152.

Cypres, R. and Bettens, B. (1989). *Pyrolysis and Gasification* (ed. G.L. Ferrero, K. Maniatis, A. Buekens, and A.V. Bridgwater). London, UK: Elsevier Applied Science.

Dai, X., Yin, X., Wu, C. et al. (2001). Pyrolysis of waste tires in a circulating fluidized-bed reactor. *Energy* 26: 385–399.

De Marco, I.R., Laresgoiti, M.F., Cabrero, M.A. et al. (2001). Pyrolysis of scrap tyres. *Fuel Process. Technol.* 72: 9–22.

Deenanath, E.D. (2014). Production and characterisation of bioethanol derived from cashew apple juice for use in internal combustion engine. PhD thesis. Faculty of Engineering and the Built Environment, University of the Witwatersrand, Johannesburg, South Africa.

Di Blasi, C. (1996). Kinetic and heat transfer control in the slow and flash pyrolysis of solids. *Ind. Eng. Chem. Res.* 35 (37): 124.

Ding, K., Zhong, Z.P., Yu, L.L. et al. (2013). Pyrolysis characteristics and kinetic study of mixed pyrolysis of municipal solid waste. *J. Southeast Univ. (Natural Science Edition)* 43: 130–135. (in Chinese).

Dombrovsky, L.A., Sazhin, S.S., Sazhina, E.M. et al. (2001). Heating and evaporation of semitransparent diesel fuel droplets in the presence of thermal radiation. *Fuel* 80: 1535–1544.

Domínguez, A., Fernández, Y., Fidalgo, B. et al. (2008). Bio-syngas production with low concentrations of CO_2 and CH_4 from microwave-induced pyrolysis of wet and dried sewage sludge. *Chemosphere* 70 (3): 397–403.

Donatelli, A., Iovane, P., and Molino, A. (2010). High energy syngas production by waste tyres steam gasification in a rotary kiln pilot plant. Experimental and numerical investigations. *Fuel* 89 (10): 2721–2728.

Fix, S.R. (1980). Microwave devulcanization of rubber. *Elastomerics* 112 (6): 38–40.

Font, R., Marcilla, A., García, A.N. et al. (1995a). Kinetic models for the thermal degradation of heterogeneous materials. *J. Anal. Appl. Pyrolysis* 32: 29–39.

Font, R., Marcilla, A., Garcia, A.N. et al. (1995b). Comparison between the pyrolysis products obtained from different organic wastes at high temperatures. *J. Anal. Appl. Pyrolysis* 32: 41–49.

Fonts, I., Gea, G., Azuara, M. et al. (2012). Sewage sludge pyrolysis for liquid production: a review. *Renew. Sust. Energ. Rev.* 16: 2781–2805.

Franco, F., Pérez-Maqueda, L.A., and Pérez-Rodríguez, J.L. (2003). The influence of ultrasound on thethermal behaviour of a well ordered kaolinite. *Thermochim. Acta* 404: 71–79.

Frigo, S., Seggiani, M., Puccini, M., and Vitolo, S. (2014). Liquid fuel production from waste tyres pyrolysis and its utilization in a diesel engine. *Fuel* 116: 399–408.

Garcia, A.N., Font, R., and Marcilla, A. (1995a). Kinetic study of the flash pyrolysis of municipal solid waste in a fluidized bed reactor at high temperature. *J. Anal. Appl. Pyrolysis* 31: 101–121.

Garcia, A.N., Font, R., and Marcilla, A. (1995b). Gas production by pyrolysis of municipal solid waste at high temperature in a fluidized bed reactor. *Energy Fuel* 9: 648–658.

Gerten, J., Fainberg, V., Hetsroni, G., and Shindler, Y. (2000). Kinetic study of the thermal decomposition of polypropylene, oil shale, and their mixture. *Fuel* 79: 1679–1686.

Gonzalez, J.F., Encinar, J.M., Canito, J.L., and Rodrıguez, J.J. (2001). Pyrolysis of automobile tyre waste. Influence of operating variables and kinetics study. *J. Anal. Appl. Pyrolysis* 58-59: 667–683.

Harrison, G. and Ross, A.B. (1996). Use of tyre pyrolysis oil for solvent augmentation in two-stage coal liquefaction. *Fuel* 75 (8): 1009–1013.

Haydary, J., Korenova, Z., Jelemensky, L., and Markos, J. (2008). Thermal decomposition of waste polymers. *ThermoPhysics* 62–68.

He, F., Yi, W.M., and Bai, X.Y. (2006). Investigation on caloric requirement of biomass pyrolysis using TG–DSC analyzer. *Energy Convers. Manag.* 47: 2461–2469.

Hu, H., Fang, Y., and Liu, H. (2014). The fate of sulfur during rapid pyrolysis of scarp tyres. *Chemosphere* 97: 102–107.

IEA (2017). *US Energy Information Administration*. International Energy Outlook.

Isayev, A.I., Yushanov, S.P., and Chen, J. (1996). Ultrasonic devulcanization of rubber vulcanizates. I: process model. *J. Appl. Polym. Sci.* 59 (5): 803–813.

Islam, M.N. and Beg, M.R.A. (2004). The fuel properties of pyrolysis liquid derived from urban solid wastes in Bangladesh. *Bioresour. Technol.* 92: 181–186.

Islam, M.R., Haniu, M.R., and Beg, M.R.A. (2007). Limonene-rich liquids from pyrolysis of heavy automotive tire wastes. *J. Environ. Eng.* 2 (4): 681–695.

Islam, M.R., Haniu, H., and Beg, M.R.A. (2008). Liquid fuels and chemicals from pyrolysis of motorcycle tyre waste: products yields, compositions and related properties. *Fuel* 87: 3112–3122.

Jahirul, M.I., Rasul, M.G., Chowdhury, A.A., and Ashwath, N. (2012). Biofuels production through biomass pyrolysis – a technical review. *Energies* 5: 4952–5001.

Jiang, H., Liu, H., Dong, J. et al. (2022). Enhancing ketones and syngas production by CO_2-assisted catalytic pyrolysis of cellulose with the Ce–Co–Na ternary catalyst. *Energy* 250: 123799. https://doi.org/10.1016/J. ENERGY.2022.123799.

Juma, M., Korenova, Z., Markos, J. et al. (2006). Pyrolysis and combustion of scrap tyre. *Petrol. Coal* 48 (1): 15–26.

Juma, M., Korenova, Z., Markos, J. et al. (2007). Experimental study of pyrolysis and combustion of scrap tyre. *Polym. Adv. Technol.* 18: 144–148.

Jung, S., Choi, D., Park, Y.K. et al. (2020a, 2020). Functional use of CO_2 for environmentally benign production of hydrogen through catalytic pyrolysis of polymeric waste. *Chem. Eng. J.* 399: 125889. https://doi.org/10.1016/J.CEJ.2020.125889.

Jung, S., Lee, S., Park, Y.K. et al. (2020b, 2020). CO_2-mediated catalytic pyrolysis of rice straw for syngas production and power generation. *Energy Convers. Manag.* 220: 113057. https://doi.org/10.1016/J.ENCONMAN.113057.

Kar, Y. (2011). Catalytic pyrolysis of car tyre waste using expanded perlite. *Waste Manag.* 31 (8): 1772–1782.

Karthikeyan, S., Sathiskumar, C., and Moorthy, S.R. (2012). Effect of process parameters on tyre pyrolysis: a review. *J. Sci. Ind. Res.* 71 (5): 309–315.

Kawakami, S., Inoue, K., Tanaka, H., and Sakai, T. (1980). Pyrolysis process for scrap tyres. In: *Thermal Conversion of Solid Wastes and Biomass ACS Symposium Series 130* (ed. J.L. Jones and S.B. Radding). Washington, DC: American Chemical Society Publishers.

Kim, S. and Lee, J. (2020). Pyrolysis of food waste over a pt catalyst in CO_2 atmosphere. *J. Hazard. Mater.* 393: 122449. https://doi.org/10.1016/J. JHAZMAT.2020.122449.

Kim, S., Byun, J., Park, H. et al. (2022). Energy-efficient thermal waste treatment process with no CO_2 emission: a case study of waste tea bag. *Energy* 241: 122876. https://doi.org/10.1016/J.ENERGY.2021.122876.

Kovo, A.S. (2010). Development of zeolites and zeolite membranes from Ahoko Nigerian Kaolin. PhD thesis. The University of Manchester, Manchester.

Kumaravel, S.T., Murugesan, A., and Kumaravel, A. (2016). Tyre pyrolysis oil as an alternative fuel for diesel engines – a review. *Renew. Sust. Energ. Rev.* 60: 1678–1685.

Kwon, D., Jung, S., Lin, K.Y.A. et al. (2021a). Synergistic effects of CO_2 on complete thermal degradation of plastic waste mixture through a catalytic pyrolysis platform: a case study of disposable diaper. *J. Hazard. Mater.* 419: 126537. https://doi.org/10 .1016/J.JHAZMAT.2021.126537.

Kwon, D., Yi, S., Jung, S., and Kwon, E.E. (2021b). Valorization of synthetic textile waste using CO_2 as a raw material in the catalytic pyrolysis process. *Environ. Pollut.* 268: 115916. https://doi.org/10.1016/J.ENVPOL.2020.115916.

Kyari, M., Cunliffe, A., and Williams, P.T. (2005). Characterisation of oils, gases and char in relation to the pyrolysis of different brands of scrap automotive tyres. *Energy Fuel* 19: 1165–1173.

Laresgoiti, M.F., Caballero, B.M., de Marco, I. et al. (2004). Characterisation of the liquid products obtained in tyre pyrolysis. *J. Anal. Appl. Pyrolysis* 71 (2): 917–934.

Lee, D.J., Jeong, K.H., Lee, D.H. et al. (2019). Catalytic pyrolysis of swine manure using CO_2 and steel slag. *Environ. Int.* 133: 105204. https://doi.org/10.1016/J.ENVINT.2019.105204.

Lee, D.J., Jung, S., Jang, Y.N. et al. (2020a). Offering a new option to valorize hen manure by CO_2-assisted catalytic pyrolysis over biochar and metal catalysts. *J CO_2 Util.* 42: 101344. https://doi.org/10.1016/J.JCOU.2020.101344.

Lee, D.J., Jung, S., Jeong, K.H. et al. (2020b). Catalytic pyrolysis of cow manure over a Ni/SiO_2 catalyst using CO_2 as a reaction medium. *Energy* 195: 117077. https://doi.org/10.1016/J.ENERGY.2020.117077.

Lee, T., Jung, S., Kim, K.H., and Kwon, E.E. (2021). Catalytic pyrolysis of pine bark over Ni/SiO_2 in a CO_2 atmosphere. *Energy* 220: 119827. https://doi.org/10.1016/J. ENERGY.2021.119827.

Leung, D.Y.C. and Wang, C.L. (1998). Kinetic study of scrap tyre pyrolysis and combustion. *J. Anal. Appl. Pyrolysis* 45 (2): 153–169.

Li, S., Yao, Q., Chi, Y. et al. (2004). Pilot-scale pyrolysis of scrap tyres in a continuous rotary kiln reactor. *Ind. Eng. Chem. Res.* 43: 5133–5145.

Lin, K.S., Wang, H.P., Liu, S.H. et al. (1999). Pyrolysis kinetics of refuse-derived fuel. *Fuel Process. Technol.* 60: 103–110.

Liompart, M., Sanchez-Prado, L., Lamas, J.P. et al. (2013). Hazardous organic chemicals in rubber recycled tyre playgrounds and pavers. *Chemosphere* 90: 423–431.

Liu, S., Xie, Q., Zhang, B. et al. (2016). Fast microwave-assisted catalytic co-pyrolysis of corn stover and scum for bio-oil production with CaO and HZSM5 as the catalyst. *Bioresour. Technol.* 204: 164–170.

Magazzino, C., Toma, P., Fusco, G. et al. (2022). Renewable energy consumption, environmental degradation and economic growth: the greener the richer? *Ecol. Indic.* 139: 108912.

MailOnline (2022). Lagos aims to use trash to make electric power. www.dailymail.co.uk/wires/ap/article-2579079/Lagos-aims-use-trash-make-electric-power.html (accessed 1 December 2022).

Malkow, T. (2004). Novel and innovative pyrolysis and gasification technologies for energy efficient and environmentally sound MSW disposal. *Waste Manag.* 24: 53–79.

Martínez, J.D., Puy, N., Murillo, R. et al. (2013). Waste tyre pyrolysis – a review. *Renew. Sust. Energ. Rev.* 23: 179–213.

Martinez, J.D., Veses, A., and Mastral, A.M. (2014). Co-pyrolysis of biomass with waste tyres: upgrading of liquid to bio-fuel. *Fuel Process. Technol.* 119: 263–271.

Mastral, A.M., Murillo, R., Callen, M.S. et al. (2000). Optimization of scrap automotive tyres recycling into valuable liquid fuels. *Resour. Conserv. Recycl.* 29: 263–272.

Midgley, T.J. and Henne, A.I. (1929). Natural and synthetic rubber I. Products of the destructive distillation of natural rubber. *J. Am. Chem. Soc.* 51: 1215–1226.

Miskolczi, N., Ates, F., and Borsodi, N. (2013). Comparison of real waste (MSW and MPW) pyrolysis in batch reactor over different catalysts. Part II: contaminants, char and pyrolysis oil properties. *Bioresour. Technol.* 144: 370–379.

Molewa, B.E.E. (2013). National Environmental Waste Management Act (59/2008): Approval of an Integrated Industry Waste Tyre Management Plan of the Recycling and Economic Development Initiative of South Africa South African National Department of Environmental Affairs 569, 35927.

Mongo, M., Belaïd, F., and Ramdani, B. (2021). The effects of environmental innovations on CO_2 emissions: empirical evidence from Europe. *Environ. Sci. Pol.* 118: 1–9.

NOAA (2022). *Carbon dioxide now more than 50% higher than pre-industrial levels.* The National Oceanic and Atmospheric Administration. https://www.noaa.gov/news-release/carbon-dioxide-now-more-than-50-higher-than-pre-industrial-levels (accessed 10 April 2023).

Olazar, M., Lopez, G., Arabiourrutia, M. et al. (2008). Kinetic modeling of tyre pyrolysis in a conical spouted bed reactor. *J. Anal. Appl. Pyrolysis* 81 (1): 127–132.

Onoji, S.E., Iyuke, S.E., Igbafe, A.I., and Nkazi, D.B. (2016). Rubber seed oil: a potential renewable source of biodiesel for sustainable development in sub-Saharan Africa. *Energy Convers. Manag.* 110: 125–134.

Osayi I.J. (2015). Production and characterisation of biocrude from used tyres and natural rubber. PhD thesis. The University of the Witwatersrand, Johannesburg.

Osayi, J.I., Iyuke, S.E., and Ogbeide, S.E. (2014). Pyrolytic oil production through pyrolysis of used tyres. *J. Catal.* 2014: 386371. https://doi.org/10.1155/2014/386371.

Pakdel, H., Roy, C., Aubin, H., and Coulombe, J.G. (1992). Formation of limonene in used tyre vacuum pyrolysis oils. *Environ. Sci. Technol.* 25: 1646.

Pakdel, H., Pantea, D.M., and Roy, C. (2001). Production of dl-limonene by vacuum pyrolysis of used tyres. *J. Anal. Appl. Pyrolysis* 57: 91–107.

Pakdel, H., Pantea, D.M., and Roy, C. (2004). Production of dl-limonene by vacuum pyrolysis of used tyres. *J. Anal. Appl. Pyrolysis* 71: 917–934.

Parekh, D.B., Parikh, P.A., and Rotliwala, Y.C. (2009). Synergetic pyrolysis of high density polyethylene and jatropha and Karanj cakes: a thermogravimetric study. *J. Renew. Sustain. Energy* 1: 033107. (1–10).

Parikh, P.A. and Rotliwala, Y.C. (2011). Thermal degradation of rice-bran with high density polyethylene: a kinetic study. *Korean J. Chem. Eng.* 28: 788–792.

Parthasarathy, P., Zuhara, S., Al-Ansari, T., and McKay, G. (2023). A review on catalytic CO_2 pyrolysis of organic wastes to high-value products. *Fuel* 335: 127073.

Phadke, A.A., Bhattacharya, A.K., Chakraborty, S.K., and De, S.K. (1983). Studies of vulcanization of reclaimed rubber. *Rubber Chem. Technol.* 56 (4): 726–736.

Pilusa, T.J., Shukla, M., and Muzenda, E. (2013). Tyre derived fuel as an alternative fuel for compression ignition engines. 2nd International Conference on Environment, Agriculture and Food Sciences (ICEAFS'2013), Kuala Lumpur, Malaysia.

Pilusa, T.J., Shukla, M., and Muzenda, E. (2014). Economic assessment of waste tyres pyrolysis technology: a case study for Gauteng province, South Africa. *Int. J. Res. Chem. Metall. Civil Eng.* 1: 1.

Pradhan, D. and Singh, R.K. (2011). Thermal pyrolysis of bicycle waste tyre using batch reactor. *Int. J. Chem. Eng. Appl.* 2 (5): 332–336.

Quek, A. and Balasubramanian, R. (2012). Mathematical modeling of rubber tire pyrolysis. *J. Anal. Appl. Pyrolysis* 95: 1–13.

Quek, A. and Balasubramanian, R. (2013). Liquefaction of waste tyres by pyrolysis for oil and chemicals – a review. *J. Anal. Appl. Pyrolysis* 101: 1–16.

Raveendran, K., Ganesh, A., Khilar, K.C., and K.C. (1996). Pyrolysis characteristics of biomass and biomass components. *Fuel* 75 (8): 987–998.

Rengga, W.D.P., Handayani, P.A., Kadarwati, S., and Feinnudin, A. (2015). Kinetic study on catalytic cracking of rubber seed (*Hevea brasiliensis*) oil to liquid fuels. *Bull. Chem. React. Eng. Catal.* 10 (1): 50–60.

Ricci, F. (2007). Channels of transmission of environmental policy to economic growth: a survey of the theory. *Ecol. Econ.* 60 (4): 688–699.

Roy, C. and Unsworth, J. (1999). Pilot plant demonstration of used tyres vacuum pyrolysis. In: *Pyrolysis and Gasification* (ed. G.L. Ferrero, K. Maniatis, A. Buekens, and A.V. Bridgwater). London, UK: Elsevier Applied Science.

Saad, J.M. and Williams, P.T. (2017). Pyrolysis-catalytic dry (CO_2) reforming of waste plastics for syngas production: influence of process parameters. *Fuel* 193: 7–14. https://doi.org/10.1016/J.FUEL.2016.12.014.

Saffarzadeh, A., Shimaoka, T., Motomura, Y., and Watanabe, K. (2006). Chemical and mineralogical evaluation of slag products derived from the pyrolysis/melting treatment of MSW. *Waste Manag.* 26: 1443–1452.

Sang, Z., Liu, P., Tian, Z. et al. (2006). Synthesis of small crystals zeolite NaY. *Mater. Lett.* 60: 1131–1133.

Schaefer, W.D. (1975). Disposing of solid wastes by pyrolysis. *Environ. Sci. Technol.* 9: 98–98.

Schnecko, H. (1998). Rubber recycling. *Macromol. Symp.* 135: 327–343.

Schure, J., Pinta, F., Cerutti, P.O., and Kasereka-Muvatsi, L. (2019). Efficiency of charcoal production in sub-Saharan Africa: solutions beyond the kiln. *Bois et Forêts des Tropiques* 340: 57–70. https://doi.org/10.19182/bft2019.340.a31691.

Senneca, O., Salatino, P., and Chirone, R. (1999). Fast heating-rate thermogravimetric study of the pyrolysis of scrap tyres. *Fuel* 78 (13): 1575–1581.

Shen, B., Wu, C., Wang, R. et al. (2006). Pyrolysis of scrap tyres with zeolite USY. *J. Hazard. Mater.* B137: 1065–1073.

Tasoff, H. (2019). *Plastic's Carbon Footprint*. Santa Barbara, CA: The Current, Science + Technology.

Trongkaew, P., Utistham, T., Reubroycharoen, P., and Hinchiranan, N. (2011). Photocatalytic desulfurization of waste tyre pyrolysis oil. *Energies* 4 (11): 1880–1896.

Ucar, S., Karagoz, S., Ozkan, A.R., and Yanik, J. (2005). Evaluation of two different scrap tyres as hydrocarbon source by pyrolysis. *Fuel* 84 (14–15): 1884–1892.

Undri, A., Rosi, L., Frediani, M., and Frediani, P. (2014). Upgraded fuel from microwave assisted pyrolysis of waste tyre. *Fuel* 115: 600–608.

Velghe, I., Carleer, R., Yperman, J., and Schreurs, S. (2011). Study of the pyrolysis of municipal solid waste for the production of valuable products. *J. Anal. Appl. Pyrolysis* 92: 366–375.

Wang, Z.H., Chen, D.Z., Song, X.D., and Zhao, L. (2012). Study on combined sewage sludge pyrolysis and gasification process: mass & energy balance. *Environ. Technol.* 33: 2481–2488.

Wang, B., Xu, F., Wang, X. et al. (2022). Comparative study on pyrolysis and gasification within CO_2 atmosphere of typical forestry biomass in Northeast Asia: thermal behavior and kinetic analysis. *Fuel* 324: 124540. https://doi.org/10.1016/J.FUEL.2022.124540.

Weisz, P.B., Haag, W.O., and Rodewald, P.G. (1979). Catalytic production of high grade fuel (gasoline) from biomass compounds by shape selective catalysis. *Science* 206: 57–65.

Williams, P.T. (2013). Pyrolysis of waste tyres: a review. *Waste Manag.* 33: 17141728.

Williams, P.T. and Brindle, A.J. (2003). Aromatic chemicals from the catalyst pyrolysis of scrap tyres. *J. Anal. Appl. Pyrolysis* 67: 143–164.

Williams, E.A. and Williams, P.T. (1997). The pyrolysis of individual plastics and plastic mixture in a fixed bed reactor. *J. Chem. Technol. Biotechnol.* 70: 9–20.

Williams, P.T., Besler, S., and Taylor, D.T. (1990). The pyrolysis of scrap automotive tyres, the influence of temperature and heating rate on product composition. *Fuel* 69: 1474–1482.

Xu, D., Lin, J., Ma, R. et al. (2022). Microwave pyrolysis of biomass for low-oxygen bio-oil: mechanisms of CO_2-assisted in-situ deoxygenation. *Renew. Energy* 184: 124–133. https://doi.org/10.1016/J.RENENE.2021.11.069.

Yang, J., Tanguy, P.A., and Roy, C. (1995). Heat transfer, mass transfer and kinetics study of the vacuum pyrolysis of a large used tyre particle. *Chem. Eng. Sci.* 50 (12): 1909–1922.

Yi, R.J. (2007). An experimental research on catalytic pyrolysis of municipal solid waste. Master thesis, Huazhong University of Science and Technology, Wuhan, pp. 32–36.

Zhang, X., Wang, T., Ma, L., and Chang, J. (2008). Vacuum pyrolysis of waste tyres with basic additives. *Waste Manag.* 28 (11): 2301–2310.

Zhang, H., Wang, G., Wang, J., and Xue, Q. (2021). Low-temperature treatment of polyethylene plastics and semi-coke mixture and CO_2 gasification of finely ground products. *Fuel* 285: 119215. https://doi.org/10.1016/J.FUEL.2020.119215.

Zhao, X., Song, Z., Liu, H. et al. (2010). Microwave pyrolysis of corn stalk bale: a promising method for direct utilization of large-sized biomass and syngas production. *J. Anal. Appl. Pyrolysis* 89 (1): 87–94.

Zi, G., Dake, T., and Ruiming, Z. (1988). Hydrothermal crystallization of zeolite Y from Na_2O-Fe_2O_3-Al_2O_3-SiO_2-H_2O system. *Zeolites* 8 (6): 453–457.

7

Climate Change Policy and Skills Education

In order to successfully address and maintain the various technologies discussed in the previous chapters to combat and control climate change, the actions of IMechE's Young Members Board (YMB) in September 2020 (Venkatesan-Crome and McDermott 2020) are brought to perspectives. Thus, the YMB presented an elegant thought-provoking inquiry on how the engineering discipline can impact and battle climate change to build a more sustainable world. It started with the definition that engineering endowed the last 200 years with industrial revolutions followed by the Information Revolution and now the evolving Cyber Revolution. To understand the role of engineering in climate change, therefore, engineering should be bearing the key to deliver a sustainable future, which is embodied with the rate of change indicating bleak implications and consequences during their lifetime. The youth should not be forced to live in a world with no fresh water, breaking down the natural wonders of the world and species becoming extinct one after another. That they want to be able to continue to enjoy the fruits of nature and live in a world of abundance not just for now, but for all lifetime and for all future generations. Even though the current dire times are challenging, it provides many opportunities for engineers to drive the change towards renewable energy sector, clean energy that is becoming a growing industry but still provides for global energy demand compared to fossil fuels. Thus, growth in the renewable energy industry should sharpen the path to a greener world that requests for all hands on deck.

It is common belief at this juncture that every facet of the engineering sector needs to embrace at least a net-zero carbon impact on the world in terms of construction, chemicals etc., such that there are no negative side-effects to every process. This is not considering things at point of use but including the entire life cycle and what the society needs to maintain that cycle. The emphasis should be the understanding that long-term solutions that are devoid of potential future side effects should be embraced. Some of the technologies and efforts recommended in order to achieve these objectives are presented in Table 7.1.

It is of note that the Negative emissions technology (NET) can contribute immensely to achieve climate targets by leading to a negative balance of carbon in the atmosphere. These technologies are met to be supported by the appropriate legislation by every country in the world committing to net-zero carbon emissions in the near future, e.g. 2050 by the United Kingdom. In this policy, projects and funding are handled in such a way that there is incentive for engineering companies to collaborate and tackle the big problems. Investments are also targeted towards achieving the most impact in order to hit the 2050 target.

The continued support and introduction of science, technology, engineering and mathematics (STEM) subjects in school and tertiary institutions should be practiced to encourage more young engineers to work on low-carbon technologies to find innovative solutions to battle climate change. The issue of equivalent or better wages for renewable energy experts as compared to the fossil fuels industry. Thus, the growing renewable sector should be marketed positively, so that the best and brightest engineers work together to combat climate change.

As the global warming on the other hand results into changes in the Earth's climate and ecosystems, every country in the world is experiencing the dramatic and negative effects on the socioeconomic, livelihoods and national economies. If not, economic growth patterns based on overexploitation of natural resources and fossil fuel energy generation have resulted into greenhouse gas (GHG) emissions and pollution of the Earth. It has been alluded to in previous chapters that global carbon dioxide (CO_2) emissions grew by 1.7% in 2018 to a record of high level. Stabilising atmospheric CO_2 concentrations requires changes in energy production and use. Effective mitigation cannot be achieved without individual agencies

Measuring Climate Change to Inform Energy Transitions: Carbon Footprint Calculations, First Edition. Sunny E. Iyuke.

Table 7.1 Possible technologies required to achieve net-zero carbon emissions.

Project/company	Location	Funding	Scope of work
SUN-to-LIQUID	ETH Zurich	European Union's Horizon 2020, Swiss State Secretariat for Education, Research and Innovation (SERI)	Produces solar kerosene from sunlight, water and CO_2
Bauhaus Luftfahrt	Germany	Airbus Group, Industrieanlagen-Betriebsgesellschaft, Liebherr-Aerospace and MTU Aero Engines as well as grants of the Bavarian Ministry for Economic Affairs, Media, Energy and Technology	The technical, economic, social and ecological aspects are considered holistically
Deutsches Zentrum für Luft-und Raumfahrt (DLR)	The German Aerospace Center, the Japanese Aerospace Exploration Agency and the Canadian Space	Germany's space agency	Its extensive research and development work in aeronautics, space, energy, transport, security and digitalisation is integrated into national and international cooperative ventures
ETH Zurich	Swiss Federal Institute of Technology, Zurich	Modern-day Switzerland created it as a centre of innovation and knowledge	From computer science through to micro- and nanotechnology and cutting-edge medicine. Every year ETH registers around 100 patents and 200 inventions on average
IMDEA Energy	Spain	Regional government of Comunidad de Madrid	IMDEA Energy Foundation Creates Carbon-Neutral Jet Fuel From Just
Abengoa	Spanish multinational company	An international company that applies innovative technology solutions for sustainable development in the infrastructures, energy and water sectors	Innovative technology solutions for sustainability
HyGear	SME based in Arnhem, The Netherlands	Technology and Services B.V.	Development and manufacturing of small-scale gas processing plants and purification systems
ARTTIC	Based in France, Belgium, Germany and Israel	ARTTIC is a set of companies	ARTTIC is an independent European provider of management services, especially in the area of large international collaborative R&D projects

Source: Adapted from Sizmann et al.

working collectively towards reduction goals. Ten stabilisation wedges can be identified as GHG reduction strategies, whereby each wedge represents 1 billion tons of carbon avoided in 2054 (Factsheet 2023).

Energy Savings: Many energy efficiency efforts require an initial capital investment, but the payback period is often only a few years. It was reported that in 2016, the Minneapolis Clean Energy Partnership planned to retrofit 75% of Minneapolis residences for efficiency and allocated resources to buy down the cost of energy audits and provide no-interest financing for energy efficiency upgrades (US EPA 2018).

Fuel Switching: Switching power plants and vehicles to less carbon-intensive fuels can achieve emission reductions. For example, switching from an average coal plant to a natural gas combined cycle plant can reduce CO_2 emissions by approximately 50% (IPCC 2014).

Capturing and Storing Emissions: CO_2 can be captured from large point sources both pre- and post-combustion of fossil fuels. Once CO_2 is separated, it can be stored underground depending on the geology of a site. With CO_2 used in enhanced oil recovery (EOR), long-term storage technologies remain expensive (Wilcox, 2020). It was also emphasised that alternatively, existing CO_2 can be removed from the atmosphere through NETs and approaches such as direct air capture and sequestration, bioenergy with carbon capture and sequestration, and land management strategies (The National Academies of Sciences, Engineering, and Medicine 2018).

As a result, many countries have committed to the implementation of the Paris Agreement adopted in 2015, which emphasised transition of energy from fossil fuels and decent work in order to minimise the severe and worsening consequences of climate change. Similarly, in 2018 COP24 in Katowice, Poland, Solidarity and Just Transition Declaration were adopted to include decent work, which are inclusive and climate-resilient types of developments. Thus, the availability of the right skills paves the way for a productive structural transformation and transition towards a greener economy and decent job creation. It is therefore urgent that skills training and development are important endeavours for a just and inclusive transition to zero emission. Skills development for green jobs would also serve as a 'buffer' against the effects of transitory disruptions and emerging challenges while economies are responding to these changes.

Many countries are challenged with influencing factors on the requisite skills due to the widening energy cost gap, socio-economic considerations, youth employment, deteriorating energy security, climate change mitigation and adaptation. Thus, the supply and demand for skills should provide a deep dive into the specifics of the jobs and corresponding skills these changes will require in the next 50 years. These reasons therefore called for the review of relevant skills training and development over five years academic curriculum in the school of Chemical and Metallurgical Engineering of the University of the Witwatersrand, Johannesburg, as a case study and presented in the next sessions.

It is pertinent to mention that one of the reasons for the case study is to provide reference information from which training bodies and academia can reference and learn as strategy and policy to combat climate change and its corresponding impacts around the world.

7.1 Case Study – Five-Year Review of Skills Training and Education in a Chemical and Metallurgical Engineering Curriculum

(Reprinted with permission from Josias van der Merwe, Head of School at the Wits University)

The importance of education to promote climate action is to help people understand and address the impacts of climate crisis, empowering them with the knowledge, skills, values and attitudes needed to act as agents of change. As a result, the UN Framework Convention on Climate Change, the Paris Agreement and the associated Action for Climate Empowerment (ACE) agenda call on governments to educate, empower and engage all stakeholders and major groups on policies. UNESCO (2017) does provide its longstanding holistic approach in combining science, education, culture, and communication and information to address the complexities of climate. At the United Nations Conference on Climate Change (COP27) at Sharm el-Sheikh, Egypt, from 6 to 18 November 2022, the organisation organised and joined a series of events and exhibitions, which provided the unique opportunity to mobilise commitments and actions to transform education to tackle climate change.

7.1.1 The Wits Senate Pledge

Wits commits to playing a transformative leadership role in South Africa's response to the climate emergency. Through research, teaching, policy engagement and its own practice, using its location in the economic hub of the country and understanding its history in mining and industry, Wits will play a pivotal role in a just and equitable transition to a net zero-carbon economy for the city, region, country, continent and the world, by 2050. At the centre of this commitment will be ongoing analyses of our own carbon footprint, with a view to transforming the university into a living laboratory for just practices of sustainability.

This policy therefore summarises the provisions from the Wits University in providing some of the required Skills and Education for Sustainable Development (ESD), which reorients education to empower learners of all ages to act for environmental integrity, economic viability and a more equitable society.

7.1.2 Curriculum Review Period of 2017–2021

The last Quinquennnial Review (QQR) of the school of Chemical and Metallurgical Engineering was conducted in 2016, which covered the period 2012–2016. The current review covers the period of 2017–2021. The review of the current cycle was scheduled to take place in 2022, but due to the COVID-19 restrictions and constraints in the previous years of 2020–2021, subsequent carryover of events to 2023 was inevitable.

The Head of School should describe how staff members were consulted and involved in the process of compiling the report. The report was drafted by the Head of School. The draft was discussed with the members of the School

Executive Committee and other members of committees within the school for comments. Hence, the final version of the report reflects the views of both senior and junior academics in the school.

Self-assessment of the school's role or place in the University, region and country. This assessment was provided within the context of the University's strategic plan and the strategic or academic plans of the school and faculty concerned.

The school of Chemical and Metallurgical Engineering was formed in 1995. It was the first of its kind in South Africa at that time. It is the combination of the former Departments of Chemical Engineering and Metallurgy and Materials Engineering. At present, the school has about 660 undergraduate students and about 280 postgraduate students. About half of the postgraduate students are employed in industry, and their research work is of interest to their employers.

The school of Chemical and Metallurgical Engineering is one of 33 schools in the university, and one of 7 schools in the Faculty of Engineering and the Built Environment. It strongly supports the faculty's role to deliver highly trained engineers for local industry and develop new knowledge and innovations through research and development. The school forms an important part of the overall university structure and supports the university's vision to be a highly research-intensive university that provides practical solutions and manpower for Africa in Africa.

After a huge surge in new first-year intakes from 2011– to 2014, current first-year intake is now around 180 (100 Chem Eng, 80 Met Eng). First-year throughput has improved due to concerted teaching efforts by an academic support programme launched by the Academic Development Unit (ADU) at Faculty and the introduction of the Common First Year (CFY).

The student enrolment statistics for the Chemical Engineering programme in 2017 decreased to 114 students, which represents a decrease over the period of 2014–2017. This increased from 2018 to 2019, and then decreased to 106 in 2021, which was below the target figure of 120 first-year students. However, one has to remember that these figures also include the repeat students. Throughput has increased encouragingly to more than 80% during the last two years.

The school's curricula in both the programmes of Chemical and Metallurgical Engineering have been benchmarked against local competitors, as well as several highly acclaimed international universities to ensure that the correct and relevant material is being taught to our students. Standard course briefs complying with Senate Standard Orders have been enforced for all undergraduate courses, and improved quality control measures to check all assessments have been introduced. Team teaching has been introduced to ensure more variation for students in their courses. All APO documentation, new course descriptions, standard credit calculations for courses and rule book descriptions have been checked and updated.

At the postgraduate level, the courses have been re-aligned to ensure that they conform to Senate Standard Orders and that their short course equivalents have been recorded correctly on the Oracle system. Postgraduate programmes have been re-organised in the school, teaching teams compiled and presentation drawn up to 2021. Where required, new Postgraduate Diploma programmes have been developed and approved for Welding Engineering, Oil and Gas Engineering, Metallurgical Engineering and the Chemical Engineering programme to date. On the research side, a target of 50% of postgraduate students compared to undergraduate students was set for 2017–2021, as well as 200 peer-reviewed journal publications in accredited journals by 2021. The school is already producing 40% of the postgraduate students graduating in the faculty, but a target of 50% is achievable with better management of the postgraduate student system.

Table of courses and programmes (where appropriate) was offered at undergraduate and post-graduate levels, including the course or programme name, number of credits, and whether the course is a service course for another school or faculty (indicate which one(s)) and/or is part of a major or field of study (specify) led by the school. The table also show the number of students enrolled over the period since the last review. Provide an analysis of the pass, failure and dropout rates, average time taken by students to complete the courses and programmes in the school, and full-time equivalents (FTEs).

Table 7.2 presents enrolments, the pass rates, failure and dropout rates and average time it takes students to complete their courses. At the undergraduate level, there was general increase of enrolment from 113 in 2017 to 146 in 2018 at level 1 before the CFY kicked in 2019. There was also general pass rates from 312 to 1,036 (79.65–91.92%) and 192 to 393 (54.08–94.55%) in 2021 across the levels (1–4) for Chemical Engineering and Metallurgy and Materials Engineering, respectively.

At the postgraduates level, there was general decrease in pass rates from postgraduate diploma (level 5) to MSc with course work (level 7) for Chemical Engineering, from 93 to 68.97% and increase from 62.96 to 83.33% for Metallurgical and Materials Engineering in 2017. However, at the full research studies of MSc and PhD by research, the reverse was the case, whereby the Chemical Engineering experienced increase from 15.44 to 21.89% in 2017 while the Metallurgical and Materials Engineering had 15.56–11.69% in 2021. The increase in pass rates in both disciplines of Chemical Engineering and Metallurgical and Materials Engineering can be attributed to the increase in research funding due to the number of SARChI Chairs in the school and industry supports. These dynamics were also seen at the course records and also presented in Table 7.3 for unit pass rates for undergraduate courses taught by the school.

Table 7.2 Summary of the total pass rates, failures and dropout rates in the school of Chemical and Metallurgical Engineering for the period 2017–2022 to reflect the status, although the review period is 2017–2021.

Characteristics	Chemical Engineering					Metallurgy and Materials Engineering				
Undergraduates										
Level	1	2	3	4	Unknown	1	2	3	4	Unknown
Enrolled	259	2018	7367	6491	340	166	1312	2524	2237	250
Passed	189	1686	6028	6142	262	104	1127	2309	2092	230
Failed	70	332	1333	345	0	61	185	214	144	0
Decision pending	0	0	6	4	77	1	0	1	1	19
Course pass %	72.97	83.55	81.82	94.62	77.06	62.65	85.90	91.48	93.52	92.00
Cancelled	46	335	292	841	43	24	88	162	56	57
Enrolled + cancelled	305	2353	7659	7332	383	190	1400	2686	2293	307
Course pass excluding failed absent and failed not qualified to write	80.43	85.02	83.87	95.83	77.06	67.97	88.74	92.62	94.45	92.00

Postgraduates

Characteristics	Chem Eng		Met and Mat Eng			
Level	**5: PG Dip**	**7: MEng**	**5: PG Dip**	**7: Meng**	**8: MSc Res**	**9: PhD Res**
Year 2017						
Enrolled	121	87	27	144	136	90
Passed	113	60	17	120	21	14
Failed	8	4	10	9	0	1
Decision pending	0	23	0	15	115	75
Course pass %	93.39	68.97	62.96	83.33	15.44	15.56
Cancelled	12	15	6	32	8	6
Enrolled + cancelled	133	102	33	176	144	96
Course pass excluding failed absent and failed not qualified to write	98.26	71.43	73.91	85.71	15.44	15.56

Table 7.3 Unit pass rates for undergraduate courses taught by the school.

	2017	2018	2019	2020	2021	2017	2018	2019	2020	2021
Characteristics	**CHMT2014A: energy balances and applications**					**CHMT2017A: introduction to extractive metallurgy**				
Enrolled	154	135	120	0	0	51	55	57	21	39
Passed	144	131	98	0	0	45	50	53	16	37
Failed	9	4	22	0	0	6	5	4	5	2
Decision pending	1	0	0	0	0	0	0	0	0	0
Course pass %	93.51	97.04	81.67			88.24	90.91	92.98	76.19	94.87
Cancelled	3	5	10	92	69	0	0	5	1	2
Enrolled + cancelled	157	140	130	92	69	51	55	62	22	41
Course pass excluding failed absent and failed not qualified to write	95.36	100.00	82.35			88.24	90.91	94.64	88.89	100.00

7.1.3 Turnover, Changes in Age Profile as It Affects Sustainability of the School

Currently, the school has 29 academic staff members and 10 visiting staff members. Two are emeritus professors who continue with research and selected teaching duties. Staff recruitment occurs via strict protocols to ensure transparency, equity and conformance to the staffing policy of the University. Posts are advertised in the press and sometimes via the internet, then applicants are short-listed by a short-listing committee, and then a selection committee is constituted. Such committees are chaired by the relevant Deputy-Vice-Chancellor for Professorial and Associate Professorial positions, and by the Dean of the Faculty for positions of Senior Lecturer and more junior positions. The details are contained in the University Human Resources Handbook.

In terms of demographics, 33% of the total academic staff component is female and 51% is Black/Indian. For the full-time staff, 31% of the academic staff are females, and 62% Black/Indian. This demographic picture compares favourably with other schools in the faculty, and Wits' transformation targets. However, there is significant room for growth of South African equity academics and we have to make opportunities for our own students to grow into these positions.

Thus, the practice on recruitment and sustainability plans in the school resolves the effects of staff retiring due to age or attrition before the due dates of retirements.

The statistics on FTE students and the Teaching Input Units (TIUs) for the review period from 2017 to 2021 are presented in Table 7.4. The number of academic staff in the school is also measured in FTEs. The figure has clearly shown the impact of the increasing student/staff ratio on TIU, FTE, Research Output Units, Publication units etc. (Table 7.5). A full-time equivalent staff member is defined as a staff member who works at the institution for a full year on a full-time basis (100% appointment and employed from 1 January to 31 December). While TIU refers to each course/unit as allocated to one CESM (classification of educational subject matter or field of study) and one level of study. The number of enrolments for the course, less the number of cancellations multiplied by the course weight, determines the student Enrolled FTE contribution of the course to a particular cell in the funding grid. The FTE contribution multiplied by the funding group factor and the factor for the funding level of that cell in the grid determines the number of TIUs generated by the course. The TIUs multiplied by the US$ value per TIU (US$1181947 for 2017 FTEs) determine the input subsidy generated by the course.

Thus, decrease in workload due to decreasing enrolment from 2017 to 2021 (507–404 Teach Enrolled FTE) is clearly shown by the ratios of Teach Input Unit per Permanent Academic Staff FTE (SFTE) Ratio, Enrolled FTE Per Permanent Academic SFTE Ratio, etc. in Table 7.4. This was achieved because the school has coped slightly by employing sessional as presented earlier in Table 7.3. The values of 15.17–14.15 from 2017 to 2021 for the Enrolled FTE Per Permanent Academic

Table 7.4 FTE and TIU in Chemical and Metallurgical Engineering.

Level	Characteristics	2017	2018	2019	2020	2021
Undergraduate	Total course regn.	4183.50	4280.00	3819.50	3866.50	3982.00
	Teach enrolled FTE	507.98	525.30	440.41	435.62	404.55
	Teach input unit	1665.80	1776.41	1496.17	1539.33	1431.82
	Teach input subsidy	22043320	24563055	21199034	25616930	21793423
Postgraduates	Total course regn.	653.05	568.85	588.22	520.47	488.76
	Teach enrolled FTE	136.05	125.74	142.39	142.88	131.51
	Teach input unit	1059.31	969.26	1077.91	1113.78	1013.11
	Teach input subsidy	14017703	13402336	15272816	18535143	15420281
Total course registration		**4836.55**	**4848.85**	**4407.72**	**4386.97**	**4470.76**
Teach enrolled FTE		**644.03**	**651.04**	**582.79**	**578.50**	**536.06**
Teach input unit		**2725.11**	**2745.68**	**2574.08**	**2653.11**	**2444.93**
Teach input subsidy (US$)		**1 933 567**	**2 035 678**	**1 955 595**	**2 367 403**	**1 995 373**

ENROLLED FTE: It represents the number of students being taught if each of them was doing a full load of courses, or pursuing his/her research full-time. TEACHING INPUT UNITS (TIU): The FTE contribution multiplied by the Funding Group factor and the factor for the Funding Level determines the number of TIUs generated by the course. INPUT SUBSIDY: The TIUs multiplied by the US$ value per TIU determines the input subsidy generated by the course.

Table 7.5 SLRE is the FTE value for the staff member, weighted by a factor relative to the rank of the staff member.

Characteristics	2017	2018	2019	2020	2021
Teach enrolled FTE	644.03	651.04	582.79	578.50	536.06
Permn. acad staff FTE univ. (excl. hosp. only)	29.08	28.93	28.28	27.57	27.42
Enrolled FTE per perm. acad. staff FTE ratio	22.14	22.50	20.61	20.98	19.55
Acad. staff FTE univ. (excl hosp. only)	42.45	43.03	39.89	38.73	37.88
Enrolled FTE per academic staff FTE ratio	15.17	15.13	14.61	14.94	14.15
Teach input unit	2725	2746	2574	2653	2445
Teach input unit per perm acad. staff FTE ratio	93.70	94.90	91.02	96.24	89.18
Teach input unit per academic staff FTE ratio	64.20	63.80	64.53	68.50	64.55
Research output units	66.85	69.75	64.15	74.14	73.50
Publication units	71	52	57	44	79
Research output units per perm. acad. staff FTE	2.30	2.41	2.27	2.69	2.68
Publication units per permn. acad. staff FTE	2.44	1.79	2.03	1.60	2.90
Total research outputs per permn. acad. staff FTE	4.74	4.20	4.30	4.29	5.58

ENROLLED FTE: It represents the number of students being taught if each of them was doing a full load of courses, or pursuing his/her research full-time. TIUs. The FTE contribution multiplied by the Funding Group factor and the factor for the Funding Level determines the number of TIUs generated by the course. Publication Units are the publications submitted in journals (staff, students both). These are unit allocated as per DHET. Research outputs units are for the PhD, Research Masters and research portion of the coursework masters which is submitted to DHET for subsidy. Total Research output = Publication units + Research output unit. SLRE is the FTE value for the staff member, weighted by a factor relative to the rank of the staff member. It is used as a measure of research productivity as in the ratio publications/SLRE.

SFTE Ratio. Since five of the professors could only give 0% to 5% of their time to teaching, because they spend more of their time for research and administration, there was general increase in the postgraduate Teach Input Subsidy from US$ 751620 – UD$ 826825 from 2017 to 2021 as shown in Tables 7.4 and 7.5.

7.1.4 Workload Model

Although the school adopted the workload model developed by the Faculty, a fair share strategy is practiced, where the workload is shared by the academics in line with expertise and course credit load. Staff may, however, be required to teach combined degree courses depending on the staffing position in the school. Thus, the practice of combined course teaching in the school has helped tremendously in removing imbalance between the Chemical and Metallurgical Engineering programmes, whereby they were seen as principles and content oriented programmes, respectively. Junior staff who are still completing their higher degrees are generally given a lighter teaching load, in order to allow them sufficient time to complete their research degrees. The workload model has been used as a guide for the allocation of teaching duties but, during a period of staff shortages and high turnover, it has been inevitable that inequalities have occurred.

However, academic staff in the school are expected to teach at both undergraduate as well as postgraduate level. At least two semester courses on undergraduate level and one postgraduate course are the norms for most staff members, although some do teach more than others. The biggest classes are the Introduction to Process Engineering I and Process Engineering 11, where classes are divided among a number of lecturers who, especially in the first-year class, rotate. Administrative help with undergraduate students, timetables and venues are supplied by Mrs Maria Modiba (who took to early retirement at the end of this review period), the school's undergraduate officer. Academic staff is allowed the freedom to take on as many postgraduate students as they feel they can comfortably handle, and no targets are set in this regard in the school. Each academic staff member is expected to serve on at least one school committee and assist with the administration and school management in that way. Short courses are offered by academic staff on a voluntary basis in a surplus share model with the school, where the largest chunk of any surplus generated goes into the research account of the particular staff member. Private work/consultation work is encouraged on condition that teaching and research duties and outputs are not compromised in any way. However, the introduction of CFY in 2019 further supported the combined teaching model.

Table 7.6 Staff profile with respect to equity targets.

		Equity information					
		Race				Gender	
Quantity	Definition	Black	White	Coloured	Indian	Male	Female
All programmes							
No. of staff with PhD's	DE	14	14		1	22	7
No. of staff with Master's degrees	ME	5	3		1	6	3
No. of staff with Bachelor's degrees	BE		1			1	
No. of staff with diplomas	Dip	2	2			2	2
No. of staff who are registered with ECSA	RE	2	4			5	1
No. of staff qualified "one higher" than programme taught	QE						
No. of full professors	Prof	3	3			4	2
No. of associate professors	A Prof	7	4			9	2
No. of senior lecturers	SL	3	3		1	5	2
No. of lecturers	L	4	3			5	2
No. of technical staff	TE	10	1			8	4
No. of administrative staff	AE	4		2			6
No of other staff	OE				1 Assoc. Lectur.		

7.1.5 Three-Year Rolling Plan of the University Employment Equity Plan

In general, the school of Chemical and Metallurgical Engineering has a favourable equity profile amongst the academics as well as support staff, which complies well with the university plan as presented in Table 7.6. The current academic staff complement in the school is 29, with 16 full-time staff members and one sessional appointment dedicated to the Chemical Engineering courses, and 13 to the Metallurgical Engineering programme. Several members of staff teach courses common to both programmes in the school.

With the recent changes and appointments, the school's position in terms of employment equity is very encouraging in comparison to other schools in the University. Currently 24.1% of the permanent academic staff falls into the employment equity/transformation category (female, and/or black), whereas another 37.9% is black from other African countries originally. The support staff, cleaners and workshop staff comprise a total number of 18 people of whom 95% are equity incumbents. More than half of the school staff is female. The school is committed to maintaining and even improving its equity position further.

It should be noted that although the number of black staff as listed in Table 7.6, includes those from African countries.

7.1.6 Staff Qualifications and Improvement Since the Last Review

Six new appointments to the school were made since 2017 while one was appointed from a 5% to a 100% position as professor in physical metallurgy and another has also joined as a senior lecturer in Physical Metallurgical Engineering at the beginning of 2021.

The academic staff has been able to go on regular sabbaticals with typically one academic from Chemical and another from Metallurgical Engineering on sabbatical at a given time. This has been helpful in terms of establishing their research work. It has helped those busy with PhD studies to give all their attention to it. In addition, new collaborations with several overseas university colleagues have been established.

Since the last QQR, one academic has completed her PhD and she was promoted to associate professor in 2022. Another female academic graduated in 2020. A male academic in the physical metallurgy department is at the point of submitting his PhD thesis for examination. While four are on the verge of submitting their PhD theses, a male on the Chemical Engineering side has submitted his PhD towards the end of 2022.

7.1.6.1 Support Staff

Support staff in the school supports both programmes. The Principal Technician in the workshop is Workshop Manager, who was appointed on 1 June 2017 to the post. The analytical supervisor resigned and her position was divided into two Laboratory Assistant positions, and the positions have been filled. The two vacant positions of Laboratory Assistant for the Chemical Engineering as well as the mineral processing labs have been filled as well. These new positions provide help to the analytical supervisor as he oversees the analytical services in the school.

The long-serving undergraduate officer went on early pension and her position has been replaced since 1 December 2021.

7.1.7 Staff Development

New staff members attend a staff induction programme, which serves as a formal introduction to the University, its mission, systems and requirements of academic staff. The university regularly offers a two-day course (The Teaching Role), which includes feedback on a video recording of a role-play lecture by course participants. The course is presented by the Centre for Learning and Teaching Development (CLTD). Other topics addressed are voice projection, large/small classes, language in teaching and postgraduate supervision. Every year a performance planning discussion is held between each staff member and the HOS, during which objectives for the next year are discussed and agreed. The University is currently rolling out a new performance appraisal system, keeping in line with developments in industry. Training requirements are identified as part of a performance planning discussion. The school encourages training of staff, academic as well as administrative and technical staff, and makes funds available for such training by staff. Academic staff serve three years of probation before being confirmed in their positions. During this probation period potential problems can be identified and solved. Staff members are reminded of the schedules, requirements and procedures for promotion. Potential promotion dates for staff are set, and staff members are encouraged to apply for promotion at a suitable time.

All staff members are encouraged to attend conferences and take sabbatical leave for the purpose of research and their personal development. New and young staff are encouraged to embark on a six-month sabbatical shortly after their confirmation to consolidate and launch their research careers.

Retention and recruitment of high-quality staff in all branches of engineering are seriously hampered by the low level of remuneration offered by the university when compared with industry. The school has unfortunately lost staff to other universities and another school in the university over the last review period. The university aims to ensure that all staff members are paid in-line with the 75th percentile of academic institutions in South Africa. The school is in a fortunate position of being able to provide subventions to most academic staff in order to bring salaries more in line with industry salaries, via supplementation from the Minerals Education Trust Fund (METF) for the Metallurgical Engineering staff and from school reserve funds for Chemical Engineering staff.

7.1.8 Regular Evaluation of Staff Teaching Performance

The impact of the programme on the students is measured by means of feedback questionnaires, class representative meetings, meetings with the School Student Representative Council (SSRC) and analyses of test and exam marks. The information received is discussed in the school teaching and learning committee and staff meetings, where action plans are put in place to curb possible problems or correct mistakes.

The impact on industry or employers of our graduates involves studies by the University Strategic Planning and Resources Committee (SPARC), who present their findings to the university and the school. Alumni students are also frequently contacted, and electronic feedback is circulated by email on a monthly basis. Alumni also commented through the school's website.

The University CLTD performs course and lecturer evaluations on a regular basis. This has been the practice for many years and forms part of staff files available for performance assessment, confirmations and promotion.

The school academic staff meets every 6–8 weeks for a staff meeting, where an agenda is set, and pertinent issues are reviewed and discussed. The School Teaching and Learning Committee, comprising senior academics from both Chemical and Metallurgical Engineering, together with the course coordinators for the different academic years in programmes, deal with academic and curriculum matters. Decisions are taken back to the Chemical and/or Metallurgical Engineering staff for further discussion and decisions, before being ratified at the staff meeting. Decisions and even concerns are referred to the Faculty Undergraduate Teaching and Learning Committee, where the school is represented by the chairperson of the School Teaching and Learning Committee.

The school has an Industrial Advisory Board (IAB) which meets once a year and provides the school with advice on curricula and industry needs and views on the training of graduates. This is fed back to the School's Teaching and Learning Committee for implementation.

7.2 Information with Respect to Activities

While it is difficult to highlight prizes won by staff and students over a period of five years, it is a regular occurrence in the school that staff members and students won nominations for exceptional performance and outstanding achievements. Suffice to say that the school of Chemical and Metallurgical Engineering regularly produces its fair share of highlights and high achievers.

Staff members in particular regularly take part in the activities of professional organisations like the Southern African Institute of Mining and Metallurgy, South African Institution of Chemical Engineers/Institution of Chemical Engineers (SAIMM, SAIChE/IChemE), the Institute of Tribology, the Welding Institute and other learned organisations, like Engineering Council of South Africa (ECSA) in terms of serving on management and conference organising committees, and represent the school on these structures.

The previous accreditation exercise in the school of Chemical and Metallurgical Engineering took place in September 2012, and the school received a clean bill of health. No deficiencies were noted and the school is in the process of preparing for the next visit in September 2023.

The school has embarked on an exercise to re-engage anew with industry and forge new relationships with industry, as well as neighbouring universities and renowned international institutions. Several early retirees were appointed in the school to assist with this exercise, and a number of MoUs/MoAs (memoranda of understanding/agreement) have also been signed. The school is talking to several local universities to collaborate on postgraduate teaching, and all the postgraduate courses have been updated for presentation from 2016 onwards. The IAB of the school has also been reactivated and reconstructed and to date two meetings have taken place.

7.2.1 Awards, Prizes or Other Forms of Recognition to Staff and Students

In 2012, an Associate Professor received C2 National Research Foundation (NRF) rating and MSc Student funding from PMSA at US$ 21984. A student received US$2145 Faculty top slice, while another academic received US$ 375335 funding from M-ERA.NET, and he is also expected to receive a further US$ 134048 from the Department of Science and Innovation (DSI). This funding helped to establish a working relationship with the Fraunhofer Institute in Germany and Ultramat SA in South Africa to perform collaborative research on carbide materials. In 2023, an Associate Professors from the chemical and Metallurgy programmes received C2 NRF rating, respectively.

While an academic was nominated for the SAEEC annual awards for 2021 (Energy Professional for 2021 and Energy Female Professional for 2021, two professor (female and male) edited a book titled- – Acid Mine Drainage – From Waste to Resources.

An Associate Professor from the mineral side received award from COALTECH to the value equivalent to support two MSc students, NRF award for the value of US$42 605 for 1 Post-Doc, 1 PhD and 1 MSc: 2022–2024 and another SASOL-NRF Award of US$64 424 for 1 PhD and 1 MSc: 2022–2024.

7.2.2 Extracurricular Involvement of Staff and School

Sillanpaa and co-workers (Karimi-Maleh et al. 2021) received 357 citations for Novel coronavirus disease 2019 (COVID-19) pandemic: From transmission to control with an interdisciplinary vision and Recent advances in using of chitosan-based adsorbents for removal of pharmaceutical contaminants: A review, respectively.

From 2017 to 2012, three full professors received Non-NRF Chairs for South African Institute of Welding Chair in Welding Science, Element 6 Ceramics Chair and Chamber of Mines Chair in Extractive Metallurgical Engineering, respectively.

As mentioned above, the Staff Wellness Week always involves the annual Lenn Smith Race: The 5-km fun run/walk that started in the school many years ago, staff from the school takes leading positions perhaps because of their interest, passion and enthusiasm on the wellness programme.

7.2.3 Professional Accreditation Assessments

Both programmes in the school were accredited by ECSA after a visit in 2022 for another five years with effect from 2022.

The ECSA visitation committee reported that during the accreditation visit in 2017, a deficiency was identified. After an interim accreditation visit in 2019, the outcome accreditation was granted until the next regular cycle of 2022. Some additional changes to the curriculum have been implemented after the last ECSA visit in 2017, where a common first year engineering programme (CFY) was introduced across all engineering disciplines in the faculty at the beginning of 2019. The subsequent changes to the Chemical Engineering and Metallurgy and Materials Engineering curricula in the second year were mainly in response to the introduction of this change. There were a few modules that were previously major full-year courses in the first-year curriculum before 2019, that could not be accommodated in the new CFY programme in their existing format and some of the aspects of those courses were incorporated into the new common first-year courses and some aspects were deemed irrelevant to other engineering disciplines. This necessitated the development and introduction of new courses in second year in 2019 to accommodate the content lost from those courses.

Their general observation of the programmes is that the programmes are well-designed and supported by various teaching and learning processes, typical for BSc Engineering in Chemical Engineering and BSc Engineering in Metallurgy and Materials Engineering with horizontal and vertical articulation possibilities. This qualification provides the educational base for the development of engineers engaged in the design, development and research related to the appropriate disciplines and environment. The learning programme consists of a coherent assembly of knowledge areas associated with engineering practice, with an appropriate balance of theoretical, practical and reflective knowledge, which should lead to achieving the competencies. Student numbers have decreased over the last nine years with most new student enrolments in the National Curriculum Statement (NCS) enrolment band of 40 and above. The dropout rate is relatively high and is about 43%. However, the overall pass-rate over a three-year average seems to be very high, but during the interview with the Head of Department (HoD) it was clarified that those dropout students did not write the final examinations, explaining the high pass rates.

As mentioned earlier, the school is committed in line with the university policy on combating climate change to accomplish the Wits Senate Pledge, highlighted in Section 7.1.1.

7.2.4 Community Engagement and Academic Citizenship

Staff give lectures at various schools from time to time to inform learners about careers in Chemical and Metallurgical Engineering. This takes place on an ad hoc basis. The school participates in the university open day where learners are informed about the schools programmes. On ad hoc basis academics in the school are called to give presentations during career weeks at schools around the country. The nanobiotechnology research group led by Prof S Iyuke participates annually in intervarsity brewing competitions. Since its inception 11 years ago, the school has always been in second position overall.

The annual Lenn Smith Race around campus takes place around September every year. The 5-km fun run/walk started in the school many years ago, which now follows the Wits Parkrun route and is open to all Wits staff and postgraduate students. The school is a major participant in this annual event.

7.2.5 Effect to This Core Function

Research performance and commitments of academic staff are negotiated once a year during a goal-setting and personal development exercise. During the previous review cycle, only two academics, who have since been assimilated into the school employment, were paid by a research entity in the school, and their performances are evaluated by the chairholder of that research entity (Prof Falcon). At this stage, however, these colleagues are now full-time staff who earn their salaries from the university.

Academic staff are encouraged to be research active through school incentive schemes, e.g. Research incentive (RINC) funding allocations for successful postgraduate supervision to graduation level, RINC for early career academics (ECA) on a competitive grant basis in the school and finally school RINC grants for interschool/faculty joint collaborative projects. Start-up funds are also granted to all new academic staff members upon their initial appointment.

All academics are informed of available funding resources by immediate circulation via e-mail to all school academic staff members. Limited time is available for research due to heavy undergraduate and postgraduate teaching loads for all staff members, so their performances described in this report are highly commendable. A lack of research funds and research equipment for postgraduate students are other contributing factors limiting research output.

7.2.5.1 School's Resourcing Strategies

For about six years, the number of postdoctoral fellows in the school has increased. Sadly, not all of them have been equally productive, but they are certainly starting to make their presence felt and having an impact on research outputs. In 2017, the school of Chemical and Metallurgical Engineering had increased number of postdoctoral fellows than the seven postdoctoral fellows in 2015. Funding permitting, probably twice this number can be accommodated in the school to increase research output.

Since 2013, the school started to conclude an increasing number of Memoranda of Understanding/Agreement (MoUs/MoAs) with other universities abroad, as well as with local organisations like Mintek, the CSIR, Necsa, Eskom and others. The aim is to start attracting postgraduate students and staff for sabbatical research visits to Wits in order to enhance our own reputation locally and abroad, but fundamentally also to increase the quality and quantity of research outputs and postgraduate student supervision in the school, which provides staff availability for community services.

7.2.5.2 Enabling Mechanisms Provided Elsewhere in the University

Funding from external funding agencies for research in the school of Chemical and Metallurgical Engineering is difficult to quantify, as it comes into the school through individual researchers from a variety of state and government agencies, as well as on an ad hoc basis from industry. Staff members in the school receive support from, among others, NRF, DST (Department of Science and Technology), WRC (Wits Research Council), TIA (Technology and Innovation Agency), SAMMRI, Sugar Research Institute, Mintek and several private companies (Cape Fencing, Thermaspray, etc.).

Most academic staff members in the school of Chemical and Metallurgical Engineering are research active. The incentive system in the school works well to encourage all staff to become research-active.

7.2.5.3 School's Contribution to Strategic Partnerships

Interaction with industry is crucial on many levels, from providing well-trained undergraduate students with skills required by industry, to collaborative research, and further educational support. It has been recognised that further interaction with industry will be beneficial for the programmes in the school (both undergraduate and postgraduate) and university, especially in terms of vacation work training. The Wits University has a number of initiatives underway to expand its research and industry relationships, such as the Wits Mining Institute, which will inevitably impact on both undergraduate programmes in the school.

7.2.6 Planned Actions

(1) The coal and carbon research group situated in the school and hosted the DST Chair of clean coal technology for many years. This Chair has now come to an end and the research group is considering other avenues of funding to support the university plan towards energy transition from fossil fuel policy. There are plans in place to recruit a new Chair and increase the research output of the group. The research group has close links with the Schools of Geoscience and Mining Engineering, and industry, including Eskom, Exxaro, Anglocoal and Sasol, to name but a few.
(2) The school could be structured to provide a home for interaction with industry on all levels, from education through to consulting. Industry board members are invited to board meetings to discuss educational and research-based requirements. Meetings take place on a bi-annual basis. Several useful suggestions have already been made and incorporated into the teaching of some undergraduate subjects.
(3) Provision of more consulting work to industry, and forming strategic long term partnerships with selected industrial concerns. It is foreseen that staff members involved in such activities will transfer appropriate knowledge in their teaching and training to undergraduate students. For example, Prof S Iyuke is currently negotiating partnerships Toyota South Africa Motors (TSAM) and First National Battery (FNB) towards net-zero carbon energy and power production.

7.3 Finance

The school's main source of income is through Governing Council funding, i.e. student fees and state subsidies for teaching and research. As shown in Table 7.7, the total teach input subsidy increased marginally from US$1933567 in 2017 to US$1995373 in 2021, with a peak amount of US$2367403 in 2020 due to the support from postgraduate students in tutorial classes and income from short-course income. These incomes were observed to be sufficient to pay for the staff salaries and

Table 7.7 Summary of enrolled FTE, TIUs and input subsidy.

Level	Characteristics	2017	2018	2019	2020	2021
Undergraduate	Total course regn.	4183.50	4280.00	3819.50	3866.50	3982.00
	Teach enrolled FTE	507.98	525.30	440.41	435.62	404.55
	Teach input unit	1665.80	1776.41	1496.17	1539.33	1431.82
	Teach input subsidy	22043320	24563055	21199034	25616930	21793423
Postgraduate	Total course regn.	653.05	568.85	588.22	520.47	488.76
	Teach enrolled FTE	136.05	125.74	142.39	142.88	131.51
	Teach input unit	1059.31	969.26	1077.91	1113.78	1013.11
	Teach input subsidy	14017703	13402336	15272816	18535143	15420281
Total course registered		**4836.55**	**4848.85**	**4407.72**	**4386.97**	**4470.76**
Teach enrolled FTE		**644.03**	**651.04**	**582.79**	**578.50**	**536.06**
Teach input unit		**2725.11**	**2745.68**	**2574.08**	**2653.11**	**2444.93**
Teach input subsidy (US$)		**1 933 567**	**2 035 678**	**1 955 595**	**2 367 403**	**1 995 373**

ENROLLED FTE: It represents the number of students being taught if each of them was doing a full load of courses, or pursuing his/her research full-time. TIUs. The FTE contribution multiplied by the Funding Group factor and the factor for the Funding Level determines the number of TIUs generated by the course. INPUT SUBSIDY The TIUs multiplied by the US$ value per TIU determines the input subsidy generated by the course.

running costs. Research income is distributed to staff members pro-rata for their research output of published papers and successfully completed MSc and PhD students. No separate allocations are made to research groups, centres or institutes. The Comps Centre, which left the school in 2013, provided their own funding and received research funding from the school on the same basis as all the other individual staff members. From the table below showing the income of the school over the past few years, it is clear that the school is a sustainable entity that is able to generate enough money for its teaching and learning activities and can support its salary budget and operational expenses budget. Money for capex remains very low and most expenses are funded through a faculty allocation annually, which is based on a principle of competitive bidding. Strict financial management through monthly controls is practiced and ensures that the school remains solvent and operates within its budget. Redirected fees are employed to subsidise student excursions, notes, and computer usage and supply, as well as laboratory consumables. The school has sufficient staff members to handle the current undergraduate student numbers, but efforts are being made to reduce the first-year intake while increasing the postgraduate numbers. A reasonable quality of teaching is still possible with the assistance of postgraduate students to help manage tutorial classes. Based on the current financial situation, the school plans to grow its postgraduate student income and short course income, as well as to optimise undergraduate throughput.

7.3.1 Foundation Accounts

There are ad hoc consultation projects from time to time by some staff members, but these are small amounts that accrue to the school. The school has several foundation accounts, mostly for prizes and donor funding, such as Donald Williams Foundation account (second year Chem Eng prize), OB Volckman Foundation account (best Chem Eng student prize), Adrian Goddard-Gencor Memorial Foundation account (best Metallurgy student prize), Alexander Lorant Foundation Account (donor funding), SAIMM Foundation Account (donor funding for student support) and a School Reserves Foundation Account (for Chem Eng staff salary subventions). Due to poor attendance and marketing (or rather lack thereof), not much money has come into the school through short courses. This is being addressed at the moment to improve this source of third-stream income in future.

7.3.2 Third-Stream Income, e.g. Donor Funding, Contract Research, Short Courses

On Third Stream Income Generation, although the school of Chemical and Metallurgical Engineering planned to increase its third-stream income generation activities to become financially more sustainable, the effects of COVID-19 from 2020 to about the first half of 2022 affected the industry partners as well as government agencies. However, several initiatives are underway to re-engage with industry to address this matter.

Table 7.8 Financial resources (note: Items may vary depending on the provider's resource allocation practice).

Item		Years up to present (US$)				
		2017	2018	2019	2020	2021
Staffing/salaries	SE	1 346 766	1 477 768	1 562 663	1 655 687	1 935 656
Capital equipment	CE	52 361	93 764	106.556	48 669	134 119
Operating/running	OE	163 526	211 779	187 529	149 976	168 482
Research		10 618	17 202	27 106	21 237	27 043
Other		0	0	0	0	0
Total		**1 573 378**	**1 613 093**	**1 670 850**	**1 875 678**	**2 265 409**
Indices	FTE/SE	39 000	42 333	50 007	53 377	67 342
	CE/FTE	1516	−2686	−3410	1569	4666
	OE/FTE	4735	6067	6001	4835	5862

7.3.2.1 Other Funding Sources

It is observed that the school's external funding and reserves that were building up over a number of years have contributed immensely to the school survival and financial sustainability. Thus, the school has decided to actively focus on generating third-stream income by consulting and running short courses.

The budget and financial allocation for the school are presented in Table 7.8. The school remains financially sustainable at current staff and student levels, but all budgets are under severe pressure as a result of subsidy uncertainties and no increases in the subsidy after COVID-19. An illustration of this is the fact that the school's operating budget for running expenses has decreased since 2018, but the research budget allocations increased slightly. Considering that the capex budget for the whole faculty of engineering and built environment (EBE) is only approximately US$ 107239 annually, it is clear to see that it is difficult to sustain any purchases of teaching equipment and the school of Chemical and Metallurgical Engineering has done remarkably well to upgrade its undergraduate teaching laboratory for Chemical Engineering despite these constraints. Very small amounts are allocated to computing equipment and library books annually.

The research component reflected represents funds from the University and Faculty Research Councils, including the SPARC initiative, and excludes all funds from research grants, chairs, NRF or contracts.

As mentioned above, let us re-emphasised that although it is difficult to sustain any purchases of teaching equipment and the school of Chemical and Metallurgical Engineering, but the school has done remarkably well to upgrade its undergraduate teaching laboratory for Chemical Engineering despite these constraints. Very small amounts are allocated to computing equipment and library books annually.

7.3.3 Sustainability, Efficiency and Effectiveness of the School

It's rewarding to undertake internationally recognised research and to transfer those ideas to industry and society as efficiently as possible. The research specifically in the area of making chemical and energy processes more efficient and environmentally friendly. In particular, it looks at sustainability and has developed novel, fundamentally based methods, to evaluate how to assess the efficiency of processes without doing detailed designs.

The fundamental ideas are also being applied to biological processes in conjunction with researchers from the school of Molecular and Cell Biology. These are towards making bio-hydrogen, bio-ethanol and bio-diesel using novel biological processes, which are funded and carried out in the school.

Furthermore, the school had two bookkeepers during the review cycle of 2017–2021. The stability of these staff members has increased the level of accountability, leading to a stable accounting system.

7.3.4 Relationship Between the School's Planning Activities and the School's Revenue Generating Activities

It is obvious that staff development would lead to increase in productivity, which will in turn increase revenue generation. As a result, the school encourages regular course attendance on CLTD courses by staff members to improve their teaching and research skills and development. In addition, the school organises a teaching colloquium once a quarter to discuss teaching matters and curriculum developments of interest to all academics. Four of the school's staff members have enrolled in the PG Diploma in Higher Education in 2015 and recently completed these qualifications between 2017–and 2021. In terms of postgraduate supervision, staff are encouraged to attend several relevant courses offered by CLTO and the postgraduate office. Younger staff are allocated co-supervisory roles together with more experienced staff members to mentor and guide them.

The NRF offers time-out grants for staff still busy with PhD studies, and other academics have so far successfully utilised this funding facility, which allowed them replacements to allow them time to concentrate on their PhD studies.

Staff are encouraged to submit to regular teaching evaluations in all the courses they teach, and in the case of confirmation of posts and promotion, student evaluations form an integral part of the process.

The Higher Education Information Management System (HEMIS) financial records for the school is presented in Tables 7.7 and 7.8. They give the student, staff and resource indices for the chemical and metallurgical programmes. The increase in the salaries expenses was as a result of new appointments and inflation-related increases on an annual basis. The increase in the operating index up to 2018, 2019 could be attributed to increasing resources required to support the increasing student numbers. The indices for the last two years reflect the financial strain in the school, which is symptomatic of the COVID situation and budget restraints in the university. This has placed pressure on the available resources. Typical first year class size is recommended to be 70 students for the Metallurgy and Materials programme and 120 students for the Chemical Engineering programme. Currently, these are the intake number targets for 2017, which have largely been achieved.

From Table 7.7, it is clear that staff has been able to be quite research active. These figures can be considered as satisfactory indices of the experience and productivity of staff in the school. It must be emphasised that the school attaches great importance to industry experience, but it is difficult to attract persons with significant experience to academia due to the salary disparity between industry and educational sector, as discussed above. Despite this, the school has been successful in attracting several staff members amongst the latest appointees who all have industrial experience. In addition, a number of visiting staff has joined the school after many years in industry, which has placed the school in a very good position to capitalise on this experience from all these sources.

The school has been through a number of reasonably large changes, including finance, during the review period. Five new academic staff members joined the school, two lecturers, two senior lecturers and an associate professor. In addition, a new Head of School (March 2021) has been appointed after the acting Head of School started in January 2020. The COVID-19 pandemic has had a significant impact for the last two years and we are only now returning to normal. However, it is truly a testament to the mettle and dedication of staff members that the academic project was kept on track under very difficult conditions.

The school of Chemical and Metallurgical Engineering, therefore, is confident in its ability to provide excellent training to both Chemical and Metallurgical Engineering students and produce competent graduate engineers complying fully with the professional attributes and requirements. The school is still attracting outstandingly high-quality applicants and is proud to maintain and strengthen the proud history in education and industry, and makes a contribution to the future of the country.

7.3.5 How Present Financial Situation of the School Inform Future Planning?

In 2012, the school commissioned the Strategic Planning office to conduct an investigation to determine the optimum size of the school in terms of undergraduate and postgraduate student numbers, within the new government framework of capping total student intake and differential TIUs allocated to postgraduate and undergraduate students. This was carried out with the intention to decrease undergraduate student intake, an increased focus on postgraduate student numbers and improved management of throughput, in both cases, and careful management of the student/staff ratio between 16 and 20. Unfortunately, as shown in Table 7.8, the indices of student/staff ration still stood between 39.00 and 67.34 from 2017 to 2021, respectively.

Thus, since the Head of School is responsible for the functional management of the school and strategic planning and implementation thereof, the head is strongly mandated to implement future plans with assistance from the existing committees, especially the school executive committee.

7.4 Teaching in the School

The school currently teaches the following undergraduate programmes:

- BSc Eng. (Chemical Engineering)
- BSc Eng. (Metallurgy and Materials Engineering)
- BSc Chemistry with Chemical Engineering, a programme taught in conjunction with the school of Chemistry, and which allows successful candidates to complete the third and fourth year of Chemical Engineering for a BSc Eng (Chemical Engineering) qualification after completion of the BSc degree.

On the postgraduate level, the school has programmes in the following specialisation areas:

(a) Metallurgical Engineering:
  1. Welding Engineering
  2. Coal science and technology
  3. Minerals processing and extractive metallurgy
  4. Pyrometallurgy
  5. Materials science and physical metallurgy

(b) Chemical Engineering:
  1. Oil and gas Engineering
  2. Chemical Engineering

Postgraduate degrees are taught on the basis of a so-called 50/50 basis for MSc degrees, which implies half coursework credits and half research report credits for the qualification, or an MSc by research only. PhD degrees are research-based only. The Welding Engineering programme has an MEng qualification associated with it, which until recently involved coursework only. A Graduate Diploma of Engineering (GDE) allows non-engineering students into the postgraduate programme on certain conditions. Upon successful completion of the GDE, non-engineering students can progress to MSc and PhD degrees. However, the GDE has been replaced by a Postgraduate Diploma (PG Dip) qualification since 2017. Examinations on both the undergraduate and postgraduate levels are quality controlled by firstly an internal examiner and secondly an external examiner. Final pass/fail decisions are made at Faculty Examination Board meetings.

The two undergraduate engineering programmes produce graduates complying with the requirements of ECSA outcome levels set by ECSA. The last ECSA accreditation visit took place in 2022 and no deficiencies or concerns were recorded for either of the school's two programmes.

Assessments in the different subjects take place through a combination of formal open and close book tests, assignments, projects, laboratory reports and tutorial homework. A number of subjects make use of continuous evaluation, especially the year subjects. The final marks obtained in each subject is a combination of a year mark and an examination mark. In subjects with ECSA outcome levels attached to it, the achievement of this outcome level(s) is/are also assessed. If a student in the final year does not achieve the ECSA graduate attributes (GA) associated with a subject, the subject or portions thereof have to be repeated. The school's aim is to produce high skills level qualified graduates who are useful in industry and able to perform competently as engineers in training when they start working.

Current undergraduate and postgraduate throughput rates are far from ideal. On the undergraduate level, it can be ascribed to a number of factors, amongst them wrong career choices by students, poor preparedness and very large classes. The school endeavours through a variety of teaching modes and tutorial classes to remedy this. On the postgraduate level, a stricter mechanism has been introduced by the Faculty of Engineering and the Built Environment to ensure timely submissions of research proposals, progress reports and final submission of theses/dissertations. These measures taken should facilitate an improvement in throughput. As Table 7.8 indicates, it has almost disappeared in the School and the two groups nowadays perform on par with each other.

7.4.1 Throughput According to Gender and Race

Table 7.9 also presents the student direct intake, enrolment graduation, and throughput. Graduation throughput, averaged around 39% from 2013 to 2018, based on direct intake and who graduated within the minimum time of four years for the Chemical Engineering programme. The school is conscious that student throughput needs to be improved, and we are investing a lot of time and energy in increasing student competency and helping students from disadvantaged backgrounds

Table 7.9 Student demographics by race and gender (2013 enrolment).

Race	Calendar instance year	Total registration	Dropout, % (moved excluded, dropped)	Graduated in 4 years, %	Graduated in 5 years, %	Graduated in 6 years, %	Graduated in 6+ years, %	% Still registered
African	2021	117	20.51	0.00	0.00	0.00	0.00	79.49
	2020	114	22.81	0.00	0.00	0.00	0.00	74.56
	2019	143	25.87	0.00	0.00	0.00	0.00	56.64
	2018	155	37.42	22.58	0.00	0.00	0.00	29.03
	2017	113	36.28	27.43	14.16	0.00	0.00	6.19
Chinese	2020	1	0.00	0.00	0.00	0.00	0.00	0.00
	2019	1	0.00	0.00	0.00	0.00	0.00	100.00
	2017	3	33.33	66.67	0.00	0.00	0.00	0.00
Coloured	2021	1	0.00	0.00	0.00	0.00	0.00	100.00
	2020	1	0.00	0.00	0.00	0.00	0.00	0.00
	2019	3	33.33	0.00	0.00	0.00	0.00	66.67
	2018	4	25.00	50.00	0.00	0.00	0.00	25.00
	2017	3	0.00	33.33	33.33	0.00	0.00	33.33
Indian	2021	17	11.76	0.00	0.00	0.00	0.00	88.24
	2020	18	27.78	0.00	0.00	0.00	0.00	66.67
	2019	23	21.74	0.00	0.00	0.00	0.00	65.22
	2018	15	13.33	53.33	0.00	0.00	0.00	13.33
	2017	22	22.73	22.73	18.18	0.00	0.00	13.64
White	2021	11	18.18	0.00	0.00	0.00	0.00	81.82
	2020	3	0.00	0.00	0.00	0.00	0.00	66.67
	2019	16	31.25	0.00	0.00	0.00	0.00	37.50
	2018	11	9.09	72.73	0.00	0.00	0.00	0.00
	2017	11	45.45	18.18	9.09	0.00	0.00	0.00

Note: This is information obtained from the University; we are aware that there are discrepancies in the total numbers which could be attributed to exclusion and readmission of students over the years.

to adapt to the multiple challenges they face during their studies in the school. It does seem though as if about 80% of a cohort eventually finishes the degree after six years as shown in Tables 7.10 and 7.11.

Table 7.10 indicates the student demographics enrolled in 2013 and completed in the minimum time for both programmes in the school. Unfortunately, the data presented here and that presented in Table 7.9 (for both programmes) is not consistent. However, Table 7.11 indicates that black males make up the largest component of the student population in the school, followed by black females. Percentage-wise the dropout and completion figures between the different population groups have narrowed considerably during the last five years. Due to the small numbers, it appears as though there is a huge discrepancy between the success rates of the white males and females, while amongst the other population groups (collectively classified as black) it is similar.

The student enrolment statistics for the Chemical Engineering programme is presented in Table 7.10. It can be seen that the enrolment in 2017 decreased to 114 students, which represents a decrease over the period of 2014–2017. This increased from 2018 to 2019, and then decreased to 106 in 2021, which is below the target figure of 120 first-year students. However, one has to remember that these figures also include the repeat students. Throughput has increased encouragingly to more than 80% during the last two years.

As presented in other records at faculty most students admitted have NSC/Wits points in the 46–48 range. This is far higher than the school minimum requirement of 42 points for Chemical Engineering. The school regularly gets more than 3000 applications per year for the available 120 places in Chemical Engineering.

Table 7.10 Cohort analysis (total numbers).

Pathway	Group						
	Level of achievement						
Year of entry	Number entered 1st year	Dropped out, %	Graduated within 4 years, %	Graduated in 5 years, %	Graduated in 6 years, %	Graduated in 6+ years, %	Still registered, %
P (2021)	106	25.0	0.0	0.0	0.0	0.0	81.0
P-1 (2020)	115	29.0	0.0	0.0	0.0	0.0	86.0
P-2 (2019)	144	41.0	0.0	0.0	0.0	0.0	103.0
P-3 (2018)	151	58.0	46.0	0.0	0.0	0.0	47.0
P-4 (2017)	114	45.0	39.0	21.0	0.0	0.0	9.0
P-5 (2016)	132	43.0	38.0	29.0	18.0	1.0	3.0
P-6 (2015)	112	33.0	35.0	20.0	18.0	6.0	0.0
P-7 (2014)	150	56.0	32.0	30.0	19.0	11.0	2.0
P-8 (2013)	162	59.0	42.0	18.0	25.0	14.0	4.0

Table 7.11 Cohort analysis (based on gender).

Gender	Calendar instance year	Total registration	Dropout, % (moved. excluded, dropped)	Graduated in 4 years, %	Graduated in 5 years, %	Graduated in 6 years, %	Graduated in 6+years, %	Still registered, %
Female	2021	59	27	0	0	0	0	73
	2020	65	25	0	0	0	0	75
	2019	74	23	0	0	0	0	77
	2018	78	35	33	0	0	0	32
	2017	55	35	35	22	0	0	9
	2016	60	25	27	30	15	2	2
	2015	60	20	37	23	17	3	0
	2014	67	45	18	18	10	6	3
	2013	77	34	26	16	12	12	1
Gender neutral	2021	1	100	0	0	0	0	0
Male	2021	46	17	0	0	0	0	83
	2020	50	26	0	0	0	0	74
	2019	70	34	0	0	0	0	66
	2018	73	42	27	0	0	0	30
	2017	59	44	34	15	0	0	7
	2016	72	39	31	15	13	0	3
	2015	52	40	25	12	15	8	0
	2014	83	31	24	22	14	8	0
	2013	85	39	26	7	19	6	4

The experience has showed that most students enrolled in the fourth year graduate (that is, there is a very low fourth year failure rate). It should also be noted that the fourth year classes enrolled from 2019 to 2021 were between 100 and 130. This has a significant impact on the resources available at a fourth-year level and affects especially courses like Design and the Research Project. In 2022, the final year numbers have dropped to around 70 (Table 7.2). The throughput rate for students completing their degrees in minimum time is currently averaging 39% (Table 7.10), which compares favourably and are in some instances better than those observed in the other engineering disciplines at Wits.

It is expedient to note that the school has enough full-time and visiting staff members to supervise up to 240 postgraduate research students, and teach the different postgraduate programmes in the school. All MSc degrees by research students are required to submit at least a draft article before being allowed to submit for examination, while in the case of PhD students, they are required to have submitted at least two papers before being allowed to submit any thesis for examination.

The biggest obstacles to increase postgraduate student numbers are the lack of money to supply bursaries to students and the lack of interest amongst South African students to engage in postgraduate studies once they have completed their undergraduate degrees. The latter fact can be ascribed to social and economic circumstances and pressure at the local communities from which many of these students originate. The school continues to engage with government agencies and private companies to obtain funding for bursaries for both undergraduate and postgraduate students to try and alleviate the problems currently experienced. The school receives a regular stream of inquiries from international postgraduate students from the rest of Africa, as well as local students from other South African universities for postgraduate studies, but there had not been any special measures in place to encourage this to a greater degree, because the local prospective students need every available spaces.

7.5 Effect of Skills Training and Education Policies on Climate Change

In conclusion and accordance with UNESCO (2021), unemployment, poverty and social exclusion are the most severe consequences that follow when skills development policies fail to respond not only to socio-economic needs but also to the demands of transitioning to more sustainable development. Just as the business success and well-being of the people in a community are very much intertwined, so also the impact of climate change on businesses as their value chains depend on the community. The essential ingredient to improve any economy is to enhance the knowledge and skills of the people to ensure their sustainable economic growth (UNESCO 2017).

As seen from the discussions and calculations in the preceding chapters, including triangulation of CO_2 projections and gradient analyses, if everyone stopped emitting GHGs today, the world would still have to face the impacts of climate change. Whether this change is positive or negative, according to scientists, each degree Celsius of global temperature increase can be expected to have impacts that maybe difficult to quantify. Thus, adapting certain economic, social and environmental perspectives should be globally practiced by all nations since clear spatial boundaries do not exist between countries and people of all nations. These are some of reasons why the Paris Agreement calls upon all nations to keep the global temperature rise well below 2°C compared to the pre-industrial revolution. Governments are therefore encouraged to pursue efforts to limit the temperature increase even further to 1.5 C (UNFCCC 2017). The climate agreements and practices should therefore focus on achieving a transitional change from carbon resources in the order of low-carbon → net-zero carbon → green growth → climate-resilient development → green hydrogen. Hence, countries that have ratified the Paris Agreement are to develop a clear policy framework and provide a strategic vision for both mitigation and adaptation activities that enable development aligned with allowable climate change and resilience.

Provision of the right skills, however, paves the way for a productive structural transformation towards a greener economy and decent job creation. Thus, an effective anticipation and development of skills are fundamental for a just and inclusive transition. Such skills development to creating green jobs also serves as a 'buffer' against the effects of transitory disruptions and emerging challenges that will surface as economies become green. A well-coordinated policy approach will ease the transition to a greener future, which again, should be inclusive. The following policy clusters highlight further the foci required for the energy transition requirements (ILO 2019a).

7.5.1 Policy Cluster of Core Skills Required for Green Jobs

Skills development for green jobs through technical and vocational education and training (TVET) is most often done by adding green components to existing qualifications or education programmes (ILO 2019a). At the moment, however, only a few countries have systems or measures in place that are fully dedicated to developing skills for green jobs. In as much as there are indications that gender issues are being addressed in the TVET sector, such that more female students are encouraged to register for science- and technology-related programmes, enrolments in universities and TVET institutions have not transitioned from the traditional gender stereotypes, with more male students in STEM fields of studies. Most affected in this regard are the developing countries owing to the lack or inefficient trained-the-trainers policy. It should therefore be emphasised that employers' and workers' organisations have an important role to play in the design, implementation and evaluation of skills development and quality-based training delivery to prioritise the needs of vulnerable groups.

Thus, clustering policy would help countries respond effectively to climate change and also advance a just transition as discussed next.

Policy cluster 1: The policy should be improved from the existing ones, coherence at planning, design and implementation stages to uphold the concept of green transition. The policy should address common features find at national sectoral or subnational governmental levels of the skills for green jobs landscape. In instances where social partner engagement is weak, such negative consequences for the coordination and relevance of policies on skills for green jobs should be enhanced for the set objectives of sustainability. Thus, coordinating combined top-down and bottom-up initiatives would provide effective and more sustainable policy support for the green transition.

Policy cluster 2: Planning and coordination at national, subnational and sectoral levels, skills for green jobs should receive government taxes and incentives in those sectors directly affected by climate change and environmental depletion in terms of energy, transport, construction and waste management practices. As a result, the private sector and trade unions should play pivotal roles in the transition to sustainable economies.

Policy cluster 3: Core and transferable skills are requirements for an inclusive transition for a greener future, which should support a wide range of both technical and non-technical skills to drive the green transition. The International Labour Office (ILO) estimates the impact of needs for occupational skill at the global energy transition and circular economy, which have revealed transferable technical skills, new skills like carbon credit, trade and marketing. These are also potentially transferable within occupations to retraining the needed workers to acquire new skills. Apart from the hard technical skills, the core soft skills can confer a comparative advantage that are transferable across occupations. Thus, Figure 7.1 presents countries that are grouped according to performance in environmental and skills policies, while Table 7.12 presents the core skills required for green jobs, by skill level of occupation (ILO 2019a).

Carbon Credits: A Climate change Mitigation Strategy

Policy cluster 4: To enable the green transition, the technique to identify, manage and forecast skills needs for green jobs can benefit from an appropriately improved labour market information and institutionalised social dialogue. Thus, the analysis and anticipation of demand for green job skills depends on well-researched information and occupational needs

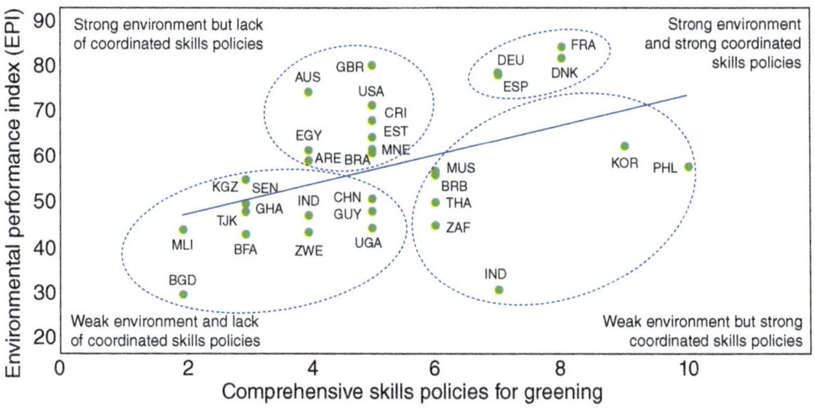

Figure 7.1 Countries grouped according to performance in environmental and skills policies (Key: *Y*-axis: the EPI uses the distance-to-target technique for indicator construction, which situates each country relative to targets for worst and best performance corresponding to scores of 0 and 100, respectively. *X*-axis: the presence of comprehensive skills policies for greening was calculated on a 0–10 scale. Country codes are used as follows: Australia (AUS), Bangladesh (BGD), Barbados (BRB), Brazil (BRA), Burkina Faso (BFA), China (CHN), Costa Rica (CRI), Denmark (DNK), Egypt (EGY), Estonia (EST), France (FRA), Germany (DEU), Ghana (GHA), Guyana (GUY), India (IND), Indonesia (IDN), Republic of Korea (KOR), Kyrgyzstan (KGZ), Mali (MLI), Mauritius (MUS), Montenegro (MNE), the Philippines (PHL), Senegal (SEN), South Africa (ZAF), Spain (ESP), Tajikistan (TJK), Thailand (THA), Uganda (UGA), the United Arab Emirates (ARE), the United Kingdom (GBR), the United States (USA) and Zimbabwe (ZWE). Source: "Skills for a greener future. Key findings", ILO (2019b).

Table 7.12 Core skills required for green jobs, by skill level of occupation (Source: ILO 2019a).

Required across the labour force	Required in medium to high-skilled occupations
Environmental awareness and protection; willingness and capability to learn about sustainable development	Analytical thinking (including risk and systems analysis) to interpret and understand the need for change and the measures required
Adaptability and transferability skills to enable workers to learn and apply the new technologies and processes required to green their jobs	Coordination, management and business skills that can encompass holistic and interdisciplinary approaches incorporating economic, social and ecological objectives
Teamwork skills reflecting the need for organisations to work collectively on tackling their environmental footprint	Innovation skills to identify opportunities and create new strategies to respond to green challenges
Resilience to see through the changes required	Marketing skills to promote greener products and services
Communication and negotiation skills to promote required change to colleagues and customers	Consulting skills to advise consumers about green solutions and to spread the use of green technologies
Entrepreneurial skills to seize the opportunities of low-carbon technologies and environmental mitigation and adaptation	Networking, IT and language skills to perform in global markets
Occupational safety and health (OSH)	Strategic and leadership skills to enable policy-makers and business executives to set the right incentives and create conditions conducive to cleaner

that will in turn address the skills challenge and the job preparedness. In a case or countries where such frameworks do not yet exist, this need brings an opportunity to create institutions and structures that support national human resources development council, involving the Government, employers, workers and providers of training and education. The arrangement will facilitate the exchange of information and establish industry groupings that could later be formalised as sectoral skills councils. In the same vein, career guidance and counselling initiatives would facilitate and boost a human-centred approach to the green transition.

Policy cluster 5: The training should target, be inclusive and prioritise the skills needs of vulnerable group of the population with well-equipped teachers to address essential and effective climate education. Invariably, availability of teachers and trainers with the appropriate knowledge on sustainable land and ecosystem management, energy efficiency and green technologies is crucial. These trainers and teachers promote environmental awareness amongst young people and in spreading environmental training beyond the formal education system into the adult population. Training these teachers and trainers is critical in any skills response strategy at national, sectoral and local levels. It is very important, therefore, that disadvantaged groups should remain largely represented and included in the country's skills development programmes for green jobs. It is also expedient to emphasise the need to develop systematic training and retraining initiatives in skills for green jobs aimed towards groups, such as youth, older workers, people with disabilities, indigenous people, women, migrant workers, unemployed people, informal workers and those living in rural areas.

Women should also be targeted for apprenticeship and skills training for environmentally sustainable jobs to overcome disparities in the labour market as well as skills shortages in certain occupations. The provision of incentives will enhance women's participation in sectors with green growth potential, notably through technical training programmes, which will achieve the double objective of solving skills shortage problems in this area while also increasing women's participation in technology-driven occupations.

Policy cluster 6: Public-private partnerships (PPP) could drive the sustainable development agenda beyond 2030 with the aid of skills funds. This need can also be met through collaboration between private-sector and multinational companies and the public sector through promotion and empowering a PPP for green jobs. The private-sector can facilitate efficient engagement and involvement of workers' organisations both in establishing a sustainable and functional TVET system and in developing skills within sectors and enterprises. While PPP can catalyse and boost diversified, innovative and new approaches in financing lifelong learning and provide a platform for combined contributions to support the TVET system systematically and independently, the government can provide financial incentives through its agencies to promote green markets. Thus, strong and inclusive PPP could be instrumental in reducing the financial constraints on effective delivery of high-quality training, as well as generating more opportunities to formulate new, innovative and data-driven skills development policies (ILO 2019a).

7.5.2 Calculation of Carbon Credits and Carbon Footprint in Banana Supply Chain

In the report from FAO (2017), the term 'carbon footprint' is defined as the quantity of GHGs generated by an item or activity. It can be expressed as CO_2 equivalent that is amount of CO_2 that would have the same impact on climate change. The carbon footprint of a product can be calculated as the sum of all materials, energy and waste across all lifecycle activities multiplied by emission factors and global warming potentials (GWPs) of the product. As a widely consumed cash crop, banana including the plantain family has broad leaves well adapted for photosynthesis by having a large surface area and contain openings called stomata to absorb CO_2 from the atmosphere and oxygen out into the atmosphere. In the primary production stage, banana farming does not produce a large contribution to GHG emissions. In studies analysing GHG emissions 'from cradle to retail', primary production contributed between 16% and 20% of the total amount of GHG emissions (FAO 2017).

The carbon footprint of the banana industry across the entire value chain can be divided into three broad areas of production and packing, transport and shipping, and ripening for export. The biggest banana-producing countries are India, Brazil, West African countries, etc. where very little quantity is exported and most of the produce is kept for domestic consumption. The international standard organisation (ISO 14064) standard on quantification of emissions and removals of GHGs represents a fundamental instrument for the estimation of carbon footprint in the banana industry. ISO standards 14040 and 14044 on Life Cycle Assessment (LCA) and PAS 2050 Carbon Footprint Verification also provide a recognised framework for carbon footprint and lifecycle analysis (FAO 2017). It is also identified that carbon sources on banana farms include:

- Fabrication and acquisition of material involve indirect use of energy for the production and transport of the plant materials such as plastics and tools used to build and maintain the farm, prepare the fields and for transport on the farm itself.
- Fabrication and use of plastics.
- Use of chemical fertilisers, especially nitrogen and potassium in monoculture production.
- Use of chemical pesticides.
- During harvest, machine-harvested crops have a higher carbon footprint.
- Energy involved in packaging for the manufacture of carton, boxes in packing stations, fumigation with fungicides etc.

7.5.2.1 Climate Change Mitigation Strategy on Banana Production

Malunjkar et al. (2015) reported a field study that was conducted on banana crop in Jalgaon district of Maharashtra to evaluate energy efficiency of two irrigation methods, namely, micro-irrigation system (MIS) and conventional flood irrigation (FI). The study revealed that water saving of 32% and electricity saving of about 36% was achieved by micro irrigation against FI. The potential of gaining carbon credits by switching from FI to MIS for 5000 ha area under banana crop in Jalgaon district was estimated at 11 750 (tCO2e or carbon credits) worth Rs. 2000 (US$33.28)/ha.

According to Kyoto Protocol and methodologies, suggested by the United Nations Framework on Climate Change (UNFCCC) for energy efficiency in agriculture, the potential of carbon credits was estimated. According to the UNFCCC, one carbon credit is equal to one tonne CO_2 reduction, which is also referred to as Certified Emission Reductions, CERs (Gold Standards, 2009). The gain of carbon credits by improving energy efficiency of banana and electricity savings in pumping of water due to drip irrigation was then calculated from the total emission reductions obtained in the study. Thus, micro-irrigation has a definite role in minimising the impacts of climate change on Indian agriculture by reducing the consumption of water and electricity as presented in Table 7.13. Since the use of fossil fuels increases the GHGs emission in traditional farming, energy efficiency in agriculture thus has huge impact on overall power scenario globally. The UNFCCC however has made it mandatory that any carbon saving in the developing world could be traded with the developed world with appropriate certification records on emission reductions. Thus, any work attaining saving of electricity by various means in the developing world can sell these certificates to any entity in the developed world, which should be done at a mutually agreed price (Malunjkar et al. 2015). Banana crop being one of the major consumers of water and energy, it is necessary to evaluate the savings of water and electricity for banana crop due to use of drip irrigation as against conventional irrigation. As a result, these pilot studies were undertaken to generate scientific information about water and power savings and standardise the same to assess the carbon credits that could be gained by adoption of drip irrigation (Malunjkar et al. 2015) and solar-powered MIS (Kumar et al. 2015) as discussed below.

Table 7.13 Irrigation water and electricity consumption in irrigation methods.

Particulars	MIS	FI	% Saving in MIS over FI	MIS	FI	% Saving in MIS over FI
	Jain R&D farm			Farmers field		
Depth of irrigation applied (mm/ha)	1455.6	2244.2	35.14	1669.7	2359.8	29.24
Total water consumption (kL/ha or m^3/ha)	14 556	22 442	35.14	16 697	23 598	29.24
Total electricity consumption (kWh/ha)	4657.8	7630.3	38.96	5343	8023.5	33.41
Total number of irrigations applied	151	40	—	126	42.1	—
Total pumping hours used for irrigation application (h/ha)	397.7	1726.3	—	456.2	1815.3	—
Hours used per irrigation	2.6	43.2	—	3.6	43.1	—

Source: Adapted from Malunjkar et al.

7.5.2.2 Climate Change Mitigation by Solar-Powered Irrigation on Banana Production

The work of Kumar et al. (2015) describes the development of a solar-powered MIS with strategies to address the limitations of drip irrigation for smallholders, such as the one reported by Malunjkar et al. (2015). These include,

(1) Development of MIS that operates with low pressure,
(2) Matching pumping scheme to draw water from small-scale water harvesting ponds or borehole,
(3) Introducing an economical and efficient alternative MIS by attaining the perfect match between water available and area under command for extensive small-scale vegetable production.

The reported MIS (Figure 7.2) is the assembly of many components that allows continuous and smooth flow of water in order to achieve wettability efficiency through microtubules. The water source, which could be surface water or borehole, is directly connected to delivery head of the pump with high pressure at the delivery head for proper functioning. The solar-powered irrigation systems against the conventional method require a huge initial investment making this a costly preposition as shown in lower part of Table 7.14. As a result, it becomes necessary to embark on design procedures to minimise cost.

Figure 7.2 The components used in the development of solar-powered irrigation system. Source: Kumar et al./with permission from Elsevier.

Table 7.14 Components and specifications for solar-powered micro-irrigation method (Kumar et al. 2015).

Components	Specification details
Solar panel	Category: high efficiency; power output: 148 W
	Minimum power voltage: 15 V; power current: 4–5 A
	Dimension: length and width preferably be less than 1200 mm (wind velocity point of view)
	Maximum wind load: 60 m/s
Charge control unit (CCU)	Normal and max input solar cell array voltage: 15 and 25 V, respectively
	Maximum charging and consumption current: 30 A and 25 mA, respectively
	Constant voltage for lead-acid battery: 14.5 ± 0.3 V
Battery	Rated voltage: 12 V; normal capacity: 10 A for 5 h
	Charging voltage and environment: 14–15 V and up to 50 °C, respectively
Inverter	AC type, microprocessor controlled 300 W continues power handling
	Input and output voltage: 10–15 and 220 V
	Rated capacity: 500 VA; output frequency: 50 Hz
	Auto shutdown in case of short circuit, raised temperature and overload
Small capacity electric pump	Type: AC; drive power: 60 W; input power: 110 W; pipe size: 13 mm
	Discharge: 15 L/min (at suction head and delivery
	head of 1.5 and 4.5 m respectively)
Overhead tank	Capacity: 1000 L, outlet: 25 mm diameter and 300 mm long mouthpiece connected to main pipeline (50 mm diameter PVC pipe) with matching flow control valve
Micro-tube irrigation system	Sub main: 40/50 mm diameter PVC pipe
	Lateral: 16 mm diameter LDPE (low-density polyethylene pipe)
	Micro-tube spacing: 50 cm row-to-row and plant to plant
	Micro-tube discharge capacity: 1 L/h at given pressure head (equivalent to 3 m hydraulic head) coverage: 200 m^2/h. 1000 m^2 per day for 5 h application
	Overall command for this system: 0.5 acre with alternate day irrigation

Cost of development for solar irrigation system (US$)	
Solar panel with cable	144
Solar charge control unit	17
Battery with accessories	83
Inverter	58
Pump	17
Overhead tank	42
Civil work	50
Total	411

7.5.2.3 Design of the Micro-Irrigation System

Water discharge from the micro-tube (MT) is calculated using orifice formula, given as,

$$q_{\text{mt}} = k \cdot k_q \cdot A_e \cdot \sqrt{2 \cdot g \cdot (H - H_{\text{m}})} \tag{7.1}$$

where q_{mt} is discharge, L/h; $k = 3.6$ for converting flow into L/h; k_q discharge coefficient taken as 0.61 for smoothly cut MT; A_e is cross-sectional area of MT, mm^2; H is the head available at the emitter inlet, m; H_{m} is the head loss in the MT and g is the acceleration due to gravity taken as 9.81 m/s^2.

The discharge from the emitting MT is designed such that the sum of head loss from different section is equal to available head of 3.0 m. The discharge, Q, is optimised using the Darcy–Weisbach and orifice formula for emitter discharge such that the flow in all the section remains laminar (Kumar et al. 2015). Variation of water flow rate is taken into account (Camp et al. 1997) as,

$$q_{var} = \frac{q_{max} - q_{min}}{q_{max}} \times 100 \tag{7.2}$$

where q_{var}, q_{max} and q_{min} are variation in emitter flow rate, maximum flow rate and minimum flow rate, respectively. The Christiansen uniformity coefficient (CUC) is described by Kang et al. (1999) as,

$$\text{CUC} = \left[1 - \frac{(1/n)\sum_{i=1}^{n} |q_i - \overline{q}|}{\overline{q}}\right] \times 100 \tag{7.3}$$

where q_i and $\overline{q}$ are ith emitter flow and average flow, respectively, n is number of emitter on the lateral. The distribution uniformity (Du) suggested by Kang et al. (1999) is the third parameter to determine the irrigation system functionally, which is given as,

$$\text{Du} = \frac{\text{Average Low Quarter Discharge}}{\text{Average Discharge}} \times 100 \tag{7.4}$$

The MT irrigation system (Table 7.14) is configured into a 'T' connection (Figure 7.2) to enable the lateral to commands two rows cropping to reduce the requirement of lateral pipe up to 50%. The arrangement of the solar power operated MIS addresses efficient use of water and energy in agriculture and the environment. In connection with operability of the design therefore, the three broad components viz, solar power generation, water lifting and water distribution using water-emitting devices are operated under gravity as presented in Figure 7.2, while the detail technical description is provided in Table 7.14. The solar power generation unit consists of solar panel, solar charge control unit, battery and inverter. The system is integrated in such a way that the components are compatible and operate optimality to achieve appropriate efficiency and economics. The size of the solar panel is 1.44 m^2 with a capacity of 148 W power from the sun, while it charges the battery through solar charge control unit. The pump is powered from the energy stored in the battery through inverter. The pump that is connected to the inverter operates with 110 W input power and 0.06 kW shaft power to lift water at the rate of 900 L/h (0.22 lps) over a 4.5 m delivery head. The water is in turn pumped to an elevated temporary 1000 L capacity water storage tank. The tank is placed at elevation such that it can deliver a gravity head of 3 m, which is equivalent to 0.3 kgf/cm^2 hydraulic pressure.

While the overall cost to the system is calculated to be equivalent of $411 and the different components are presented in Table 7.14, the detailed operational scheme can be found elsewhere (Kumar et al. 2015).

References

Camp, C.R., Sadler, E.J., and Busscher, W.J. (1997). A comparison of uniformity measures for drip irrigation systems. *Trans. ASAE* 40 (4): 1013–1020.

Factsheet, 2023. Center for Sustainable Systems, University of Michigan. 2023. "Climate Change: Policy and Mitigation Factsheet. https://css.umich.edu/sites/default/files/2023-10/Climate%20Policy_CSS05-20.pdf [Accessed 17/12/2023].

FAO (2017). Carbon footprint of the banana supply chain. In: *Promoting the Sustainability of the Banana Industry*. FAO I6842EN/1/02.17.

ILO (2019a). *Skills for Employment*, Policy Brief. International Labour Office.

ILO (2019b). *Skills for a Greener Future: Key Findings*. Geneva.

Kang, Y., Yuan, B.Z., and Nishiyama, S. (1999). Design of microirrigation laterals at minimum cost. *Irrig. Sci.* 18 (3): 125–133.

Karimi-Maleh, H., Ayati, A., Davoodi, R. et al. (2021). Recent advances in using of chitosan-based adsorbents for removal of pharmaceutical contaminants: A review. *J. Clean. Prod.* 291: 125880.

Kumar, M., Reddy, K.S., Adake, R.V., and Rao, C.V.K.N. (2015). Solar powered micro-irrigation system for small holders of dryland agriculture in India. *Agric. Water Manag.* 158 (2015): 112–119.

Malunjkar, V.S., Deshmukh, S.K., and Balakrishnan, P. (2015). Carbon credits: a climate change mitigation strategy. *Int. J. Sci. Res. Publ.* 5 (3): 1. ISSN 2250-3153.

Sizmann, A., Steinfeld, A., Brendelberger, S., et al. 2019. *Solar Fuel Breakthrough*. SUN-to-LIQUID Press Release.

Gold Standards, 2009. First Upgrade of Gold Standard Version 2.0. Source: https://globalgoals.goldstandard.org/standards/PRE-GS4GG-Energy/gsv2.1_toolkit_clean-11.pdf [Accessed 17/12/2023].

IPCC (2014) Climate Change 2014: Mitigation of Climate Change. The Intergovernmental Panel on Climate Change (IPCC), UNEP.

The National Academies of Sciences, Engineering, and Medicine (2018) *Negative Emissions Technologies and Reliable Sequestration: A Research Agenda*.

UNESCO (2017). *Education for Sustainable Development Goals. Learning Objectives*. Paris: UNESCO https://unesdoc.unesco.org/ark:/48223/pf0000247444.

UNESCO (2021). *Skills Development and Climate Change Action Plans Enhancing TVET's Contribution*. Paris, France: United Nations Educational, Scientific and Cultural Organization.

UNFCCC 2017. UN Climate Change Annual Report 2017. Source: https://unfccc.int/documents/66311 [Accessed 17/12/2023].

US EPA (2018). *2016 Climate Leadership Award Winners*. EPA Center for Corporate Climate Leadership. EPA https://www.epa.gov/climateleadership/2016-climate-leadership-award-winners [Accessed 17/12/2023].

Venkatesan-Crome, C., McDermott, E. 2020. *The Role of Engineering in Combating Climate Change*. https://www.imeche.org/news/news-article/the-role-of-engineering-in-combating-climate-change (accessed 23 March 2023).

Wilcox, J. (2020). *The Challenge of Scaling Negative Emissions*. Kleinman Center for Energy Policy, Source: https://kleinmanenergy.upenn.edu/podcast/the-challenge-of-scaling-negative-emissions/ [Accessed 17/12/2023].

Notation

Acronyms

BARON	Branch and Reduced Optimisation Navigator
CCU	charge control unit
CHMT	Chemical & Metallurgical Engineering
CLTD	Centre for Learning, Teaching and Development
COP	conference of the parties
CPF	centralised production facility
Dip	diploma
DME	dimethyl ether
FCC	fluid catalytic cracking
FEBE	Faculty of Engineering & the Built Environment
FI	flood irrigation
FTE	full-time equivalent
GAMS	general algebraic modelling system
GB	gigabyte
GHG	greenhouse gas
HAZOP	hazard and operability study
HEFC	hybrid electrolyser fuel cell
HOMER	hybrid optimisation of multiple energy resource
JFF	Johannesburg food farmers
LPG	liquefied petroleum gas
LSR	light naphthas
MATLAB	MATrix LABoratory
MIS	micro irrigation system
NLP	non-linear programming
PEME	proton exchange membrane electrolyser
PEMFC	proton exchange membrane fuel cell
PG	postgraduate
PNGV	Partnership for a New Generation of Vehicles
PV	photovoltaic
QQR	Quinquennial Review
RAM	random access memory
TIU	teaching input units
TVET	technical and vocational education and training
UG	undergraduate

Measuring Climate Change to Inform Energy Transitions: Carbon Footprint Calculations, First Edition. Sunny E. Iyuke.

Parameters

A_{H2O}	activity of water
α	charge transfer coefficient
AI	annual interest rate
C	concentration, mol/m^3
CF	cost factor
P_{cr}	critical pressure, MPa
T_{cr}	critical temperature, K
D	diffusion coefficient, m^2/s
j_o	exchange current density, A/m^2
F	Faraday's constant, C/mol
IR	inflation rate
δ_{mem}	membrane thickness, μm
MW	molar mass, kg/mol
$MW_{mem,dry}$	molar mass of the dry membrane, kg/mol
n_{cell}	number of cells
z	number of electrons transferred
P	pressure, MPa
P_{ref}	reference pressure, MPa
T_{ref}	reference temperature, K
R	ideal gas constant, J/mol K
Tax_{rate}	tax rate
D_W	water diffusion coefficient, m^2/s
K_{darcy}	water permeability coefficient, m^2
Y_{plant}	plant life
$Y_{purchase}$	purchase year
Y_{rep}	replacement year

Variables

Cash	cash flow, US$
Cost	cost, US$
C_{cap}	capital cost, US$
I	current, A
J	current density, A/m^2
Dep	depreciation, US$
DF	discount factor
n_d	electro-osmotic drag
FCI	fixed capital investment
D_{eff}	effective diffusion coefficient, m^2/s
IRR	internal rate of return
C_{main}	maintenance cost, US$
$\dot{M}$	mass flow rate, kg/hr
$\dot{N}$	molar flowrate, mol/s
n	molar flux, mol/m^2 hr
y	molar fraction
OP_{hours}	operating hours, hour
NPV	net present value, US$
PAT	profit after tax, US$
PATPD	profit after tax plus depreciation, US$
PBT	profit before tax, US$

PBP	payback period, year
Power	power, kW
PV	present value, US$
A	reaction area, m^2
C_{rep}	replacement cost, US$
Rev	revenue, US$
P_{sat}	saturation pressure, MPa
T	temperature, K
TAC	total annualised cost, US$
TDC	total direct cost, US$
TIC	total indirect cost,
TP	total production, US$
V	voltage, V

Greek symbols

α	charge transfer coefficient
ρ	density
$\rho_{mem,dry}$	density of dry membrane, kg/m^2
λ	humidity degree
σ_{mem}	membrane conductivity, $ohm^{-1}\ m^{-1}$
β	partial pressure, MPa
ε_p	percolation
ε	electrode porosity
δ	thickness, m
μ	viscosity, $kg\,s/m^3$

Sets

M	number of cells in the fuel cell stack
N	number of cells in the electrolyser stack
Q	number of plant operating years

Subscript

act	activation
AL	assembly labour
an	anode
annual	annual
BOP	balance of plant
cat	cathode
cap	capital
ch	channel
ch,o	channel at reference conditions
construction	construction
cons	consumed
cs	control and sensors
diff	diffusion
DC	direct cost
dms	delivery management system

DS	distribution and selling
ED	engineering and design
elect	electrolyser
eo	electro-osmotic drag
fc	fuel cell
gms	gas management system
H_2	hydrogen
H_2O	water
H_2-H_2O	hydrogen-water mixture
IF	installation factor
INS	insurance
LC	legal and contract fees
land	land
mem	membrane
me	membrane-electrode interface
me,o	membrane-electrode interface at reference conditions
MBOP	mechanical balance of plant
ohm	ohmic
ocv	open circuit voltage
other	other
O_2	oxygen
O_2-H_2O	oxygen-water mixture
PC	project contingency
pe	pressure effect
PR	patents and royalties
pelect	power electronics
prod	produced
RD	research and development
stack	stack
sys	system
tms	thermal management system

Index

Note: Page numbers with "*f*" denote figures and "*t*" denote table, respectively.

C

d

e

Z